13th International Scientific-Technical Conference on Electromechanics and Robotics "Zavalishin's Readings" (ER(ZR) –2018)

MATEC Web of Conferences Volume 161 (2018)

St. Petersburg, Russia
18-21 April 2018

Editors:

A. Ronzhin **V. Shishlakov**

ISBN: 978-1-5108-6196-1

Printed from e-media with permission by:

Curran Associates, Inc.
57 Morehouse Lane
Red Hook, NY 12571

Some format issues inherent in the e-media version may also appear in this print version.

Printed by Curran Associates, Inc. (2018)

For additional information, please contact EDP Sciences – Web of Conferences at the address below.

EDP Sciences – Web of Conferences
17, Avenue du Hoggar
Parc d'Activité de Courtabœuf
BP 112
F-91944 Les Ulis Cedex A
France

Phone: +33 (0) 1 69 18 75 75
Fax: +33 (0) 1 69 28 84 91

contact-edps@webofconferences.org

Additional copies of this publication are available from:

Curran Associates, Inc.
57 Morehouse Lane
Red Hook, NY 12571 USA
Phone: 845-758-0400
Fax: 845-758-2633
Email: curran@proceedings.com
Web: www.proceedings.com

TABLE OF CONTENTS

Preface

The conference "Zavalishin's Readings" was established as a tribute to the memory of Dmitry Aleksandrovich Zavalishin (1900-1968) – a Russian scientist, corresponding member of the USSR Academy of Sciences, founder of the school of valve energy converters based on electric machines and valve converters energy. The first conference was organized by the Institute of Innovative Technologies in Electromechanics and Robotics of the St. Petersburg State University of Aerospace Instrumentation in 2006.

The purpose of the conference is the exchange of information and progressive results of scientific research work of scientific and pedagogical workers, young scientists, graduate students, applicants and students in the field of: automatic control systems, electromechanics, electric power engineering and electrical engineering, mechatronics, robotics, automation, technical physics and management in the electric power industry.

The 13th International Conference on Electromechanics and Robotics "Zavalishin's Readings" - 2018, ER(ZR)-2018 was held on April 18-21, 2018 in St. Petersburg, Russia by the St. Petersburg State University of Aerospace Instrumentation (SUAI) and St. Petersburg Institute for Informatics and Automation of the Russian Academy of Sciences (SPIIRAS) in cooperation with German Centre for Research and Innovation Moscow (DWIH) and Russian Foundation for Basic Research (grant № 18-08-20012).

During the conferenc the invited talks were given by Prof. Michael Beetz (University Bremen, Germany), Prof. Karsten Berns (TU Kaiserslautern, Germany), Prof. Christian Ott (DLR, Germany), Prof. Sebastian Stueker (KIT, Germany), Prof. Oliver Jokisch (HfTL, Germany), Prof. Ilshat Mamaev (KIT, Germany).

More then 150 papers were submitted to the conference and each paper was reviewed by several scientists. Around 30% of the best papers were published in English proceedings by EDP Sciences on EI Serial Journal of MATEC web of Conferences indexed in SCOPUS, Thomson Reuters (Web of Science), Inspec, etc. Due to great efforts of reviewers and scientific editors Prof. Andrey Ronzhin and Prof. Vladislav Shishlakov this issue was carefully prepared and consists of 45 contributions of authors from Armenia, India, Israel, Japan, Germany, France, Mexico, Portugal, Russia, Serbia, Taiwan, Thailand, Turkey and Viet Nam. Special thanks are due to the members of the Local Organizing Committee for their tireless effort and enthusiasm during the conference organization.

We express our deepest gratitude to all participants for their valuable contribution to the successful organization of ER(ZR)-2018, hope for and look forward to your attention to the next International Conference on Electromechanics and Robotics "Zavalishin's Readings" in 2019. The conference website is located at: http://suai.edu.ru/conference/zav-read/.

General Chair
of 13th International Scientific-Technical Conference
on Electromechanics and Robotics "Zavalishin's Readings" – 2018,
Rector of the St. Petersburg State University of Aerospace Instrumentation
Prof. Yulia A. Antokhina

MATEC Web of Conferences **161**, 01001 (2018) https://doi.org/10.1051/matecconf/201816101001
13th International Scientific-Technical Conference on Electromechanics and Robotics "Zavalishin's Readings" - 2018

Emotion based human-robot interaction

Karsten Berns* and *Zuhair* Zafar

Robotics Research Lab (RRLAB), Department of Computer Science, TU Kaiserslautern, Germany

Abstract. Human-machine interaction is a major challenge in the development of complex humanoid robots. In addition to verbal communication the use of non-verbal cues such as hand, arm and body gestures or mimics can improve the understanding of the intention of the robot. On the other hand, by perceiving such mechanisms of a human in a typical interaction scenario the humanoid robot can adapt its interaction skills in a better way. In this work, the perception system of two social robots, ROMAN and ROBIN of the RRLAB of the TU Kaiserslautern, is presented in the range of human-robot interaction.

1 Introduction

An increasing interest in emotion can be seen in the behavioural, biological and social sciences but also more and more in advanced robotics. Research over the last two decades suggests that many phenomenon, ranging from individual cognitive processing to social and collective behaviour, cannot be understood without taking into account affective determinants (i.e., motives, attitudes, moods, and emotions) [1]. The emerging field of affective science seeks to bring together the disciplines which study the biological, psychological and social dimensions of affect [2]. The fact that emotions are considered to be essential to human reasoning suggests that they might play an important role in autonomous robots as well.

Providing a robot with social skills has been studied for many years and still an active research area. Several challenges have been recorded in this field due to the complexity of interpersonal communication. A social robot should work in the human's daily life environment and perceive human's feelings, intentions, and demands. Furthermore, a social robot needs to recognize and understand human's emotions to interact naturally. However, the perception of human emotions is not an easy task even for humans themselves which is a challenge in humanoid robotics.

How social robots express emotions is equally significant along with perceiving emotional states of a person. Like humans, social robots should be able to express different emotional states using their facial expressions or body movements in order to express their feelings on any event. Currently, there exists few robots that are able to generate different emotions, however these systems are quite primitive and perform poorly in real-time.

The work in this paper discusses the importance of perception system that enables a social robot to analyse complex human behaviours and interactive mode in order to interact naturally for intelligent human-robot interaction. The rest of the paper is organized as follows: previous research has been discussed in Section 2, in Section 3 we describe the low-level features and high-level perception system of social robot, ROBIN in detail. In Section 4 we discuss experiments conducted and conclude the paper in Section 5.

2 Previous research

In the last decade, numerous researchers developed task specific robots. However, few social service robots have been developed with the aim to develop robust human-robot interaction [19-23]. One such robot, ROMAN, has been developed by RRLAB as a test platform for human-robot interaction [3]. It consists of an upper body, two arms, neck, and an expressive head. ROMAN can generate facial expressions, gestures, and expressive body postures. It can generate nonverbal expressions using 47 DOF. Furthermore, it is also equipped with an expressive speech synthesizer. Figure 1 shows the ROMAN robot.

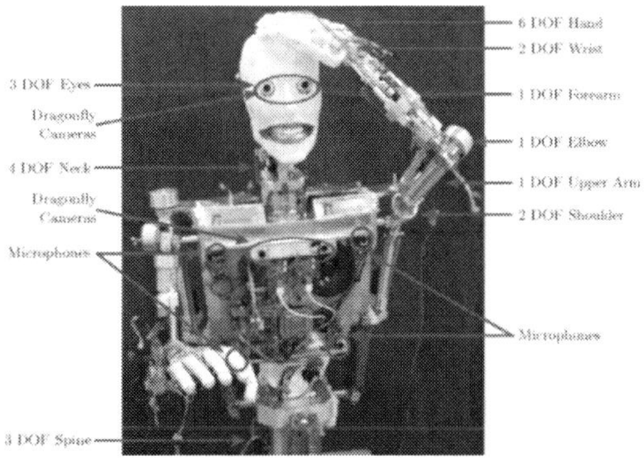

Fig. 1. Interactive Humanoid Robot, ROMAN.

* Corresponding author: berns@cs.uni-kl.de

The control architecture of the robot, emotion-based architecture, is developed by Hirth [4]. The goal of the implementation is to realize the functions of emotions: regulative (emotion), selective (percepts), expressive (habits), motivational (motives), and rating (emotion), as well as the secondary functions.

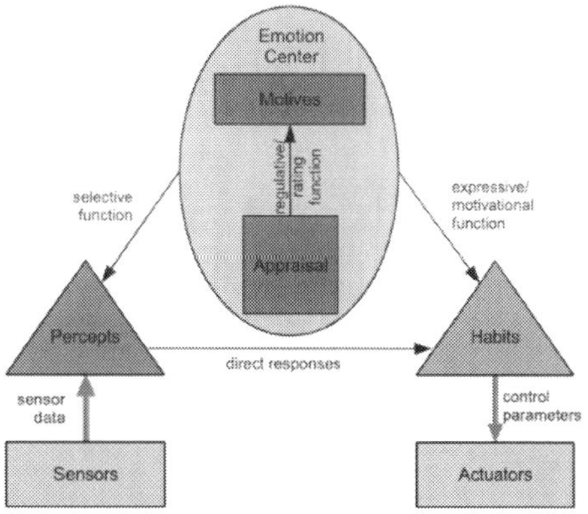

Fig. 2. Emotion-based control architecture, consisting of four main groups, motives, emotional state, habits of interaction, and percepts of interaction.

Compared to previous publications [3] where the rating and the regulative function were not included, all five functions of emotion mentioned in [5] are realized, since the rating and the regulative function are very important with a view to more cognitive or learning applications. The perception system perceives and interprets information of the environment. Depending on this information, direct responses, performed by the habits, are activated and the motives calculate their satisfaction. This satisfaction changes the current emotional state of the robot. Besides this, the motives activate several habits to change the robot's behaviour in order to reach a satisfied state. They also determine which information of the percepts is needed in the current situation. That way, the selective function is realized. This architecture has been examined using Tangram game [6] playing scenario as shown in the Figure 2.

Fig. 3. ROMAN playing Tangram game with human.

Beside the control architecture, ROMAN has been equipped with expressive function which enables it to produce different emotional states. Figure 4 depicts the facial expression of six basic emotions. A silicon skin has been glued to ROMAN's head. By moving small metal plates the different expressions have been generated [7].

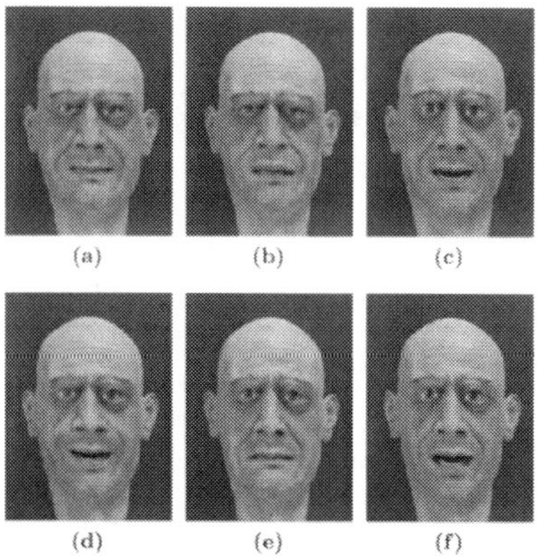

Fig. 4. Facial expressions of ROMAN realized with a silicone skin (anger, disgust, fear, happiness, sadness and surprise).

However, generating expressions accurately using silicon skin is quite challenging and cannot be generalized to different facial expressions. Moreover, ROMAN lacks in robustness when performing different gestures in real-time. Apart from emotion-based control architecture, important interactive modules like human feedback, context-aware perception and personality traits inference are missing in ROMAN.

In order to address these shortcomings, Technical University of Kaiserslautern has developed a social humanoid robot, ROBot-human-INteraction (ROBIN). It is equipped with a backlit projected face, arms, hands and torso. It can speaks via its built-in speech synthesis module in English and German language. The face makes use of projective technology to express almost any facial expression using action units. The head is able to move sideways for about ±45 degrees. RGB-D sensor is installed on the chest of robot. Additionally, a high definition camera is also installed on the head. The whole arm has 14 degrees of freedom, where hands are able to perform nearly all gestures. ROBIN has its own processor that can handle all the movements of joints. Figure 5 shows the ROBIN robot. In the following section, we describe the perception system of ROBIN.

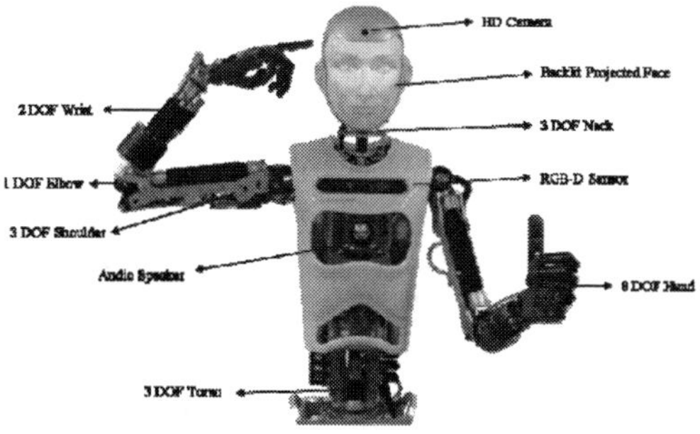

Fig. 5. Interactive humanoid robot, ROBIN, developed by TU Kaiserslautern, Germany.

3 Perception system of ROBIN

The human perception system is a part of the central nervous system, which is responsible for detecting and interpreting the surrounding environment. It consists of two eyes and the perception part of the brain. Eyes play the role of the camera to perceive the environment whereas the brain does all the complex image processing. Human, in general, depends on the visual system more than all other sensory systems. This is not only due to the quantity of information it provides, but also the robustness and efficiency in processing the information being a key factor.

Perception of the environment includes analysing and interpreting human behaviours. Most of the human behaviours are realized using verbal and nonverbal cues. Psychologically, nonverbal cues convey a lot of information and sometimes more than what verbal cues can express. Analysing and interpreting the nonverbal cues is a challenging task even for humans. ROBIN perception system can be divided into 2 subcategories: low-level perception and high-level perception system. Former deals with low-level features, e.g., facial expressions, posture, head pose, etc. while latter deals with complex human behaviours, e.g., feedback perception, context-aware perception and personality traits perception. We discuss each of these in the following subsections.

3.1 Low-level perception system

Before recognizing complex human behaviours and gearing a robotic system with emotional intelligence, recognition of low-level features accurately in real-time is highly critical. In this level, information from sensors are processed and analysed on the frame basis. The sensor used is an RGB-D sensor that provides depth information as well as RGB images. The function of low-level perception is to extract *percepts*, which are low-level information such as face location, head pose estimation, posture recognition, hand recognition, facial expressions recognition etc. In the section, we discuss major low-level features implemented in ROBIN.

3.1.1 Face detection

Face detection is an essential skill that a social robot should possess. It is the core process for many other skills like face tracking, facial expression recognition, face recognition, gender/age recognition, and head pose estimation. The present work utilizes a Haar cascade classifier to find candidate faces in 2D RGB images. The faces detected by the Haar cascade classifier is not always representing a real face. This is due to the used 2D features that can be found in any variations of intensities (colors) analogous to a human face. To reduce the false positives that Haar cascade classifier produces, the corresponding depth information acquired by an RGB-D sensor is used [8] as shown in Figure 6.

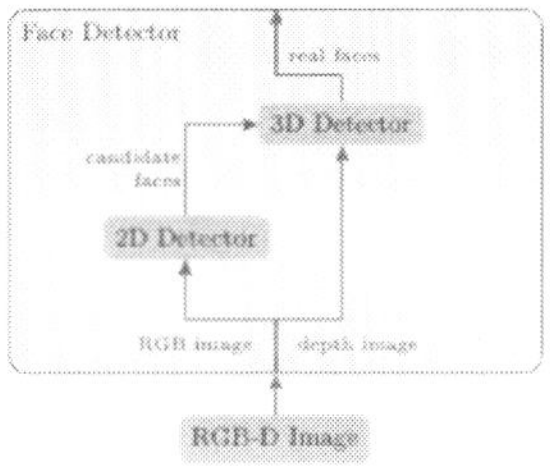

Fig. 6. Working of face detection module.

3.1.2 Facial expression recognition

Facial expressions of humans play a significant role in the interpersonal interaction. They are the primary source of expressing emotions and feelings. The work at hand uses action units to recognize the facial expression of the interaction partner [9]. It uses a deep convolutional neural network (CNN), or deep learning to estimate the activation of each of the related action units and combinations. The architecture comprises six layers (excluding input layer) as shown in Figure 7. For each action unit, CNN receives a human face as a 32×32 gray image and outputs the confidences of two classes active and not active of the corresponding action unit. It uses two convolution layers each with its own subsampling layer (max-pooling). 6 basic and neutral facial expressions have been recognized using action units results with an overall accuracy of 90.2%.

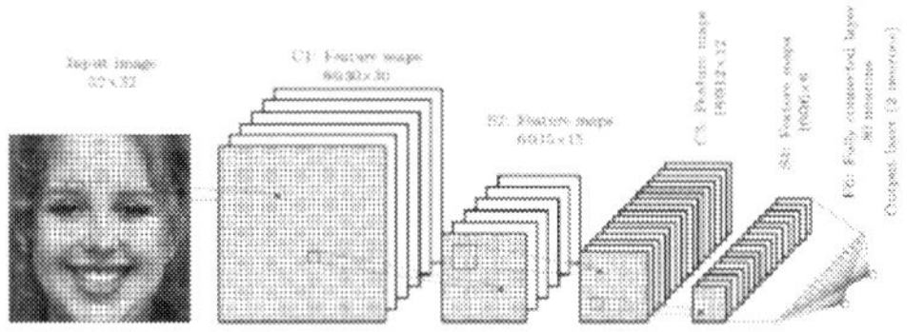

Fig. 7. Deep architecture of facial expression recognition module.

3.1.3 Posture recognition

Human pose also shows a whole lot of information about human emotional state. Human pose can convey information about human internal state whether he/she is nervous, or confident, or thinking, or not interested etc. In order to realize this on the robot, we use an approach proposed by Zafar et al. [10] that uses depth data along with NiTE library to detect human joint positions and then convert them into meaningful angles for feature vector generation task. The resultant feature vector is quite unique for each posture and is invariant to height, body shape, illumination, proximity and appearance of human. The working schematics of the proposed approach is shown in figure 8. The system is able to recognize overall 21 gestures real-time when classified by using multi-class SVM with an accuracy of 98%.

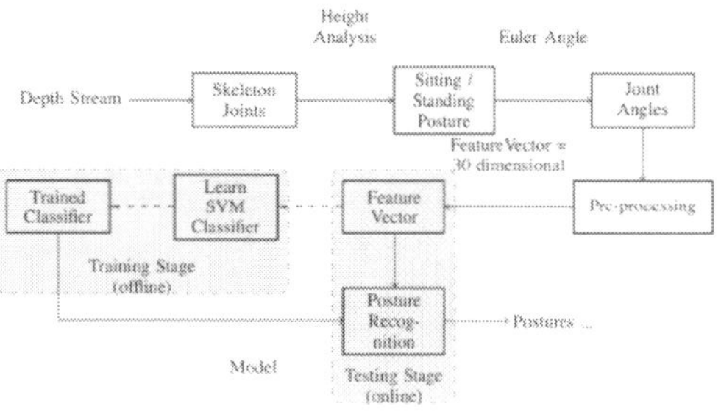

Fig. 8. Working of Posture Recognition Module.

3.1.4 Static hand gesture recognition

Like other nonverbal cues, hand gestures are also used in interpersonal face-to-face communications. They also reveal some of the human's emotions and feelings. In this work, the hand gesture recognition approach proposed by Zafar and Berns [11] is used as shown in Figure 9. This approach has been implemented to recognize 18 different hand gestures and shown high efficiency and robustness. It uses bag-of-features (BoF) of scale invariant feature transform (SIFT) key points and support vector machines (SVMs) to recognize hand gestures. Interest point approaches extracts useful information from an image and bag of feature approach is used to represent them in terms of a feature vector. A total of 15 different hand gestures are recognized with more than 94% accuracy in real-time.

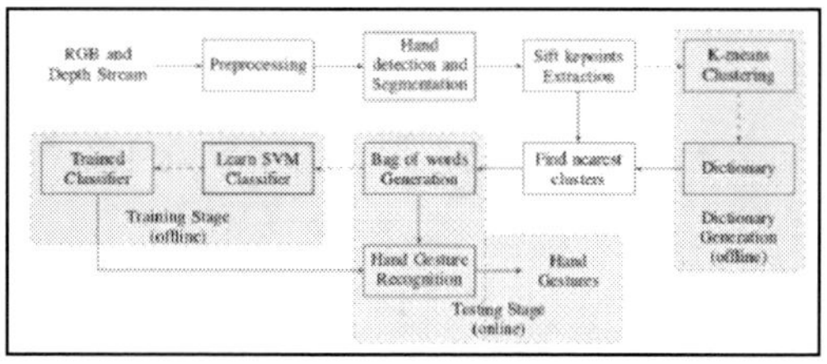

Fig. 9. Schematic flow of hand gesture recognition module.

3.1.5 Dynamic hand gesture recognition

Similar to static, dynamic hand gesture also shows complex human behaviour when interpret accurately along with the context. The system uses the approach presented by Zafar et al. [12]. The recognizer uses only depth image information, and the hand position provided by a hand tracker library, in order to construct its feature vectors. The recognizer builds two types of feature vectors to increase accuracy; the frame feature vectors that describe a static hand, and the sequence feature vectors that describe a contiguous segment of frames. The recognizer also uses two statistical classifiers. The frame feature vectors are utilized by the frame classifier. The results of the classifier, then become part of the sequence feature vector, which in turn are utilized by the sequence classifier as shown in Figure 10. A total of 16 different interactive dynamic hand gestures has been recognized with an approximate accuracy of 87%.

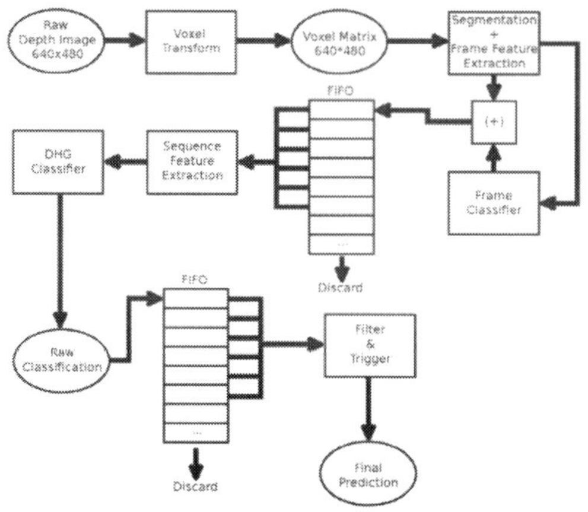

Fig. 10. Working flow of dynamic hand gesture recognition module.

3.1.6 Head pose estimation

Head movements convey a lot of information in inter-human communication. Humans have the ability to interpret these movements quickly and effortlessly, while it is regarded as a difficult challenge in computer systems and robotics. Detecting the human head movement requires estimating the head pose (position and orientation) over time. This work uses depth information for head pose estimation. This approach is called Direction-Magnitude Pattern (DMP). These features are suitable for depth information because they result in a surface map of the face. Three linear Support Vector Machines for Regression (ε -SVRs) are trained to detect the pose angles roll, pitch and yaw as presented in [8] and extended in [13]. The input of these SVRs is the feature descriptor of the face under test which is DMP features. Dynamics of the approach is shown in Figure 11. The reports a mean error of ± 3 degrees when tested in real-time. Head gestures include nodding, shaking, tilting, looking up, down, right, left, forward are realized in this work.

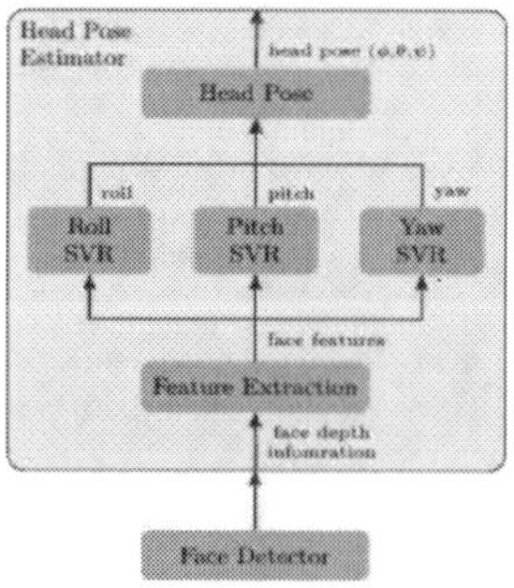

Fig. 11. Working flow of dynamic hand gesture recognition module.

3.1.7 Human identity recognition

Recognizing human identity is an important aspect in intelligent human-robot interaction. In human-human interaction, humans recognizes other human's identity and according to his past experiences all together changes his whole behaviour. According to some psychologists, recognition of identity is one of the basic and equally important step on which we build our interaction. In order to recognize human identity, we use Local Binary Pattern Histogram approach to extract local feature from human face and classify them [14]. More than 20 different subjects have been stored in the system and robot with an accuracy of 97.3% recognizes human identity correctly.

3.2 High-level perception system

High level perception deals with more complex human behaviour which uses combination of low level features stated above. Human behaviour can be recognized by recognizing some basic low level features and realize it over longer period of time to extract some useful cues.

In order to realize robust and natural human-robot interaction, we have developed higher level human perception system that extracts human feedback features, context aware features and lastly the human personality features. These features are all equally important to realize a natural and efficient multimodal human-robot interaction. The following subsections discuss the details of high-level features.

3.2.1 Human feedback perception

Humans use feedback in their conversation to exchange information about four communicative functions. These functions are continuation, perception, understanding and attitudinal reactions. Any expression, verbal or non-verbal, can be regarded as a feedback if it serves one or more of these functions.

Human feedback perception by a robot requires many low-level perception processes. These processes include a frame-based perception of human articulators such as the face, head, hands, body, and so on. In this work, the outputs of these processes are called percepts. Head pose, facial expression, hand static gestures, and posture are examples of the percepts. These low-level percepts then can be accumulated over time to form higher-level information. High-level percepts include dynamic head gestures, dynamic hand gestures, body movements, and so on.

In order to communicate successfully, according to Allwood [15], two participants should establish a contact with each other. It is necessary that participants show their ability and willingness to continue in the interaction. Continuation signal tells the interlocutor about the desire of continuing in the interaction regardless of the contents of the message. Two types of continuation signal can be distinguished: You go on and I go on [16].

Once the contact is established, one participant possibly produces a message. The receiver should be able and willing to perceive the message and he may or may not understand the message. Understanding the perceived message means that the contents of the message is interpreted in the same way that the speaker intends to. Once the receiver understands the message, he/she may give attitudinal and behavioural reaction according to the acceptance of the message. Attitudinal reaction can be positive, negative or neutral. The implementation of feedback perception is described in [16] in detail.

3.2.2 Context-aware perception

One of the key aspect in human-human interaction is the knowledge of context. Humans use this feature intelligently in order to reduce the number of possible interactive topics whenever they interact with each other. As soon as person A meets person B, person A automatically recalls every information regarding person B starting from some basic things how person B looks like, his interests, his views and opinions about different things to some more subjective things like his personality and his personal relation with him etc. In short all that information acquired over past interactions becomes active. This information plays a huge role during interaction. Human after an interaction with other person know which things are more interested to him and which are not and hence ask questions or talks on those topics.

Similarly, in human-robot interaction, this feature can make the whole interaction intelligent. In order to implement this context aware interaction feature, robot has to store all the information of interaction with each person in a separate file. This file is accessed only when that person is recognized using face recognizer module. Robot can access the information acquired in the past interactions and use it to bring topics that are more relevant to that person for interaction.

How this file is being generated? The process is quite similar to how human acquires knowledge. According to psychological studies, humans examine each other and ask questions starting from some basic questions like "What is your name?", "What do you do?", "Where do you come from?" etc. and with time they ask about their interests, their hobby's, their goals and so on. Similarly, the robot recognizes the person in the start and checks whether there is any information regarding that person is present in its database. If not, then robot activates the introductory question script where, it started asking about person whereabouts and his/her occupation and then moved more towards asking his/her goals, interests, favourite things, etc. Currently, the system has more than 60 questions which robot can ask and saves the responses in a separate file. Based on the information stored in this file, robot finds relevant things, topics or ask questions.

This type of interaction is intelligent as it is not scripted or fixed. Robot can talk on topics which are relevant to person interacting with him. We have conducted experiments where robot first asks questions in the start to know about the person. The person responses are saved in a file. This file is stored with a name of the person in the system. Next time when that person again interacts with robot, robot recognizes him and brings all the information regarding that person for context aware human-robot interaction. Experiments have shown the effectiveness of the system and reports promising results.

3.2.3 Personality traits perception

Knowledge of human personality is a bonus in human-robot interaction. Robot can behave differently according to human personality type. However, recognizing human personality is a very challenging task. Even for humans, sometimes it is difficult to know the personality of our counterparts. Numerous psychologists have presented their theories which reports some features that represent human personality type. Most of these features are not the defining features and hence do not points directly towards a personality type. However there exists some studies which maps different non-verbal cues to different personality types. According to psychologists, it is almost impossible to derive personality type from nonverbal cues which makes the problem difficult. Since tone of human speech conveys a lot about human personality, recognizing this feature is highly critical in precise recognition of human personality. However, recognizing tone of human speech is not realized in this project.

Other methods include recognizing some cues about human personality type from human body gestures and appearance, and asking questions regarding different situations and analyse how the person responds. Though these methods may not be precise alone but combining both of these methods provide us a good hint about the personality type of human. According to psychological survey conducted by Jensen [17], different personality types can be recognized using non-verbal patterns namely: (a) Extraversion, Introversion (b) Open to new experiences, Traditionalist, (c) Conscientiousness, Careless, (d) Agreeableness, Self-centred and (e) Neuroticism, Emotionally stable.

Some of the features the exploits personality type (E) including proximity feature, gesticulation features, facial expression features etc. Individuals high on extraversion prefer to stand close to conversation partner and the also like to sit close to co-communicator. Individuals low on extraversion, prefers to stand and sit at a distance when in a conversation. Similarly extraversion is more related to frequent use of rapid body movements while neuroticism in involved in more self-touching behaviour. Extraversion is also associated with more intense facial expressions, e.g., smiling a lot during conversation. Agreeableness personality type is positively correlated to laughter and sympathetic facial display. These and more features represent (roughly) some of these personality type.

The five-factor model dealing personality traits plays a critical role in the field of communication studies, psychology and philosophy. Based on the verbal and nonverbal cues, the personality trait of a human can be guessed and to some extent the specific trait can be detected. We require a human to judge in which dimension the personality of a human lies. The judgmental process is extremely fuzzy as there are so many facets ingrained in every human. Interestingly, there is no quantitative standard to judge the severity of each and every facet. Intuitive and perceptive skills do the trick for the person judging personality of another person. However, as mentioned earlier, researchers have come up with some theories and methods which can help to recognize some of these personality traits.

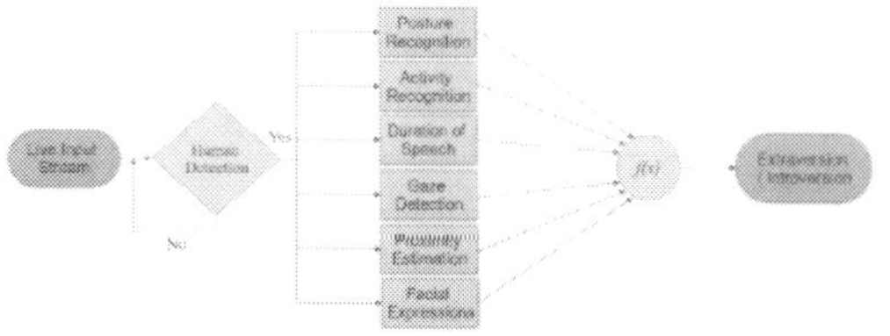

Fig. 12. Working flow of extroversion-introversion trait detection.

Figure 12 shows the work flow of our approach which uses human nonverbal cues for recognition of extroversion-introversion trait. Robot uses ASUS Xtion sensor along with OpenNI and NiTE Library to detect humans in the scenario and recognize major facets relevant to extroversion-introversion dimension. A score is estimated based on weighted function f(x) which determines extroversion-introversion personality trait.

Our approach uses nonverbal features, i.e., human posture, rapid body movements (activity), duration of speech, proximity, gaze and facial expressions as presented by Zafar et al. [18]. Integration of these nonverbal cues in a meaningful way is highly critical for overall outcome. Simplest approach is to add all the binary outcome of each facets to generate a score. However, the problem lies in the situation when one or more nonverbal feature(s) would either be detected wrongly or not detected at all. This would cause misleading outcome. In order to address this, a weighted outcome strategy has been employed. This ensures that the outcome is not sensitive to any particular nonverbal facet, rather it considers the appropriate impact of each facet in general. According to empirical studies and literature survey, it has been found out that the activity facet plays the most vital role in the recognition of extroversion-introversion. Similarly, duration of speech and posture are also considered significant in the recognition process. Equation (1) shows the weighted function f(x), which can have a value ranging from 0 - 1.

$$f(x) = 0.3\,(A) + 0.2\,(S) + 0.15\,(P) + 0.15\,(G) + \\ + 0.1\,(D) + 0.1\,(E), \tag{1}$$

where A is the activity detection, S is the duration of speech, P is the posture, G is the gaze detectors, D is the distance detector and E is the facial expressions detector. Since facial expressions are quite dynamic and vary from person to person especially when a person is speaking, they are prone to false recognition. Hence, this facet has low weight. Similarly, proximity of a person happens only for a short time during the interaction. Hence it also has low weight. The score generated by the function f(x) is used to recognize extroversion/introversion. Higher value indicates strongly extrovert person and lower value shows extremely introvert person.

4 Experimentation

For experimentation of feedback system, 5 different subjects play 20 questions game with the robot. As few subjects don't know how to play the game, robot explains the game and game-play at the beginning. To answer multiple choice question, basic hand gestures are used. An information card is placed beside the robot for the user to consult which hand gesture represents which answer option. For evaluation of the experiments, subject plays the game with the robot. System writes the activation of all the behaviours in a separate file along with head, hand, face gestures and the question for every frame. A Human expert analyses the subject behaviours in the recorded videos. He observes the feedback behaviours whenever a question has been asked by the robot and fills in the questionnaire. He also takes notice of subtle changes in the behaviour, for example lack of interest, understanding or not understanding behaviours etc. At the end of the experiments, the recorded file by the system is compared with expert questionnaire.

Figure 13 shows the plot of extroversion-introversion score of a job interview experiment. Initially, candidate acts as an introvert which can also be visualize from the plot. Candidate has crossed his arms and looking downwards with minimal body movements. This behaviour results in low score for extroversion. After period of time, candidate behaves like an extrovert as seen in the figure. Different features like open arms, body movements, facial expressions, mutual eye gaze, etc. are active and lead to high score for extroversion. Images at each time-stamp are used to validate the weighted score.

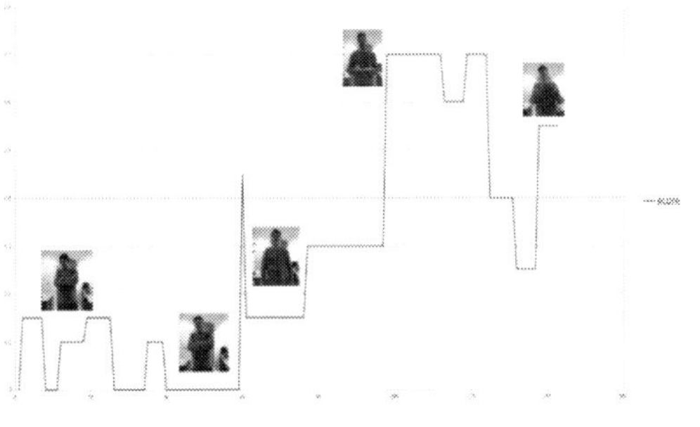

Fig. 13. Weighted score plot of extroversion-introversion trait.

5 Conclusion

In this work, we have discussed the perception system of humanoid robot ROBIN. In order to achieve emotion-based human-robot interaction, there is a need of robust perception system that is able to perceive and understand some basic human nonverbal behaviour as well as some high-level perception tasks which include understanding human feedback communication channel, context-aware interaction and personality traits inference. The system is tested in real-time with active low-level and high-level perception features.

References

1. Affective Science. http://www.ncbi.nlm.nih.gov
2. A.G. Francis, M.J. Mehta, A. Ram, *Handbook of Research on Synthetic Emotion and Sociable Robotics* (Hershey, NewYork, 2009)
3. J. Hirth, K. Berns, Humanoid robots, 97–116 (2009)
4. J. Hirth, N. Schmitz, K. Berns, International Journal of Social Robotics, **3** (3), 273–290 (2011)
5. H.R. Lueckert, I. Lueckert, *Einfuehrung in die Kognitive Verhaltenstherapie: Allgemeine Grundlagen* (Ernst Reinhardt Verlag, Munich 1994)
6. J. Elffers, M. Schuyt, *Tangram: The Ancient Chinese Puzzle* (Evergreen, 1999)
7. K. Berns, J. Hirth, Proc. Of IROS, 3119–3124 (2006)
8. S. Saleh, A. Kickton, J. Hirth, K. Berns, Proc. of ACHI (2013)
9. S. Al-Darraji, K. Berns, A. Rodic, Proc. of RAAD, 413–420 (2016)
10. Z. Zafar, R. Venugopal, K. Berns, Real-Time Recognition of Human Postures for Human-Robot Interaction, Proc. of ACHI, Rome, Italy, (2018)
11. Z. Zafar, K. Berns, Proc. of ACHI, 333–338 (2016)
12. Z. Zafar et al., International Conference on Robotics in Alpe-Adria Danube Region, 649–656 (2017)
13. S. Saleh, K. Berns, Proc. of PETRA, 5 (2015)
14. T. Ahonen, A. Hadid, M. Pietikainen, Proc. of ECCV, 469–481 (2004)
15. J. Allwood, International Conference on Cooperative Multimodal Communication, **2155**, 113–124 (2001)
16. S. Al-Darraji, Z. Zafar, K. Berns, Proceedings of the 30th International BCS Human Computer Interaction Conference: Fusion!, 27 (2016)
17. M. Jensen, Personality, J. of Social Science, **4**, 57–70 (2016)
18. Z. Zafar, S.H. Paplu, K. Berns, Proc. of RAAD, (2018)
19. T. Hashimoto, S. Hiramatsu, T. Tsuji, H. Kobayashi, Proc. of RO-MAN, 326–331 (2007)
20. M. Noma, N. Saiwaki, S. Itakura, H. Ishiguro, Proc. of Humanoids, 163–168, (2006)
21. M. Bennewitz, F. Faber, D. Joho, S. Behnke, Proc. of RO-MAN, 1072–1077 (2007)
22. K. Papoutsakis et al., Proc. of PETRA, 56 (2013)
23. T. Asfour, K. Welke, P. Azad, A. Ude, R. Dillman, Proc. of Humanoids, 447–453 (2008)

Speech interaction strategies for a humanoid assistant

Sebastian Stüker*, *Stefan* Constantin, *Jan* Niehues, *Thai-Son* Nguyen, *Markus* Müller, *Ngoc Quan* Pham, *Robin* Rüde and *Alex* Waibel
Institute for Anthropomatics and Robotics, Karlsruhe Institute of Technology, Karlsruhe, Germany

Abstract. The goal of SecondHands, a H2020 project, is to design a robot that can offer help to a maintenance technician in a proactive manner. The robot is to act as a second pair of hands that can assist the technician when he is in need of help. In order for the robot to be of real help to the technician, it needs to understand his needs and follow his commands. Interaction via speech is a crucial part of this. Due to the nature of the situation in which the interactions take place, often the technician needs to speak to the robot when under stress performing strenuous physical labor, the classical turn based interaction schemes need to be transformed into dialogue systems that perform stream processing, anticipating user intentions, correcting itself as more information become available, in order to be able to respond in a rapid manner. In order to meet these demands, we are developing low-latency streaming based automatic speech recognition systems in combination with recurrent neural network based Natural Language Understanding systems that perform slot filling and intent recognition in order for the robot to provide assistance in a rapid manner, that can be partly based on speculative classifications that are then being refined as more speech becomes available.

1 Introduction

The goal of the project SecondHands, sponsored within the European Union's 8th framework programme *Horizon 2020*, is to develop a humanoid robot that acts as an assistant to maintenance technicians. It is supposed to either proactively support the technician in performing routine maintenance operations, or to offer support upon request from the human. Within the project a consortium of five partners are collaborating on this task: Ocado, Karlsruhe Institute of Technology (KIT), École Polytechnique Fédérale de Lausanne (EPFL), Ocado, and Universitá di Roma Sapienza.

Conceptually, the robot's task is to provide a second pair of hands to the maintenance technician, such that once the robot has been trained, it can predict when it can usefully provide help and knows which actions to take to provide it. In order to reach this goal the project aims to advance abilities and key technologies relevant to industrial robotics, such as cognition, human-robot interaction, mechatronics, and perception.

To operate within environments designed primarily for industrial efficiency in cooperation with a mostly human workforce, a robot needs a rich repertoire of human-like skills, and, in the opinion of the project's partners a humanoid or human-like form, specifically in order to use the same methods of access and to be able to efficiently cooperate with humans when they are performing their tasks. One key perceptual and cognitive capability in this cooperation is the communication with the human co-workers via natural language. Traditionally this means the construction of a spoken dialog system that consists of an automatic speech recognition system, a natural language understanding system, a dialog manger, a natural language generation system and a speech synthesis system. The construction of goal oriented dialogues in this manner often leads to heavily turn-based systems that require several turns to resolve ambiguities and correct errors in an explicit manner, in order to finally reach the goal.

However this approach is ill-suited in the human assistance scenario, as the human does not have the time nor the cognitive capacities to conduct this kind of dialogue while performing his strenuous task.

KIT has therefore set-up to develop a slightly different kind of dialogue system that works in a stream-based manner. That is to process the human speech in a run-on manner to feed this continuous low-latency stream of words from it into a stream-based natural language understanding component that passes information on to the planner of the robot as it becomes available, correcting itself if necessary, when more information has become available.

Also, the dialogue is supposed to be become more "social" in order to improve the acceptance of the robot by the human technicians.

In addition we are pursuing novel, alternative approaches for end-to-end dialogue modelling that are based on memory-networks, and which in their attributes seem to be specifically suited for the use case at hand.

The dialog systems constructed in this manner have become part of the robot that has been constructed as a proof-of-concept within the project: ARMAR VI. It will be verified against a series of real maintenance tasks that

*e-mail: sebastian.stueker@kit.edu

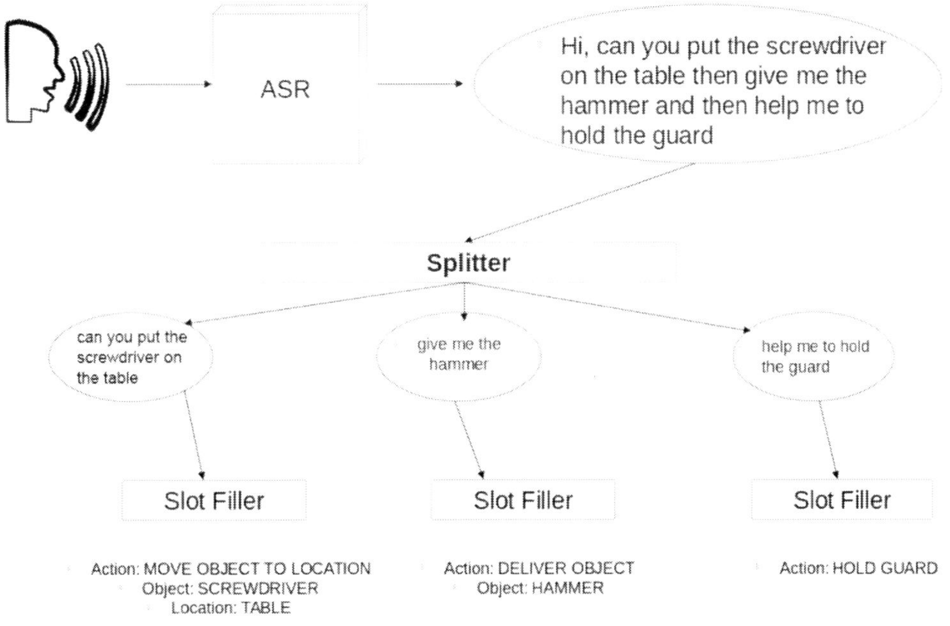

Figure 1. Natural language understanding pipeline.

occur within the Ocado's highly automated Customer Fulfilment Centres (CFCs).

2 Low latency speech recognition

Currently, in the design of modern online automatic speech recognition (ASR) systems, there is usually an audio segmenter component which seeks to detect complete segments in input audio streams, often based on detected silence. Later, an ASR component is in-charge of decoding the detected segments into text. In order for the system to keep up with the audio stream it has to run in real-time or faster. The faster the decoding runs the lower the latency which is bounded by the length of the detected segments.

In our system [1], the ASR component was built to process an audio stream and output a stream of transcripts which is the input to the subsequent components (e.g., Natural Language Processing). By using a dynamic decoding framework for speech recognition, we can avoid the detection of audio segments, and directly perform the decoding as soon as a small part of speech is recorded. Specifically, the decoding framework allows us to compute acoustic feature vectors every 0.25 seconds of the speech data and dynamically extend the search network by processing the new frames. This so called run-on recognition helps us to avoid the latency caused by waiting for the end of the current segment.

Normally, only at the end of an utterance the most probable hypothesis is determined. However, since waiting until the end of the utterance leads to a long latency, we detect when a part of the hypothesis becomes stable and can be kept. This is possible due to the fact that our search implements a viterbi beam search, so that for earlier portions of a hypothesis it is possible and frequently happens, that all competing alternatives get pruned away later in the search.

We also introduced a novel protocol in which the ASR component and the subsequent components follow to dramatically reduces the latency while still maintain the recognition accuracy. Basically, we output probable parts of the unstable hypothesis and send them to the subsequent component. Later, the recognizer can revise its decision and overwrite the previous output if necessary. In this way, the recognition component does not need to wait until a stable portion or end-of-segment, instead it finds the most probable hypothesis every iteration of the incremental decoding, detects and sends the update portions to the subsequent component.

3 Natural language understanding in stream mode

We designed a component that takes the output of the ASR counterpart and extract the information required from the utterance. Specifically, the main task of the Natural Language Understanding (NLU) component is to identify one of the predefined actions, and the corresponding objects and locations. The actions are revolved around transporting the items from and to the technicians, as well as helping the technicians in handling the guards. The included objects are the tools required in the working environment, such as screwdriver, hammer, and ratchet-spanner. Our pipeline is illustrated in Figure 1.

We use a standard recurrent neural network with Long-Short Term Memory (LSTM) units [2] to encode the whole utterance. The main NLU model is showed in Figure 2. The network runs in a bidirectional manner to get the contextual information from both sides of each word [3]. For predicting the objects and the locations, we used the local context which are the states at each time step. However, for predicting the actions, we found that it is beneficial to utilize the global context by joining the states of the first and the last steps. We also design our system to be able to detect multiple actions in one utterance. In order to solve this problem, we train an addition sentence splitter which is also fundamentally a bidirectional recurrent neural network. The sentence splitter annotates the splitting point of the long sentence. For example, "can you put the screwdriver on the table then give me the hammer and then help me to hold the guard" is split into "can you put the screwdriver on the table", "give me the hammer", and "help me to hold the guard". We generate the training data for this splitter simply by randomly concatenating sentences in our training data.

Our NLU component currently performs recognition whenever the segment is sent by the ASR component. The segment splitting of the stream is considered a subtask in the ASR component. Potentially, our component can also be minimally modified to be able to process the stream in a token basis. Such design allows us to be able to produce the prediction sooner than waiting for the whole utterance to finish.

4 End-to-end dialog modelling with memory networks

The NLU component, described in section 3, process the splitted sentence parts individually without regarding previous parts or asking questions. However, in the human-robot interaction, it is often necessary to regard the previous parts of a dialog. For example, the human asks the robot to redo the last action for another object. Often, it is also necessary to have multiple interactions between the human and the robot before an action can be executed by the robot. For example, a human asks a robot to hand over a screwdriver. In this example, the information, what size the screwdriver should have, is missing. Therefore, a robot should be able to ask for this information and after getting the information, it can hand over the user a suitable screwdriver. The robot should also be able to ask questions to resolve ambiguities. To be able to regard previous parts of the dialog and ask questions, a Dialog Manager (DM) and a Natural Language Generation (NLG) component are necessary. A dialog system with a DM, NLU, and NLG component that can be trained end-to-end make it possible to change or extend the domain of the dialog system without changing the dialog system. It is only necessary to train the system with an extended or changed dataset.

The end-to-end trainable Memory Network [4] is an interesting neural network architecture for using it as NLU, NLG, and DM component that can be trained end-to-end [5]. It is a neural network architecture that has access to the memory. All utterances before the current utterance of the user are saved in the memory as utterance-embeddings. To every utterance-embedding in the memory, it is calculated how relevant it is to the last utterance of the user. All memory entries are multiplied by their relevance and summed up. This sum plus the utterance-embedding of the last utterance of the user is the output of the Memory Network. This conforms to the NLU and DM component of a dialog system. The output of the Memory Network is multiplied by a trainable matrix. Each vector representation of a possible output candidate is multiplicated by that result. The candidate with the highest value is the output of the system. This conforms to the NLG component of a dialog system. An alternative to this kind of NLG is to generate the answers word by word [6]. The trainable matrices are trained by the backpropagation algorithm.

The Memory Network achieved good results in experiments with the Dialog bAbI Tasks. In the first task of the Dialog bAbI Tasks a dialog system has to request for four restaurant reservation parameters and should be able to generate after receiving all parameters an API call with them. The dialog system has achieved more than 99 % accuracy in the first task of the Dialog bAbI Tasks. To test if the system is suitable for the robot domain, we made a dataset similar to the bAbI Dialog tasks for the robot domain. In this dataset, the dialog system achieved an accuracy of more than 95 %.

5 Non-verbal cues as feedback for the user

Backchannel (BC) responses like "uh-huh", "yeah", "right" are used by the listener in a social dialog as a way to provide feedback to the speaker. In the context of human-computer interaction, these responses can be used by an artificial agent to build rapport in conversations with users and make the human-robot interaction more social. In the past, we proposed an approach by training artificial neural networks on acoustic features such as pitch and power and also attempted to add word embeddings via word2vec [7]. We now refined this approach by evaluating different methods to add timed word embeddings via word2vec. Comparing the performance using various feature combinations, we observed that adding linguistic features improved the performance over a prediction system that only uses acoustic features [8]. The most commonly used acoustic features in related research are fast and slow voice pitch slopes and pauses of varying lengths. A neural network is able to learn advantageous feature representations on its own. By inputting the absolute pitch and power (signal energy) values for a given time context, the network is able to automatically extract the pitch slopes and pause triggers by subtracting the neighboring values in the time context for each feature. We train the network on the outputs [1, 0] for BCs and [0, 1] for non-BCs. The placement of BCs is dependent on previous BCs: If the previous BC utterance was a while ago, the probability of a BC happening shortly is higher and vice versa. After each BC, the

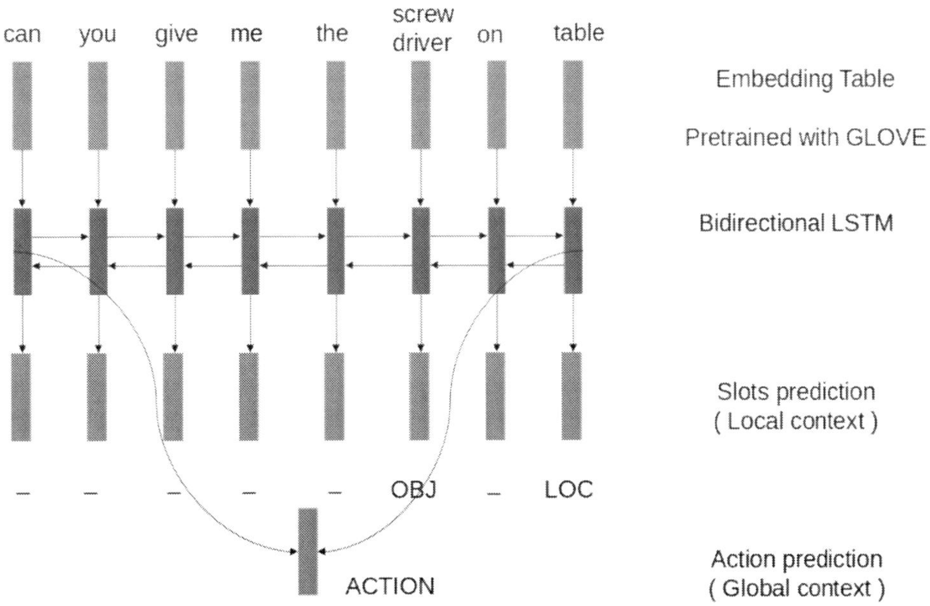

Figure 2. Natural language understanding model details.

probability of a new BC rises over time. To accommodate for this, we want the neural network to also take its previous internal state or outputs into account. We do this by using LSTMs instead of dense feed forward layers. Our goal is to generate an artificial audio track containing utterances such as "uh-huh" or "yeah" at appropriate times. The neural network outputs a noisy value between 0 and 1. To generate an audio track from this output, we need to convert this noisy floating probability value into discrete trigger time stamps: We first run a low-pass filter over the network output, which removes all the higher frequency components and yields to a less noisy and more continuous output function. To ensure our predictor does not use any future information, this low-pass filter must be causal. The commonly used gaussian filter is symmetric, which in our case means it uses future information as well as past information. To prevent this, we cut the right side of the filter off asymmetrically at some multiple c of the standard deviation σ. Then we shift the filter to the left so the last frame it uses is ± from the prediction target time. This means the latency of our prediction increases by $c \cdot \sigma ms$. If we choose $c = 0$, we cut off the complete right half of the bell curve, meaning we do not need to shift the filter, which keeps the latency at 0 ms, but at the cost of accuracy of the low-pass filter.

All of the results in Table 1 use the following setup: two LSTM layers (with 70 and 35 neurons), a context width of 1500 ms, a context frame stride of 2, and the margin of error is 0ms to +1000 ms. We initialized the weights using Glorot uniform initialization and used Adam for optimization. These parameters were chosen to give the best results on the validation data set. A more detailed comparison of training methods, feature combinations, context widths and network layouts can be seen in our previous work [7]. The experimental setup is slightly different there, so the absolute F1-values should not be directly compared. Precision, recall and F1-Score are given for the completely independent evaluation data set. The performance with both linguistic and acoustic features is significantly better than with just acoustic features ($p < 1$ %). The improvement of adding the ICSI meeting corpus to word2vec training is not statistically significant.

6 Conclusion

In this paper we have given an insight into the dialogue modelling activities that we are performing within the H2020 project *SecondHands*. The goal of the project is the construction of a humanoid robot that acts as an assistant to the human technicians performing maintenance tasks.

In order for the robot to collaborate well with the technician, we have shown several ways for building a stream-based spoken dialogue system that offers low-latency and does not rely on turns. We thereby made use of stream-based speech recognition that works with speculative output, of recurrent-neural-network based natural language understanding that can operate in a stream-based manner, or of end-to-end dialogsystems based on memory networks.

In order for the dialog to be more socially acceptive to the human user, we also demonstrated a non-verbal backchannel system that can support the communication with the user.

Table 1. Objective results with various input feature combinations.

Features	Precision	Recall	F1-Score
power only	0.244	0.516	0.331
accoustic (power, pitch, ffv)	0.279	0.515	0.362
linguistic (word2vec$_{dim=30}$)	0.244	0.478	0.323
accoustic and linguistic (power, pitch, ffv, word2vec$_{dim=30}$):			
· word2vec trained on Switchboard	0.298	0.510	0.376
· word2vec trained on Switchboard	0.305	0.519	0.385

This work has been conducted in the SecondHands project which has received funding from the European Union's Horizon 2020 Research and Innovation programme (call:H2020- ICT-2014-1, RIA) under grant agreement No 643950.

References

[1] J. Niehues, T.S. Nguyen, E. Cho, T.L. Ha, K. Kilgour, M. Müller, M. Sperber, S. Stüker, A. Waibel, Interspeech 2016 2513–2517 (2016)

[2] S. Hochreiter, J. Schmidhuber, Neural computation **9**, 1735 (1997)

[3] G. Mesnil, X. He, L. Deng, Y. Bengio, *Investigation of recurrent-neural-network architectures and learning methods for spoken language understanding.*, in *Interspeech* (2013), 3771–3775

[4] S. Sukhbaatar, A. Szlam, J. Weston, R. Fergus, *End-To-End Memory Networks* (2015), 2440–2448, `http://papers.nips.cc/paper/5846-end-to-end-memory-networks`

[5] A. Bordes, Y.L. Boureau, J. Weston, *Learning End-to-End Goal-Oriented Dialog*, in *Proceedings of the International Conference on Learning Representations, ICLR 2017, Toulon, France, April 24-26, 2017* (2017), `http://arxiv.org/abs/1605.07683v4`

[6] S. Constantin, J. Niehues, A. Waibel, *An End-to-End Goal-Oriented Dialog System with a Generative Natural Language Response Generation*, in *Proceedings of the International Workshop on Spoken Dialogue Systems, IWSDS 2018, Singapore, Singapore, May 14-16, 2018 (forthcoming)* (2018)

[7] R. Ruede, M. Müller, S. Stüker, A. Waibel, *Yeah, Right, Uh-Huh: A Deep Learning Backchannel Predictor*, in *Proceedings of the Eight International Workshop on Spoken Dialogue System (IWSDS 2017)* (Pittsburgh PA, U.S., 2017)

[8] R. Ruede, M. Müller, S. Stüker, A. Waibel, *Enhancing Backchannel Prediction Using Word Embeddings*, in *Proceedings of INTERSPEECH* (Stockholm, Sweden, 2017)

Advances in the development of a cognitive user interface

Oliver Jokisch [1, 2, *] and *Markus* Huber [2]

[1] Leipzig University of Telecommunications (HfTL), Institute of Communications Engineering, 04277 Leipzig, Germany
[2] InnoTec21 GmbH, 04105 Leipzig, Germany

Abstract. In this contribution, we want to summarize recent development steps of the embedded cognitive user interface UCUI, which enables a user-adaptive scenario in human-machine or even human-robot interactions by considering sophisticated cognitive and semantic modelling. The interface prototype is developed by different German institutes and companies with their steering teams at Fraunhofer IKTS and Brandenburg University of Technology. The interface prototype is able to communicate with users via speech and gesture recognition, speech synthesis and a touch display. The device includes an autarkic semantic processing and beyond a cognitive behavior control, which supports an intuitive interaction to control different kinds of electronic devices, e. g. in a smart home environment or in interactive respectively collaborative robotics. Contrary to available speech assistance systems such as Amazon Echo or Google Home, the introduced cognitive user interface UCUI ensures the user privacy by processing all necessary information without any network access of the interface device.

1 Research background

The joint research project Universal Cognitive User Interface (UCUI) 2015–2018 is developing methods, data and a prototype [1] to easily manage connected home appliances (as example scenario) by corresponding intuitive actions of the users. In the framework of our contribution, some preliminary results from the UCUI project shall demonstrate the potential for a novel class of interfaces for the human-machine or human-robot interaction.

With UCUI, the user can control the system via speech, gestures or even a virtual keyboard. The system is designed to operate autonomously, neither using an extensive database nor a network connection. Recent speech dialog systems and cognitive user interfaces allow a verbal, natural human-machine-interaction and achieve an excellent performance. However, the leading commercial solutions heavily rely on transmitting sensitive user information such as personal data or voice recordings through public networks and on processing, storing and analyzing the data on servers of external service providers.

The UCUI demonstrator is realizing a cognitive user interface for intuitive interaction with arbitrary electronic devices and ensures privacy by design. The system collects user-specific data which are processed by a cognitive behavior control to allow an adaptation to the users' communication style and to improve the strategy in problem solving. The underlying paradigm requests a systems' adaptation to the user and not vice versa, assuming the fact, that such systems are mainly used by human beings who are less trained in the use of complex technical devices. In addition, user-specific data are not delivered to other users to avoid possible conclusions from these data. In order to achieve an appropriate system behavior, a variety of possible human-machine interactions needs to be integrated into the UCUI system, since alternative input phrases may have an identical meaning in speech control. Therefore, all input and output modalities are fused on a semantic processing level.

For the data preparation we conducted Wizard-of-Oz (WoZ) experiments to collect typical user inputs [2]. In further steps, the user behavior shall be analyzed and integrated into the system model.

The analysis and classification software in the system is based on the Unified Approach to Signal Synthesis and Recognition [3, 4], hosted by the BTU Cottbus-Senftenberg and Fraunhofer IKTS including a speech recognizer and synthesis engine, both ported to the hardware.

The project partners are mainly focusing on the cognitive processing of meanings and knowledge about the user habits. For the representation of semantic data feature-values-relations [5-8] are used, processed by Petri net transducers (PNT) [9, 10]. Feature-values-relations are treelike non-sequential structures, where a feature has a set of values which themselves can be features again. Petri net transducers are used to translate input signals into such structures and also for translating them into output signals.

The system shall be capable of learning from the behavior of users in order to improve its function. Multiple devices will be able to cooperate (distributed microphone array, task assignment, etc.) over a strongly encrypted wireless connection. The system design is based on studies of user-machine interactions in a real home-automation scenario and takes into account relevant legal and ethical aspects.

For the demonstrator, the project partners reduced the task to the domain of controlling a heating installation.

* Corresponding author: jokisch@hft-leipzig.de

The semantic processing transforms all inputs (speech, gestures and touch screen) into a unified representation. By the cognitive behavior control, the representation can be transformed into any output channel (speech, acoustic signals, display). Figure 1 shows the first version of the UCUI demonstrator.

Fig. 1. Demonstrator of UCUI, first version 2017 [1].

2 User-centered experience

2.1 Data retrieval by Wizard-of-Oz experiments

The described cognitive interface is developed user-driven, which poses a challenge, as the overall system is still under construction. The project partners need to evaluate and to optimize some system functions before their implementation. For this purpose, the Wizard-of-Oz (WoZ) method is used in the UCUI project [2]. The main component in Wizard-of-Oz experiments is a human being (the wizard), who simulates the final systems' behavior. During the experiment, the test user interacts with the interface of a simulated technical system. All system reactions to the user are pretended by the wizard.

Wizards have to react accurately and in short time on user inputs. This can be supported with predefined, frequent responses in rapid access, e.g. "please wait, your inquiry is processed" or similar statements, and by a suitable training of wizards to achieve constantly accurate responses. User scenarios and tasks require a known goal of the actions which can be only achieved by the means of the system without restricting the user in his solution strategy, verbal utterances or gestures. The task construction has to consider the interaction variety of the user and should communicate the options to the user.

Within the UCUI tests, the user is receiving written instructions beforehand, and the interface system is demonstrated on the basis of a simple vendor machine application by an investigator (not identical with the wizard). Furthermore, the task assignment is based on hypotheses with regard to the expected user and system behavior.

Finally, a successful, user-driven system construction includes a series of WoZ experiments, whereby the tested system states should increasingly interact in autonomous mode with the user, i.e. less-controlled by the wizard. Consequently, the UCUI project involves three consecutive test runs, followed by the overall evaluation of the optimized system.

A Wizard-of-Oz Framework (WoOF) was built to support the user-driven construction. It allows for the creation of different evolving simulators and serves as an execution environment for these simulators.

The formulated requirements include general ones on frameworks supporting WoZ experiments as well as special ones following from the project specifications. Since the task is to simulate a real system which gives visual and audible feedback to the user, there has to be some mechanism to present visual objects on a monitor and to route audio data to the user. Eventually the system should be controllable via speech and touch-input (among other inputs) which imposes the necessity to interact with the visual objects and to route audio data from the user to the framework. Besides these basic functionalities, an adequate support for the realization of the experiments has to be included. This covers creation of simulators and experiments as well as supporting the wizard during the experiments.

For the UCUI project the first experiments [2] consisted of a series of scenarios which are seen as tasks to the participants. The wizard was able to switch between scenarios. A single scenario is understood as a unit of a user task, the aim of the task, and possible visual and audible feedback.

The outcome of the experiments included a collection of user behaviors. Therefore all interactions with the system were recorded which included the monitor content the participants saw, all touch-events they triggered on it and all spoken input during the experiments. To support the integration of the gesture control in following project phases, the participants were additionally recorded by camera. To ease the evaluation of the recorded data the audio output of the system was recorded as well.
A form of session management to allow for data per participant was also included. To respect the privacy of the participants was explicitly not a requirement on the framework. This had to be assured by the experimenters. The motivation behind was that there cannot be any algorithmic solution appropriate to all applications of the framework. So this burden was left to the users.

A session as the execution of an experiment with a distinct participant included some well-defined events. Also it was possible to add new types of events to the framework.

The preparation of the collected data was semi-automated. All data of a session were cut into pieces corresponding to the scenarios. Audio data were transliterated and phonetically transcribed. All steps were advised by a human being. The construction of the feature-values-relations and Petri net transducers (cf. next section) is an ongoing research.

Figure 2 shows the WoOF graphical interface as seen by the Wizard, mainly on the left side. We replaced the original smart home task in UCUI (the heating control feedback) on the right side by the planned interactive cooperative robot task to collect fruits and to hand them over to a user, suggested by the AI/robotic company 7Bot in [11]. In the future human-robotic experiments, we will use such a low cost platform (in this case ca. 350 USD) with open source software interface to collect our user inputs. Of course, for the WoZ experiments, the robotic arms will be manually controlled by the wizard. This will include actions to surprise users in order to collect more realistic cognitive user data in such a cooperative control task.

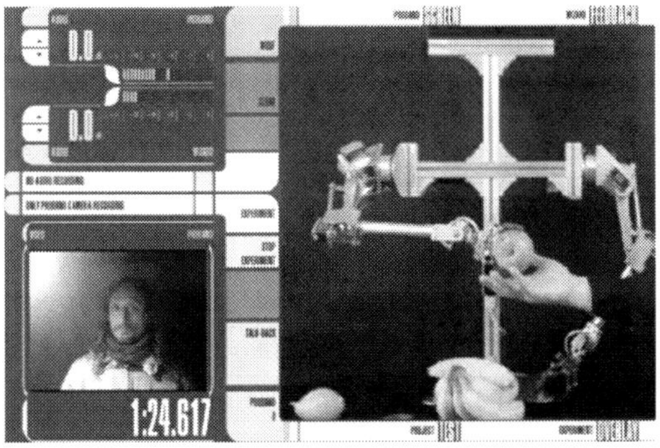

Fig. 2. Wizard panel from [2], conceptually extended by an interactive cooperative robot task (right, excerpt from [11]).

2.2 Employment of image schemas

The paradigm shift from mainly technology-centered devices – as described in introduction and preliminary WoZ experiments – towards human-centered devices requires the implementation of basic human models for a variety of human experiences.

In [12], we reported amongst others on an experimental investigation of image schemas as basic features of human knowledge that could help to support the development of more intuitive interfaces that should enable effective interactions with a system based on the subconscious application of basic prior knowledge according to Mohs et al. [13] and Turner [14]. One of such a fundamental aspect of human experience, and the focus of this investigation, was the impact of image schemas on human knowledge and language. Image schemas (e.g. up-down, center-periphery) are basic pre-conceptual, universal patterns of human experience that integrate information from multiple generic and perceptual modalities such as visual, acoustic and haptic information, i.a. suggested by Lakoff and Johnson [15]. They serve to structure human knowledge, behavior and experience. As basic building blocks of human knowledge, generated in earliest childhood interactions, image schemas may be available to all potential users. While previous research seems to support this hypothesis (e.g. Hurtienne et al. [16]), surveys on concept formation suggest that some image schemas occur earlier in infancy compared to other image schemas. Mandler and Pagán Cánovas [17] considered path schemas such as up-down, container, location, blockage, into, out of, open as basic image schemas occurring early in infancy at preverbal stages. On the contrary, center-periphery, scale, balance, cycle and other process schemas, near-far, multiplicity/unity as well as attributional image schemas (e.g. big-small, warm-cold) are built upon these basic primitives and, thus, should exert less influence on the development of human thought and knowledge [17].

To investigate whether developmental occurrence of image schemas influences their appliance in human speech interaction with machines/computers, we applied the WoZ paradigm and tested two hypotheses in [12]: First, we expected early, basic image schemas to be employed more often than later image schemas. Second, we expected no impact of individual difference variables on the frequency of applied image schemas, since these basic building blocks of human knowledge should be equally available to all users regardless of age, gender and technical experience. Forty-three German native speakers (20 men, 23 women; mean age 29.2 years, SD 9.9 years) participated in the speech interaction study. To calculate technical experience (TE), participants were asked to indicate the frequency with which they used a variety of technical devices at home (e.g. smartphone, laptop) and in public (e.g. ticketing machine, self-service banking). For seventeen items, participants had to select response options from 0 (I don't know this device) via 5 (almost daily) to 7 (more than once per day). The mean of all responses was then calculated for each participant with a higher score indicating greater technical experience. The TE score in our test persons ranged from 2.35 to 4.24 (mean TE score: 3.34). Both, age and gender were surprisingly not correlated with TE.

The speech interaction task took place in a quiet and moderately illuminated room. All participants were asked to complete the questionnaire ascertaining their technical experience as well as demographic data. Participants were then asked to stand in front of a multi-touch panel. They were presented with twenty test scenarios, investigating the image schemas-underlying free speech interaction with a heating device. They were not informed about image schemas in any way and were told that there are no right or wrong answers, since they have to test the functionality of a newly-developed heating device. Participants were asked to respond to the various scenarios, as they would also do at home and to simply tell the heating device which changes, if any, they would like the device to carry out. A content analysis was carried out to identify image schemas underlying utterances (see Table 1). Image schemas underlying speech and gesture responses were analyzed by two independent coders. To ensure reliability of coding, one coder coded the entire speech and gesture dataset. A second coder that was blind to the study hypotheses coded 25% of both datasets. In the speech interaction study, the two coders agreed 88% of the time.

Ten different image schemas (up-down, verticality, horizontality, balance, scale, warm-cold, process, near-far, multiplicity/ unit and container) were identified in all scenarios.

Each participant employed on average 6.7 different image schemas (SD 1.0) in speech interaction. The four image schemas up-down, verticality, horizontality and container were regarded as early, basic image schemas, whilst the six image schemas balance, scale, warm-cold, process (including cycle), near-far and multiplicity/ unity were regarded as later image schemas [17]. There were no significant effects for gender, age and technical experience (all p's > .05) on frequencies. In general, basic image schemas were employed significantly more often (67.8%) than image schemas occurring later in development (32.2%). In line with the second hypothesis mentioned, gender, age and technical experience were not correlated with both the frequency of early and late image schemas. This suggests that image schemas as basic building blocks of human knowledge are available to users of all ages and technical experience independently from gender.

Table 1. Coding of speech utterances into image schemas [12].

Image schema	Example speech utterances
Basic	up/down, to rise/lower, higher/lower
Path: up-down	(to move/drive) up/down, to rise/to lower, higher/lower
Vertical path	to set (German: *stelle ein; stelle auf*)
Horizontal path	from…to, along, back/front, backwards/forward
Container	to put in, to hold, in the room, in the house, in daytime
Later	
Balance	to regulate, to adjust
Attribute: warm-cold	make it warmer/colder, to heat (up or down), to temper
Process (cycle, iteration)	to turn (up or down), to deactivate
Near-far	near xx degrees, next week, next day
Unity or multiplicity	whole day, whole week, a few (degrees)
Scale	to reduce, to cut, to increase or decrease, how much

With regard to [13, 14] interaction should be based on (automatically retrievable) prior knowledge available to all potential users. The illustrated survey from [12], however, demonstrated that not all image schemas are equally intuitive in human computer interaction: The developmental occurrence of image schemas impacts upon the frequency of applied image schemas when interacting with technical devices. Early image schemas should, thus, be given preference over late image schemas in the interface design.

3 Concepts of semantic processing and behavior control

3.1 Feature-value-relation (FVR)

Cognitive user interfaces require a bidirectional translation between input signals and representations of meaning. While low-level signals are sequential, semantics is, in general, non-sequential. In [5], feature-values-relations (FVR) for representation and processing of semantic information were introduced. An example is depicted in Figure 3 showing an FVR for the speech input "Increase the temperature to 23 degrees on Saturday." where the relevant values of the input are related to semantic categories relevant for the system.

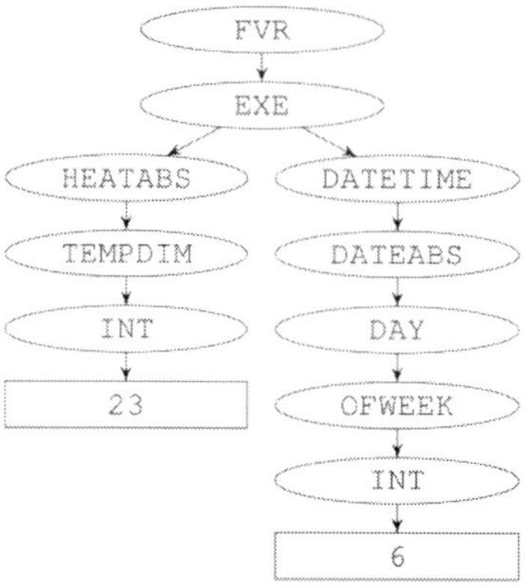

Fig. 3. FVR example as part of a heating control from [1].

These categories depend on the available actions of the system and the domains of usage, and are collected in a world model.

In [6, 7] FVRs are equipped with weights – which are omitted in fig. 2 – and related to language modelling, whereas [8] defines several operations on FVRs. A description of a behavior control – in an instinctive and an adapting version – building upon these operations is given in [12]. In our multi-modal system any input signals are transformed into FVRs representing the individual semantics (cf. exemplary discussion regarding a touchscreen in [12]). All FVRs lead to joint input semantics, which can then be unified with the current state (another FVR). This state serves as memory between dialog turns and contains all data gathered during an ongoing dialog. By comparison of the new state with the world model – and thus identifying the goal of a dialog – the semantics of an appropriate system action can be computed. The world model encodes what data is needed to execute a specific action. Whenever there is not enough data, another dialog turn requesting more data is initiated until execution of an action is possible. Such requests can be routed to different parts of the system, e.g. available sensors, a user model holding the user's habits, initiating a visual or auditory prompt for user input, any module from where the system can get and incorporate missing data. The flexibility of the approach arises from the fact that all processing is done in terms of FVRs and thus independently from the concrete system and any modalities of input and output. So from a behavior point of view it does not matter if the task is controlling a heating or taking part in a collaborative human-robot interaction.

3.2 Petri net transducer (PNT)

For the technical realization, the so called Petri net transducers (PNTs) were proposed [9, 10], that process labelled partial orders (LPOs), which in turn can represent FVRs. The application of PNTs to the bidirectional translation between sequences and partial orders allows us to build a seamless signal-to-semantics recognition network. Moreover we are able to prime this network by composition with a semantic structure representing an expectation on the next input. This expectation is a truly semantic one but adjusts the recognizer down to all low-level parts. On the other hand we use the same techniques for the synthesis side, where we can for example inject syntactical restrictions to adapt speech output to the users' wording. As with multi-modality on the input side, we can use different translation units to present the same semantics on different output channels.

3.3 Fock space

Based on a further concept described in [18], we are currently entering into a novel theory for mapping FVRs into a so-called Fock space from quantum mechanics. Given V as the set of nodes of an FVR there exists a Hilbert space V with the dimension $|V|$ since every element of V is used as one basis vector. The Fock space is then defined as $\bigoplus_{n=0}^{\infty} V^{\otimes n}$ where $\oplus$ is the direct sum and $\otimes$ denotes the tensor product. This allows us to use a different branch of mathematics for the processing of semantics. As a side effect we gain new insights as a first approach to discover semantic structures from data or using them for action planning as described in [18].

4 Realization as a hardware prototype

As described in [1], the first demonstrator was realized in August 2017 as an integrated circuit device, which still involves external power supply and a RS232 interface for the communication with arbitrary electronic devices. Four microphones, a loudspeaker and the touch panel were integrated. Figure 4 shows the main board.

Fig. 4. First version of UCUI circuit board from [1].

The board (100 x 130 mm) includes two digital signal processors (DSPs), one Field Programmable Gate Array (FPGA), four RAMs, a flash memory, an audio codec and a motion sensor. The FPGA performs acoustic signal analysis, some other algorithms for speech recognition as well as signal and data routing. One DSP finalizes speech recognition and runs cognitive processes based on FVRs. Beyond it executes speech synthesis and controls the display. In next steps, our project partners Javox Solutions GmbH and XGraphic Ingenieurgesellschaft mbH will run their algorithms for beam forming, noise and echo cancellation as well as the gesture recognition on the second DSP.

5 Conclusions

In highly user-adaptive scenarios of human-machine and human-robot interactions, we suggest a decided cognitive modelling including semantic and behavior processing. The introduced UCUI prototype is able to communicate with users in parallel via speech, gesture recognition, speech synthesis and a touch screen. The device operates autarkic and supports a widely intuitive interaction to control different kind of *machines* surveyed with early and late image schemas. It can ensure user privacy by its system design and does not rely on network access.

Conventional interfaces cannot benefit from semantic prior knowledge. By using PNTs, semantic structures – corresponding to input signals within a multimodal hierarchical signal processing system – can be computed without premature decisions.

The current demonstrator is supporting the tasks of speech recognition, synthesis, semantic processing and a simplified cognitive behavior control on the embedded platform. After our next development steps, an extended behavior control model will enrich the interaction opportunities. Furthermore, the ultrasonic gesture control will be implemented.

This contribution has been developed in the project Universal Cognitive User Interface (UCUI) which is partly funded by the German Federal Ministry of Education and Research (BMBF) within the research program IKT2020 (grant #16ES0297). As associated partners we thank the ministry and the main project partners Fraunhofer IKTS, BTU Cottbus-Senftenberg, Javox Solutions GmbH, XGraphic Ingenieurgesellschaft mbH and Agilion GmbH for the decided work and for their previous publications as the foundation of this article.

References

1. F. Duckhorn et al, Proc. ISCA Interspeech, 3435–3436 (2017)
2. M. Huber, O. Jokisch, Proc. of Knowledge Management Conference, Novo Mesto (Slovenia), June 2017. International Institute for Applied Knowledge Management, 31–40 (2017)
3. R. Hoffmann, M. Eichner, M. Wolff, Verbal and Nonverbal Communication Behaviors, 200–218 (2007)
4. M. Wolff, UASR: Unified Approach to Signal Synthesis and Recognition (2000) Available at: http://www.b-tu.de/en/fg-kommunikationstechnik/research/ projects/uasr
5. M. Huber et al, Proc. of Elektronische Sprachsignalverarbeitung (ESSV), series Studientexte zur Sprachkommunikation, **53**, 25–32 (2009)

6. G. Wirsching, Proc. IEEE 3rd Intern. Conference on Cognitive Infocommunications (CogInfoCom), 71–76 (2012)
7. G.Wirsching, R. Lorenz, Proc. of IEEE 4th Intern. Conf. on Cognitive Infocommunications, 369–374 (2013)
8. P. Geßler, *Kognitive Gerätesteuerung. Master thesis* (BTU Cottbus-Senftenberg, 2017)
9. R. Lorenz, M. Huber, G. Wirsching, Proc. of 35th Intern. Conference on Application and Theory of Petri Nets and Concurrency, **8489**, 233–252 (2014)
10. M. Huber, R. Römer, and M. Wolff, Proc. Elektron. Sprachsignalverarbeitung (ESSV), Saarbrücken, ser. Studientexte zur Sprachkommu-nikation, **86**, 122–129 (2017)
11. 7Bot: a low cost Robotic Arm that can See, Think and Learn! (www.7bot.cc, retrieved on 10/03/2018: https://www.kickstarter.com/projects/1128055363/7bot-a-powerful-desktop-robot-arm-for-future-inven).
12. M. Huber et al, Proc. of Knowledge Management Conference, (to be published)
13. C. Mohs, J. Hurtienne, M.C. Kindsmüller, J.H/ Israel, H.A. Meyer, IUUI Research Group, MMI-Interaktiv, **11**, 75–84 (2006)
14. P. Turner, Behavior & Information Technology, **27**, 475–482 (2008)
15. G. Lakoff and M. Johnson, *Philosophy in the Flesh: The Embodied Mind and Its Challenge to Western Thought* (New York: Basic Books, 2011)
16. J. Hurtienne, et. al, Interactive Computing, **22**, 475–484 (2010)
17. J. Mandler, C. Pagán Cánovas, Language and Cognition, 1–23 (2014)
18. P. beim Graben, M. Huber, R. Römer, W. Wolff, Proc. Elektronische Sprach-signalverarbeitung (ESSV), **90**, 167–174 (2018)

Adaptive particle filter for localization problem in service robotics

Alexander Heilig[1], *Ilshat* Mamaev[1 *], *Björn* Hein[1] and *Dmitrii* Malov[2]

[1]Karlsruhe Institute of Technology, Institute for Anthropomatics and Robotics - Intelligent Process Control and Robotics Lab (IAR-IPR), Engler-Bunte-Ring 8, 76131 Karlsruhe, Germany
[2]SPIIRAS, Laboratory of Autonomous Robotic Systems, 199178, Saint-Petersburg, 14-th line of V.I., 39, Russia

Abstract. In this paper we present a statistical approach to the likelihood computation and adaptive resampling algorithm for particle filters using low cost ultrasonic sensors in the context of service robotics. This increases the efficiency of the particle filter in the Monte Carlo Localization problem by means of preventing sample impoverishment and ensuring it converges towards the most likely particle and simultaneously keeping less likely ones by systematic resampling. Proposed algorithms were developed in the ROS framework, simulation was done in Gazebo environment. Experiments using a differential drive mobile platform with 4 ultrasonic sensors in the office environment show that our approach provides strong improvement over particle filters with fixed sample sizes.

1 Introduction

Service robotics market is estimated to almost triple by 2022 from the 2016 level [1]. In comparison to industrial robotics, service robots are inexpensive, built from low-cost hardware and need to fulfil softer requirements for accuracy, repeatability and reliability. This also applies to the sensors. For Simultaneous Localization and Mapping applications (SLAM) in the industry traditionally optical systems, such as 2D/3D cameras and LIDAR sensors, are used. Both types of the sensors are not beneficial in the light of the above requirements, additionally cameras potentially threatening customer privacy. On the other hand, ultrasonic range sensors or sonars are quite appealing for service robotics, the price and power consumption are better than those of optical sensors. Moreover, sonars have no problems with transparent materials like glass and do not acquire any personal data. In this paper we focus on the localization problem of a mobile platform with sonar sensors.

To be able to accomplish service tasks, robots must be able to estimate its pose with respect to a fixed reference frame. This process is known as localization problem. Moreover, localization task can be divided into two subclasses: weak and strong localization. The first one deals only with a qualitative pose estimation such as high level spatial reasoning – for example, determining the room in which the robot is located. On the other hand, strong localization provides numerical estimation of the robot pose. We present an approach to strong localization on the global context using sonars as exteroceptive sensors and odometry as proprioceptive ones. The localization problem is solved using particle filter with a statistical approach to the computation of the likelihood and adaptive resampling algorithm. Modelling and simulation implemented in the ROS framework and Gazebo simulation environment. The evaluation of the proposed approach will be also conducted on an experimental setup using differential drive mobile robot in the institute/office environment.

2 Particle Filter Concept

Uncertainty is a linchpin of the most problems robotics faces in the real world applications. Probabilistic robotics is a relatively new approach, which represents uncertainty using the calculus of probability theory [2]. Bayes filter is the most general approach to compute beliefs based on observations, action data and prior probability. However, in terms of the computational tractability, the general Bayes filter is not trackable for continuous state spaces [3]. The classical trackable solution in probabilistic methods is the Kalman Filter [4], this recursive approach to the discrete-data linear filtering problem based on Gaussian filters has been the subject of extensive research in the SLAM. However, due to the use of the normal distribution assumption, Kalman filter comes to its limits dealing with situations with high ambiguity [5]. The latter is especially relevant for ultrasonic sensors due to low resolution, multiple reflexions and low sampling rate.

In 1999, particle filters were introduced in the localization context under the name of Monte Carlo Localization [6]. In this context, particle filters offer a probabilistic approach on how to estimate the state of a robot in a known map. They are based on the assumption that the measurement of external values is

* Corresponding author: ilshat.mamaev@kit.edu

related to the internal state of the robot, i.e. its pose. From there on, the particle filter creates semi-random states in the world – particles – which will give an estimation on their respective states likelihood to represent the actual state. This approach is needed, because the robot's odometry diverges over time and loses its accuracy. The estimation over external sensors corrects this error. A particle filter mainly consists of three phases: motion applying, likelihood estimation, particle resampling (Figure 1).

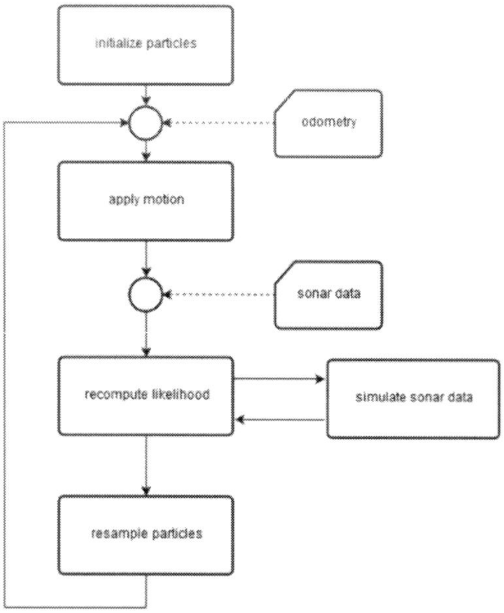

Fig. 1. Particle filter concept and its phases.

This work aims to improve the quality of the particle filter by modifying the computation of likelihood and the resampling algorithm. Particle initialization is uniformly sampled around the map, so the initial state of the robot is unknown. Motion from odometry data is applied in a frequency of 100 Hz to each particle.

2.1 Ultrasonic sensors

In comparison with laser scanners and 3D camera systems, ultrasonic range finders exhibit several shortcomings such as low resolution, cross-talk, multiple reflexions and low sampling rate. Nevertheless, sonars are quite appealing in the context of the service robotics by virtue of the low price, small size and low power consumption. Different research groups have validated the usage of the ultrasonic sensors in localization task [7-9] and even in SLAM [10, 11]. In this paper we exploit the widely used ultrasonic sensor in the robotics community – HC-SR04. The example of the range data from the sensor captured in the test environment is shown in Figure 2.

For the sonar sensor simulation, the hector sonar plugin has been used in Gazebo [12]. This plugin uses the Gazebo integrated ray sensor plugin which scans a cone, originating from the sensor. The plugin then returns the smallest distance it found to simulate sonar data. The simulated sensors were configured to work similar to the real sensors which have been shown to produce reliable results with small variance.

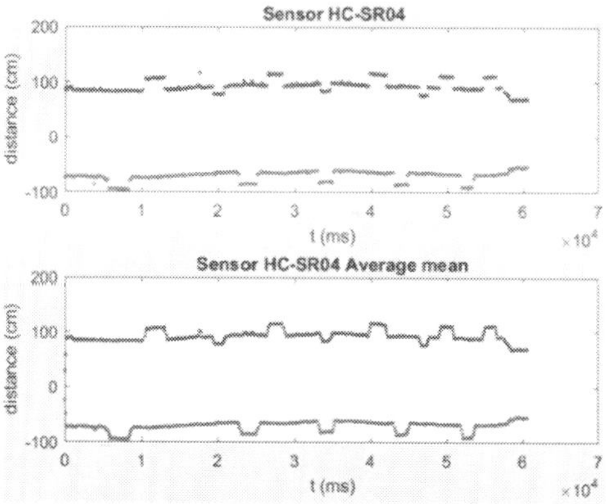

Fig. 2. Raw range measurements (above) and average mean filtered data (below) from HC-SR04 in the test environment.

In the particle filter, sonar sensor measurements must be estimated for each particle. As the number of particles varies from a few hundred to a few thousand, this estimation needs to have low computational cost. The particle filter contains a black and white image in a PNG format representing the real map. Black pixels indicate objects, while white ones indicate free space. Each particle's position is projected onto this bitmap. Using the robot's universal robot description and the sensor attributes, three rays are shot from each particle's sensors. The rays cover the sonar sensor cone's edges as well as the center line. The breadth-first search ray sweep is interrupted, as soon as an object is found by one of the rays, imitating the real sonar's ability to detect only the smallest distance as well as reducing the computational cost. The distance measured for each sensor is then translated from pixels to meters and saved for each particle.

2.2 Likelihood update

After having applied the robot's motion model to all particles, their likelihoods must be updated by including the system observations into the particle filter. The algorithm must assess how well each particle's estimated system observation coincide with the real observation or measurements. The sonar sensors used in this work exhibit a Gaussian noise with standard deviation of 0.005m. Taking this distribution into account, an estimation of likelihood of one estimated sensor measurement to the true measurement can be calculated by

$$f(x) = \frac{1}{\sqrt{2\pi\sigma^2}} e^{-\frac{(x-\mu)^2}{2\sigma^2}}, \quad (1)$$

where μ denotes the actual sensor reading and σ the standard deviation. Taking the arithmetic mean over all probabilities gives a measurement on how likely a particle is to represent the robot state. This calculation has been slightly altered to accommodate differences between particles. Often enough, one sensor reading differs from one particle to another, which contributes only a quarter to the actual likelihood calculated in the end. However, one sensor's reading can be enough to discern a particle with actually high likelihood from a particle with actually low likelihood. For this matter, a particle which has a single sensor reading deviating more than two times the standard deviation σ receives a penalty and has its overall likelihood reduced by 10% for every penalty it got. The resulting likelihood of each sensor of one particle is then saved in a set containing the likelihoods of this particle's ancestors' likelihoods for a fixed size of generations as the most current likelihood. To get a measurement of the particle's final estimation towards the robot state, a weighted average is calculated over the set $\bar{l}$ of current and ancestor likelihoods. A particle's ancestor is defined as the particle it originated from during previous resampling steps. The weighted average is calculated as

$$mean(\bar{l}) = \frac{2 * \sum i * l_i}{n * (n+1)}, \quad (2)$$

$$\bar{l} = (l_1, .., l_n), i \in [1, n).$$

2.3 Resampling algorithm

Resampling is the last step in particle filter localization algorithms. Particles are deleted, newly generated or altered by taking into consideration previous system observation. This procedure aims to compensate the steadily increasing error of odometry feedback. However, traditional resampling runs into several issues whose consequences can be detrimental for the accuracy of the particle filter.

One of these issues is kidnapping, where the robot is placed somewhere else by an external force which is not measured and from which it must recover to find its actual state again. The traditional particle filter does not recover from such a scenario while the proposed method finds the next possible states, leading to good estimations, but not excluding ambiguities (Fig. 2). Another problem one encounters is the local optimum, i.e. places where the particles get stuck. This usually occurs when estimating the wrong state at a given point in time due to the sensor readings giving high probabilities for this particular state. As the robot moves on, the particles should diverge, as they follow the movement and escape the apparent optimum. However, resampling may lead the particles back to the previous state, resulting in erroneous estimations again.

In this work, the resampling part is based on a systematic resampling approach and extended to increase the accuracy of the state estimation [13]. The particles, sorted by ascending likelihood, are used to generate a set of normalized accumulated likelihoods over all particles, where each accumulation is associated with the respective particle. This approach then generates n ordered equidistant numbers where n equals the size of the resampled particles.

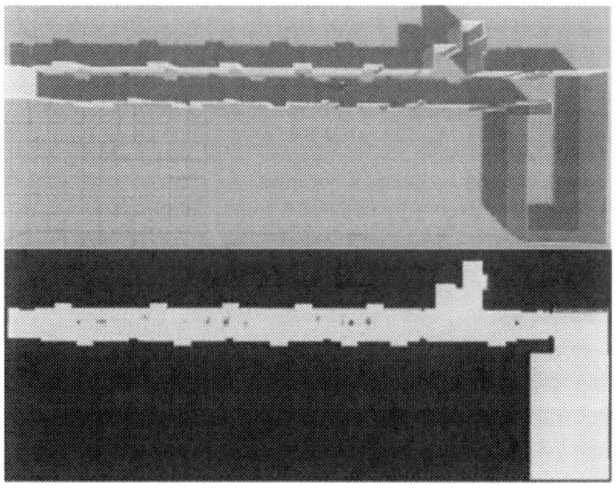

Fig. 3. Ambiguity of states: sensor observations correspond to multiple possible states in the map.

With systematic resampling, one random number $u \in [0, n^{-1}]$ is generated and then used to systematically compute equidistant numbers

$$u_i = \frac{i}{n} + u, i = 2. \quad (3)$$

For each u_i the first particle whose accumulated likelihood satisfies

$$u_i < acc_j, i, j = 1 \quad (4)$$

is drawn for resampling. This accomplishes three things:

1. Prevents sample impoverishment, as the number of resampled particles will be the same as the numbers of previous particles.
2. If there are high likely particles, the algorithm will converge towards them.
3. Systematically ensures that less likely particles will still be picked in contrary to e.g. multinomial resampling.

Drawn particles receive a resampling noise in x, y and θ and are then placed into the new set of particles. The noise added to the drawn particles is restricted to not overstep more than half the distance the motion model had applied to it before. This measurement is taken to ensure the advance of particles along the line of the applied velocity, which in turn guarantees that particle cannot get stuck in local minima. Lastly, the most likely particle is added once more without generating additional noise. This prevents diverging of particles when the odometry error is low. After resampling, the variance of resampling noise and the number of particles is increased, if the best particle's likelihood is underneath a certain threshold. By spreading the particles further out, yet keeping a high density, the particle filter searches for better estimations of the robot state. Starting in the vicinity of the previous particles, small errors in odometry can be detected and corrected. This approach further increases the chance to recover from the kidnapping problem, especially if the robot has not been moved very far from its previous state, i.e. when replacing the robot, because he was in the way of the user. As soon as a state which promises a good estimation is discovered, the number of particles as well as the variance is reduced again. As a result of this approach, only a small number of particles can be used as the default size for the particle filter, thus decreasing the computational cost.

3 Mobile robotic platform "Slamdog"

A low-cost mobile robotic platform "Slamdog" was developed within several student lab projects for teaching and research purposes. This medium sized platform was built upon the commercial children toy vehicle "Feber Dareway 12V" with differential drive kinematics, which was heavily modified: new wheels from BLICKLE for better grip, 14-bit rotatory position sensors AMS AS5047P for odometry, IMU Invense MPU-9150, ultrasonic sensors HC-SR04, IBT H-Bridge etc. Raspberry Pi 3 Model B was used as a low level controller. Furthermore, the platform was expanded with NVIDIA Jetson TX2 development kit as a high level controller, which is mainly used for machine learning tasks and for interfacing with 3D ToF cameras. In this paper, only ultrasonic sensors and the low level controller were used, the localization node was running either on a normal desktop PC or Jetson TX2.

The software for the platform was developed in C++11 under the Robot Operating System (ROS) framework. A ROS hardware driver was implemented as a "Slamdog" ROS node, which provides interfaces for the motor control, ultrasonic sensors, encoders and a buzzer. For simulation of the environment, the robot and the sensors, this work relies on Gazebo 7, for the visualization of particles RViz is used.

4 Evaluation

The high level test scenario for the "Slamdog" was a postman's task: to deliver package inside an office space. In order to be able to do this successfully, it was necessary to navigate in a weakly structured space. The localization problem, presented in this paper, complements the path planning problem within this task. As it was already mentioned one of issue for the localization in this task is the kidnapping of the robot. Figure 4 shows a scenario, where the robot is pushed from its original pose into a different state. This example of small distance kidnapping can be solved by the resampling algorithms dynamically adapting variance and particle numbers.

To further analyze the capabilities of this particle filter approach, different scenarios were simulated multiple times. For each scenario, the distance from the particle filter's best particle as well as the mean distance for all particles and the standard deviation will be reported. Convergence of the particles will be measured by the standard deviation σ and is further defined as $\sigma \leq 0.25$ meters, which roughly equals the radius of the "Slamdog" robot. Measurements were taken every second over twenty seconds. For all scenarios, the initial particle orientation was uniformly distributed over $[-\pi, \pi]$.

Fig. 4. Slamdog robotic platform in the test environment.

The first scenario "Converged" initialized the particles with a uniformly distribution of 0.5 meters centered at the robot pose to measure the filter's ability to correctly keep a converged state. The second scenario "Kidnapped" is similar to the scenario in Figure 5. The particle filter was initialized with the same distribution as "Converged", but the centered particle pose was offset by 1.0 meter to simulate kidnapping. The third scenario "Ambiguity" was supposed to measure the filter's ability to deal with ambiguities. The robot was placed in the long corridor as depicted in Figure 3 and the particles were initialized over the whole corridor by using a uniformly distribution of 20.0 meters in x- and 8.0 meters in y-direction. The fourth scenario "Diverged" was meant to measure the filter's overall ability to converge to the robot's pose once non-ambiguous measurements exist. The robot was placed at the right end of the long corridor (Fig. 3) and drove to the right. The particles were initially uniformly distributed like in the third scenario. Each experiment was simulated ten times. The results are presented in Table 1.

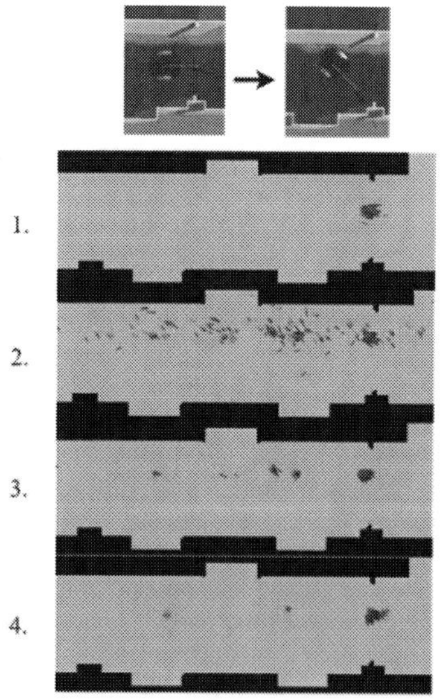

Fig. 5. A kidnapping problem: Robot is forcibly translated and reoriented a small distance from the previous converged state (1). Resampling diverges the state and searches the closer vicinity (2). The particles converge (3) and reach their new best estimation.

Table 1. Results of the particle filter evaluation. Values are measured in distance to the correct robot state in meters for 'best', mean (μ) and standard deviation (σ) or in the time t (in seconds) it took the filter to converge.

Scenarios		1.	2.	3.	4.	5.	6.	7.	8.	9.	10.	Mean
Converged	best (m)	0.39	0.21	0.38	0.22	0.42	0.50	0.17	0.30	0.37	0.44	0.34
	μ (m)	0.43	0.41	0.43	0.37	0.48	0.50	0.34	0.41	0.51	0.54	0.44
	σ (m)	0.14	0.14	0.19	0.14	0.12	0.15	0.19	0.13	0.12	0.11	0.14
	t (sec)	0.0	0.0	0.0	0.0	0.0	0.0	0.0	0.0	0.0	0.0	0.0
Kidnapped	best (m)	0.26	0.08	0.46	0.36	0.15	0.42	0.04	0.34	0.19	0.30	0.26
	μ (m)	0.30	0.34	0.28	0.289	0.21	0.39	0.36	0.38	0.34	0.35	0.32
	σ (m)	0.14	0.12	0.12	0.15	0.11	0.11	0.16	0.12	0.14	0.14	0.13
	t (sec)	3.0	3.0	3.0	4.0	4.0	4.0	2.0	1.0	3.0	3.0	3.0
Ambiguity	best (m)	1.95	1.56	0.91	0.20	1.42	3.45	0.43	1.51	0.20	0.60	1.22
	μ (m)	2.11	1.68	0.59	0.42	2.61	3.23	0.45	1.47	0.37	0.83	1.38
	σ (m)	0.41	0.18	0.88	0.14	1.22	0.16	0.16	0.38	0.14	0.84	0.45
	t (sec)	-	13.0	-	14.0	-	10.0	14.0	-	12.0	-	12.6
Diverged	best (m)	0.38	0.32	0.50	0.24	0.38	0.48	0.50	0.30	0.48	0.22	0.38
	μ (m)	0.38	0.45	0.50	0.43	0.49	0.51	0.63	0.39	0.32	0.35	0.45
	σ (m)	0.14	0.16	0.12	0.14	0.18	0.11	0.13	0.12	0.15	0.17	0.14
	t (sec)	12.0	11.0	11.0	11.0	11.0	7.0	11.0	11.0	12.0	4.0	10.1

The results show that the particle filter is able to keep a fitting state once converged as depicted in "Converged" (see Table 1.). The filter did not expand and converged to a small standard deviation of 0.14 meters overall. The "Kidnapped" scenario illustrated the filter's ability to expand and search its vicinity. The filter needed less than 4 seconds to converge and did so correctly all ten tries. The largest errors were measured in the third scenario. Five times out of ten the particle filter converged in multiple places which shared the same characteristics, i.e. ambiguities (s. Table 1 "Ambiguity" where $\sigma \geq 0.25$). However, three times it has converged correctly. The "Diverged" scenario displays how the particle filter keeps searching the map for a fitting state until at around 11 seconds where salient measurements are taken.

One aspect which can be noted over all the experiments is the range of the best particle found when converged correctly, whose distance to the actual robot spans from 0.08 meters ("Kidnapped", second set) to 0.5 meters (e.g. "Converged", sixth set). This relatively large difference can be explained by the beam angle of the sonar sensors and the distance to the walls on each side in the map. This increases the possible states near the correct state, especially in x-direction and negatively affects finding the best particle. Overall, the particle filter manages to converge correctly in 33 of the 40 experiments undertaken to depict the correct robot state.

5 Conclusions and future work

In this paper we have presented a resampling method, which increases the probability to recover from problems such as stuck particles or kidnapping. This was accomplished by dynamically adapting the configuration of the particle filter to the quality of the state estimation. The algorithm prevents sample impoverishment and converges to high likely states once they are found. The usually high computational cost of estimating the sensor reading for each particle has been simplified while keeping satisfying estimations. Ambiguity of robot states is reduced unless the robot is kidnapped far away to a region that resembles the previous one.

As this particle filter works on comparatively low cost and computational power, the robot system is kept inexpensive. Furthermore, as the main focus lies on applications in the service market, the sonar sensors which have minimal invasiveness regarding privacy and safety, present desired attributes for the sector of service robotics.

In our current implantation we deliberately use only sonar sensors for the localization. This approach can be extended using sensor fusion methods for the sonars, IMU and visual odometry data based on the 2D/3D cameras.

To further analyze the capabilities of this particle filter approach, different scenarios were simulated multiple times. For each scenario, the distance from the particle filter's best particle as well as the mean distance for all particles and the standard deviation will be reported. Convergence of the particles will be measured by the standard deviation σ and is further defined as $\sigma \leq 0.25$ meters, which roughly equals the radius of the "Slamdog" robot. Measurements were taken every second over twenty seconds. For all scenarios, the initial particle orientation was uniformly distributed over $[-\pi, \pi]$.

References

1. *Service Robotics Market and Volume* (Type (Professional Service Robots, Personal and Domestic Service Robots) and Key Players Analysis - Global Forecast to 2022," iGATE RESEARCH, 2017)
2. S. Thrun, W. Burgard, D. Fox, *Probabilistic Robotics* (Cambridge, MA: MIT Press, 2005)
3. A. Burguera, Y. Gonzales, G. Oliver, Advances in Sonar Technology, 213–2 32 (2009)
4. R.E. Kalman, Transactions of the ASME--Journal of Basic Engineering, **82**, 35–45(1960)
5. J. Neira, J. Tardós, IEEE Transactions on Robotics and Automation, **17** (6), 890–897(2001)

6. F. Dellaert, D. Fox, W. Burgard, S. Thrun, IEEE International Conference on Robotics and Automation, **2**, 1322–1328 (1999)
7. A. Burguera, Y. Gonzalez, G. Oliver, Sensors, **9**, 10217–10243 (2009)
8. D. Fox, Robotics Research, **22** (12), 985–1003 (2003)
9. A. Großmann, R. Poli, Robotics and Autonomous Systems, **37**, 1–18 (2001)
10. L. D'Alfonso, A. Grano, P. Muraca, P. Pugliese, *16th* International Conference on Advanced Robotics (ICAR), 1–6 (2013)
11. A. Pandey, K. Krishna, H. Hexmoor, International Conference on Integration of Knowledge Intensive Multi-Agent Systems, 283–288 (2007)
12. Hector_gazebo_plugins, Available: http://wiki.ros.org/hector_gazebo_plugins.
13. R. Douc, O. Cappé, Image and Signal Processing and Analysis (ISPA), 64–69 (2005)

Control of shimmy vibration in aircraft landing gears based on tensor product model transformation and twisting sliding mode algorithm

Suwat Kuntanapreeda*

King Mongkut's University of Technology North Bangkok, Department of Mechanical and Aerospace Engineering, Bangkok 10800, Thailand

Abstract. Shimmy vibration is a common phenomenon in landing gear systems during either the take-off or landing of aircrafts. The shimmy vibration is undesirable since it can damage the landing gear and discomforts the pilots and passengers. In this work, tensor product model transformation (TPMT) and twisting sliding mode algorithm (TSMA) are utilized to design a robust controller for suppression of the shimmy vibration. The design has two steps. First, the TPMT is applied to determine the first part of the controller to suppress the vibration of the undisturbed system. After that, the TSMA is adopted to obtain another part of the controller to eliminate the remaining vibration caused by disturbances. By integrating these two parts, the proposed controller is obtained. Simulation studies are provided to demonstrate the effectiveness of the controller.

1 Introduction

Shimmy vibration is a common phenomenon in landing gear systems during either the take-off or landing of aircrafts. It is the state of self-excite oscillations in lateral and torsional directions caused by the interaction between the tires and the runway. The energy input is provided by the kinetic energy of the forward motion of the aircraft [1, 2].

The shimmy vibration is undesirable since it can damage the landing gear and discomforts the pilots and passengers. The damage of the landing gear can also lead to accidents. Thus, suppression of the shimmy vibration is needed. Passive dampers are normally used to suppress the shimmy vibration due to their simplicity [3]. However, damping requirements often conflict with good high-speed direction control [4]. Thus, active shimmy vibration control strategies have drawn attention recently [5-7]. In [5], an active control strategy based on model predictive control and tensor product model transformation is proposed. It was shown that the proposed active control method can effectively suppress shimmy vibration. In [6], robust and nonlinear optimal control of shimmy vibration is developed. The controller is designed by integrating sliding mode control together with State-Dependent Riccati Equation. In [7], an active shimmy vibration controller based on a filtered PID scheme is proposed. The parameters of the controller are tuned by means of a population decline particle swarm optimizer. Results under different scenarios were investigated and stochastic robustness verification was addressed to verify the controller effectiveness.

Tensor product model transformation (TPMT) is an effective numerical technique based on the recently developed high order singularity value decomposition (HOSVD) [8–10]. It transforms a linear parameter varying (LPV) system into a tensor product (TP) model form, which is described by a convex combination of linear time invariant (LTI) systems. Various types of convex hulls also can be derived. The TPMT has been successfully applied in many challenging problems [5, 11–14]. In [11], the TPMT was adopted for designing a controller to stabilize shared impedance/admittance-based bilateral telemanipulation under varying time delay. The controller fulfilled the stability requirement within the time delay domain considered in the design. A TPMT-based control and synchronization scheme for fractional-order chaotic systems was proposed in [12]. Numerical results of the fractional-order Lorenz and Liu chaotic systems illustrated the effectiveness of the scheme. In [13], a TPMT-based modelling and control design approach for morphing aircraft in transition process was investigated. Its effectiveness was shown by simulations of a variable-sweep morphing aircraft. In [14], a TPMT approach for level control of a three tanks system was presented. Simulation results were obtained to validate the controller design. A generalization of the TPMT for control design can be found in [15].

High order sliding mode (HOSM) control was first introduced by Levant [16, 17]. Twisting sliding model algorithm (TSMA) is one of the best choices among other HOSMs for the stabilization of nonlinear systems under external disturbances or uncertainties. Examples demonstrating applications of the TSMA can be found in [18–22].

This paper presents a method based the TPMT and the TSMA to design a robust controller to supress shimmy vibration in aircraft landing gears. The rest of the paper is

* Corresponding author: suwat.k@eng.kmutnb.ac.th

organized as follows. In the next section, some preliminaries are provided, including the mathematical model of the shimmy vibration, the TPMT and the TSMA. The proposed robust control design is presented in Section 3, followed by simulation results in Section 4. The conclusion is drawn in Section 5.

2 Preliminaries

2.1 Mathematical model

Consider an aircraft landing gear system as shown in Figure 1. The mathematical model describing shimmy vibration in the landing gear can be written as [5]:

$$\dot{\mathbf{x}} = \mathbf{A}(V, e)\mathbf{x} + \mathbf{B}u + \mathbf{D}, \quad (1)$$

where $\mathbf{x} = \begin{bmatrix} \psi \\ \dot{\psi} \\ y \end{bmatrix}$ is the system state,

ψ is the yaw angle (rad),
y is the lateral deflection (m),
u is the control command (volt),

$$\mathbf{A}(V,e) = \begin{bmatrix} 0 & 1 & 0 \\ -\dfrac{K}{I_z} & -\dfrac{c}{I_z} + \dfrac{\kappa}{VI_z} & \dfrac{F_z}{I_z\sigma}(C_{M\alpha} - eC_{F\alpha}) \\ V & e-a & -\dfrac{V}{\sigma} \end{bmatrix},$$

$$\mathbf{B} = \begin{bmatrix} 0 \\ k_e \\ 0 \end{bmatrix},$$

V is the taxiing velocity (m/s),
e is the wheel caster length (m),
$T = k_e u$ is the input torque (N·m) and
$\mathbf{D} = [0, d, 0]^T$ is the bounded disturbance torque (N·m).

Here, V and e are considered as variable parameters. Thus, the system (1) is a linear parameter varying (LPV) system in which V and e are the time-varying parameters.

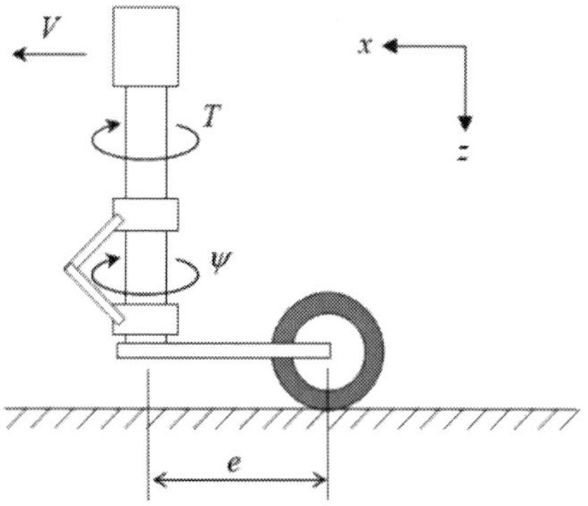

Fig. 1. Aircraft landing gear system.

The values of the system parameters used in this paper are summarized in Table 1. Examples of uncontrolled system (i.e., $u = 0$) are shown in Figures 2 and 3. Note that the system is stable in velocity 25 m/sec while it is unstable in velocity 75 m/sec.

Table 1. System parameters [5].

Symbol	Parameter	Value
V	Taxiing velocity	20–80 m/s
e	Wheel caster length	0.1–0.5 m
a	Haft contact length	0.1 m
I_z	Moment of inertia	1 kg·m^2
F_z	Vertical force	9 kN
K	Torsional spring constant	20 N·m/(rad·s)
c	Torsional damping constant	20 N·m/(rad·s)
$C_{F\alpha}$	Side force derivative	20 rad^{-1}
$C_{M\alpha}$	Moment derivative	-2 m/rad
κ	Tread width moment constant	-270 N·m^2/rad
σ	Relaxation length	$3a = 0.3$ m
k_e	Input torque constant	10 kN·m/volt

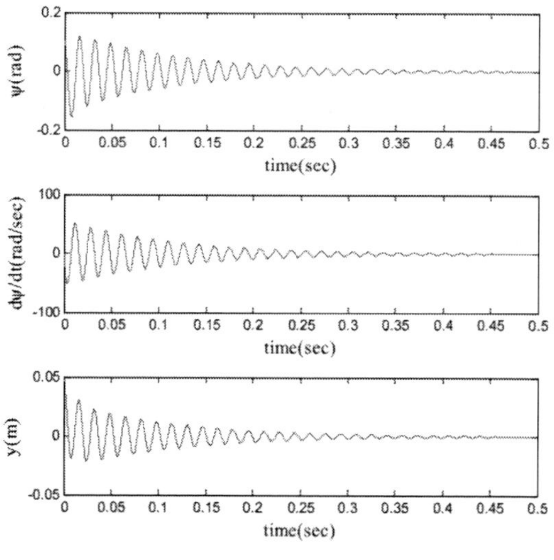

Fig. 2. Shimmy vibration at $V = 25$ m/sec and $e = 0.3$ m.

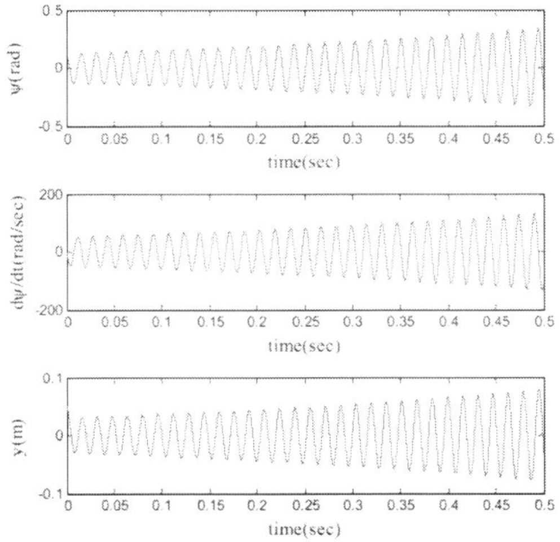

Fig. 3. Shimmy vibration at V = 75 m/sec and e = 0.3 m.

2.2 Tensor product model transformation

Tensor product model transformation (TPMT) is an effective numerical methodology to transform linear parameter varying (LPV) systems into convex tensor product (TP) model representations [8, 9]. In this paper, the LPV system is written as:

$$\begin{aligned}\dot{\mathbf{x}} &= \mathbf{A}(\mathbf{p}(t))\mathbf{x} + \mathbf{B}(\mathbf{p}(t))\mathbf{u} \\ &= \mathbf{S}(\mathbf{p}(t))\begin{bmatrix}\mathbf{x} \\ \mathbf{u}\end{bmatrix},\end{aligned} \tag{2}$$

where

$$\mathbf{S}(\mathbf{p}(t)) = [\mathbf{A}(\mathbf{p}(t)) \quad \mathbf{B}(\mathbf{p}(t))] \tag{3}$$

is the system matrix, $\mathbf{x}$ is the state vector, $\mathbf{u}$ is the input vector and $\mathbf{p}(t)$ is a time varying vector in a bounded space. The system is called quasi LPV (qLPV) if $\mathbf{p}(t)$ includes some elements of $\mathbf{x}$.

The TPMT converts the system matrix (3) into a convex combination of R constant linear time invariant (LTI) system matrices

$$\mathbf{S}_r = [\mathbf{A}_r \quad \mathbf{B}_r], \quad r = 1, 2, ..., R \tag{4}$$

as

$$\mathbf{S}(\mathbf{p}(t)) = \sum_{r=1}^{R} \omega_r(\mathbf{p}(t))\mathbf{S}_r, \tag{5}$$

where $\omega_r(\mathbf{p}(t))$ is the weighting function. Thus, the system (2) can be written as

$$\dot{\mathbf{x}} = \sum_{r=1}^{R} \omega_r(\mathbf{p}(t))(\mathbf{A}_r\mathbf{x} + \mathbf{B}_r\mathbf{u}). \tag{6}$$

Based on a parallel distributed compensation (PDC) controller design framework, the control law can be expressed as

$$\mathbf{u} = -\left(\sum_{r=1}^{R} \omega_r(\mathbf{p}(t))\mathbf{K}_r\right)\mathbf{x}, \tag{7}$$

where the matrix gains $\mathbf{K}_r$ can be obtained by solving the following LMIs [8, 9]

$$-\mathbf{X}\mathbf{A}_r^T - \mathbf{A}_r\mathbf{X} + \mathbf{M}_r^T\mathbf{B}_r^T + \mathbf{B}_r\mathbf{M}_r > 0 \tag{8}$$

for all r and

$$\begin{aligned}&-\mathbf{X}\mathbf{A}_r^T - \mathbf{A}_r\mathbf{X} - \mathbf{X}\mathbf{A}_s^T - \mathbf{A}_s\mathbf{X} \\ &\quad + \mathbf{M}_s^T\mathbf{B}_r^T + \mathbf{B}_r\mathbf{M}_s + \mathbf{M}_r^T\mathbf{B}_s^T + \mathbf{B}_s\mathbf{M}_r \geq 0\end{aligned} \tag{9}$$

for all $r < s \leq R$, where $\mathbf{K}_r = \mathbf{M}_r\mathbf{X}^{-1}$. The control law (7) renders the closed loop system of the system (2) asymptotically stable. The reader refers to [8, 9] for more details. Note that the TPMT and the control law (7) can be determined by using the TP tool [10].

2.3 Twisting sliding model algorithm

Twisting sliding model algorithm (TSMA) is one of the popular high order sliding mode (HOSM) control. The TSMA was first introduced in [16, 17].

Consider the second order system described by

$$\begin{aligned}&\dot{x} = y, \quad \dot{y} = u + \delta(t,x,y) \\ &|\delta(t,x,y)| \leq D, \quad D > 0,\end{aligned} \tag{10}$$

where x, y are the state variables, u is the control input and $\delta(t,x,y)$ is the bounded disturbance. The TSMA control law can be written as [16]

$$\begin{aligned}u &= -M_1\text{sign}(x) - M_2\text{sign}(y) \\ &= -M(\text{sign}(x) + 0.5\text{sign}(y)),\end{aligned} \tag{11}$$

where $M_1 = M$ and $M_2 = 0.5M$. The control law renders the closed loop system a finite time stable equilibrium, provided that $M > 2D$.

Note that if the bound D is unknown, the fixed gain M can be replaced by an adaptive gain M(t) [23].

3 Control design

The control command u is expressed as

$$u = u_{TPMT} + u_{TSMA}. \tag{12}$$

The control design has two steps. In the first step, u_{TPMT} is determined by assuming that the system (1) is undisturbed (i.e., d = 0). Thus, the system (1) can be expressed as

$$\dot{\mathbf{x}} = \mathbf{A}(V,e)\mathbf{x} + \mathbf{B}u_{TPMT}, \tag{13}$$

which is the same form as the system (2). Then, by applying the TPMT and solving the LMIs (8) and (9), it yields

$$u_{TPMT} = -\left(\sum_{r=1}^{R} \omega_r(\mathbf{p}(t))\mathbf{K}_r\right)\mathbf{x}. \tag{14}$$

Thus, the closed loop system containing the system (13) and the control law (14) is asymptotically stable. By substituting (12) into (1), it results in

$$\begin{aligned}\dot{\mathbf{x}} &= \mathbf{A}(V,e)\mathbf{x} + \mathbf{B}(u_{TPMT} + u_{TSMA}) + \mathbf{D} \\ &= \mathbf{A}(V,e)\mathbf{x} + \mathbf{B}u_{TPMT} + \mathbf{B}u_{TSMA} + \mathbf{D}.\end{aligned} \quad (15)$$

In the second step, since $\dot{\mathbf{x}} = \mathbf{A}(V,e)\mathbf{x} + \mathbf{B}u_{TPMT}$ is asymptotically stable, x_1, x_2 and x_3 are bounded since d is bounded. Then, we simplify the system (15) as

$$\begin{aligned}\dot{x}_1 &= x_2 \\ \dot{x}_2 &= k_e u_{TSMA} + \delta(t),\end{aligned} \quad (16)$$

where $\delta(t)$ includes d and any unmodelled parts of the system. Note that $\delta(t)$ is bounded since x_1, x_2, x_3 and d are bounded. From (11), we obtain

$$u_{TSMA} = -M(\text{sign}(x_1) + 0.5\text{sign}(x_2)). \quad (17)$$

Therefore, the control command (12) can be written as

$$u = -\left(\sum_{r=1}^{R} \omega_r(\mathbf{p}(t))\mathbf{K}_r\right)\mathbf{x} - M(\text{sign}(x_1) + 0.5\text{sign}(x_2)). \quad (18)$$

4 Simulation results

The Runge–Kutta method with the time step of 0.0001 sec was used in all simulations. The initial condition was set as $\psi(0) = 0.1$ rad, $d\psi(0)/dt = 0$ rad/sec and $y(0) = 0.05$ m.

The time varying vector $\mathbf{p}(t)$ contains V and e. From Table 1, the space of $\mathbf{p}(t)$ is selected as [20, 80] × [0.1, 0.5]. By executing the TPMT of the system (13), using the TP Tool [10] with 100 × 100 sampling grid points, the ranks of the sampled tensors were found to be 3 and 2 on V and e, respectively. Thus, six vertex systems can exactly represent the system. The weighting functions are shown in Figure 4.

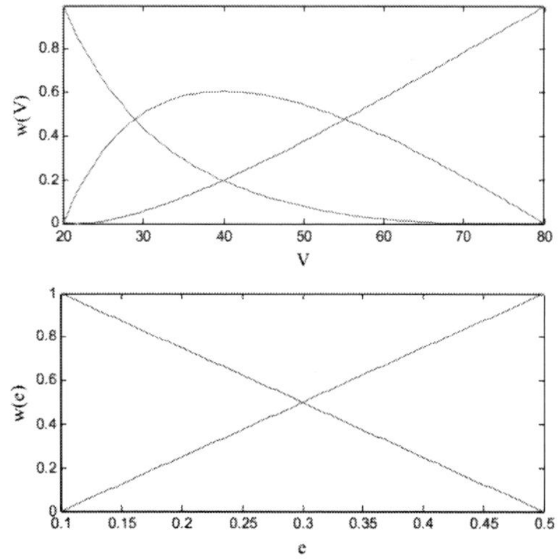

Fig. 4. Weighting functions on V and e.

By solving the LMIs (8) and (9), the following six linear feedback gains were obtained:

$$\begin{aligned}\mathbf{K}_1 &= [6.5065 \quad 0.0222 \quad -15.7799], \\ \mathbf{K}_2 &= [6.7727 \quad 0.0223 \quad -11.2479], \\ \mathbf{K}_3 &= [7.4875 \quad 0.0226 \quad -10.5572], \\ \mathbf{K}_4 &= [8.1582 \quad 0.0239 \quad -13.6185], \\ \mathbf{K}_5 &= \mathbf{K}_6 = [0.0 \quad 0.0 \quad 0.0].\end{aligned} \quad (19)$$

The closed loop control result of the system (1) using only u_{TPMT} for $V = 75$ m/sec, $e = 0.3$ m and $d = 0$ kN·m is shown in Figure 5. It shows that the shimmy vibration was suppressed within 0.07 sec.

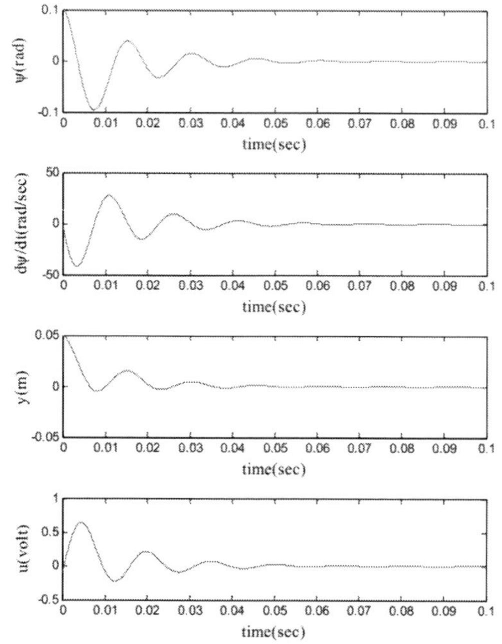

Fig. 5. Control responses of the undisturbed system using only the TPMT controller.

Next, the following disturbance torque was imposed into the system:

$$d(t) = 5\sin(1000t) \text{ kN·m}. \quad (19)$$

The result is shown in Figure 6. It is observable that additional vibration due to the disturbance was occurred although the controller had continuously responded to suppress the vibration.

To eliminate the effect of the disturbance, u_{TSMA} was integrated into the control law. Here, $M = 1$ was selected. In summary, the control law becomes:

$$u = -\left(\sum_{r=1}^{R} \omega_r(\mathbf{p}(t))\mathbf{K}_r\right)\mathbf{x} - (\text{sign}(x_1) + 0.5\text{sign}(x_2)). \quad (20)$$

where the weighting functions are given in Figure 4 and the linear feedback gains in Eq.(19).

The closed loop control result is shown in Figure 7. The controller was able to suppress the vibration as desired. However, due to the discontinuity nature of the TSMA, chattering was observed.

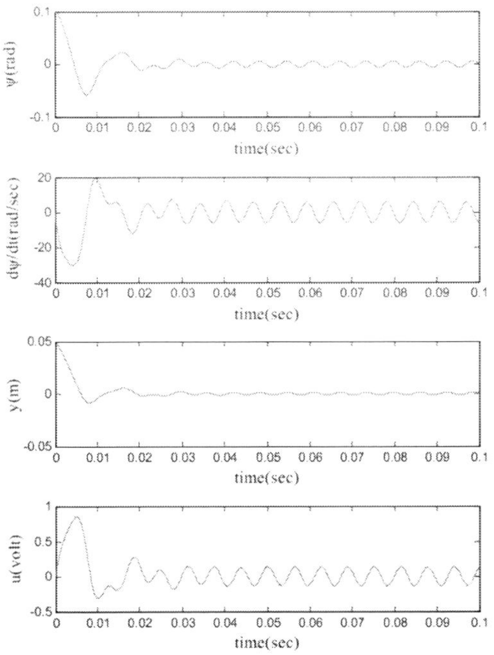

Fig. 6. Control responses of the disturbed system using only the TPMT controller.

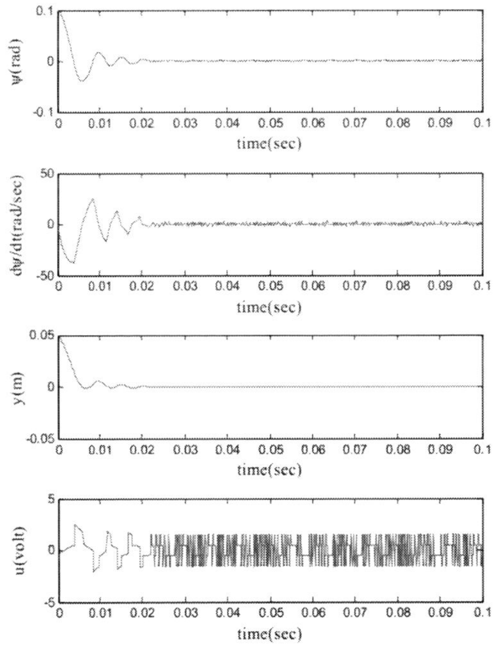

Fig. 7. Control responses of the disturbed system using both TPMT and TSMA controllers.

Finally, by replacing the discontinuous function sign(x) with the function tanh(5x) to solve the chattering problem, smoother control responses were achieved, as shown in Figure 8.

Similar results were also obtained for different values of V and e.

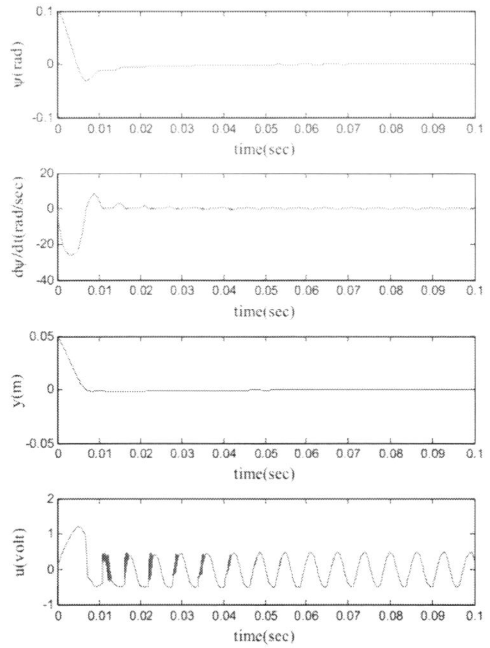

Fig. 8. Control responses of the disturbed system using both TPMT and TSMA controllers (replacing sign(x) with tanh(5x)).

5 Conclusions

The tensor product model transformation (TPMT) and twisting sliding mode algorithm (TSMA) are utilized to design a robust controller for suppression of the shimmy vibration in aircraft landing gears. The shimmy vibration is undesirable and it can lead to accidents. Simulation results illustrated that the proposed controller effectively suppressed the vibration even though the system was subjected to an external disturbance torque.

References

1. G. Somieski, Aerospace Science and Technology, **8**, 545–555 (1997)
2. E. Esmailzadeh, K.A. Farzaneh, Journal of Vibration and Control, **5**, 45–56 (1999)
3. Y. Li, C. Howcroft, S.A. Neild, J.Z. Jiang, Journal of Vibration and Control, **408**, 234–251 (2017)
4. J. Pritchard, Journal of Aircraft, **38**, 130–137 (2001)
5. A. Hajiloo, W.F. Xie, Asian Journal of Control, **17**, 476–485 (2015)
6. H. Tourajizadeh, S. Zare, Aerospace Science and Technology, **50**, 1–14 (2016)

7. C. Orlando, A. Alaimo, Mechanical Systems and Signal Processing, **84**, 551–569 (2017)
8. P. Baranyi, IEEE Transactions on Industrial Electronics, **51**, 387–400 (2004)
9. P. Baranyi, Y. Yam, P. Várlaki, *Tensor Product Model Transformation in Polytopic Model-Based Control*, (Florida: CRC Press, 2013)
10. S. Nagy, Z. Petres, P. Baranyi, 8th International Symposium of Hungarian Researchers on Computational Intelligence and Informatics, 483–495 (2007)
11. P. Galambos, P. Baranyi, G. Arz, Proc. IMechE Part C: J Mechanical Engineering Science, **228**, 765–777 (2014)
12. S. Kuntanapreeda, Asian Journal of Control, **17**, 371–380 (2015)
13. Z. He, M. Yin, Y. Lu, Proc. IMechE Part G: J Aerospace Engineering, **230**, 378–391 (2016)
14. E.-L. Hedrea, C.-A. Bojan-Dragos, R.-E. Precup, T.-A. Teban, 21st International Conference on Intelligent Engineering Systems, 113–118 (2017)
15. J. Kuti, P. Galambos, P. Baranyi, IFAC PaperOnline, **50**, 5604–5609 (2017)
16. A. Levant, International Journal of Control, **58**, 1247–1263 (1993)
17. A. Levant, Automatica, **43**, 576–586 (2007)
18. W. Xu, Y. Xu, X. Zheng, 5th Annual IEEE International Conference on Cyber Technology in Automation, Control and Intelligent Systems, 1379–1384 (2015)
19. H. Chen, Z. Sun, W. Sun, T. W. Yeow, 34th Chinese Control Conference, 8331–8336 (2015)
20. T. Han, Q. Hu, Aerospace Science and Technology, **70**, 428–434 (2017)
21. P. Khamsuwan, S. Kuntanapreeda, T. Sangpet, 12th IEEE Conference on Industrial Electronics and Applications, 798–803 (2017)
22. R. Luo, H. Su, Chinese Journal of Physics, **55**, 2199–2207 (2017)
23. Y.B. Shtessel, J.A. Moreno, L.M. Fridman, Automatica, **75**, 229–235 (2017)

Energy-efficient control of a multi-section supercapacitor power supply of an electric drive

Vladimir Mozzhechkov*

Tula State University, Department of Robotics, 300012 Tula, Russia

Abstract. A method for synthesizing the control laws of a multi-section supercapacitor power supply of an electric drive is developed. The synthesized control law for an electric drive realizes the prescribed motion and minimizes the required capacitance of the power source. It is achieved through optimal disconnection and connection to the power line of the drive at designated times of one of the power supply sections. Reduction of the required capacitance of the power supply is achieved through a fuller discharge of some of its sections in motion conditions requiring a low level of electrical voltage and saving high voltage in other sections for the respective motion conditions. A mathematical formulation of the problem and a method of its solution is proposed. An example of the implementation of the proposed method is considered.

1 Introduction

Currently, electric drives with capacitor power supply (CPS) are widely used in industrial applications. Electric energy in such power supplies is stored in conventional capacitors and/or supercapacitors (ultracapacitors). Capacitors in the CPS can be periodically or continuously recharged, but when the drive performs energy-intensive motions, the role of the charging line is negligible, the main part of the energy in the drive comes from the capacitors.

Let us consider a class of problems assuming that the CPS must provide the energy necessary for the prescribed (programmed) motion of the electric drive for a given finite time interval. This class includes, for example, some tasks of robotics, the tasks of power supply for the electric vehicle cyclically moving along the designated route with a specified recharge point, the tasks of opening and closing valves on pipelines using electric drives powered by CPS.

CPS can be structurally divided into sections, with each containing several electrically connected capacitors and/or supercapacitors. This type of CPS is called multi-section capacitor power supply (MCPS). The choice of the capacitance of the sections and their number is carried out on the basis of design considerations and known energy costs for the realization of the assigned motion. The electrical connections of the MCPS sections to each other and to the power line of the drive are traditionally considered fixed.

The capacitance of the MCPS required for the implementation of the prescribed motion of the electric drive can be reduced by connecting of one of the MCPS sections to the power supply line at pre-scheduled times. Reduction of the required capacitance in this case is due to the possibility of more complete discharge of the MCPS sections in motion conditions requiring low voltage level and saving of high voltage charge for the respective motion conditions, which may be only in one section or a few MCPS sections.

Control laws for electric drives, which take into account the possibility of optimal disconnection and connection to the power line of one of the MCPS sections, lead to the increase of the energy efficiency of the considered drives, reduction of capacitance and, as a consequence, weight, size and cost of MCPS.

The topic of energy efficiency of electric drives with an autonomous power source is given considerable attention in modern scientific literature. In the paper [1] an adaptive energy-efficient control is developed for planar motion control of electric ground vehicles with four in-wheel electric drives. The article [2] presents a plug-in hybrid electric vehicle energy management control strategy that aims to improve the real energy usage efficiency. In the publication [3], the authors consider the possibility of increasing the energy efficiency of actuators based on the use of the load torque predictor in the drive control law. The paper [4] describes a supervisory energy efficiency control strategy for electrical energy transfers in multisource renewable energy systems. In the paper [5] an efficient energy-management system for a hybrid electric vehicle, using ultracapacitors and neural networks, was developed. In the publication [6], in order to improve energy efficiency, the authors propose design and new control of DC/DC converters to share energy between supercapacitors and batteries in hybrid vehicles. The article [7] considers the problem of energy-efficient coordinated use of such energy sources as electric batteries, ultracapacitors and fuel cells in the onboard energy-storage system of the vehicle. The paper [8] describes the energy-efficient application of batteries and

* Corresponding author: itc@tulaprivod.ru

ultracapacitors in electric energy storage units for hybrid-electric vehicles.

However, the available literature on the subject, despite the considerable interest in the topic of energy efficiency of electric drives with an autonomous power source, does not provide methods of synthesis of the above-mentioned control laws. This fact makes the development of such methods a necessity.

2 Formulation of the problem

We specify and mathematically describe the problem of synthesizing the control law of an electric drive, which assumes a rational shutdown and connection to the power supply line of the drive of one of the MCPS sections.

Let us assume that the motion of the drive output link on a limited interval of time is given, i.e., we assume a known function $\alpha^*(t)$, $t\in[0,T]$, where α^* is the coordinate of the output link of the drive, t time, T the upper boundary of the time interval.

Along with the motion program, we shall assume that the law of variation of the load torque $M(t)$ exerted on the motor shaft is given.

We will consider drives with DC motors with independent excitation. The dynamics of this motor is given by the following differential equations:

$$di/dt=(u-iR-\omega C_e)/L, \quad (1)$$

$$d\omega/dt=(iC_m-M)/J, \quad (2)$$

$$d\varphi/dt=\omega, \quad (3)$$

where i is the current in the motor winding; U voltage across the motor winding; R, L resistance and inductance of the rotor winding; ω, φ rotational speed and angular position of the motor rotor; C_e back EMF constant; C_m motor torque constant; J is the moment of inertia of the moving parts of the drive transposed to the motor rotor.

The mechanical transmission (gearbox) in the drive system converts the rotation of the motor shaft into a movement of the drive output shaft. The following equality holds:

$$\alpha=\varphi/k_g, \quad (4)$$

where α is the linear or angular coordinate of the output link of the drive (angle of rotation or linear travel depending on the type of drive), k_g gear ratio.

MCPS is divided into several sections. The MCPS section represents several permanently connected capacitors. The capacitance of the section is denoted by C_j, $j=1,n$. The number of sections n is assumed to be given. The maximum voltage of the charged MCPS section should not exceed the value of U_{max}.

At any one time, only one MCPS section must be connected to the drive power line. It is possible to disconnect and connect any section of the MCPS to the power line. At this time the voltage in the supply line is equal to the voltage of the MCPS section connected to it.

The voltage of the j-th section of the MCPS $U_j=q_j/C_j$, where q_j - electric charge of the j-th section. In the differential form, we get: $dU_j/dt=dq_j/dt(1/C_j)$, or

$$dU_j/dt=i/C_j, \quad (5)$$

where i - discharge current, in the considered problem equal to the current in the armature winding of the drive motor.

Electronic switches in the electric drive control system allow to disconnect and connect the anchor winding of the motor to the supply main with a switching polarity determined by the sign of the control signal ν, i.e.

$$u=U_j\,\mathrm{sign}(\nu), j(t)\in\{1,..., n\}, \quad (6)$$

where ν - is the signal, that determines the polarity of connecting the motor winding to the power line; sign() is a sign function, takes values from the set {-1, 0, + 1} in accordance with the sign of its argument.

Without a significant loss of generality, we specify electric drive control law as follows. In order to minimize the deviation of the realized $\alpha(t)$ from the desired $\alpha^*(t)$ law of motion, the value of ν at each moment of time will be chosen as follows:

$$\text{if } (\alpha^*-\alpha)+(\omega^*-\omega)\kappa>0, \text{ then } \nu=+1, \text{ otherwise } \nu=-1, \quad (7)$$

where κ is a constant (speed feedback coefficient).

The system of equations (1-7) is a mathematical description (model) of an electric drive with a multi-section controlled source of power. In it, equations (1-5) describe the control object, and equations (6, 7) - control law.

For the electric drive described by the system of equations (1-7), it is required in the control law (6, 7) that the function $j(t)$ and the coefficient κ must be determined to provide a sufficiently accurate implementation of the prescribed motion $\alpha^*(t)$, $t\in[0,T]$ of the drive output shaft with the minimum total capacitance of the MCPS sections.

3 Method of synthesizing

The formulated problem of synthesizing the control law does not allow for an exact analytical solution. Therefore, the following numerical method for finding approximate solutions to the problem under consideration is proposed.

The proposed method is based on specifying a finite set of time moments, in which the values of the control function $u(t)$ can change. Thus, the desired control law $u(t)$ is approximated by a function that changes values at certain instants of time from a given finite set. In this case, the specified time sampling does not extend to the other variables. As a result, the problem reduces to analyzing a finite set of control law variants.

A significant part of control law variants is excluded from the subsequent analysis based on the application of simple logical rules, set out below. For the next analysis version of the control for the simple analytical dependencies found, the optimal values of the capacitance of the sections and the total required capacitance of the source are calculated.

The computationally simple dependencies and logical rules allow you to quickly analyze a large volume of possible control options. As a result, among all possible control law variants, the optimal variant is identified, to which the minimum required capacitance of the source corresponds. In a more detailed presentation, the proposed method is reduced to performing the following actions.

1. Through the solution of equations (1)-(4), the inverse problem of dynamics [9, 10] is solved, and functions $u^*(t)$, $i^*(t)$, $t\in[0,T]$ are found, which determine the values of voltage and current in the motor winding for each moment of time, corresponding to the given motion of the output link of the drive $\alpha^*(t)$ and the law of change moment of the load force $M(t)$.

2. The time interval $[0,T]$ is divided into N equal intervals of length Δt, hereinafter referred to as time step. The time moment corresponding to the beginning of the k-th time sample ($k = 1, \dots N$,) is denoted by t_k.

For each time sample, we assign the MCPS section number that is connected to the drive power line for the entire duration of the given time step. As a result, the value of the vector s consisting of N components is determined. The k-th component of this vector s_k takes values from the set $\{1, \dots, n\}$, thereby determining the number of the MCPS section connected to the power line during the k-th time step.

The vector s contains for each instant $t\in[0,T]$, the values of the control variable j in the expression (6). Thus, the vector s is a discrete approximation of the function $j(t)$ used in the formula (6).

The search of all possible vectors s is equivalent to the combinations of numbers from 0 to n^N-1. Only one vector s corresponds to each number from the specified range. It is representation of an integer from the range from 0 to n^{N-1} if the sections are assigned numbers from 0 to n-1. In terms of combinatorics, the set of all possible values of the vector s is equivalent to the set of placements with repetitions of n different objects at N different places.

Having carried out a search and analyzing all possible values of the vector s, you can find the optimal value that determines the optimal control of the source.

The search can be shortened by using the rule: in the following stages, only those vectors s are analyzed, in which the first occurrences of the unique components increase with the increase in their serial number.

All other options for management are excluded from the subsequent analysis as identical to those options that are subject to analysis in accordance with this rule.

Let us prove the expediency of using this elimination rule for variants, that is, we show that the optimal value of the vector s coincides with its optimal value found among the set of vectors s, in which the first values of the unique components increase with the increase in their ordinal number.

Indeed, the vectors s, equivalent up to the values of its components, are equivalent in the framework of the problem being solved. For example, the vector s = (1 1 2 3 1 2) and the vectors (2 2 1 3 2 1), (3 3 2 1 3 2), (2 2 3 1 2 3), (*a a b c a b*), (*b b a c b a*) only the designation of the applied sections and are equivalent in order of their application, i.e. are equivalent in accordance with the source control law corresponding to them and, therefore, are equivalent in terms of the result achieved.

In the framework of the problem being solved, the vector s' is equivalent to the vector s'' if, as a result of changing the designations of the source sections, one can transform the vector s'' into the vector s'. Therefore, it is expedient to define the principle of section designation, which makes it possible to isolate vectors s that are subject to further analysis, and all other vectors (equivalent up to the notation of its components) are excluded from consideration.

As such principle, we accept the following: by using the next section of the source, we assign it a number one more than the maximum number among the numbers of the previously used sections, start with 1. In this case, the values of the unique components (i.e., the component, the values of which are different from each other) for the first time occurring in the vector s increase with increasing their ordinal number in the vector s. All vectors that do not comply with this principle are excluded from further consideration as equivalent to one of the vectors to be analyzed. For example, the vector (1 1 2 3 1 2) corresponds to this principle, and the vectors (3 3 2 1 3 2), (2 2 1 3 2 1), (3 3 2 1 3 2), (2 2 3 1 2 3) does not correspond and can be excluded from further consideration. It follows from the foregoing that the value of the optimal vector among all possible values coincides with its optimal value found among the set of vectors s in which the first occurrences of the unique components increase with the increase in their ordinal number.

3. For each j-th section of MCPS, the time function $I_j(t)$, $t\in[0,T]$, is produced, which determines the current requested from the j-th section by the formula:

$\forall k\in\{1,\dots,N\}$: if $s_k=j$, then $\forall t\in[t_k, t_k+\Delta t]$: $I_j(t) = i^*(t)$, otherwise $I_j(t) = 0$;

and the time function U^*_j, $t\in[0,T]$, determining the voltage requested from the j-th section by the formula:

$\forall k\in\{1,\dots,N\}$: if $s_k=j$, then $\forall t\in[t_k, t_k+\Delta t]$: $U^*_j(t) = u^*(t)$, otherwise $U^*_j(t) = 0$.

4. The problem of minimizing the capacitance C_j, $j\in\{1,\dots, n\}$ is solved. To this goal, from equation (6) we find:

$$C_j = Q_j(0)/(U_j(0)-U_j(T)), \tag{8}$$

where $Q_j(t) = \int I_j(\tau)d\tau$, $\tau\in[t, T]$ is the electric charge picked up from the j-th MCPS section on the time interval $[t, T]$.

It follows from formula (8) that one of the necessary conditions for minimizing the value of C_j is the equality of the initial voltage in each MCPS section to the maximum voltage:

$$U_j(0)= U^{max}, j\in\{1,\dots, n\}. \tag{9}$$

From equalities (8), (9) it follows that the desired optimal capacitance value of the MCPS section is:

$$C_j^{\min} = Q_j(0)/(U^{\max} -U_j(T)) , j\in\{1,\dots, n\}. \tag{10}$$

In the resulting formula, the value $U_j(T)$ is to be chosen. The optimal value of $U_j(T)$, as follows from equation (10), should be as small as possible in order to minimize the corresponding capacitance of MCPS, and simultaneously, it must be large enough that the following conditions are fulfilled:

$$U_j(t) \geq U^*_j(t), j \in \{1,..., n\}, \forall t \in [0,T].$$

We obtain the calculated relations ensuring the optimal choice of $U_j(T)$. From equation (6) with account of (10) we find:

$$U_j(t)=U_j(T)+(U^{max}-U_j(T))Q_j(t)/Q_j(0), t \in [0, T]. \quad (11)$$

The conditions $U_j(t) \geq U^*_j(t), j \in \{1,..., n\}, \forall t \in [0,T]$ after substituting expressions (11) in them take the form:
$U_j(T)+(U^{max}-U_j(T))Q_j(t)/Q_j(0) \geq U^*_j(t), _{j \in \{1,..., n\}}, \forall t \in [0,T]$,
or, equivalently,
$[1-Q_j(t)/Q_j(0)]U_j(T) \geq U^*_j(t)-U^{max}Q_j(t)/Q_j(0), j \in \{1,..., n\}, \forall t \in [0,T]$.

The obtained expression for the initial time moment (at this moment $Q_j(t)/Q_j(0)=1$) leads to the obvious requirement: $U^*_j(0) \leq U^{max}$. For the remaining instants of time $t \geq 0+\varepsilon$, $\varepsilon>0$, we obtain:
$U_j(T) \geq [U^*_j(t)-U^{max}Q_j(t)/Q_j(0)]/[1-Q_j(t)/Q_j(0)]$,
$j \in \{1,..., n\}, \forall t \in [0+\varepsilon, T]$.

Starting from minimizing the value of $U_j(T)$, we find

$$U_j(T)=\max [U^*_j(t)-U^{max}Q_j(t)/Q_j(0)]/[1-Q_j(t)/Q_j(0)], \; t \in [0+\varepsilon,T], j \in \{1,..., n\}. \quad (12)$$

Having determined the optimal value of $U_j(T)$ by formula (12), we calculate the optimal values of the capacitances of MCPS sections according to equation (10).

5. The total minimum capacitance of the MCPS sections $\sum_{j=1,n} C_j$, corresponding to the analyzed variant of the vector s, is computed and stored.

6. After the search of all possible vectors s generated at the second stage of this algorithm is completed, the option s^{opt} is chosen among them, which corresponds to the minimum capacitance of the MCPS sections calculated at the fifth stage of this algorithm.

As a result of the implementation of the listed stages of the method, we find: a discrete approximation of the desired control law determined by the optimal value of the vector s found, the capacitance values of the C_j^{min} sections and the initial values of the potentials $U_j(0)=U^{max}$ in them, which ensure the minimization of the required capacitance of the MCPS.

7. For the found program for connecting MCPS sections, defined by the vector s^{opt}, following the known methods of control theory [11, 12], we determine the coefficient κ, ensuring a fairly accurate realization of the prescribed motion $\alpha^*(t)$.

The value of the coefficient κ does not significantly affect the minimum allowable total capacitance of the MCPS sections. This is because the optimal capacitance of the sections depends mainly on the trajectory of the motion α(t). It determines the profile of the functions $u^*(t)$, $i^*(t)$, which in turn determine the optimal connection law and the capacitance of the MCPS sections.

If the coefficient κ is chosen so that the accuracy of realization of the prescribed motion is high, this means that in this case the function $\alpha(t)$ is close to the function $\alpha^*(t)$ and the subsequent actions described in the above method lead to the result that is invariant to the value of the coefficient κ.

This circumstance makes it possible to realize the separation of the synthesis of the control law realized in the proposed method into two problems: the problem of choosing the optimal dependence $j(t)$ in formula (7), which determines the sequence of use of the MCPS sections, and the problem of choosing the optimal value of the coefficient κ in formula (6).

The accuracy of the solution obtained can be estimated on the basis of modeling a synthesized control system using equations (1-7).

If the control law found ensures the achievement of a specified accuracy of the implementation of the specified motion, then the process of solving the required task is successfully completed.

Otherwise, it is possible to improve the accuracy of the implementation of the specified motion through an increase of the amplitude of the control action determined according to expression (7), i.e., on the basis of an increase of U_j. To achieve this, the values of C_j are increased to the levels ensuring the required accuracy.

The above method assumes that the number of source sections is specified. However, it can be extended to a more general case where the number of sections is to be determined. In such a case, it seems rational to consistently increase the number of sections of the source from one to a value, when a further increase in their number does not lead to a sufficiently significant decrease in the required capacitance of the source. Thus, a reasonable compromise is achieved between the complication of the source design and the decrease in its capacitance.

4 Example of the control law synthesis

Consider the implementation of the proposed method in relation to the task described above, in which the program of motion of the output link of the drive is given in the form of a polynomial: $\alpha^*(t)=(\gamma_0 t+\gamma_1 t^2/2+\gamma_2 t^3/3+\gamma_3 t^4/4)/k_g$, $t \in [0, T]$.

Therefore, given equality (4), the program for changing the coordinate of the motor shaft is described by a polynomial: $\varphi^*(t)=\gamma_0 t+\gamma_1 t^2/2+\gamma_2 t^3/3+\gamma_3 t^4/4$, and the speed change program is given by the polynomial: $\omega^*(t)=\gamma_0+\gamma_1 t+\gamma_2 t^2+\gamma_3 t^3$.

Let us consider an application in which the drive opens a safety valve. At the initial time t=0, the valve is closed.

When the valve has "closed" status, all variables of the system under consideration have zero values. During the opening process, the valve is moved by the drive from the "closed" state to the "open" state in a given time T.

At time $t=T$, the coordinates α^*, φ^* take values corresponding to the state of the valve "closed": $\alpha^*(T)=\alpha_T$, $\varphi^*(T)=\alpha_T k_g$.

When switching to the valve state "closed", it is necessary that the valve plug comes into contact with the seat smoothly.

Mathematically, this requirement is described by the conditions $\omega(T)=0$, $\alpha^*(T)=\alpha_T$, $\alpha(t)\leq\alpha_T$, $t\in[0, T]$.

Since at the initial and final moment of time the program value of the speed $\omega^*=0$, we get: $\gamma_0=0$, $\gamma_1+\gamma_2T+\gamma_3T^2=0$, or $\gamma_0=0$, $\gamma_1= - (\gamma_2T+\gamma_3T^2)$.

We take into account that $\varphi^*(T)= \alpha_Tk_g$.

Then: $\alpha_Tk_g=\gamma_0T+\gamma_1T^2/2+\gamma_2T^3/3+\gamma_3T^4/4$.

Using the values of γ_0, γ_1, found above, we obtain: $\alpha_Tk_g=-(\gamma_2T+\gamma_3T^2)T^2/2+\gamma_2T^3/3+\gamma_3T^4/4$, from which the following value is found: $\gamma_2 =-6(\alpha_Tk_g/T^3+\gamma_3T/4)$.

Let us select the following parameters of the system: $\alpha_Tk_g=12\pi$, $k_g= 1200\pi$ *rad/m*, i.e. the motor shaft makes 6 full revolutions resulting in the stroke of the plug of $\alpha_T=10^{-2}$ *m*.

Let us select $T=1$ s,$\gamma_3=-450$, then: $\gamma_0=0$, $\gamma_1=1.20$ *rad/s*^{2}, $\gamma_2=448.81$ *rad/s*^{3}.

We assume that the law of the change in the load torque $M(t)$ is given by the polynomial $M(t)=\mu_0+\mu_1t+\mu_2t^2+\mu_3t^3$ in which: $\mu_0=15$ *Nm*; $\mu_1=0.1$ *Nm/s*; $\mu_2=0.07$ *Nm/s*^{2}; $\mu_3=-0.1$ *Nm/s*^{3}.

The graphs of the functions $\alpha^*(t)$, $\omega^*(t)$, $M(t)$, $t\in[0,T]$ are shown in Figure 1.

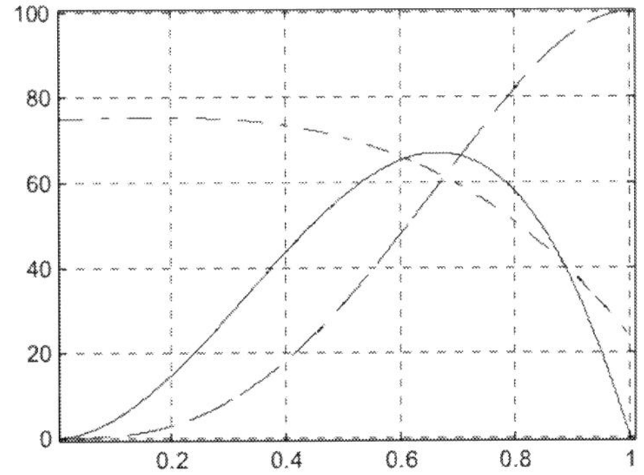

Fig. 1. The graphs of the functions $\alpha^*(t)\cdot10^4$, $\omega^*(t)$, $M(t)\cdot5$ (respectively - dashed, solid, dot-dash line).

The motor parameters have the following values: $C_e=1.50$ *V·s/rad*; $C_m=1.32$ *Nm/A*; $L=15\cdot10^{-4}$ *H*; $R=1$ Ω; $J=10^{-2}$ *kg·m*^{2}; $U^m=150$ *V*.

Further we proceed with the computations according to the method proposed above.

Solving the inverse problem of dynamics [9, 10] for the above-mentioned values of the drive parameters and the functions $\alpha^*(t)$, $M(t)$, with respect to the equations (1)-(4), we find the functions $u^*(t)$, $i^*(t)$, $t\in[0,T]$: their graphs are shown in Figure 2.

In the case of an MCPS consisting of one section ($n=1$), we determine by formula (12) the optimum value of the voltage at the output of the MCPS at the final instant of time $U_1(T)=100{,}86$ *V*.

Then, according to equation (10), we find the optimum capacitance of a single-section MCPS: C_{min} = 195.8 *mF*. The processes that characterize the functioning of the drive when the valve is open in the case of an MCPS consisting of a single section are shown in Figure 3. The coefficient κ in the control law (7) is chosen to be 50.

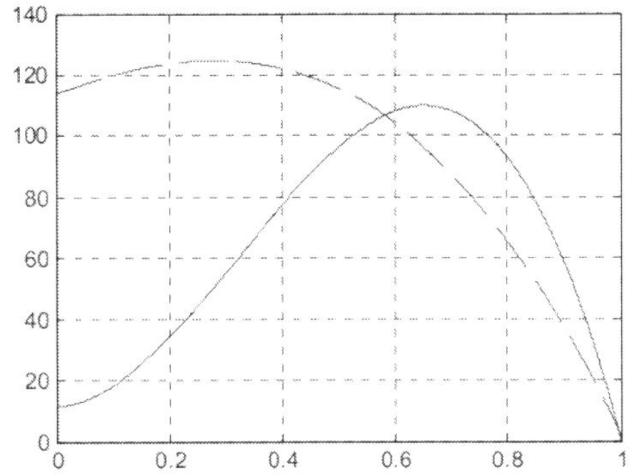

Fig. 2. The graphs of the functions $u^*(t)$, $i^*(t)$ 10, (respectively - solid, dashed line).

The optimal choice of the MCPS capacitance is characterized by the point of contact of the graphs of the functions $u^*(t)$ and $u(t)$ observed in Figure 3.

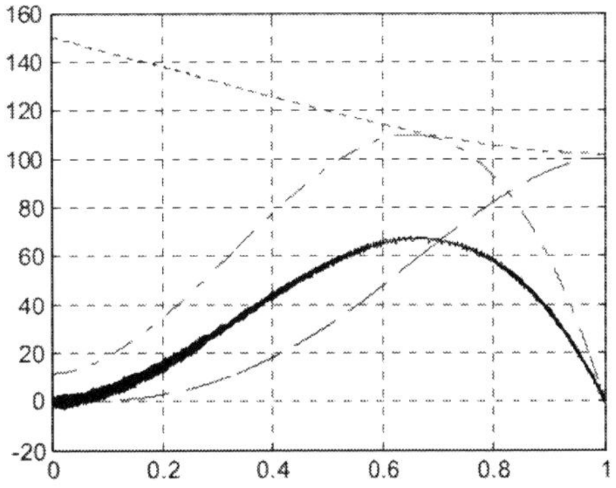

Fig. 3. The graphs of the functions $\alpha^*(t)$ 10^4, $\alpha(t)$ 10^4, $\omega^*(t)$, $\omega(t)$, $u^*(t)$, $u(t)$ in MCPS with the number of sections n=1 (dashed, dashed, solid, solid, dot-dash and dotted lines).

Assign the number of MCPS sections to $n=3$.

The interval of time [0,1] is divided into N = 10 equal parts. We generate and analyze vectors *s*, taking into account the above recommendations.

As a result of the search of vectors s, we find the optimal value s_{min}=(0 0 1 1 2 2 2 2 2 1). The coefficient κ in the control law (7), as in the case of a single-compartment MCPS, was chosen to be 50.

Figure 4 shows the processes corresponding to the solution found, characterizing the operation of the actuator when the valve is open in the case of an MCPS consisting of three sections.

Analyzing the obtained s_{min} value and the graph of $u(t)$, we see that the charge of the 0-th section is first consumed, then the 1st, 2nd, and the charge of the 1st section is again consumed at the final time step.

The optimum capacitance values of the MCPS sections are: $C_0^{min}=20.55$ *mF*; $C_1^{min}=34.04$ *mF*; $C_2^{min}=79.06$ *mF*. Residual voltages in MCPS sections: $U_1(T)=33.66$ *V*, $U_2(T)=71.23$ *V*; $U_3(T)=92.46$ *V*. The total capacitance is $C_0^{min}+C_1^{min}+C_2^{min}=133.7$ *mF*.

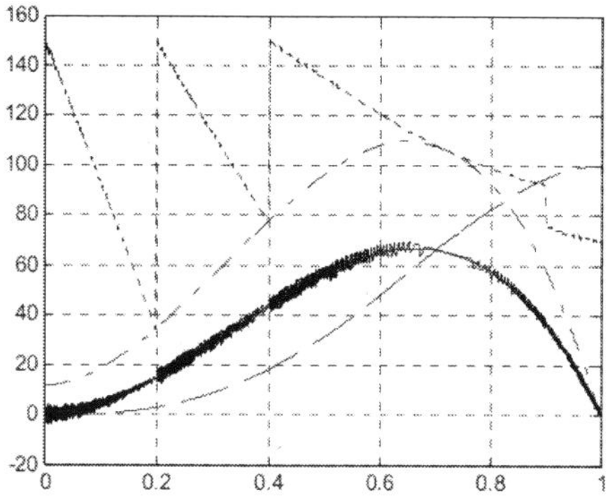

Fig. 4. The graphs of the functions $\alpha^*(t)\ 10^4$, $\alpha(t)\ 10^4$, $\omega^*(t)$, $\omega(t)$, $u^*(t)$, $u(t)$ in MCPS with the number of sections n = 3 (dashed, dashed, solid, solid, dot-dash and dotted lines).

In the example considered, the capacitance of the three-section MCPS is 1.46 times smaller than in the single-section MCPS.

5 Conclusions

A method for synthesizing the optimal for energy efficiency control law for a multi-sectional capacitor power supply of an electric drive is proposed. As a criterion of energy efficiency, the total required capacitance of MCPS sections is considered. The synthesized control law for electric drive realizing the prescribed motion, ensures the minimization of the required capacitance of the power source on the basis of rational disconnection and connection to the power line of the drive at designated times of one of the power supply sections.

Reduction of the required capacitance of the source is achieved as a result of a more complete emptying of some of its sections at time intervals with a low level of requested potential and saving high potential for upcoming time intervals in only some part of the source sections.

The task is formalized. The proposed method for solving it is based on specifying a finite set of time instants, into which the values of the control function can change. Thus, an approximation is made for the desired control law for a function that changes values at certain times from a given finite set. In this case, the specified time sampling does not extend to the other variables.

As a result, the problem reduces to analyzing a finite set of control options. A significant part of the management options is excluded from the subsequent analysis based on the application of simple logical rules, set out below. For the next analysis version of the control for the simple analytical dependencies found, the optimum values of the capacitance of the sections and the total storage capacitance of the source are calculated. Simple in the computational plan, dependencies and logical rules allow you to quickly analyze a large number of possible control options.

Among all possible variants of controls, an optimal variant is identified, to which the minimum required capacitance of the source corresponds.

An example of the implementation of the proposed method is considered for the electric drive of a safety valve. This example demonstrates the possibility of a significant reduction in the required MCPS capacitance when using the control law synthesized by the proposed method.

References

1. Y. Chen, J. Wang, IEEE Trans. on Control Systems Technology, **22** (4), 1362–1373 (2014)
2. H. Yu, M. Kuang, R. McGee, IEEE Trans. on Control Systems Technology, **22** (4), 1323–1336 (2014)
3. C. Du, A.R. Plummer, D.N. Johnston, Control Engineering Practice, **60**, 87–98 (2017)
4. F. Guerin, D. Lefebvre, V. Loisel, Control Engineering Practice, **20** (12), 1310–1324 (2012)
5. J. Moreno, M.E. Ortuzar, J.W. Dixon, IEEE Trans. on Industrial Electronics, **53** (2), 614–623 (2006)
6. M.B. Camara; H. Gualous, F. Gustin, A. Berthon, IEEE Trans. on Vehicular Technology, **57** (5), 2721–2735 (2008)
7. A. Khaligh, Z. Li, IEEE Trans. on Vehicular Technology, **59** (6), 2806–2814 (2010)
8. A. Burke, Proc. IEEE, **95** (4), 806–820 (2007)
9. H.K. Khalil, *Nonlinear systems* (Prentice Hall, 1996)
10. P.D.Krutko, Journal of computer and systems sciences international, **35** (6), 857–885 (1996)
11. V.A. Mozzhechkov, Automation and Remote Control, **64** (1), 23–36 (2003)
12. I.Boiko, *Non-parametric Tuning of PID Controllers: A Modified Relay-Feedback-Test Approach* (Springer-Verlag, 2012)

Multi-alternative control of large systems

Semen Podvalny* and *Eugeny* Vasiljev

Voronezh State Technical University, 394026 Voronezh, Russia

Abstract. The article deals with analysis of evolutionary concept of control of large systems – multi-alternativeness. Based on the biocybernetical approach to the task we are revealing information components of this concept, which is formulated in the following 3-M principles: multilevel structure is implementing in complex systems as the property of homeostasis; multi-version of algorithms and separation of control functions for the adaptive systems control; modularity as guarantee for this diversity.

1 Definition of a problem

The solution of the problem of large systems control within the robust-adaptive algorithms meets the difficulties of physical implementation of these algorithms arising due to big size of the systems themselves, their internal interrelations, nonstationarity of their parameters and dynamic uncertainty.

At the same time, a cybernetic approach to the analysis of this task indicates the condition that in biological systems the contrary phenomenon is observed: i.e., the growth of biosystems complexity causes the increase of their ability to the balanced development under the conditions of unpredictable change of the external environment and pressure of natural selection.

A specified property of «non-complicating simplicity» of live organisms and systems gives the grounds to assume that their activity is the cornerstone of certain evolutionary principles and their reproduction in anthropogenous systems (i.e., generated by human intellect) gives the way to realize a truly degree of stability and balance in them in the form of a direct analog of a biological homeostasis [1].

In this work the following principles are offered [2]:

- the principle of variety and division of functions;
- the principle of a multilevel control structure;
- the principle of a modularity,

all of them forming the general concept of multi-alternative control (from Latin alternare - to alternate) [3, 4].

Actual trends of usage of the of multi-alternative control principles, their initial biological content and the particular examples of their technical realization are shown below.

2 Actuality of multi-alternative control

Multi-choice control was actively used as early as 1957 by A.M. Lyotov to solve the problems of automatic control of dynamic objects, who offered to change the feedback sign in the regulator, i.e. to change its structure, choosing one of two alternatives depending on the current state of an object. Further studies of this problem have led to the development of systems with variable structure [5]. Presently, this trend developed into the theory of hybrid (switched) control systems combining the properties of continuous and discrete systems [6-8]. There are the potentialities of the formation of complex nonlinear control rules in the switched systems, which change depending on the changing external factors of the environment.

The principle of variety is included in the mandatory recommendation of the International Power Commission to develop the emergency protection systems for nuclear power plants. Such systems were identified as diversity system. The most commonly used forms of diversity in the protection systems are: technical, functional, parametrical, software and design forms. An example of technical diversity is the realization of the active protection which transforms the power unit to the state of safety in the presence of an electric power source, and the passive, nonvolatile protective option providing this safety only by force of gravity.

Due to the expansion of microprocessor control and the unprovability of nonfailure software operation of the automation equipment, the actuality of implementation of the diversity principle is the realization of the software programmable functions using the non-programmable means. Presently, the principle of diversity is being developed as the main method to increase the software reliability.

In the spaceships automatic control systems the latest tendency is to avoid the hardware reservation and the transition to reconfigured systems providing the flight safety with the artificially formed structural redundancy, i.e., the variety of control algorithms.

* Corresponding author: vgtu-aits@yandex.ru

3 Biological sources of multi-alternativity

3.1 Principle of variety and division of functions

The most important prerequisite of existence and evolution of all biological systems on Earth under the influence of the environment is the mandatory variety of its taxons due to which in the situations when the change of habitat becomes adverse for the existence of an arbitrary element of the system its function is taken by the other one (or many) biological groups that are capable of restoring the broken balance in the system, e.g., the continuity of a ffiood chain.

At the level of individual organisms the diversity of adaptive mechanisms is exposed in the form of differentiation and division of functions of their cells and organs for use in various environments within the natural habitat. The nature chose do develop the diversity, but not combination of functions in «universal» organs and systems of organisms. Animals vision is the example of the result of evolution which led to the formation of color-sensitive vision, but with low photosensitivity of the visual receptors functioning in the daylight and the highly sensitive receptors that do not sense color and function when the luminance is low.

Another more amazing example of the diversity principle and the division of functions is the presence of the so-called stem cells in the highest mammal forms which initially were not functionally focused, however, being capable to get narrow functional specialization depending on the current requirements of an organism.

This diversity provides stability of existence of the biological systems and their effective consumption of energy from both the external environment and internal resources.

3.2 Principle of multilevel control structure

In biosystems the multi-alternativity is implemented not only in the variety of the live world, but also as a multilevel structure, the hierarchy of this variety. Hierarchy of a structure is the basis of the most important property of all live systems, i.e. the principle and phenomenon of homeostasis [1] which activates higher levels of protection or reaction of an organism in the process of exhaustion of lower protection levels.

This property is implemented generally at the expense of a control system with two or more hierarchical levels one of which continuously supports the preset condition of an organism under the low-scale (insignificant) environmental changes, whereas the other levels react only to critical deviations of these conditions and as soon as these deviations manifest themselves, these protection levels start the qualitatively new mechanisms of essential reorganization of an organism, particularly, the spasmodic, uneven mechanism generating the alternative control structures, as shown in Figure 1.

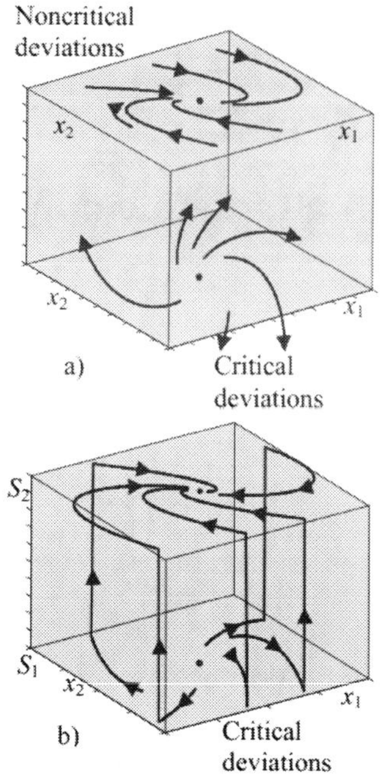

Fig. 1. Principle of multi-alternative control: a) parametrical level; b) structural level – the transition from structure S_1 to structure S_2.

W. Ashby [1], distinguished five (sic!) independent mechanisms of stabilization of glucose levels in the blood of a human.

The hierarchy of control forms, as a result of accumulation, the expanded set of tools, the multilayered protective «belt» which protects the system from unpredictable changes of the environment.

3.3 Principle of a control structure modularity

The source of diversity is the principle of a modularity which manifests itself in the formation of new live structures from the already available and rather small number of biologically stable modules [9]. The modularity of a structure is the major prerequisite of the «non-complicating simplicity» of the live world.

In particular, the known variety of proteins is realized as combinations of 20 basic amino acids the placement of which in protein is encoded in the genetic code in the form of a chain of triplets using only four nucleotides: adenine, guanine, thymine, cytosine. This set of blocks provides all proteinaceous variety of fauna: the chain comprising only 100 triplets can form 20100 variants of protein; the actual size of such chains may be several thousand triplets and even tens of thousand triplets. Discussing the development mechanisms, E. M. Galimov [10] notes that «the most economical way of the formation of a low-entropy product is the combination of already available low-entropy structures, and the evolution, as we should understand it should comprise not only and, moreover, not all possible small changes but the spasmodic changes caused by the new combinations of existing structures», thus emphasizing the combinatory nature of evolution. The discussed multi-alternative approach to the control in complex systems is formulated by W.Ashby as the information principle of the essential variety: «Variety can destroy variety».

The list of biological examples of multi-alternative control shown below brings us to the conclusion that complex biological systems have reached the high level of adaptive interaction with the environment by means of selective specialization of this interaction on the basis of: a set of considerably simple mechanisms of the division of functions; modularity and hierarchy of structure, thus they avoid the restrictions imposed by the growing functional complexity of living organisms. These adaptive mechanisms are united with the universal concept of multi-alternativity.

4 Examples of realization of multi-alternative control

4.1 Process of counterflow heat exchange

Let's review an example of realization of the multi-alternativity principles in the control system of two-circuit counterflow heat exchange process, which is shown schematically in Figure 2.

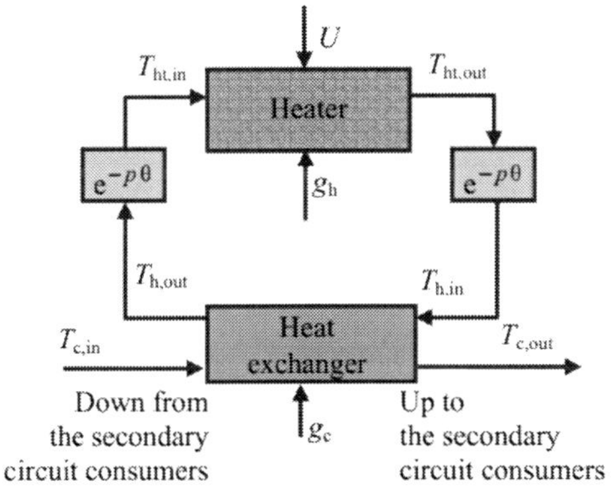

Fig. 2. Counterflow heat exchange process.

Here: $T_{c.in}$, $T_{c.out}$ are the temperatures of the "cold" secondary circuit heat-transfer agent at the input and output of the heat exchanger, respectively; $T_{h.in}$, $T_{h.out}$ are the temperatures температуры of the "hot" primary circuit heat-transfer agent at the input and output of the heat exchanger, respectively; $T_{ht.in}$, $T_{ht.out}$ – are the temperatures circuit heat-transfer agent at the input and output of heater; θ is the time required for the heat exchanger to pass the distance between the heat exchanger and the heater; $e^{-p\theta}$is the transfer function of the delay element; g_c, g_h are the mass flow rates of the heat-transfer agent (per second) in the secondary circuit (the consumers' circuit) and the heater circuit, respectively; U is the temperature of the heater's external surface. Proceeding from the aprioristic uncertainty and the nonstationarity of the parameters of an object, the controllers may be synthesized in the class of the fuzzy control, having the potential to form independent individual control rules for various conditions of a controlled object. This opportunity corresponds nicely to the declared multi-alternativity principles. According to the multilevel principle we shall construct the block diagram of an automatic control system with five control levels in each channel of impact g_h and U, as shown in Figure 3.

Basing on of the current value of controlled temperature $T_{ht.out}$ the linguistic variable $\tilde{T}_{ht.out} = \gamma$ is formed where γ contains the fuzzy variables γ: $\gamma=\{\gamma_1, \gamma_2, \gamma_3, \gamma_4, \gamma_5\}$, using the measures of comparison with the required average value of $T_{ht.out.r}$.

The increment Δg_h of a mass flow rate g_h is put in compliance with the linguistic variable α: $\alpha=\{\alpha_1, \alpha_2, \alpha_3, \alpha_4, \alpha_5\}$ with the values using the extent of a change α in comparison with its current level. For fuzzy variables $\alpha_1,\ldots, \alpha_5$ separate control channels are added, the contribution of each such channel into the resulting fuzzy impact β developed by the controller shall be defined by a measure of proximity of a current state γ to the required state γ_r:

$$\gamma_r = [\mu_{\gamma 1}(T_{ht.out.r}), \mu_{\gamma 2}(T_{ht.out.r}),\ldots, \mu_{\gamma 5}(T_{ht.out.r})] \quad (1)$$

Using the Cartesian product of γ and γ_r, we obtain the transition matrix of $M(\gamma\times\gamma_r,\beta)$ from the current state γ to the required state γ_r under the influence of the sought for control β:

$$M(\gamma\times\gamma_r,\beta) = \gamma\times\gamma_r. \quad (2)$$

For the definition of the general control β for each of n channels of the system the transition matrices of $M(\gamma_i\times\gamma_j, \alpha_s)$, $i = 1 \ldots n$; $j = 1 \ldots n$; $s = 1 \ldots n$; $n = 5$ are formed and the values of their elements characterize the possibility of a transition of an object from the condition γ_i to the condition γ_j under the influence of control α_s. As a measure of proximity of the required control β to the known controls $\alpha_1,\ldots, \alpha_5$ the extent of inclusion of a transition matrix $M(\gamma\times\gamma_r,\beta)$ in the corresponding matrices is applied to each channel:

$$\nu_s = \nu(\beta,\alpha_s) = \nu[M(\gamma\times\gamma_r,\beta), M(\gamma_i\times\gamma_j, \alpha_s)] \quad (3)$$

The received values of the proximity measures of ν_s are used for logical conclusion:

$$\mu_s(\Delta g_h) = \&_{\Delta gh}(\nu_s, \mu_{\alpha s}(\Delta g_h)) \quad (4)$$

the result of which $\mu_s(\Delta g_h)$ are the functions of accessory $\mu_{\alpha s}(\Delta g_h)$ truncated at the ν_s level. Control β can be calculated using the operation of association applied to all fuzzy subsets $\mu_s(\Delta g_h)$:

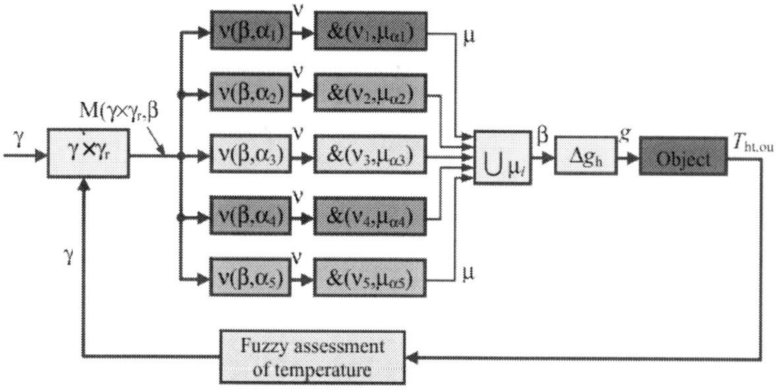

Fig. 3. Block diagram of a 5-level g_h-controlled control channel.

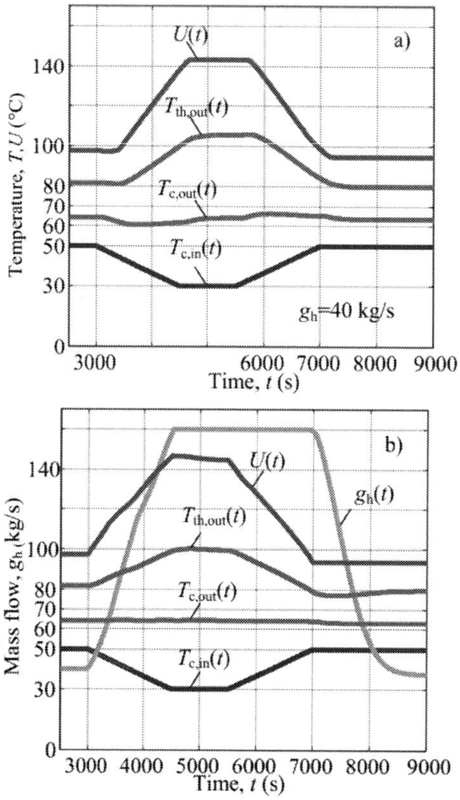

Fig. 4. Transition processes in the heat exchange control system with: (a) one control channel, and (b) three control channels with five control levels in each channel.

$$\beta = \{\mu_s(\Delta g_h)\} = \{\cup_{\Delta gh}\mu_{\alpha s}(\Delta g_h))\}. \tag{5}$$

Procedures (1)-(5) define the multi-alternative algorithm utilizing the g_h control channel. The control channel is constructed similarly utilizing the heater temperature U.

The content of this algorithm has clear physical meaning: at a shortage of resources of low level control the higher hierarchical level of the system is started providing the required quality of control as a result. The modular structure of each level of a system allows to create the control β in the form of multidimensional function $\beta(v_1,v_2,v_3,v_4,v_5)$, i.e., generally, in the form of a n-dimensional vector, providing high flexibility of control in the full range of current object states. In Figure 4a the result of work of control system with a single control channel by g_h is shown. The system keeps the value $T_{c,out}$ in the preset limits 60 ... 70°C, however the value of temperature $T_{ht,out}$ of the heat-transfer agent at the output of the heater exceeds the critical value 100°C.

To eliminate this shortcoming according to the diversity principle the two additional control channels are added to the system, one of them utilizing the heater temperature U, and the other channel using the disturbance input $T_{c,in}$. The quality of $T_{c,out}$ and $T_{th,out}$ control in the final multi-alternative system is improved considerably as confirmed by Figure 4b.

4.2 The space station power supply system

The space station power supply system includes three levels of subsystems, shown in Figure 5a: solar batteries (SB) subsystem; external of high voltage source (ES) subsystem; electrochemical storage batteries (EB) subsystem. Interaction principle of these subsystems results in the hierarchical structure in which the transfer of power supply function to each subsequent level occurs only in the condition of completely exhausted power resources of the previous level.

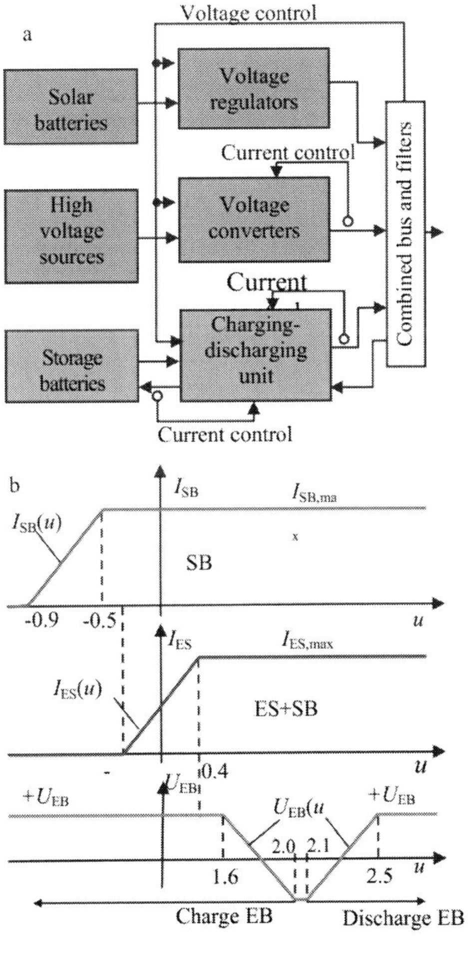

Fig. 5. An example of the multi-alternativity principles in the structure and operating modes of space station power supply system.

In Figure 5b it is shown that at the shift of the operating signal u outside the control range [-0.9;-0.5] the solar batteries subsystem provides its full current $I_{SB,max}$, to the load thus starting the external power source subsystem (control range [-0.4; 0.4]). With the solar battery completely loaded and selection of an external power source ($I_{SB}=I_{SB,max}$ and $I_{ES}=I_{ES,max}$) power supply from storage batteries is started. Similar transfer of control is carried out within each subsystem which contains a large number of identical modules. The modularity and hierarchy of a structure exclude any possibility of cascade (technologically conditional) increase of failures in the system whereas at any current in the load active control is provided for just a part of power supplied to power consumers corresponding to one single module of this either one or some other subsystem. All other modules are in the uncontrolled mode, some generates maximum current, and the others that are not required at the moment are disconnected.

As a result achieved in this «non-complicating simplicity» system, there is a possibility to unify the control circuits and the development of modular subsystems from interchangeable modules thus resulting in the reliable functioning of the whole power supply system in the wide range of deviations of its parameters, loads and at failures of various individual modules.

These examples adequately demonstrate that multi-alternative control of complex systems provides high quality of their functioning not just as a result of formation of the general algorithm of either robust or adaptive control, but by means of independent, autonomous synthesis of the corresponding hierarchical subsystems.

5 Conclusions

The solution of control tasks of complex systems by means of synthesis of uniform universal robust-adaptive algorithms is characterized by the accruing contradiction between the theoretical results and the extent of their realization achieved in this direction in practice.

Focusing on the biological origins of the problem being considered shows that the variety of conditions of the environment in the course of evolution of live organisms have led to the development of their corresponding adaptive mechanism: the multi-alternativity of functioning.

The concept of multi-alternativity offered in this work as a way of understanding the mechanisms of evolution and adaptation in living organisms gives the opportunity of purposeful reproduction of these mechanisms into the control systems of complex objects. This concept determines by itself the «non-complicating simplicity» principle as a reply to the problems of large systems development. Examples of realization of this mechanism in the automatic control systems utilized for various purposes confirm its high efficiency.

The offered way of a solution of the problem of large systems control follows in the course of the quickly developing scientific direction, viz., the biological cybernetics, which studies along with the basic feedback principle, the deeper analogies of anthropogenous systems with evolutionary principles of the biological communities functioning, which determine their final existence and the development.

References

1. W.R. Ashby, *Design for a brain* (London: Chapman & Hall, 1966)
2. S.L. Podvalny, E.M. Vasiljev, Autom. Remote Control, **8**, 1471–1499 (2015)
3. S.L. Podvalny, E.M. Vasiljev, Autom. Remote Control, **2**, 311–317 (2015)
4. S.L. Podvalny, E.M. Vasiljev, V.F. Barabanov, Autom. Remote Control, **10**, 1886–1891 (2014)
5. S.V. Emelianov, S.K. Korovin, I. G. Mamedov, *Variable-structure control systems: discrete and digital*, (Inc. Boca Raton, FL, CRC Press, 1995)
6. D. Liberzon, *Switching in systems and control,* (Boston, MA, Birkhäuser, 2003)
7. S.N. Vassilyev, A.A. Kosov, Autom. Remote Control, **6**, 1163–1183 (2011)
8. S.L. Podvalny, E.M. Vasiljev, International conference «Stability and control processes in memory of V.I. Zubov» (SCP), 120–122 (2015)
9. V.A. Krassilov, *Theory of evolution: a new synthesis is needed, evolutionary research. Macroevolution,* (Vladivostok, Dal'nevost. Nauchn. Tsentr, 1984)
10. E.M. Galimov, *Phenomenon of life, between equilibrium and non-linearity. Origin and principles of evolution*, (Moscow, Editorial URSS, 2006)

Stability of digital feedback control systems

Eugene Larkin*, *Alexey* Bogomolov and *Sergey* Feofilov

Federal State Budget Institution of higher Education "Tula State University", Lenina av., 92, Tula, Russia

Abstract. Specific problems arising, when Von Neumann type computer is used as feedback element, are considered. It is shown, that due to specifics of operation this element introduce pure lag into control loop, and lag time depends on complexity of algorithm of control. Method of evaluation of runtime between reading data from sensors of object under control and write out data to actuator based on the theory of semi-Markov process is proposed. Formulae for time characteristics estimation are obtained.
Lag time characteristics are used for investigation of stability of linear systems. Digital PID controller is divided onto linear part, which is realized with a soft and pure lag unit, which is realized with both hardware and software. With use notions amplitude and phase margins, condition for stability of system functioning are obtained. Theoretical results are confirm with computer experiment carried out on the third-order system.

1 Introduction

Including Von-Neumann type computer to digital control loops born many problems, main of which is the problem of gap from emergence a situation, which requires an adequate control system response, till the real action, affected onto the controllable object [1-3]. Time lag depends on a number of factors, such as computer architecture, clock frequency, instructions structure, operating environment, scheduling discipline, transactions order, mathematical foundation of control algorithms, etc.

Control algorithms have next specific features, which had been studied by several authors [4-6]:

Control algorithms have next specific features, which had been studied by several authors [4-6]:

algorithms are a cyclic ones, i.e. they have start operator, but does not have the end operator;

quest of peripherals is realized by means of inclusion into algorithm special transaction management operators;

for an external observer selection of branch in places of algorithm ramification is a stochastic one, and probabilities of branching depend on a distribution of data processed;

for an external observer algorithm operators' run-time is, a random one, distribution function of time of operator execution depends on distribution of data processed.

Thus, a process of a deterministic algorithm interpretation with Von-Neumann type controller for external observer by several authors [5, 7, 8] is regarded as semi-Markov process with continuous time. Operators of an algorithm are considered as states of semi-Markov process. Interpretation of algorithm may be considered as sequence of state switches or wandering through the states of semi-Markov process.

Among the algorithm operators there are operators, which request sensors and operators, which request actuator of the object under control. Time intervals between semi-Markov process states, which are abstract analogues of operators mentioned and influence of intervals onto the quality of control in linear control systems is subject of following investigations.

2 General semi-Markov model of the control algorithm

The flowchart of digital feedback control system is shown on the Figure 1 [1, 2, 3, 5].

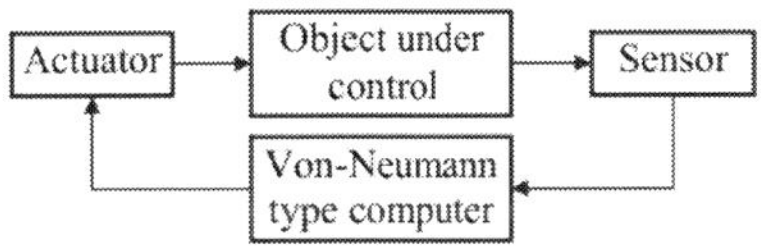

Fig. 1. Digital feedback control system.

The system includes object under control and the Von-Neumann type computer as a controller. The physical state of object under control is estimated by sensor, which generates data for computer processing. Output of Von-Neumann type computer through actuator affects on the object, so the feedback is closed.

Model of a system soft is the semi-Markov process [8-11]

$$\mu = \{A, \mathbf{h}(t)\}, \quad (1)$$

where $A = \{a_1, ..., a_j, ..., a_n, ..., a_J\}$ is the set of J states; $\mathbf{h}(t) = [h_{j,n}(t)]$ is the semi-Markov matrix of size $J \times J$; $h_{j,n}(t) = p_{j,n} \cdot f_{j,n}(t)$ is the weighed time density of residence in the state a_j before its switch into the state a_n; $p_{j,n}$ is the probability of switch from a_j into a_n;

* Corresponding author: elarkin@mail.ru

$f_{j,n}(t)$ - is the pure time density of residence in the state a_j with further switch into the state a_n.

Common structure of semi-Markov process, which simulates the soft of computer, is shown on the Figure 2 a. The process is ergodic, and is performed by the full graph with loops. From this structure all possible structures with J states may be obtained.

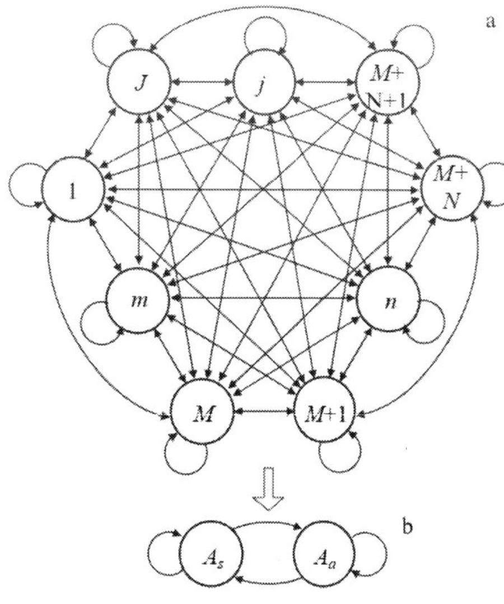

Fig. 2. Structures of Semi-Markov process; a - initial, b - after simplification.

In context of the task under solution the set of states A may be divided onto three disjoint subsets (classes):

$$A = A_s \cup A_a \cup A_o \quad A_s \cap A_a = \varnothing;$$

$$A_a \cap A_o = \varnothing; \; A_s \cup A_o = \varnothing, \tag{2}$$

where $A_s = \{a_1, ..., a_m, ..., a_M\}$ are states, which simulate sensor quests; $A_a = \{a_{M+1}, ..., a_n, ..., a_{M+N}\}$ are states, which simulate actuator quests; $A_o = \{a_{M+N+1}, ..., a_j, ..., a_J\}$ are states, which simulate other operators of cyclic algorithm.

As it is follows from the structure of semi-Markov process, quests to Von-Neumann computer peripherals, shown on the Figure 1, may be generated in the next cases:

when there are direct switches from states of subsets A_s or A_a to states of subsets A_s or A_a;

when there are wanderings from states of subsets A_s or A_a to states of subsets A_s or A_a through states of subset A_o.

To avoid second case one should to simplify the semi-Markov process (1) with use method, described in [12], as follows:

$$\mu \rightarrow \mu' \tag{3}$$

$$\mu' = \{A', \mathbf{h}'(t)\}, \tag{4}$$

$$A = A_s \cup A_a =$$
$$= \{a_1, ..., a_m, ..., a_M, a_{M+1}, ..., a_n, ..., a_{M+N}\}; \tag{5}$$

where $\mathbf{h}'(t) = [h'_{j,n}(t)]$ is the semi-Markov matrix of size $M+N \times M+N$; $h'_{j,n}(t) = p'_{j,n} \cdot f'_{j,n}(t)$.

In turn, simplified process (3) may be transformed into the process μ'', shown on the Figure 2 b as follows:

$$\mu' \rightarrow \mu''; \tag{6}$$

$$\mu'' = \left\{ \{A_s, A_a\}, \begin{bmatrix} h_{ss}(t) & h_{sa}(t) \\ h_{as}(t) & h_{aa}(t) \end{bmatrix} \right\} \tag{7}$$

where $h_{ss}(t) = p_{ss} f_{ss}(t)$ is the weighted time density of residence the process μ'' in one of subset A_s states with further switch into a state of the subset A_s; $h_{sa}(t) = p_{sa} f_{sa}(t)$ is weighted time density of residence the process μ'' in one of subset A_s states with further switch into a state of the subset A_a; $h_{as}(t) = p_{as} f_{as}(t)$ is weighted time density of residence the process μ'' in one of subset A_a states with further switch into a state of the subset A_s; $h_{aa}(t) = p_{aa} f_{aa}(t)$ is weighted time density of residence the process μ'' in one of subset A_a states with further switch into a state of the subset A_a.

Time density $f_{sa}(t)$ gives soughed-for lag in digital feedback control system. To define $f_{sa}(t)$ one should to evaluate probabilities of residence of semi-Markov process (4) in states of set (5) as follows [5, 7, 13]:

$$\pi_m = \frac{T_m}{\tau_m}, 1 \le m \le M, \tag{8}$$

where T_m is the time of a residence the process (4) in the state a_m; τ_m is the time of a return the process (4) at the state a_m;

$$T_m = \int_0^\infty t \sum_{j=1}^{M+N} h_{mj}(t) dt; \tag{9}$$

For estimation of τ_m one should split the state a_m onto ${}^b a_m$ and ${}^e a_m$. This may be done by means of transposition of m-th column of matrix $\boldsymbol{h}'(t)$ to $(M+N+1)$-th column. Emptied m-th column and $(M+N+1)$-th row should be fulfilled with zeros. As the result, matrix $\tilde{\mathbf{h}}'(t)$, of size $(M+N+1) \times (M+N+1)$ should be formed, The expectation of the time of return the process into state a_m is as follows:

$$\tau_m = \int_0^\infty t \cdot L^{-1}\left({}^r\mathbf{I}_m \cdot \sum_{k=1}^{\infty} \{L[\tilde{\mathbf{h}}'(t)]\}^k \cdot {}^c\mathbf{I}_{M+N+1} \right) dt, \tag{10}$$

where ${}^{r}\mathbf{I}_m$ is the row vector of size $M+N+1$, m-th element of which is equal to one and other elements are equal to zeros; ${}^{c}\mathbf{I}_{M+N+1}$ is the column vector of size $M+N+1$, ($M+N+1$)-th element of which is equal to one, and all other elements are equal to zeros; $L[...]$, $L^{-1}[...]$ are direct an inverse Laplace transform correspondingly.

The time density $f_{sa}(t)$ among τ_m, $1 \le m \le M$ also depends on weighed time densities of switch from states of subset A_s into stets of subset A_a:

$$f_{sa}(t) = \frac{\sum_{m=1}^{M} \pi_m \sum_{n=M+1}^{M+N} h'_{m,n}(t)}{\sum_{m=1}^{1} \pi_m \sum_{n=M+1}^{M+N} p'_{m,n}}. \qquad (11)$$

Next characteristics of lag are of interest when investigation of digital control systems stability: expectation T_{as}, dispersion D_{as}, minimum $T_{as\,\min}$ and maximum $T_{as\,\max}$ values [14]. They are as follows:

$$T_{as} = \int_0^\infty t f_{as}(t)dt\,; \quad D_{as} = \int_0^\infty (t - T_{as})^2 f_{as}(t)dt\,;$$

$$T_{as\,\min} = \min_{m,n} \left[\arg h_{m,n}(t)\right];$$

$$T_{as\,\max} = \max_{m,n} \left[\arg h_{m,n}(t)\right]. \qquad (12)$$

3 Stability of linear systems

Let us consider an influence of lag time $f_{sa}(t)$ on the stability of digital feedback control system, shown on the Figure 1 for the case, when model of object under control is the linear one. Control action is calculated as linear combination of sensor data, derivatives of the first and higher orders and integrals of the first and higher orders of sensor data (PID controller). Structural scheme of the linear system is shown on the Figure 3 [1, 2].

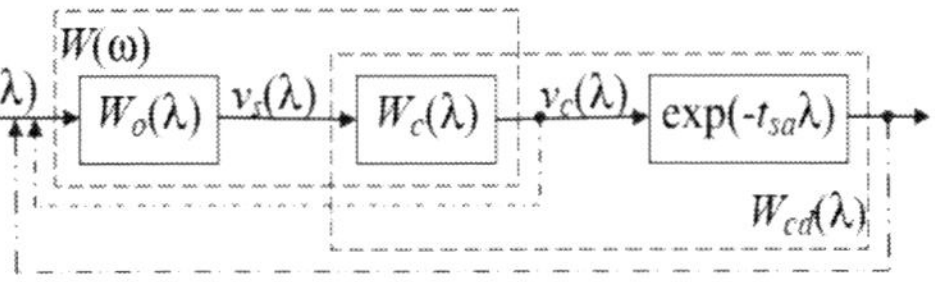

Fig. 3. Structural scheme of linear system.

On the Figure 3: $v_{cd}(\lambda)$ is the quasi-analogue computer output signal Laplace image; $v_s(\lambda)$ is the sensor signal Laplace image; $v_c(\lambda)$ is the quasi-analogue computer output signal Laplace image without lag; $W_o(\lambda)$ is the controllable object transfer function, including transfer functions actuator, object itself and sensor; $W_{\tilde{n}}(\lambda)$ is PID controller transfer function without lag [4, 15]; $\exp(-t_{sa}\lambda)$ is the transfer function of pure lag; $W(\lambda)$ – transfer function of system linear part; $W_{cd}(\lambda)$ is the transfer function of computer λ is the Laplace variable (the differentiation operator); t_{sa} is a random parameter, in accordance with section 2.

Substitution into transfer function $W_o(\lambda)$ $\lambda = i\omega$, where ω is the circular frequency; $i = \sqrt{-1}$ gives next form of the transfer function from the input of object under control till the output of computer without taking into account pure lag:

$$W(i\omega) = \frac{v_c(i\omega)}{v_{cd}(i\omega)} = \alpha(i\omega) \cdot \exp[i\varphi(\omega)], \qquad (13)$$

where $\alpha(\omega) = |W(\omega)|$ is the amplitude-frequency response; $\varphi(\omega) = \arccos \frac{\mathrm{Im}[W(i\omega)]}{A(i\omega)}$ is the phase-frequency response.

Both amplitude-frequency response and phase-frequency response are shown on the Figure 4.

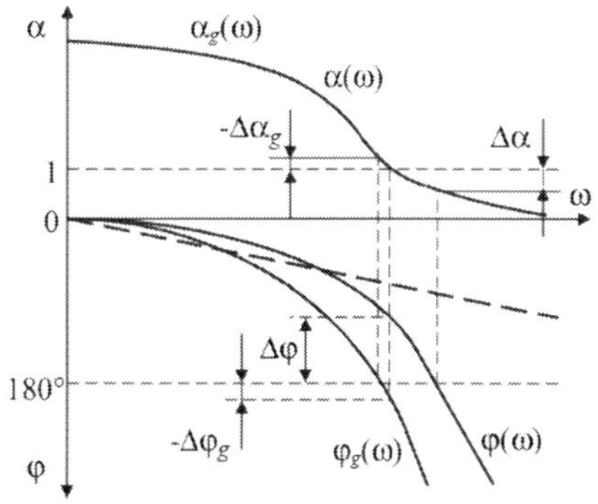

Fig. 4. Amplitude and phase frequency responses.

Differences $\Delta\alpha = 1 - \alpha(\omega)$ at the point $\varphi(\omega) = 180^\circ$ and $\Delta\varphi = 180 - \varphi(\omega)$ at the point $\alpha(\omega) = 1$ give for a system without lag amplitude and phase margins respectively [1, 2].

Let us insert the unit $\exp(-t_{sa}\lambda)$ into the control loop. The amplitude response of the unit is equal to 1 in all frequency range [16, 17, 18, 19]. Phase response has the form of an inclined straight line

$$\varphi = t_{sa}\omega, \qquad (14)$$

with the slope factor equal to t_{sa} (shown on the Figure 4 with dashed line).

In the system with lag values of straight line (14) are summed with values of phase response $\varphi(\omega)$. So amplitude $\alpha_g(\omega)$ and phase $\varphi_g(\omega)$ response of the system with a lag are as follows;

$$\alpha_g(\omega) = \alpha(\omega)\,; \quad \varphi_g(\omega) = \varphi(\omega) + t_{sa}\omega. \qquad (15)$$

Summing the values of straight line with values of the curve increases a steepness of the curve. Increase of a steepness leads to decrease of amplitude $\Delta\alpha_g$ and phase $\Delta\varphi_g$ margins. If margins became negative the system losses stability.

So, to provide stability of the system one should to operate with use next simple method.

Method of a digital feedback control system stability estimation.

1) For the system $W(\lambda)$ define amplitude and phase margins $\Delta\alpha_g$ and $\Delta\varphi_g$.

2) For the control algorithm define $T_{as\max}$.

3) Check the next criteria for the system with a lag:

$$\varphi\{\arg[\alpha(\omega)=1]\} - T_{as\max}\cdot\arg[\alpha(\omega)=1] \le 180^\circ; \quad (16)$$

$$\alpha\{\arg[\varphi(\omega)+T_{as\max}\omega=180^\circ]\} \le 1. \quad (17)$$

If both conditions are satisfied, the system is stable, if not, there is necessary to change system hardware or software.

4 Example

As an example one would consider the system, in which linear part is performed by the transfer function

$$W(\lambda)=\frac{v_c(\lambda)}{v_{cd}(\lambda)}=\frac{0{,}2}{0{,}01\lambda^3+0{,}1\lambda^2+0{,}2\lambda+0{,}01}. \quad (17)$$

System response on the Heavisaid function is shown on the Figure 5.

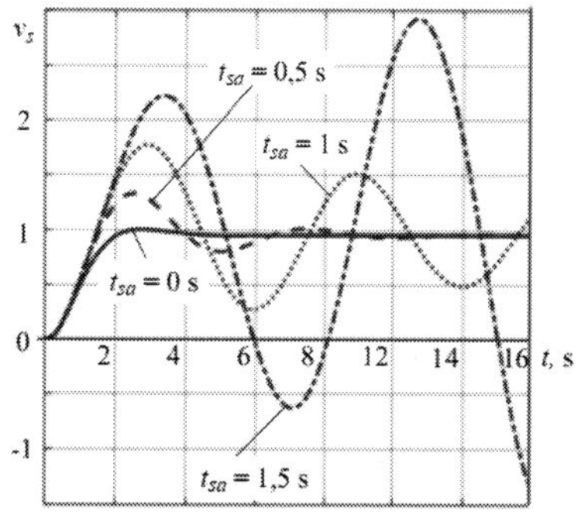

Fig. 5. Response of the system on the Heaviside function.

Response the system with closed loop without lag ($t_{sa}=0$) is shown with solid line. Response when $t_{sa}=0{,}5$ s is shown with dashed line< response when $t_{sa}=1$ s is shown with dotted line, and response when $t_{sa}=1{,}5$ s is shown with dash-dotted line. Criteria (15), (16) are met when $t_{sa}=0$, $t_{sa}=0{,}5$, $t_{sa}=1$ and are not met when $t_{sa}=1{,}5$. This is corfirmed by the graphs of transition processes. Process diverges when $t_{sa}=1{,}5$, and converges in all other cases.

5 Conclusion

Von-Neumann type computer lag time is the important factor, which should be taken into account when working out a digital control system, both hardware and software. Working out approach to evaluation a behavior of the system with pre-determined configuration opens new page in the theory and practice of system design. With use results expounded above one can to establish demands to software, especially to its time characteristics. This, in turn permits to limit computational complexity of control algorithm and facilitate decision making by software developers.

Further investigation in this area should be directed to working out practical recommendations for working out algorithm satisfied obtained limitations, and to finding links with other methods of software design.

The research was carried out within the state assignment of the Ministry of Education and Science of Russian Federation (No 2.3121.2017/PCH).

References

1. I.D. Landau, G. Zito, *Digital Control Systems, Design, Identification and Implementation*, (Springer, 2006)
2. J. Aström, B. Wittenmark. *Computer Controlled Systems: Theory and Design*, (Tsinghua University Press, Prentice Hall, 2002)
3. S.G. Tzafestas *Introduction to Mobile Robot Control*, (Elsevier, 2014)
4. K.J. Astrom, T.Hagglund, *Advanced PID Control. ISA* The Instrumentation, Systems, and Automation Society (2005)
5. E.V. Larkin, A.N. Ivutin, V.V. Kotov, A.N. Privalov, Interactive Collaborative Robotics (ICR 2016), 189–198 (2016)
6. A. Visioli *Practical PID control*, (Springer: London, 2006)
7. E.V. Larkin, A.N. Ivutin, 3-rd Meditteranean Conference on Embedded Computing (MECO-2014), 236–239 (2014)
8. N. Limnios, A. Swishchuk, Adv. in Appl. Probab., **45** (1), 214–240 (2013)
9. V. Korolyuk, A. Swishchuk, *Semi-Markov random evolutions*. (Springer-Science+Buseness Media, 1995)
10. T.R. Bielecki, J. Jakubowski, M. Niewęgłowski, Stochastic Processes and their Applications, **127**, (4), 1125–1170 (2017)
11. J. Janssen, R. Mancaб *Applied Semi-Markov processes*. Springer US (2005)
12. E.V Larkin., A.N. Ivutin, D.O. Esikov. MATEC Web of conferences: Australia, Melbourne (2016)
13. G.R. Grimmett, D.R. Stirzaker, *Probability and Random Processes*, (Oxford: Clarendon Press, 2001)
14. A.N. Shiryaev *Probability*, (Springer Science+ Business Midia, 1996)

15. A. Visioli *Practical PID control*, (Springer: London, 2006)
16. B. Wittenmark, J. Nilsson, M. Torngren. Proceedings of 1995 American Control Conference – ACC'95, American Autom Control Council, 2000–2004 (1995)
17. R. Krtolica, U. Ozguner, H. Chan, H. Goktas, J. Winckelman, M. Liubakka, Proceedings of American Control Conference, 2648–2653 (1991)
18. L. Wu, B. Ma, J. Cheng, J. Yin. Chinese Control and Decision Conference, 1273–1276 (2010)
19. C, Zheng, H, Zhang, Z Wang. Neurocomputing, **72**, 1744–1754 (2009)

Development trends of combined inductance-capacitance electromechanical energy converters

Hamlet Karayan and *Seyran* Gandilyan*

Department of Optics, Yerevan State University, Armenia

Abstract. In the article the modern state of completely new direction of electromechanical science such as combined inductive-capacitive electromechanics is considered. The wide spectra of its possible practical applications and prospects for further development are analyzed. A new approach for mathematical description of transients in dualcon jugate dynamic systems is proposed. On the basis of the algorithm differential equations for inductive-capacitive compatible electromechanical energy converters are derived. The generalized Lagrangian theory of combined inductively-capacitive electric machines was developed as a union of generalized Lagrangian models of inductive and capacitive electro-mechanical energy converters developed on the basis of the basic principles of binary-conjugate electrophysics. The author gives equations of electrodynamics and electromechanics of combined inductive-capacitive electric machines in case there are active electrotechnical materials of dual purpose (ferroelectromagnets) in the structure of their excitation system. At the same time, the necessary Lagrangian for combined inductive-capacitive forces was built using new technologies of interaction between inductive and capacitive subsystems. The joint solution of these equations completely determines the dynamic behavior and energy characteristics of the generalized model of combined machines of any design and in any modes of interaction of their functional elements

1 Introduction

Back in 1980-90-ies in the writings of Academician A.G. Iosifyan and Professor I.P. Kopylov on the basis of a critical analysis of the physical and technical foundations of the processes of generation, transmission, conversion and consumption of electromagnetic energy in inductive (magnetically induction) and capacitive (electro-induction) electric machines and apparatuses the possibility of implementing a third type of electromechanical energy converter has been installed – combined inductive-capacitive (magneto-electric induction), in which flux-chains (magnetic and electric) play an important role during the energy-exchange processes [1, 2].

The development of modern electromechanical science at the junction of the latest achievements of classical mechanics, electrodynamics, solid state physics, electrotechnical materials science, semiconductor physics, atomic physics, chemistry, cybernetics, etc., shows that the possible practical embodiment of the idea of inductive capacitive electromechanics currently predominates and can provide a new qualitative leap in the development of those branches of human activity, the basic value for which has a complex application ix systems electromechanical energy converters (from medical robotics to create a means of space exploration) [3-5].

Practical implementation of the idea of inductive capacitive electromechanics can be strongly influenced by the development of the main direction of nanoscience – nanostructural material science (nanocrystalline, nanocomposite, nanofibrous, etc.) [6, 7].

System analysis of fundamental research conducted at the turn of the 20th and 21st centuries in the field of physico-mathematical modeling, automated design and manufacturing of inductive-capacitive components of a microsystem electric drive is given by some foreign and Russian scientific publications, for example in [8-14].

In this paper, which is an organic continuation of the work of the authors [14,15], a new approach to the solution of some nodal questions of the generalized physico-mathematical modeling of combined inductive-capacitive energy converters allowing to take into account a large number of interrelated factors that determine their dynamic and energy characteristics is proposed.

2 Equations of electromechanical connections in combined inductive capacitive electromechanical energy converters

In [16-18], the generalized Lagrangian theory of generalized inductively-capacitive electric machines was developed as a union of generalized Lagrangian

* Corresponding author: angastroi@mail.ru

models of inductive and capacitive electromechanical energy converters, developed on the basis of the basic principles of binary-conjugate electrophysics.

Let us consider a more correct formulation of this problem. Without loss of generality, we can assume that the generalized combined machine, which is under consideration, consists of four contour chains: two stator symmetric three-phase circuits ($\hat{S}$ – inductive, $\hat{R}$ - capacitive phase, denoted by $\hat{a}, \hat{b}, \hat{c}$ and $\hat{m}, \hat{n}, \hat{h}$), respectively, two rotor excitation circuits of a constant magnetic – $\hat{F}$ and electric $\hat{P}$ field (spatially located between systems $\hat{S}$ and R) (Figure 1).

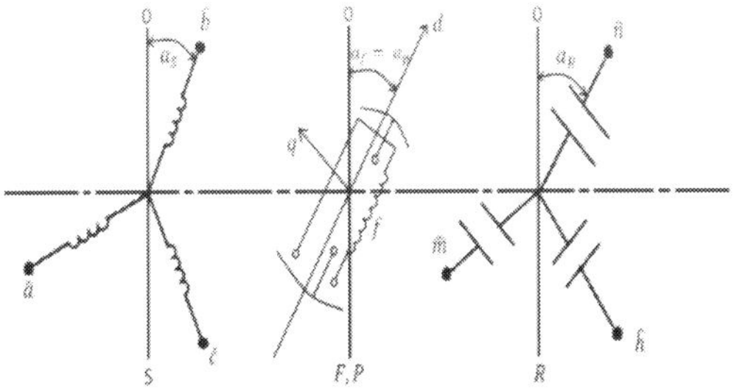

Fig. 1. Scheme of contour circuits of a combined inductive-capacitive machine with a dual excitation system.

Excitation of a constant magnetic flux $\Psi_f = \int Bds$ of the inductive part is carried out with the help of an excitation winding f, DC powered, the excitation of a constant electric flow $Q_p^* = \int Dds$ of the capacitive part is carried out by means of dipole electrodes, the power of which is carried out through special contact rings from a constant-voltage source. Between the phase circuits of the two subsystems there is a series (or parallel) galvanic coupling: respectively between $\hat{a}$ and $\hat{m}$ through the resistance R_{am}; $\hat{b}$ and $\hat{n}$ through the resistance R_{bn} ; $\hat{c}$ and $\hat{h}$ through the resistance R_{ch} (Figure 2).

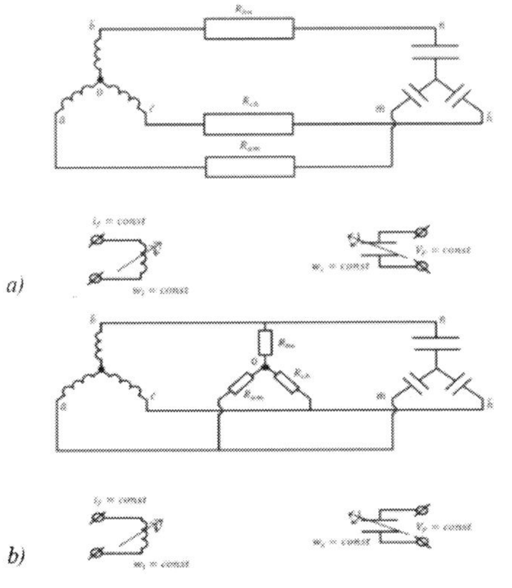

Fig. 2. The electrical circuit of the combined inductive-capacitive alternator: a – for series connection of phase circuits; b – for parallel connection of phase circuits.

The energy state of a combined machine in the multidimensional phase space of generalized variables is described by means of generalized coordinates - q (q_a, q_b, q_c, q_f) (as an electric charge in the current cords of an inductive subsystem), ψ^* (ψ_m^*, ψ_n^*, ψ_h^*, ψ_p^*) (as a magnetic flux "charge" inductive working electric field of a capacitive subsystem) corresponding generalized pulses Ψ^* (Ψ_a^*, Ψ_b^*, Ψ_c^*, Ψ_f^*) (working magnetic flux), Q^* (Q_m^*, Q_n^*, Q_h^*, Q_p^*) (working electric flux). As a mechanical generalized coordinate, we take the rotor rotation angle $\gamma = \int_0^t \omega dt$ or linear deviation $x = \int_0^t Vdt$.

The output of the equations of the generalized model of a combined machine is based on the Lagrange equations of the second kind describing the behavior of a nonconservative system with lumped parameters:

$$\frac{d}{dt}\left(\frac{\partial \mathcal{L}}{\partial \dot{a}_i}\right) - \left(\frac{\partial \mathcal{L}}{\partial a_i} + \frac{\partial D_0}{\partial \dot{a}_i}\right) = B_i; \quad (1)$$

$$i = a, b, c, f, m, n, h, p,$$

where $a\ (a_i) = a$ (q_a, q_b, q_c, q_f; ψ_m^*, ψ_n^*, ψ_h^*, ψ_p^*) – combined symbolic vector of generalized coordinates; $\dot{a}\ (\dot{a}_i) = \dot{a}\ (i_j = \frac{dq_i}{dt};\ V_j^* = \frac{d\psi_j^*}{dt})$, $(i = a, b, c, f;\ j = m, n, h, p)$ – generalized velocity vector; $\beta\ (\beta_i) = \beta\ (e_1 = -\frac{d\Psi_i}{dt},\ i_i^* = -\frac{dQ_i}{dt})$ generalized forces; $\mathcal{L}$ – Lagrangian energy function; $D_0 = -\frac{1}{2}\ (\sum_{j,i}(R_i i_i^2 + G_j V_j^{*2}))$ – loss or dispersion function.

From the point of view of energy conversion processes, in principle the following cases can be realized.

1. The mechanical energy is supplied to the system $\widehat{FP}$, the system $\hat{F}$ is excited. Then the combined machine operates in the inductive generator mode.
2. Mechanical energy is supplied to the system $\widehat{FP}$, the system $\hat{P}$ is excited. Then the combined machine operates in the mode of a capacitive generator.
3. Mechanical energy is supplied to the system $\widehat{FP}$, the system $\widehat{FP}$ is excited by a constant current. Then the system $\widehat{FS}$ operates in the inductive generator mode, $\widehat{PR}$ – in the mode of the capacitive generator.
4. Electromagnetic energy brings systems $\hat{S}$, the system $\hat{F}$ excited by a direct current, the system $\hat{P}$ – by a constant voltage. Then, the system $\widehat{FS}$, operates in the inductive motor mode, and the system $\widehat{PR}$ – in the mode of the capacitive generator.

If in the structure of a combined inductive capacitive electric machine there are no active working materials (such as ferroelectrics and magnets), then it is represented as a simple linear sum of two machines - inductive and capacitive, operating on the same shaft and interconnected by anchor electric circuits.

With a linear connection between the inductive and capacitive subsystems, the Lagrange function $\mathcal{L}$ is an additive function.

$$\mathcal{L} = \mathcal{L}_L(x, p, \Psi, q) + \mathcal{L}_c(x, p, Q^*, \psi^*). \quad (2)$$

In this case, the equations of electrodynamics of a combined machine can be expressed in the following vector-matrix form:

$$\left.\begin{aligned} \frac{d}{dt}\begin{pmatrix}\psi^*\\ Q^*\end{pmatrix} + \begin{pmatrix}R_L & 0\\ 0 & G_c\end{pmatrix} * \begin{pmatrix}i_L\\ V_c^*\end{pmatrix} = \begin{pmatrix}V_L\\ i_c^*\end{pmatrix} \\ \begin{vmatrix}\psi^*\\ Q^*\end{vmatrix} = \begin{vmatrix}L & 0\\ 0 & C\end{vmatrix} * \begin{vmatrix}i_L\\ V_c^*\end{vmatrix} \end{aligned}\right\} \quad (3)$$

where ψ, R_L, L – submatrices of magnetic flux linkages, active resistances and inductances of a magnetic machine, Q^*, G_c, C – submatrices of electric flux linkages, active conductivities and capacitances of electrical; i_L, V_L - vectors of currents and voltages of magnetic circuits, V_c^*, i_c^* - vector of voltages and currents of electrical circuits.

The equation of dynamics is supplemented by the equation of motion of the rotor

$$M = (J_m + J_e)\frac{d\omega}{dt} + \frac{K\omega}{P_0} + \frac{P_0}{2}\left(i_L^{\tau}\frac{dL}{dt}i_L + V_c^{*\tau}\frac{dC}{dt}V^*\right)\frac{1}{\omega}, \quad (4)$$

where J_m, J_e – moments of inertia of the rotors of inductive and capacitive subsystems, P_0 – number of pole pairs.

When the phase start circuits are connected in series, the inductive subsystem as a voltage source is a "carrier system" with an active-capacitive load, and with a parallel connection, the "carrier" system is a capacitive subsystem as a current source with an active-inductive load.

In this case, the equations of electrodynamics of a combined machine can be written in the following form: equations for stator system $\widehat{SR}$ for the serial coupling of phase stator circuits

$$V_{SP} = \frac{d\Psi}{dt} + R_L i_L + C_0^{-1}\int i_c dt, \quad (5)$$

where the vector i_c^*- is determined from the systems of equation

$$\frac{dQ^*}{dt} + G_c V_c^* = i_c^*, \quad (6)$$

matrix C_0 – is expressed in the form of

$$C_0 = \begin{vmatrix}C_{mm} & 0 & 0\\ 0 & C_{nn} & 0\\ 0 & 0 & C_{hh}\end{vmatrix}, \quad (7)$$

equations for stator system $\widehat{SR}$, with a parallel connection of phase stain chains

$$i_{RS}^* = \frac{dQ^*}{dt} + G_c V_c^* + L_0^{-1}\int V_L dt, \quad (8)$$

where the vector V_L – is determined from the system of equation

$$\frac{d\Psi}{dt} + R_L i_L = V_L, \quad (9)$$

and matrix L_0 is expressed in the form of

$$L_0 = \begin{vmatrix}L_{aa} & 0 & 0\\ 0 & L_{bb} & 0\\ 0 & 0 & L_{cc}\end{vmatrix}, \quad (10)$$

3 Resonance in an inductive capacitive machine

Resonance in the electrical circuit can be obtained both in parallel connection of the inductive part with the capacitive part – the resonance of the currents, and in their subsequent connection – the resonance of the voltages [19].

In a combined machine that does not contain external power sources, the reactive power is determined by the partial alternating current – $\omega = \omega_L = \omega_C$ and the maximum values of the energy accumulated in the inductive and capacitive subsystems of the machine:

$$\left.\begin{aligned} P_L = V_L i_L \sin\frac{\pi}{2} = \frac{Li_L^2}{2} = \omega W_{L(\max)}, \\ P_c = -V_C^* i_C^* \sin\frac{\pi}{2} = \frac{Li_L^2}{2} = -\omega\frac{CV_C^{*2}}{2} = -\omega W_{C(\max)} \end{aligned}\right\} \quad (11)$$

where $W_{L(\max)}$ and $W_{C(\max)}$ – the maximum value of energy periodically accumulated in the magnetic field of the inductive machine and in the electric field of the capacitive machine.

On the basis of the law of conservation of energy, the reactive power of a combined machine is equal to the algebraic sum of the reactive powers of inductive and capacitive machines, i.e.

$$P^0 = P_L + P_C = \omega\,(W_{L(\max)} - W_{C(\max)}). \quad (12)$$

When $P^0 = 0$, there is a resonance of the voltages (for a series connection) and a current resonance (with a parallel connection).

In this expression for the resonance of the voltage and current have the form

$$\left.\begin{aligned} V = [V_R^2 + (V_L - V_C)^2]^{1/2} = i_L[R^2 + (\omega L - 1/\omega C)^2]^{1/2} \\ I = [I_R^2 + (I_L - I_C)^2]^{1/2} = V_C^*[(1/2)^2 + (1/\omega L - \omega C)^2]^{1/2}. \end{aligned}\right\} \quad (13)$$

At resonance, the energy characteristics of the combined machine are the best, since the reactive power exchange between the inductive and capacitive subsystems takes place, and to create magnetic and electric fields the machine ceases to consume reactive power from external sources of energy.

If there are active electrotechnical materials of dual purpose – ferroelectromagnets, in the structure of the excitation system a combined inductive-capacitive electrical machine, then between the inductive and capacitive subsystems the electrophysical bond is nonlinear.

The Lagrange function - $\mathcal{L}_{LC}$ is not an additive function and can be expressed as:

$$\mathcal{L}_{LC} = \mathcal{L}_L + \mathcal{L}_C + \Delta\mathcal{L}_{LC}, \quad (14)$$

where $\Delta\mathcal{L}_{LC}$ a member of the Langrangian, takes into account the effect of one subsystem on another.

In our opinion, the necessary Lagrangian for combined inductive-capacitive EMF energy can be constructed by introducing a new interaction matrix (mutual influence) between the inductive and capacitive subsystems – $\widehat{K_0}(q,\psi^*)$, which is expressed in the form

$$\widehat{K_0}(q,\psi^*) = \begin{vmatrix} K_{am} & 0 & 0 & 0 \\ 0 & K_{bn} & 0 & 0 \\ 0 & 0 & K_{ch} & 0 \\ 0 & 0 & 0 & K_{fp} \end{vmatrix}, \quad (15)$$

where $K_{am} = K_0(q_a,\psi_m^*)$, $K_{bn} = K_0(q_b,\psi_n^*)$, $K_{ch} = K_0(q_c,\psi_h^*)$ – express mutual influence between phase chains, K_{fp} – determines the energy state of the combined inductive excitation system.

Since the active materials - ferroelectromagnets are completely installed in the rotor of a combined inductive capacitive EPM, in the general case K_{fp} is a continuous functional of the thermal, electrophysical and mechanical characteristics of this material, expressed in the form

$$K_{fp} = K_{fp}(\beta(i_f), \varepsilon(V_p), V_o, T), \quad (16)$$

where $\beta(i_f)$ – magnetic permeability, $\varepsilon(V_p)$ – electrical permeability, V_o– volume, T – temperature of active material.

Basic analytical properties of matrix elements $\widehat{K_0}(q,\psi^*)$ can be summarized as:
the elements K_{am}, K_{bn} и K_{fp} are continuous periodic functions, have a continuous periodic derivative in the interval $0 \le \gamma < \infty$ and satisfy the conditions

$$\left.\begin{aligned} & 0 \le |K_0(q_i,\psi_j^*)| \le 1 \\ & \lim_{\substack{q_i \to 0 \\ \psi_j^* \to 0}} |K_0(q_i,\psi_j^*)| = 00 \\ & \int_0^{q_{max}} \int_0^{\psi_{max}^*} T_m \, dq_i d\psi_j^* \le K^0 \\ & K^0 > 0; i = a,b,c; j = m,n,h \end{aligned}\right\} \quad (17)$$

When the phase circuits are connected in series, the inductive subsystem as a voltage source is a "carrier system" with an active-capacitive load, and with a parallel connection, the "carrier" system is a capacitive subsystem as a current source with an actively inductive load:

$$\left.\begin{aligned} & \mathcal{L}_L = T_L - \Pi_L; \ \mathcal{L}_c = T_c - \Pi_c \\ & T_L = \frac{1}{2} i^\tau L i + i^\tau \widehat{K_0}(q,\psi_j^*) V^*; \ \Pi_L = \frac{1}{2} V_я^* C_я V_я^*; \\ & T_c = \frac{1}{2} V^{*\tau} C V^* + i^\tau \widehat{K_0}(q,\psi_j^*) V^*; \ \Pi_c = \frac{1}{2} i_я^\tau L_я i_я \end{aligned}\right\} (18)$$

where T_L, T_c – the effective electrokinetic energy of the inductive and capacitive subsystems, respectively; Π_L и Π_c – energy of the electric field of the capacitive and magnetic fields of the inductive load; $i^\tau \widehat{K_0}(q,\psi_j^*) V^*$- the volume energy between the two subsystems (the superscript τ denotes transposed matrices), L – matrix of inductances, $i\ (i_a, i_b, i_c, i_f)$ – vector of currents of inductive machines, C – capacitance matrix, $V_я^*(V_m^*, V_n^*, V_h^*, V_p^*)$ – armature voltage armature vector, $L_я$ – matrix of inductances, $i_я(i_a, i_b, i_c, 0)$ – vector of the armatures of the inductive machine:

$$L_я = \begin{bmatrix} L_{SS} & L_{SF} \\ 0 & 0 \end{bmatrix}; \quad C_я = \begin{bmatrix} C_{RR} & C_{RP} \\ 0 & 0 \end{bmatrix}, \quad (19)$$

On the basis of equations (1), (2), (14) - (18), the equations of electrodynamics and electromechanics for stationary modes of combined inductive capacitive EMF can be written in the following vector-matrix form: the equation of electrodynamics

$$\left.\begin{aligned} & \frac{d\Psi}{dt} + R_L i_L + K_0(q_i,\psi_j^*) \frac{dV_C^*}{dt} = V_{LC} \\ & \frac{dQ^*}{dt} + G_c V_c^* + K_0(q_i,\psi_j^*) \frac{di_L}{dt} = i_{CR} \end{aligned}\right\} \quad (20)$$

the equation of electromagnetic moments

$$M_{LC} = -\frac{P_0}{2}\left[i_L \frac{\partial L}{\partial \gamma} i_L^\tau + V_c^* \frac{\partial C}{\partial \gamma} V_c^{*\tau} + i_L \frac{\partial K_0(q,\psi_j^*)}{\partial \gamma} V_c^{*\tau} \right] \quad (21)$$

The simultaneous solution of the system of equations (20) and (21) completely determines the dynamic behavior of the generalized EMF model in any modes of interaction of their functional elements.

To study the combined EMF of a specific design, it is necessary to take into account all the structural features of their functional elements in the problems of computer-aided design in these equations. In this case, the dependencies of the elements of the interaction matrix on the physical characteristics and design parameters of the combined inductive capacitive EMF allow us to establish the use of the theory of experimental design [20].

4 Conclusion

1. The article presents differential controls of inductive capacitive electromechanical energy converters, on the basis of which, in a particular case, the equations of inductive and capacitive electric machines of alternating current are presented.
2. The use of achievements in inductive electric machines for the development of capacitive electric machines by means of inverse-coupled electrodynamics can significantly accelerate the development of the latter. The successful development of capacitive electric machines can become the basis for the creation of an inductive-capacitive machine with a working electromagnetic field.

3. The article has established the possibility of optimizing the operating mode of an inductive generator with the help of a capacitive generator (or vice versa), with the formation of a parallel (or sequential) resonant circuit (the controlling parameter is the interaction matrix). A combined machine is a resonant machine and has the best energy performance.

A deeper application of the developed theoretical principles for generalized physico-mathematical, computer simulation, calculation of the optimal energy indices of combined inductively-capacitive EMF energy and their systems will be given in separate studies by the authors.

References

1 A.G. Iosifyan, Electricity, **12**, 19–21 (1987); **9**, 7–14 (1989)
2 I.P Kopylov, S.V. Gandilyan, Electrical Engineering, **9**, 25–40 (1998)
3 I.P. Laptsevich, L.N. Grechikhin, N.G. Kuts, Energetika (Izv., Higher Educational Institutions of the CIS), **6**, 64–77 (2012)
4 G.V. Nosov, S.V. Pustynnikov, Izvestiya Tomsk Polytechnic University, **4**, 111–115 (2012)
5 V.A. Chelukhin, Izvestiya Tomsk Polytechnic University, **325** (4), 76–80 (2014)
6 A.A. Fraerman, UFN. **182** (12), 1345–1351 (2012)
7 N.V. Ter-Oganessin, Ferroelectrics. **438** (1), 101–106 (2012)
8 B.J. Gallacher, J.S. Burdess, A.J. Harris, M.E. McNie, Proc. on Progress in Phys, 10–20 (2001)
9 P. Muralt, Rep. on Progress in Phys, **64**, 1339–1388 (2001)
10 *Nanotechnology in electronics* (M.: The techno sphere, 2005)
11 V.E. Turkov, S.A. Zhukova, D.D. Riskin, Nano- and microsystem equipment. **12**, 7–21 (2014)
12 J. Altman, *Military Nanotechnologies: Possibilities for the Application and Preventive Control of Arms* (M.: Technology, 2006)
13 A.N. Kovshov, Yu.F. Nazarov, I.M. Ibragimov, *Fundamentals of nanotechnology in technology* (M.: Academy, 2009)
14 H.S. Karayan, S.V. Gandilyan, Armenian J.Phys. **9** (3), 244–259 (2016)
15 G.S. Karayan, S.V. Gandilyan, Vestnik MPEI, **5**, 63–70 (2017)
16 A.G. Iosifyan, G.L. Areshian, Dokl. AN Arm. SSR. **73** (1), 54–61 (1981)
17 S.V. Chelukhin, Izv. supreme. educational institutions (Electromechanics), **5**, 48–52 (1987)
18 S.V. Gandilyan, Electricity. **8**, 22–28 (1988)
19 S.V. Gandilyan, Izvestiya AN RF (Energy and Transport), **2**, 50–62 (1993)
20 N.I. Sidnyaev, N.T. Vilisova, *Introduction to the theory of experimental planning* (M: Izd.-in MSTU him NE. Bauman, 2011)

Synthesis of control laws of electromechanical systems under polynomial approximation of characteristics of nonlinear elements

Vladislav Shishlakov*, *Elizaveta* Vataeva, *Natalia* Reshetnikova and *Dmitriy* Shishlakov

SUAI, Department of Electromechanics and Robotics, 190000, St. Petersburg, Bolshaya Morskaya Str., 67, Russia

Abstract. The paper presents the solution of the problem of synthesis of the laws of control of electromechanical automatic control system under polynomial approximation. As a mathematical apparatus of the decision the generalized Galerkin's method is applied. This method is extended to a new class of objects with nonlinear characteristics approximated polynomially.

1 Introduction

Providing the set indicators of quality of work of the nonlinear systems of automatic control (ACS) in the transitional modes represents very difficult task which correct solution is in many respects connected with creation of adequate mathematical model.

As the elements and devices having nonlinear characteristics enter into structure of electromechanical systems and complexes, when developing mathematical models, degree of their adequacy is directly connected with the accepted way of approximation of not linearities [1-3].

The most widespread method of approximation of nonlinear characteristics is piecewise and linear approximation. However, the models received thus not always can with sufficient degree of accuracy reproduce nonlinear characteristics, especially, if it is about extreme characteristics such as mechanical characteristic of the asynchronous engine or dependence of coefficient of coupling on the size of relative slipping. Similar dependence by means of piecewise and linear approximation can be reproduced rather precisely at very large number of piecewise line sections. Reduction of number of piecewise line sections for approximation allows to receive quickly enough result, however the error of calculation can be unacceptably high. At the same time complication of the approximating expressions for nonlinear characteristics can create considerable difficulties at a research as dynamic properties of nonlinear control systems in itself are rather difficult.

Therefore it is advisable to use polynomial (algebraic) approximation in such cases which unlike the theory of automatic control is very widely used and applied in electrical engineering. In this case, the approximating polynomial is written in the form

$$y(x) = y(x_0) + a_1(x - x_0) + a_2(x - x_0)^2 + K + a_n(x - x_0)^n \quad (1)$$

where coefficients of $a_1, a_2 ..., a_n$ are defined by expressions

$$a_1 = \left(\frac{dv}{dx}\right)_{x=x_0}$$
$$a_2 = \frac{1}{2!}\left(\frac{d^2v}{dx^2}\right)_{x=x_0} \quad (2)$$
$$a_n = \frac{1}{n!}\left(\frac{d^nv}{dx^n}\right)_{x=x_0} ;K$$

Approximation by a power polynomial consists in finding the coefficients of the series $a_1, a_2 ..., a_n$. At the set characteristic form these coefficients significantly depend on the choice of a working point and also on width of the used site of the characteristic.

An effective method for solving the problem of synthesizing the parameters of the control law for electromechanical and robotic systems and complexes is the generalized Galerkin's method allowing to solve from uniform mathematical, methodological and algorithmic positions similar problems for ACS of a wide class which dynamic properties are described by the nonlinear differential equations randomly of a high order.

2 Mathematical description of the synthesis problem

From a mathematical point of view, this approach makes it possible to approximately solve a nonlinear differential equation of arbitrary form of any order, describing the dynamics of the automatic control system [4, 5].

$$Q(c_k, p)x(t) + R(c_k, p)y(t) = S(c_k, p)f(t) ;$$
$$y(t) = F\left[x(t), \dot{x}(t)\right], \quad (3)$$

* Corresponding author: kaf31guap@gmail.com

where $x(t)$ – coordinate at the input of the non-linear link, relative to which the equation of motion of the synthesized ACS is written; $f(t)$ – the external input; $y(t)=F\left[x(t),\dot{x}(t)\right]$ – a nonlinear function representing a relationship approximating the static or dynamic nonlinear characteristic of an ACS element or device.

Polynomials of the differentiation operator p with real constant coefficients of powers of n,u,v respectively, c_k are the desired parameters of the control operator.

In the course of the solution, it is required to find the values of unknown parameters that will provide the system with specified quality indicators of its operation during the transition from one steady state to another.

As a rule, the problem is solved with constraints on the values of the coefficients of the differential equation (which are functions of the parameters of the control operator), while the solution may be close to the given, but not accurately reproduce it. This is due to the fact that in any ACS there is an unchangeable part (the executive motor, the transmission mechanism, the mismatch meters, etc.), the transmission coefficients and the time constants of the elements of which are given. In addition, the parameters of the management operator are constrained by the possibility of their technical implementation

$$c_k^- \le c_k \le c_k^+ \quad k=1,2,K,m, \tag{4}$$

where c_k^+ – the maximum permissible value of the variable parameters; c_k^- – the minimum permissible value of the variable parameters.

And also by the restriction on the roughness of the solution found for given deviations of the parameters from the values found, which is related to the technological spread of the element base

$$\Delta=\frac{\delta c_k}{c_k}\le\Delta^0 \tag{5}$$

where Δ^0 is a given value of system roughness; δc_k – variations of parameters within which the system is stable.

When determining the parameters of the control operator of a nonlinear automatic control system in the first approximation, the desired software motion is given in the form of a solution of a linear differential equation.

Since the design of automatic control systems, regulated indicators, are usually indicators of the quality of the transient process, then. there must be a one-to-one correspondence between these indicators and the parameters of the desired program motion (the damping coefficients and the natural vibration frequencies of its components).

Such a correspondence is defined for linear differential equations of the first and second orders. So the attenuation coefficient α is related to the duration of the transient process $T_{n.n.}$ by the relation [4, 5]

$$\alpha=\frac{3\div4}{T_{nn}} \tag{6}$$

The solution of a second-order linear differential equation has the form

$$x^0(t)=\left[x_y+H^*\cos(\beta t-\varphi_0)e^{-\alpha t}\right]1(t) \tag{7}$$

where x_y - the value of the desired process $x^0(t)$ at $t=\infty$; and H^* and φ_0 are defined by relations of the form

$$H^*=\sqrt{\left(x_0-x_y\right)^2+\left[\frac{\alpha\left(x_0-x_y\right)+\dot{x}_0}{\beta}\right]^2}$$
$$\varphi_0=arctg\left[\frac{\alpha\left(x_0-x_y\right)+\dot{x}_0}{\beta\left(x_0-x_y\right)}\right] \tag{8}$$

here $x_0,\dot{x}_0$ – the initial values of the coordinate under consideration, with respect to which the equation of motion of the synthesized ACS is recorded and it's derivative, respectively, at time $t=+0$.

The oscillation of the process μ is related to overshoot σ which in turn determines the natural oscillation frequency β. Thus, in the particular case (when = 0), we have

$$\sigma_m=\frac{\mu}{\sqrt{\mu^2+1}}e^{-\frac{\pi}{\mu}} \tag{9}$$

Thus, proceeding from the required values of the performance indicators of the ACS in the transient mode (time of the transient and overshoot), it is possible to determine the values of the parameters of the program motion of the form (7) that provide them.

For more complex software movements (for example, the third fourth order), some of the process parameters must be assigned, and the rest must be determined, as shown in [5].

After the desired program motion (the solution of the nonlinear differential equation) is specified and its parameters are determined, a discrepancy is formed

$$\psi\left(c_k,t\right)=Q\left(c_k,D\right)x^\circ(t)+ +R\left(c_k,D\right)F\left[x^\circ(t),D\left\{x^\circ(t)\right\}\right]- -S\left(c_k,D\right)f(t), \tag{10}$$

where $x^0(t)$ – the desired software motion (the solution imposed on the nonlinear differential equation), D – the operator of generalized differentiation.

The values of the unknown parameters c_k are determined from the orthogonality condition of the discrepancy (10) to the coordinate functions

$$\int_0^\infty \psi(c_k,t)\varphi_q(t)dt=0 \quad k,q=1,2,K,m \tag{11}$$

where $\varphi_q(t)$ – system of m continuously differentiable linearly independent coordinate functions

$$\varphi_1(t), \varphi_2(t), K, \varphi_q(t), K, \varphi_m(t) \quad (12)$$

Coordinate functions should be a complete system of functions. In this method of solving a nonlinear differential equation of general form, real exponents are used (in contrast to the Laplace transform, in which complex exponentials are used)

$$e^{-\rho_1 t}, e^{-\rho_2 t}, K, e^{-\rho_q t}, K, e^{-\rho_m t}, q = 1, 2, K, m \quad (13)$$

The experience of solving practical problems shows [4-7] that in order to best approximate the desired program motion $x^\circ(t)$ to the actual process occurring in the system with synthesized parameters, the attenuation coefficient ρ_1 of the coordinate functions is expediently chosen in the form

$$\rho_1 = \alpha \quad (14)$$

where α - the attenuation coefficient of the component of the desired program motion, which has the greatest influence on the duration of the transient process.

The remaining attenuation coefficients of the series p_{m-1} should be chosen in the form of a geometric progression (with the denominator of the progression $r = 2$), that is

$$\rho_q = \rho_1 r^{q-1} = \rho_1 2^{q-1}, q = 1, 2, K, m \quad (15)$$

this provides a shorter fade time for each of the $m-1$ exponentials compared to the decay time of the first coordinate function.

The solution of equation (11) leads to the following system of algebraic equations

$$\sum_{i=0}^{n} a_i(c_k) A_{qi} + \sum_{i=0}^{u} b_i(c_k) B_{qi} - \sum_{i=0}^{v} e_i(c_k) C_{qi} = 0, \quad q = 1, 2, \ldots, m, \quad (16)$$

where

$$A_{qi} = \int_0^\infty D^i\left\{x^\circ(t)\right\} e^{-\rho_q t} dt = A_q \rho^{i-1}, \quad i = 0, 1, \ldots, n,$$

$$B_{qi} = \int_0^\infty D^i\left\{F\left[x^\circ(t), \dot{x}^\circ(t)\right]\right\} e^{-\rho_q t} dt = B_q \rho^{i-1}, \quad i = 0, 1, \ldots, u, \quad (17)$$

$$C_{qi} = \int_0^\infty D^i\left\{f(t)\right\} e^{-\rho_q t} dt = C_q \rho^{i-1}, \quad i = 0, 1, \ldots, v.$$

In [4-6] the method of calculating Galerkin integrals is considered in detail and recurrent analytical expressions A_q, B_q, C_q for differential equations with various nonlinear characteristics are given for their piecewise linear and algebraic approximations for various types of program motions.

Solving the system of m algebraic equations (16), we determine the values of the variable parameters of the control operator, and, consequently, the coefficients of the nonlinear differential equation of motion of the ACS.

However, the problem is solved with limitations on the values of the required parameters imposed on the basis of the possibility of their technical implementation, limitations on the stability and roughness of the ACS with synthesized parameters, and also because, as a rule, there is a nonlinear relationship between the varied parameters, then the strict equality (16) will not be performed.

Therefore, the definition of the parameters of the control operator providing a given solution of a nonlinear differential equation in the computational plan is a non-linear programming problem with an objective function constructed on the basis of equations (16) and having the form

$$J = \sum_{q=1}^{m}\left\{\int_0^\infty \psi(c_k, t)\varphi_q(t) dt\right\}^2, min_{c_k} J \to 0, \quad (18)$$

The optimum of which is determined under the constraints noted above using the known methods for finding the extremum of a functional.

To develop the generalized Galerkin method for a new class of approximating nonlinear characteristics of functions, it is required to determine the values of the integrals B_{qi} on a family of elementary functions.

That is, for the solution of a problem of synthesis of ACS at polynomial approximation of characteristics it is required to calculate integral

$$B_{qi} = \int_0^\infty D^i\left\{F\left[x^0(t)\right]\right\} e^{-\rho_q t} dt = \int_0^\infty D^i\left\{\sum_{g=0}^{l} z_g \left(x^0(t)\right)^g 1(t)\right\} e^{-\rho_q t} dt \quad (19)$$

where

$$F\left[x^0(t)\right] = \sum_{g=1}^{l} F_g\left[x^0(t)\right] = \sum_{g=0}^{l} z_g\left(x^0(t)\right)^g, g = 0, 1, \ldots, l$$

what represents distribution of the principle of equivalent transformations [5, 6] on nonlinear characteristics at their analytical approximation and significantly simplifies calculation of integrals.

The experience of solving practical problems in the synthesis of linear and nonlinear mathematical models of control systems by the generalized Galerkin method shows that it is sufficient to take the solution of a linear differential equation of no higher than second order [5,7] which, as a rule, is less than the order of the differential equation describing the motion of the synthesized system. This provision seems to be completely justified, since the problem of synthesis, rather than identification, is being solved. Therefore, emphasizing the higher harmonic components in the program movement, whose influence they seek to reduce, has no practical meaning.

In addition, in the design of automatic control systems, regulated indicators, usually indicators of the quality of the transient process. Therefore, there must be a one-to-one correspondence between these parameters and the parameters of the desired program motion (damping coefficients and natural frequencies of oscillations of its components), which is defined for linear differential equations of the first and second orders.

As a result of the calculations, we obtain

$$B_{qi} = \int_0^\infty D^i\left\{F\left[x^0(t)\right]\right\} e^{-\rho_q t} dt = B_q \rho_q^{i-1}, i = 0,1,\dots,u, \quad (20)$$

where the ratios defining B_{qi} are calculated for process

$$x^0(t) = [H^* e^{-\alpha t}\cos(\beta t - \varphi_0)]1(t) \quad (21)$$

corresponding to the recording of the equation of motion of the system relative to the input and process

$$x^0(t) = \left(x_y - H^* e^{-\alpha t}\cos(\beta t - \varphi_0)\right)1(t) \quad (22)$$

corresponding to the recording of the equation of motion of the system relative to the error signal.

We get:

- for odd degree

$$B_q = \sum_{g=0}^{l} a_g \sum_{k=0}^{g} H^{*k} C_g^k \left\{ \frac{k}{2^{k-1}(k\alpha+\rho_q)}\left[1-\left(k-2E\left(\frac{k}{2}\right)\right)\right] + \frac{(k\alpha+\rho_q)\cos(k\varphi_0)+k\beta\sin(k\varphi_0)}{2^{k-1}\left[(k\alpha+\rho_q)^2+(k\beta)^2\right]} + \frac{k\begin{bmatrix}(k\alpha+\rho_q)\cos((k-2)\varphi_0)+\\(k-2)\beta\sin((k-2)\varphi_0)\end{bmatrix}}{2^{k-1}\left[(k\alpha+\rho_q)^2+((k-2)\beta)^2\right]} \right\} \quad (23)$$

- for an even degree

$$B_q = \sum_{g=0}^{l} a_g \sum_{k=0}^{g} H^{*k} C_g^k \left\{ \frac{k}{2^{k-1}(k\alpha+\rho_q)}\left[1-\left(k-2E\left(\frac{k}{2}\right)\right)\right] + \frac{(k\alpha+\rho_q)\cos(k\varphi_0)+k\beta\sin(k\varphi_0)}{2^{k-1}\left[(k\alpha+\rho_q)^2+(k\beta)^2\right]} + \frac{k\begin{bmatrix}(k\alpha+\rho_q)\cos((k-2)\varphi_0)+\\(k-2)\beta\sin((k-2)\varphi_0)\end{bmatrix}}{2^{k-1}\left[(k\alpha+\rho_q)^2+((k-2)\beta)^2\right]} \right\} +$$

$$+2\sum_{g=0}^{l}(-1)^j \left(\sum_{g=0}^{l} a_g \sum_{k=0}^{g} H^{*k} C_g^k \left\{ \frac{k}{2^{k-1}(k\alpha+\rho_q)}\left[1-\left(k-2E\left(\frac{k}{2}\right)\right)\right] + \frac{(k\alpha+\rho_q)\cos(k\varphi_0)+k\beta\sin(k\varphi_0)}{2^{k-1}\left[(k\alpha+\rho_q)^2+(k\beta)^2\right]} + \frac{k\begin{bmatrix}(k\alpha+\rho_q)\cos((k-2)\varphi_0)+\\(k-2)\beta\sin((k-2)\varphi_0)\end{bmatrix}}{2^{k-1}\left[(k\alpha+\rho_q)^2+((k-2)\beta)^2\right]} \right\} \right) +$$

$$+(-1)^j \left(\sum_{g=0}^{l} a_g \sum_{k=0}^{g} H^{*k} C_g^k \left\{ \frac{k}{2^{k-1}(k\alpha+\rho_q)}\left[1-\left(k-2E\left(\frac{k}{2}\right)\right)\right] + \frac{(k\alpha+\rho_q)\cos(k\varphi_0)+k\beta\sin(k\varphi_0)}{2^{k-1}\left[(k\alpha+\rho_q)^2+(k\beta)^2\right]} + \frac{k\begin{bmatrix}(k\alpha+\rho_q)\cos((k-2)\varphi_0)+\\(k-2)\beta\sin((k-2)\varphi_0)\end{bmatrix}}{2^{k-1}\left[(k\alpha+\rho_q)^2+((k-2)\beta)^2\right]} \right\} \right) \quad (24)$$

here E is the integer part of a number.

Summarizing the expressions (23) and (24), we obtain

$$B_q = \sum_{g=0}^{l} a_g \sum_{k=0}^{g} x_y^{g-k} H^{*k} (-1)^k C_g^k \left\{ \frac{k}{2^{k-1}(k\alpha+\rho_q)}\left[1-\left(k-2E\left(\frac{k}{2}\right)\right)\right] + \frac{(k\alpha+\rho_q)\cos(k\varphi_0)+k\beta\sin(k\varphi_0)}{2^{k-1}\left[(k\alpha+\rho_q)^2+(k\beta)^2\right]} + \frac{k\begin{bmatrix}(k\alpha+\rho_q)\cos((k-2)\varphi_0)+\\(k-2)\beta\sin((k-2)\varphi_0)\end{bmatrix}}{2^{k-1}\left[(k\alpha+\rho_q)^2+((k-2)\beta)^2\right]} \right\} \quad (25)$$

3 Conclusion

Thus, the recurrence relations defining the Galerkin integrals B_{qi} for polynomial nonlinear characteristics make it possible to apply the method for solving the problem of synthesis of the control laws of electromechanical and robotic systems and complexes for a given approximation of the characteristics of nonlinear element.

References

1. L. Bessonov, *Theoretical fundamentals of electrical engineering : Electric circuits* (Higer school, Moscow, 1978)
2. K. Polivanova, D. Zhukhovitsky, I. Negnevitsky *Theoretical bases of electrical* (Energy, Moscow, 1972)
3. G. Zeveke, P. Ionkin, A. Netushil, S. Strakhov, *Fundamentals of the theory of chains* (Energoatomizdat, Moscow, 1989)
4. V. Shishlakov, *Synthesis of nonlinear automatic control system with various types of modulation : monograph* (SPbSUAI, SPb, 1999)
5. A. Nikitin, V. Shishlakov, *Parametric synthesis of nonlinear automatic control system: monograph* (SPbSUAI, SPb, 2003)
6. V. Shishlakov, S. Tsvetkov, D. Shishlakov, *Synthesis of the parameters of continuous and pulsed multiply connected automatic control system: monograph* (SPbSUAI, SPb, 2009)
7. V. Shishlakov, A. Shishlakov, S. Timofeev, *Synthesis of automatic control system for various types of approximation of nonlinear characteristics : theory and practise* (SPbSUAI, SPb, 2017)
8. M.V. Burakov, V.F. Shyshlakov, SPIIRAS Proceedings, **51**(2), 60–77 (2017). doi:10.15622/sp.51.3

Investigation of the structural characteristics of the electrodes of energy storage devices used in power plants based on renewable energy sources

Elena Kiseleva* and *Fedor* Lelin

Joint Institute for High Temperatures of the Russian Academy of Sciences, 125412, Izhorskaya st. 13 Bd.2, Moscow, Russia

Abstract. The porous structure of electrode materials was studied and an experimental analysis of the role of macropores in the electrodes of model double layer supercapacitors was carried out. It is shown that the excess value of the volume of macropores in the electrodes plays a negative role and, on the whole, in a complex manner, decreases the specific electrochemical characteristics of these devices. As a result of the research, high-efficiency activated carbons from wood waste, designed for use in supercapacitors, have been created. Correlations have been established between the parameters of the regime of thermochemical activation of wood, the porous structure of synthesized activated carbons and electrodes based on them and the specific characteristics of supercapacitors.

1 Introduction

At present, the problem of energy and resource saving both in Russia and in the world is quite acute. The growing demand for electrical energy makes it necessary to seek new sources of energy and optimize the energy consumption received from traditional systems. It is known that most renewable energy sources have significant daily and seasonal generation non-uniformities, and wind power installations are characterized by fluctuations with lifetimes from fractions of a few seconds [1]. If long peaks and recessions of power generation can be compensated by electrochemical accumulators, the double layered electrochemical supercapacitor (EDLC) is the optimal solution to improve the quality of the power supplied. The problem of suppressing fluctuations is also characteristic for diesel-generator sets, which form the basis of modern distributed generation

In some cases, an increase in the service life and cost reduction can be achieved by combining the batteries with EDLC in the power plant. Similar schemes are also used in electric vehicles [2].

In this scheme, the batteries provide power to the consumer in the basic mode in the absence of RES, and the transient processes are compensated by the EDLC [3]. Supercapacitors can also be used in power systems for those consumers whose electrical load is clearly impulsive [4]

The supercapacitor is a pulsed electrochemical device, which is primarily a power source. It differs from batteries of various types by a significantly lower energy intensity (units W • h / kg) and increased specific power (2-10 kW / kg). The absence of Faraday processes in the operation of EDLC leads to an increase in its lifetime in comparison with accumulators, and also to the absence of restrictions on the depth of its discharge [5]. EDLCs are used to smooth out impulse loads, regenerate braking energy, or other similar processes that require high values of charge or discharge currents. For the satisfactory functioning of EDLC, it is necessary to have micro and mesopores in its active layers, which together result in a large value of the specific surface and, correspondingly, of the electric capacity. But with a charge (discharge) of high currents for a limited time, a situation may occur where a large fraction of the surface of the micropores turns out to be uncharged (undischarged).

What is the role of the macroporous electrode structure in solving this problem? It is generally believed that macropores create a channel in the active layer for the supply of electrolyte ions. But mesopores, and even micropores, also contribute to the value of the effective conductivity of the electrolyte. If we remove the macropores, then, as the estimates show, conductivity will decrease, but not catastrophically. In general, all three types of pores are usually present in EDLC electrodes: macropores, mesopores and micropores. In [6], structures with three types of pores were studied by the methods of the theory of porous electrodes and computer simulation. It follows from the theoretical model proposed in these studies that with a decrease in the proportion of macropores in the structure of the electrodes, the specific capacitance can increase substantially, even if the specific surface does not increase, and the effective conductivity of the electrolyte decreases. This result was analyzed in the present work on the basis of experimental data.

* Corresponding author: kanna787@mail.ru

Electrodes mainly determine the properties of supercapacitors. Therefore, the choice of electrode material and the technology of its preparation is extremely important for achieving high values of specific power and energy [7-8]. When using activated carbons (AC) for double-layer super-capacitor electrodes, the task of optimizing the porous structure of coal is actual. By varying the conditions for chemical activation, one can control the volume and distribution of pores by size, the nature and the number of functional groups on the surface. Earlier, the authors investigated the influence of the time and temperature of activation, the mass ratio of the activating agent and the raw material on the parameters of the porous structure, activated carbon: specific surface area, pore volume and radial distribution, and electrochemical characteristics of EDLC with aqueous (4.9 M sulfuric acid) and organic electrolytes (1M tetraethylammonium tetrafluoroborate) on the basis of these materials [9].

2 Technique of the experiment

From what has been said above, an important, as yet undefined problem, describing the effect on the value of the specific characteristics of EDLC of the content of macropores in the porous structure of AC is common to describe the principles of EDLC functioning. In order to demonstrate how the presence of a significant excessive amount of macropores in the AC affects the values of the specific characteristics of EDLC, the total pore volume was estimated in the work. In the electrode, it was evaluated using a bulk-weight method, and also from the mass and geometric volume of the dry electrode. In addition, the electrolyte volume in the electrodes was measured after measuring the characteristics of the model cells of the supercapacitors. Accordingly, the volume of macropores in the electrode was determined from the difference in the total measured volume, and the volume of micro- and mesopores determined from the LI method data.

When determining the volume of macropores in coal samples, it was found that the total volume of the measured pores of radius less than 100 nm by the method of limited evaporation is in close agreement with the data obtained by the volumetric-weight method. In addition, microphotographs of the corresponding increase in the particles synthesized in coal work do not observe macropores, unlike, for example, commercial coals from coconut shell.

In Figures 1 and 2, images of samples of coals Kuraray YP-50F and YP-80F obtained from the coconut shell are shown in addition to the images of one of the samples of the left-hand AC (Fig. 1 on the left and 2 on the left).

On this basis, we believe that in our activated coals the share of macropores is negligible. At the same time, in electrodes, the macropores formed by the interstices between the coal particles (and, to a lesser extent, the binder) are practically unavoidable. The total amount of electrolyte in the electrode based on AC from alder, concentrated in micro-, meso- and macropores, according to our estimates, should be 1.4 cm3 / g for maximum performance. In this case, the meso- and macropores of the electrode, which are the interstices between the particles, ensure the transport accessibility of the ions to each particle. Excessive amount of macropores and, accordingly, electrolyte in them are ballast. These pores do not contribute to the formation of DES, and the mass and volume of the electrolyte worsen the specific mass and volume characteristics of supercapacitors.

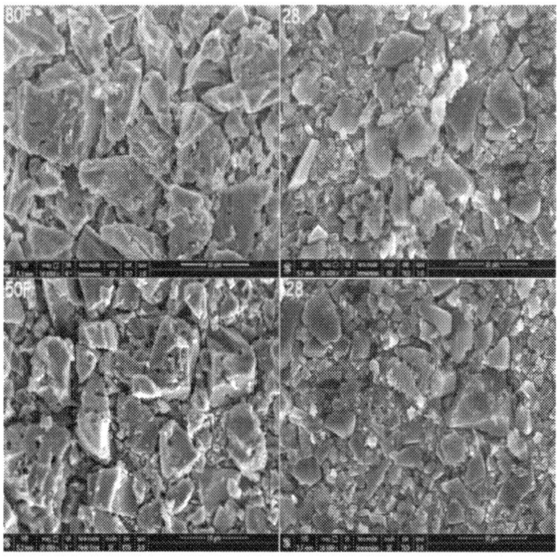

Fig. 1. Microphotographs of Kuraray YP-50F, YP-80F coal particles on the left, AC from waste wood on the right.

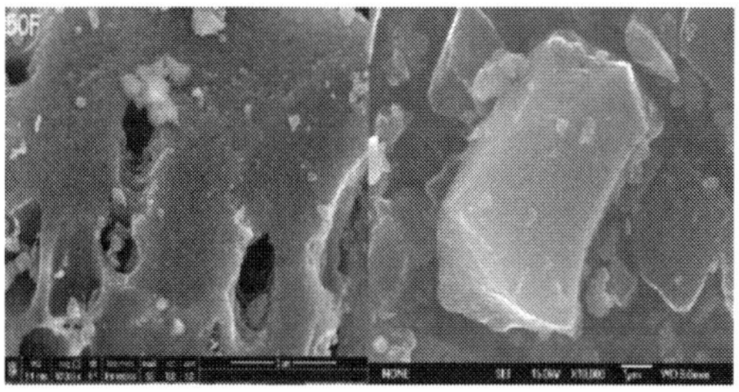

Fig. 2. Microphotographs of the Kuraray YP-50F coal particle on the left, AC from the wood waste on the right.

The conclusions drawn for the electrodes can be confirmed experimentally. Below (Figure 3) presents the experimental results for a large series of different samples of electrodes and activated coals due to the effect of the presence of macropores on the electrochemical characteristics of DSC. As mentioned above, the electrodes were made by calendering. This method of making electrodes does not completely eliminate macropores. These results represent a statistical treatment of a large number of independent experiments conducted over a long period of time, mainly to study the effect of synthesis parameters on the electrochemical characteristics of coals. In addition, the features of the electrode manufacturing technology could vary, the size of the coal particles, the binder content, the soot additives, etc., could vary. Thus, the volume of macropores in each electrode was a consequence of heterogeneous and often random causes.

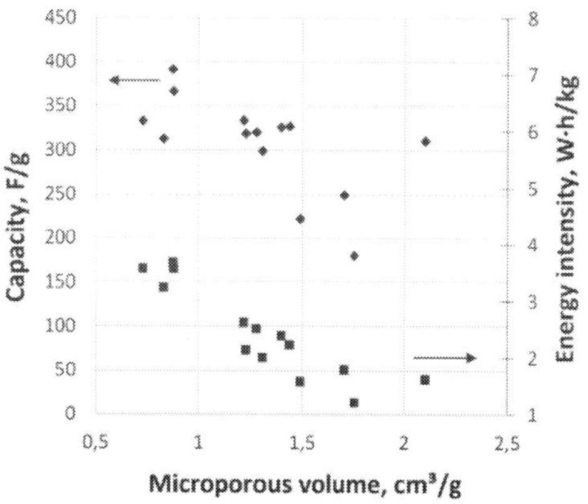

Fig. 3. Dependence of ■ the specific energy intensity in relation to the total mass of electrodes and electrolyte and ♦ – the electrical capacitance from the volume of macropores in the DSC electrode.

In Figure 3 shows the experimental correlation dependences: along the X axis, the values of the volume of macropores in the electrode are plotted, and along the Y axis the specific energy intensity (right axis), calculated on the total mass of electrodes with electrolyte and specific electric capacity (left axis) per gram of dry activated carbon . Despite the scatter in the light of the foregoing, the graphs clearly show that an increase in the volume of macropores is accompanied by a decrease in specific electrochemical characteristics.

A partial decrease in the energy intensity in Figure 3 is easy enough to explain. Since this value was calculated based on the total mass of the electrodes with the electrolyte, the more the macropores in the electrode, the greater the amount of electrolyte it contains, the higher the denominator, and the lower the specific characteristics. After all, the contribution to the surface and, accordingly, to the capacity of the DPP, macropores practically do not contribute. Nevertheless, all the decrease in the specific energy intensity (up to three times) only due to the increase in the denominator of the electrolyte mass (about 2.5 times), taking into account the mass of the electrode, can not be explained.

As for the electric capacity per dry weight of coal, here, based on general considerations, the connection with the macropores would have to be more inverse, since macropores increase the effective conductivity of the electrolyte. Apparently, such a porous system is realized in which the excess volume of macropores prevents the charge and discharge of a part of the coal particles, contributing to their isolation from other particles. In view of the fact that in the system besides macropores there are filaments and aggregates of fluoroplastic particles, more isolated particles may have poor electrical contact over the solid phase of the carbon material. If, during a charge, these difficulties can be at least partially compensated by the exposure at the maximum voltage (which is realized in practice when the supercapacitors in the working circuit are under voltage), then it is impossible to influence the discharge for this purpose.

Thus, as a working hypothesis, it can be assumed that, with an excess of macropores, some of the micropores in the more isolated particles remain uncharged or under-discharged. As a consequence, there is a noticeable drop in capacity per dry weight of coal with an increase in the proportion of macropores. In this case, the more correct characteristic, on which the capacity depends, should be not the volume, but the surface of macropores and fluoroplastic aggregates.

3 Conclusion

As a result of the research, high-efficiency activated carbons from wood waste, designed for use in supercapacitors, have been created. Correlations between the parameters of the regime of thermochemical activation of wood, the porous structure of synthesized activated carbons and electrodes based on them and the specific characteristics of supercapacitors are established.

The main part of the work is devoted to the determination of the optimal porous structure of the activated activated coals. It is shown that the main pore volume falls to a radius of about 1 nm. At the same time, the volume of the pores of these sizes at the level of 0.6 cm3/g is sufficient to ensure the high energy characteristics of supercapacitors. A high specific capacitance of a double electric layer at a level of 390 F/g in supercapacitors with sulfuric acid as an electrolyte was obtained. This value is a very high indicator, corresponding and even surpassing the world level.

In our opinion, obtaining such high specific characteristics became possible mainly due to the possibility of operative control of the parameters of the porous structure of the synthesized carbon materials achieved by the method of Limited Evaporation. At the same time, two main goals were pursued when choosing the main parameters of synthesis of AC: first, it was necessary to achieve the highest possible saturation of the specific volume of AC with micropores and mesopores, and secondly, it was required to reduce in every possible way the amount of excess ballast electrolyte in the AC.

This research has been supported by Russian Scientific Foundation (project No.11-50-00124) in part of activated carbon synthesis, electrode preparation routine, scanning electron microscopy.

References

1. W.F. Pickard, A.Q. Shen, N.J. Hansing, J. Renew. Sustain. Energy, **13**, 1934–1945 (2009)
2. A.W. Stienecker, T. Stuart, C. Ashtiani, J. Power Sources, **154** (2), 561–566 (2006)
3. A. Andreotti, F. Mottola, M. Pagano, G. Velotto, Electric Power Systems Research, **78** (6), 1038–1046 (2008)

4. A. Hande, T. Polk, W. Walker, D. Bhatia, Microprocessors and Microsystems, **31**, 420–432 (2007)
5. K.K. Dyshchikov, *Optimization of the interaction of nanostructured carbon materials and electrolytes based on ionic liquids to improve the electrical characteristics of supercapacitors*, (RUSNANOFORUM, Moscow, 2008)
6. Yu.G. Chirkov, V.I. Rostokin, J. Electrochem, **50**, 208–222 (2014)
7. M.A Zhurilova, I.A. Yanilkin, E.A. Kiseleva, I.N. Atamanyuk, E.I. Shkolnikov, Izv. Vyssh. Uchebn. Zaved., Khim. Khim. Tekhnol, **60** (4), 82–87 (2017)
8. E.A. Kiseleva, F.V. Lelin, M.A. Zhurilova, E.I. Shkolnikov, RUSS J APPL CHEM, **90** (5), 712–715 (2017)
9. E.I. Shkolnikov, E.A. Kiseleva, D.E. Vervikishko, S.A. Kochanova, and E.V. Sidorova, RUSS J APPL CHEM, **90** (4), 547–552 (2017)

Estimation of spark protection device's reliability influence on the fire safety of apartment's network with harmonic currents and voltage up to 1 KV

Pavel Kuznetsov[1,*], *Sergej* Solyonyj[2], *Yuriy* Sychev [1], *Vladislav* Shishlakov[2] and *Oksana* Solenaya[2]

[1]Saint-Petersburg Mining University, Department of Power Industry and Electromechanics, 199106, Saint Petersburg, Russia
[2]Saint-Petersburg State University of Aerospace Instrumentation, Department of Electromechanics and robotics, 190000 Saint Petersburg, Russia

Abstract. This article analyses the statistical level of fire danger in living premises. Its influence on the reliability of apartments' electrical system is estimated and role of spark protection devices in this process is designated. In addition, a role of harmonic currents in the process of dangerous contact joints detection shown and described. In the conclusion, main requirements for diagnostic of apartments' low-voltage electrical networks are formulated.

1 Introduction

Based on the data received from [1-3] it was estimated that during the last decade there were 523833 fires. However, 43.7% of them were from exogenous sources. In other words, 228915 fires happened because of problems with electricity and 90% of this amount happened in living premises because of hot, sparkling disassembling power contacts' appearance. This number equals to 206023 fires during the past 10 years. This is why development of protective and preventive methods, organisation of technical inspections from hot, sparkling disassembling contacts is an actual problem nowadays.

2 Condition of the problem

Nowadays Automatic Circuit Breakers (ACB) and residual cut-off device (ISS) exist and are widely applied in the apartments' power networks. These gadgets have some weak sides. For example, they do not react for appearance of weakened sparkling power contact joints that leads to the burnout of the conductor's isolation and this, in turn, leads to the fires in living premises. Researches [4-6] propose technical solution with the help of which it is possible to disable weakened sparkling power contact joints. However, real prototype's tests have revealed the following important disadvantages:

• spectrum converter of high-frequency electromagnetic oscillations that is applied in the prototype was manufactured as a current transformer with the air gap in magnetic core. This core was produced from ferrite magnets (very hard, but fragile material very similar to ceramics in its mechanical properties). Due to this fact core's manufacturing becomes a hard-to-perform task that leads to prime cost increase and rise in price of the protective device;

• transformer's secondary winding is connected to the current relay directly (executive part). Thus, the entire device's sensitivity depends on the relay's operating current and all high-frequency current's oscillations that are induced by the sparkling weakened power joints or by the electric arc through the transient resistance and that will not reach relay's operating current value will not be detected by the device;

• installation of device in living premises may be performed simultaneously with the voltage regulators that are widely used for lightning intensity control or power systems' productivity control. Operating schemes of voltage regulators are based on the use of thyristors or semistors. These semiconductor devices distort voltage and current sin waves' shapes during their operation cycles and generate high-frequency distortions in the power network. Protective devices interpret them as sparkling of the weakened contact joints and false alarm happens;

• during power switches of the loads a transient mode can occur with the help of circuit breakers and automatic circuit breakers. These transient modes are often accompanied by the high frequency oscillations in the power network that also lead to the false alarm.

The above described disadvantages were removed in the device shown on Figure 1 described and proposed in [3]. This unit solves the problem of high-frequency oscillations determination. It divides the oscillations from dangerous sparkling weakened contact joints and all other safe semiconductor equipment based on the time of oscillations' existence in the power network. This guarantees modernized unit's reliable operation.

This modernization is achieved due to successful implementation of high-frequency signals auto selection

* Corresponding author: kuznetsovpavel@inbox.ru

unit between logical element OR and selective detector unit. Network parameters are measured via current transformers that are connected separately to each phase. The unit also contains a high-frequency filter. This simple scheme allows to separate dangerous oscillations that can cause fire from the simple ones and send an emergency signal to the fire or local department.

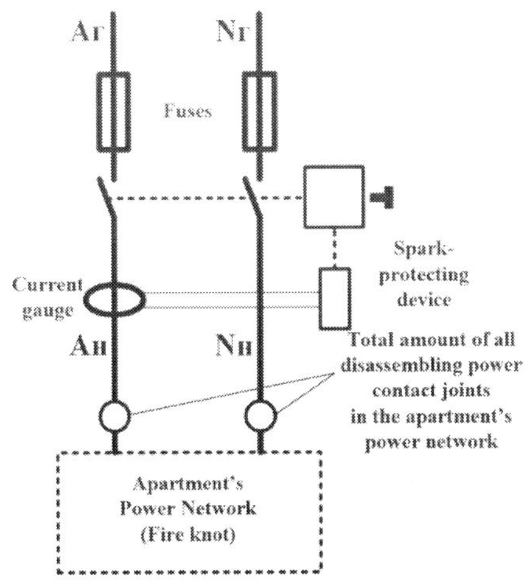

Fig. 1. Apartments' spark-protecting device.

3 Research results

The main aim of this article is to assess the fire safety level in living premises' power networks. It is also important to reveal influence of the sparkling protection system's reliability and its maintenance terms on the loads fire danger.

There was made an assumption that under the "fire node" definition we understand total sum of all contact joints in the apartment's power network.

In general, 19183000 danger fire nodes (apartments) are being controlled in [1-3] yearly.

The probability of yearly fire occurrence that depends on fire node must not exceed the value $Q(t) = 1 \cdot 10^{-6}$, where $t = 1$ year [4].

It is known that time intervals between fires in living premises do not conflict with exponential probability distribution function [7-9]. At that time, the year's fire rate in the premises because of weakened sparkling disassembling contact joints can be determined with the following formula:

$$\lambda_1 = \frac{n}{N \cdot t}, \tag{1}$$

where n=50578 is the mean fires value per year in premises because of sparkling weakened disassembling power joints; N=19183000 is the mean value of dangerous fire nodes in living premises (number of flats); t=1 year.

Probability of fire occurrence per year:

$$Q^*(1) = 1 - e^{-\lambda_1 t} \tag{2}$$

i.e. $Q^*(1) = 7.12 \cdot 10^{-4}$ that is 712 times higher than standardized value $Q(t) = 1 \cdot 10^{-6}$.

Suppose that isolation combustion happens each time when the following accident events will coincide in time and space: appearance of weakened sparkling power joint in the apartment's electrical network with voltage up to 1 KV and the sparkling protective device has failed to react [10].

Change of state during time in the contact joint can be presented as a homogenous random Markov process $\xi(t)$ [11-15] with two states: 0 – there are no weakened dangerously sparkling disassembling contact joints in the protected network; 1 – a weakened dangerously sparkling disassembling contact joint has appeared.

Values λ_1 and μ_1 are the parameters of the process $\xi(t)$.

Sparkling protection's status change in time can be described using the same homogenous Markov process $\eta(t)$ [16-18]: 0 – sparkling protection operates normally; 1 – sparkling protection has failed. Process $\eta(t)$ has the following parameters: λ_2 and μ_2.

Totality of the processes $\xi(t)$ and $\eta(t)$ can be observed as a homogenous Markov process with 4 discreet states and continuous time.

As it follows, the "contact joint – sparkling protection" system can be in one state of finite set of states:

$e_1(0,0)$ there are no weakened dangerously sparkling disassembling contact joints in the protected network, sparkling protection device operates normally;

$e_2(1,0)$ a weakened dangerously sparkling disassembling contact joint has appeared in the protecting network, sparkling protection device operates normally;

$e_3(0,1)$ there are no weakened dangerously sparkling disassembling contact joints in the protected network, sparkling protection device has failed;

$e_4(1,1)$ a weakened dangerously sparkling disassembling contact joint has appeared in the protecting network, sparkling protection device has failed.

When system occasionally comes to the state $e_4(1,1)$ apartment's conductors isolation combustion happens in the electrical network that leads to fire in most cases.

It is possible to calculate conductors isolation's combustion probability in time:

$$Q(t) = 1 - [P_1(t) + P_2(t) + P_3(t)], \tag{3}$$

where

$$\begin{cases} \dot{P}_1(t) = -(\lambda_1 + \lambda_2) \cdot P_1(t) + \mu_1 \cdot P_2(t) + \mu_2 \cdot P_2(t) \\ \dot{P}_2(t) = \lambda_1 \cdot P_1(t) - (\mu_1 + \lambda_2) \cdot P_2(t) + \lambda_1 \cdot P_1(t) \\ \dot{P}_3(t) = \lambda_2 \cdot P_1(t) - (\lambda_1 + \mu_2) \cdot P_3(t) \end{cases} . \quad (4)$$

Equation system (4) can be solved with the following initial conditions:

$$P_1(0) = 1, P_2(0) = P_3(0) = 0, \quad (5)$$

where $\lambda_1 = \frac{1}{\overline{g}_1}, \mu_1 = \frac{1}{g_1}, \lambda_2 = \frac{1}{\overline{g}_2}, \mu_2 = \frac{1}{g_2}$,

where

$\overline{g}_1$ is an average time interval between dangerous sparkling disassembling contact joint occurrence;

g_1 is an average duration of contact joint being in dangerous condition (arcing between contact joints);

$\overline{g}_2$ is an average time interval between failures of the spark protection system

g_2 is an average duration of the spark protection system being in an undetected failed condition.

From the equation system (4) with the help of known methods [19] we can find $P_1(t)$, $P_2(t)$, $P_3(t)$ and after their substitution in (3) we can receive combustion possibility of conductor in the apartment's power network:

$$Q(t) = 1 - \frac{G(S_1)}{Z'(S_1)} \cdot e^{S_1 \cdot t} + \frac{G(S_2)}{Z'(S_2)} \cdot e^{S_2 \cdot t} + \frac{G(S_3)}{Z'(S_3)} \cdot e^{S_3 \cdot t}, \quad (6)$$

where

$$G(S) = S^2 + aS + b_1 = 0, \quad (7)$$

$$Z(S) = S^3 + aS^2 + bS + C = 0. \quad (8)$$

Roots S_1, S_2, and S_3 can be found from the solving of cubic equation (8) with the known methods.

$$S^3 + aS^2 + bS + C = 0$$

where

$$a = 2\lambda_1 + 2\lambda_2 + \mu_1 + \mu_2;$$
$$b = \lambda_1\lambda_2 + (\lambda_1 + \lambda_2 + \mu_2)(\lambda_1 + \lambda_2 + \mu_1);$$
$$c = \lambda_1\lambda_2(\lambda_1 + \lambda_2 + \mu_1 + \mu_2);$$
$$b_1 = 2\lambda_1\lambda_2 + b.$$

After substitution of the found roots from equation (8) we can obtain:

$$Z'(S_1) = 3S_1^2 + 2aS_1 + b;$$

$$Z'(S_2) = 3S_2^2 + 2aS_2 + b;$$

$$Z'(S_3) = 3S_3^2 + 2aS_3 + b.$$

After substitution of $Z'(S_1), Z'(S_2), Z'(S_3)$ and $G(S_1), G(S_2), G(S_3)$ values into equation (6) we obtain combustion possibility of the conductor's isolation in time *t*.

If time interval between diagnostics of spark protection system Θ_2 is set, it is possible to find μ_2 using equation from [20]:

$$\mu_2 = \frac{1}{\Theta_2 - \frac{1}{\lambda_2}\left[1 - e^{-\lambda_2\Theta_2}\right]}. \quad (9)$$

In case if $\lambda_2\Theta_2 < 0{,}1$:

$$\mu_2 = \frac{2}{\lambda_2\Theta_2^2}. \quad (10)$$

Average time before first combustion of conductor's isolation can be defined from the system of linear algebraic equations written in matrix form. This is valid only under condition that in the initial time moment system was in one of the following conditions: $e_1(0{,}0); e_2(1{,}0)$ or $e_3(0{,}1)$:

$$\begin{bmatrix} \tau_1 \\ \tau_2 \\ \tau_3 \end{bmatrix} = \frac{1}{\lambda_1\lambda_2(\lambda_1 + \lambda_2 + \mu_1 + \mu_2)} \times \begin{bmatrix} (\mu_1 + \lambda_2)(\lambda_1 + \mu_2) & \lambda_1(\lambda_1 + \mu_2) & \lambda_2(\mu_1 + \lambda_2) \\ \mu_1(\lambda_1 + \mu_2) & \lambda_1(\lambda_1 + \mu_2 + \lambda_2) & \lambda_2\mu_1 \\ \lambda_2(\mu_1 + \lambda_2) & \lambda_2\mu_2 & \lambda_2(\lambda_1 + \mu_1 + \lambda_2) \end{bmatrix} \begin{bmatrix} 1 \\ 1 \\ 1 \end{bmatrix}. \quad (11)$$

From the system (11) it can be obtained:

$$\tau_1 = \frac{(\mu_1 + \lambda_2)(\lambda_1 + \mu_2) + \lambda_1(\lambda_1 + \mu_2) + \lambda_1(\mu_1 + \lambda_2)}{\lambda_1\lambda_2(\lambda_1 + \lambda_2 + \mu_2 + \mu_1)}, \quad (12)$$

$$\tau_2 = \frac{\mu_1(\lambda_1 + \mu_2) + \lambda_1(\lambda_1 + \mu_2 + \lambda_2) + \lambda_2\mu_1}{\lambda_1\lambda_2(\lambda_1 + \lambda_2 + \mu_2 + \mu_1)}, \quad (13)$$

$$\tau_3 = \frac{\lambda_2(\mu_1 + \lambda_2) + \lambda_2\mu_2 + \lambda_2(\lambda_1 + \mu_1 + \lambda_2)}{\lambda_1\lambda_2(\lambda_1 + \lambda_2 + \mu_1 + \mu_2)}, \quad (14)$$

where

τ_1 is an average time before first combustion of the conductor's isolation, if the system in the initial time moment was in the $e_1(0{,}0)$ condition;

τ_2 is an average time before first combustion of the conductor's isolation, if the system in the initial time moment was in the $e_2(1{,}0)$ condition;

τ_3 is an average time before first combustion of the conductor's isolation, if the system in the initial time moment was in the $e_3(0{,}1)$ condition.

Time dispersion before first combustion of conductor's isolation if the system was in one of the following conditions $e_1(0,0)$, $e_2(1,0)$ or $e_3(0,1)$ can be found from the solution of the following algebraic equations' system written in the mathematical form:

$$\begin{bmatrix}\sigma_1^2\\ \sigma_2^2\\ \sigma_3^2\end{bmatrix}=\left[\begin{pmatrix}2&0&0\\0&2&0\\0&0&2\end{pmatrix}\times\begin{bmatrix}a_{11}&a_{12}&a_{13}\\a_{21}&a_{22}&a_{23}\\a_{31}&a_{32}&a_{33}\end{bmatrix}-\begin{pmatrix}1&0&0\\0&1&0\\0&0&1\end{pmatrix}\right]\times \quad (15)$$

$$\times\begin{bmatrix}\tau_1\\ \tau_2\\ \tau_3\end{bmatrix}-\begin{bmatrix}\tau_1^2\\ \tau_2^2\\ \tau_3^2\end{bmatrix}$$

where

$$a_{11}=\frac{(\mu_1+\mu_2)(\lambda_1+\mu_2)}{\lambda_1\lambda_2(\lambda_1+\lambda_2+\mu_1+\mu_2)};$$

$$a_{12}=\frac{\lambda_1(\lambda_1+\mu_2)}{\lambda_1\lambda_2(\lambda_1+\lambda_2+\mu_1+\mu_2)};$$

$$a_{13}=\frac{\lambda_2(\mu_1+\lambda_2)}{\lambda_1\lambda_2(\lambda_1+\lambda_2+\mu_1+\mu_2)};$$

$$a_{21}=\frac{\mu_1(\lambda_1+\mu_2)}{\lambda_1\lambda_2(\lambda_1+\lambda_2+\mu_1+\mu_2)};$$

$$a_{22}=\frac{\lambda_1(\lambda_1+\mu_2+\lambda_2)}{\lambda_1\lambda_2(\lambda_1+\lambda_2+\mu_1+\mu_2)}; \quad (16)$$

$$a_{23}=\frac{\lambda_2\mu_1}{\lambda_1\lambda_2(\lambda_1+\lambda_2+\mu_1+\mu_2)};$$

$$a_{31}=\frac{\lambda_2(\mu_1+\lambda_2)}{\lambda_1\lambda_2(\lambda_1+\lambda_2+\mu_1+\mu_2)};$$

$$a_{32}=\frac{\lambda_2\mu_2}{\lambda_1\lambda_2(\lambda_1+\lambda_2+\mu_1+\mu_2)};$$

$$a_{33}=\frac{\lambda_2(\lambda_1+\mu_2+\lambda_2)}{\lambda_1\lambda_2(\lambda_1+\lambda_2+\mu_1+\mu_2)}.$$

From the equation system (15) it can be obtained:

$$\sigma_1^2=(2a_{11}-1)\tau_1+2a_{12}\tau_2+2a_{13}\tau_3-\tau_1^2; \quad (17)$$

$$\sigma_2^2=2a_{21}\tau_1+(2a_{22}-1)\tau_2+2a_{23}\tau_3-\tau_2^2; \quad (18)$$

$$\sigma_3^2=2a_{31}\tau_2+2a_{32}\tau_2+(2a_{33}-1)\tau_3-\tau_3^2, \quad (19)$$

where

σ_1^2 is time dispersion before the first combustion of conductor's isolation, if the system was in $e_1(0,0)$ condition;

σ_2^2 is time dispersion before the first combustion of conductor's isolation, if the system was in $e_2(1,0)$ condition;

σ_3^2 is time dispersion before the first combustion of conductor's isolation, if the system was in $e_3(0,1)$ condition.

Formulas (6, 12-14, 17-19) fully characterize the fire node (apartment's network). In case, if during calculation it can be obtained that $\tau_1\approx\sigma_1$, $\tau_2\approx\sigma_2$, $\tau_3\approx\sigma_3$, then probability of conductor's isolation combustion can be determined with the following formula:

$$Q_i(t)=1-e^{-\frac{t}{\tau_i}}, \quad (20)$$

where i=1, 2, 3.

4 Calculation example

4.1 Input data

The calculations were performed for the following initial data: $\lambda_1=7.12\cdot10^{-4}$ 1/year, $\lambda_2=0.2$ 1/year, $\mu_1=1.576\cdot10^{8}$ 1/year, $\mu_2=1.425\cdot10^{3}$ 1/year. It is necessary to determine possibility of conductor's isolation combustion because of weakened, sparkling disassembling power contact joint appearance in the apartment's power network during the year *Q(1)*, average time before the first conductor's isolation combustion τ_1 and dispersion σ_1^2 using formula (17), on condition that system was in condition $e_1(0,0)$ in the initial time moment.

4.2 Solution

Average time τ_1 before the first conductor's isolation combustion can be calculated with the help of formula (12), dispersion σ_1^2 can be found using formula (17), in other words $\tau_1=1.0008215\cdot10^7$ years and $\sigma_1=1.00083\cdot10^7$ years. Because our result corresponds to the condition where $\tau_1\approx\sigma_1$, possibility of conductor's isolation combustion can be found using approximate formula (20) i.e. $Q_1(1)=0.99918\cdot10^{-6}$ In our example $Q_1(1)$ value calculated by approximation formula coincides with $Q_1(1)$ value calculated by exact formula (6), i.e. $Q(1)=0.98245\cdot10^{-6}$.

5 Conclusion

Mathematical formulas (6, 12-14, 17-20) that were synthesized in this article allow to assess apartment's fire safety during appearance of weakened sparkling disassembling dangerous contact joint. In addition, this methodology helped to reveal a fact, that usage of spark-protecting devices with $\lambda_2\le0.2$ 1/year and $\Theta_2=0.084$

year allows to guarantee a robust and good fire safety for power network.

This methodology has wide range of possible applications. It can be used for assessments of different spark protection systems during their development or installation. Moreover, technical maintenance of existing anti-fire systems made with the help of the proposed methodology can help to reveal more dangerous zones and prevent thousands of possible fires in living premises cause by sparkling in power networks.

The presented results were obtained as a part of scientific research according to the contract №11036GU/2016 "The development of power network condition controller with blackout prevention possibility" provided by Foundation for Assistance to Small Innovative Enterprises in Science and Technology (FASIE).

The presented results were obtained as a part of scientific researches according to the contract № 13.3746.2017.8.9 "The designing on the basis of systematic and logic probability evaluations of rational and economically proved structure of centralized, autonomous and combined power supply systems with high reliability and stability level with usage of alternative and renewable power sources for uninterrupted power supply of enterprises with continuous technological cycle".

References

1. S.V. Solyonyj O.P. Kovaliov, Scientifical papers of DonNTU, **10** (2011)
2. A.A. Vasin, et al. Shevchenko, *FIPS* **22** (2001)
3. P.A. Kuznetsov, A.V. Yudin, S.V. Solyonij, O.Ya. Solenaya, Herald of the Kharkiv National Technical Agricultural Univercity named after P. Vasilenko, **164** (2015)
4. S.V. Solyonyj, O.P. Kovaliov, G.V. Demchenko, M.O. Nagorniy, *Unit for Power Network Protection* (Ukrpatent, 2010)
5. B.N. Abramovich et al., Oil Industry, **6**, 130–132 (2011)
6. GOST 12.1.004-91, *Occupational safety standards system. Fire safety. General requirements*, (Standards Publishing, 1992)
7. A. Dasgupta, P. Senasama, *Design and Control of Matrix Converters* (Springer, 2017)
8. H. Walter, B. Epple, *Numerical Simulation of Power Plants and Firing Systems* (Springer, 2017)
9. O.A. Shevchenko, Labor Protection, **20** (2003)
10. V. Khadkikar, M. Singh, A. Chandra, B. Singh, Joint International Conference on Power Electronics, Drives and Energy Systems (PEDES) 2010 Power India, 1–6 (2010)
11. S. Golestan, M. Monfared, and J. Guerrero, 4th Power Electronics, Drive Systems & Technologies Conference (PEDSTC2013), 510–517 (2013)
12. Je.Ju. Brazilovich et al. *Radio and Connection,* **376** (1983)
13. A.P. Kovaliov, A.V. Shevchenko, I.V. Belousenko, Industrial Power Energy, **6** (1991)
14. Yu. Huang, P.M. Pardalos, Q.P. Zheng, *Electrical Power Unit Commitment*, (Springer, 2017)
15. B.N. Abramovich, Yu.A. Sychev, A.S. Mingazov, V.V. Polishuk, Oil Industry, **10**, 126–127 (2013)
16. P.A. Kuznetsov, O.A. Stepanov, Aerospace MAI Journal, **4**, 157–163 (2017)
17. F.Z. Peng, J.S Lai, IEEE Trans. on Inst. and Meast, **45** (1), 293–297 (1996)
18. B. Singh, K. Al-Haddad and A. Chandra, IEEE Trans. on Power Systems, **46** (5), 133–138 (1999)
19. A. Ferrero and G. Supert-Furga, IEEE Trans. Instrum. Meas., **40**, 568–577 (1991)
20. B.N. Abramovich, Y.A. Sychev, International Journal of Applied Engineering Research, **11** (4), 2640–2645 (2016)

Intellectual decision-making system in the context of potentially dangerous nuclear power facilities

Alexander Danilov[1], *Vladimir* Povarov[2], *Viktor* Burkovsky[1], *Semen* Podvalny[1] and *Konstantin* Gusev[1,*]

[1]Voronezh State Technical University, 394026, Voronezh, Russia
[2]Novovoronezh Nuclear Power Plant, 396072, Novovoronezh, Russia

Abstract. The article deals with intelligent operation decision support system under condition of potentially hazardous nuclear facilities. The proposed system is referred to the class of advising systems and does not make final decisions in case of deviations of parameters to be analyzed, but generates general ways to solve an encountered problem and issues a set of recommendations for the plant personnel. In the article a fuzzy logic tool is used as mathematic tool. Lessons learnt from operation of nuclear facilities demonstrate that existing critical components (parts, areas, welding joints) are subject to increased failure under conditions of high operational loads, including beyond design loads and negative environmental impact. Usually in that situation there is probability of equipment integrity failure, when the unit is at power, with severe defect downing. For instance, the coolant leak and potential development of initial penetration defect to critical dimensions. In other words, in fact, the final observable result is always one – formation and development of operational crack which jeopardizes design integrity of the component and, accordingly, seriously compromises the nuclear power unit operation. The proposed situational model is linked with real knowledge data base where generated situational pairs are stored. The expert system is used for knowledge data base formation. Actually the proposed system consists of two independent fuzzy systems. From mathematical tool point of view, the advantage of such systems combination is lack of defuzzification unit in the first system and fuzzification unit in the second one.

1 Introduction

Accidents happened over the past decades consistently lead to the conclusion that following the way of technical progress a man is exposed to higher risk. Thanks to enhancement of technosphere and medicine average longevity increased significantly. Technosphere, created to protect people against external impacts, nowadays becomes a source of danger: commercially operated units capacity is increasing, technologies are more sophisticated, there is increase of inter-effect between facilities, equipment operation more depends on human factor. Risks and accidents scale significantly extended.

With that in mind it is required to take measures to protect people and environment against hazard generated by technosphere – accidents on hazardous nuclear facilities. These measures are important along the whole life cycle of the plant but specifically at the stage of design. At this particular stage it is possible to significantly improve nuclear facility safety without hefty increase of costs. Usually hazard level of facility operating under standard conditions, stipulated by the design, is much lower than the hazard level under emergency conditions.

Provided that different accidents at highly dangerous facilities were analyzed, it is possible to identify a number of common causes: design errors, incorrect decisions about siting and plant operation modes, underestimation of personnel training, negligence and carelessness. But accidents analysis does not solve all problems. It is required not only to find out weakness in technological cycle but forecast development of events caused by an accident at hazardous facilities, shape management decisions required for risk minimization.

Present day decision-making support systems at nuclear facilities are based on mathematical statistics data with low degree of adaptation to changing parameters. Such systems are intended predominantly for analysis of data backlog and ineffectively manage with situations forecasting tasks. Based on analyses of modern decision-making support systems at nuclear facilities it is proposed to implement a comprehensive intelligent decision making support system based on fuzzy logics tool distinguished by availability of alternative actions of personnel at hazardous nuclear facilities.

2 Parameters generation

For intelligent operation decision support system implementation fuzzy situational pattern "situation-situation" type is proposed.

For input matrix elements data from multi-parameters continuous monitoring system of the plant critical

* Corresponding author: gussev_konstantin@mail.ru

components, subject to risk of operational damage, is used [1].

The situational model incorporates three main blocks:
- conditions evaluation block (fuzzification);
- decision making block;
- block of output action (defuzzification) [2].

Provided that two linguistic variables for input and output data streams are introduced:

X_i <«*Concentration*», T_x, *[0, P_i]*>,

Y_i <«*Level of emergency situation occurrence* »,

T_y, *[0, 1]*>,

where T_x = {«small», «mean», «rather high», «high»}, T_y = {«low», «mean», «rather high», «high», «dangerous»}, and semantic rule, which introduces compliance with each fuzzy variable out of the set T_x, T_y fuzzy set $\tilde{C}_{xi}$ and $\tilde{C}_{yi}$, shown on Figure 1. Type of the membership function is selected by data approximation results and in the form of the Gaussian function [3]:

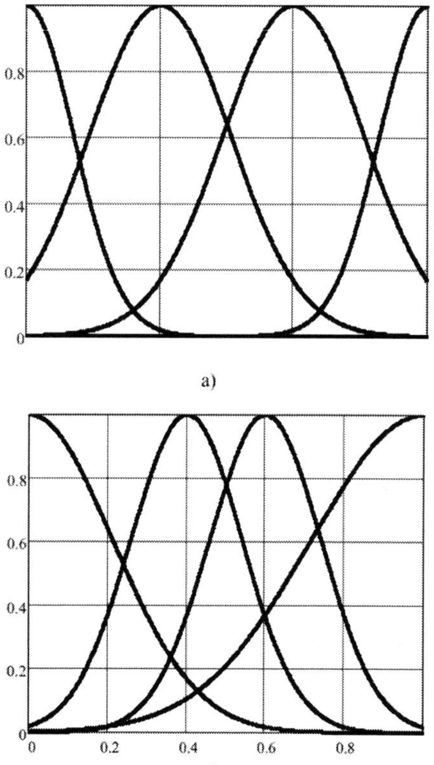

Fig. 1. a) Membership function of fuzzy set $\tilde{C}_{xi}$; **b)** Membership function of fuzzy set $\tilde{C}_{yi}$.

$$\mu(a_i) = \exp\left[-\left(\frac{a_i - x_i^k}{s_i^k}\right)^2\right],$$

where $i = 1, \ldots, n$; n – number of fuzzy model inputs while function parameters have the following interpretation: x is a center, and s is the width is the Gaussian curve.

3 Model of decision making system

Knowledge data -base. In case of situational model implementation knowledge data base is a key block in making proper operational decision. The work uses static knowledge data base where pairs of matrix of the following type are stored:

<Current fuzzy situation>-<Forecast fuzzy situation>.

Fuzzy situation represents a matrix identical to matrix S_i *(n x k), i=1,..,l,* obtained in defuzzification block. Matrix S is composed in the same way by means of the unit mentioned before and membership functions. The second element of the pair is a vector S_i*(1xk), i=1,..,l,* obtained by means of linguistic variable X and correspondent membership functions (Fig. 2). Thus, knowledge data base is in explicit form and presents a number of records in units of measurements *(l x 2)*.

In knowledge data base generation two main approaches were applied [4]:

1) generation of pairs in the knowledge data base based on historical data. At this stage reference situations, used in future for forecasting, are generated;

2) generation of pairs in the knowledge data base based on expert review of status of parameters being analyzed.

Decision making block. Within decision making procedure possible ratios for fuzzy situations are implemented. Equality extent is used in the work as a criteria for selection of closest situation from knowledge data base and, accordingly, corresponding outcome vector S:

$$\mu(S_0, S_i) = \nu(S_0, S_i) \& \nu(S_i, S_0),$$

where $i=1,\ldots,k$, μ – equality degree of fuzzy situations, ν – degree of fuzzy situations introduction. In other words, situation obtained from input data and reference situation in knowledge data base are compared in this block.

It is proposed to apply the procedure of output vector S generation depending on calculated equality degrees as follows:

$\mu_{max} > 0{,}9$	Accept S with the highest degree of equality
$0{,}5 < \mu_{max} < 0{,}9$	Select situation with three highest degrees of equality and receive result set S by means of selected $S = S_{\mu max} \cap S_{\mu max-i}, i = 1, \ldots, 2$ overlapping
$\mu_{max} < 0{,}5$	Select situation with five highest degrees of equality and receive result set S by means of selected $S = S_{\mu max} \cap S_{\mu max-i}, i = 1, \ldots, 4$ overlapping

Block of output action. As a rule, algorithms of defuzzification are conducted in this block and a solution or likely recommendation is generated at the output. To improve quality of intelligent operation decision support system it is proposed to implement defuzzification process based on two situations: current and forecast. For this purpose it is proposed to use the block for assessment of the current state parameters. In the proposed block input parameters are changed into fuzzy set by means of linguistic variable X_i <«*Concentration*», T_x, *[0, P_i]*>.

Consequently, two fuzzy sets S_i and S_0 enter decision making block input from forecasting block and correct status parameters evaluation block accordingly. At this stage for recommendations and likely solutions generation a set of fuzzy rules is applied:

$$IF\ S_0 = A\ \ AND\ \ S_i = B\ \ THEN\ R = Y.$$

Data base of rules for the decision making process is generated based on expert evaluation of parameters and looks like in the Table 1.

Table 1. Data base of rules.

Level of emergency occurrence risk	Personnel possible actions
1	2
Low	Steady conditions
Mean	Continuous monitoring of the situation and diagnostics of power installation critical damage parameters
1	2
Rather high	Diagnostics, calculated and experimental forecasting of event development, maintenance and restoration activities on operating power unit
High	Unit shutdown, maintenance activities
Dangerous	Unit shutdown and decision making about taking out of service

Thus, the result of the block of output action operation is a group of specific recommendations to be selected by the system depending on current and forecast values of the system parameters.

In the work defuzzification method with mean center is used:

$$\bar{Z} = \frac{\sum_{i=1}^{k} \mu_R(Z_i) Z_i}{\sum_{i=1}^{k} \mu_R(Z_i)},$$

where Z_i is a point where function μ_R takes the maximum value [5].

Continuous multi-parameter monitoring system is able not only to analyze and generate parameters but also forecast the facility status. To improve quality of intelligent operation decision support system it is proposed to add a block for comparison of forecast values. Figure 2 [6] presents functional diagram of fuzzy situational model.

3 Algorithm of decision making system operation

Initial data of the algorithm represent a matrix of parameters to be analyzed. After identification of liguistic variables the pairs of fuzzy situations are generated.

First of all, based on available historical data and, secondly, relying on obrained expert data. As a result all pairs are combined in the common knowledge data base.

The structure of intelligent operation decision support system consists of two fuzzy models (Fig. 3):

- fuzzy situiational system represents fuzzy control «situation-action» [7]. Following formation of rules data base and knowledge data base fuzzy input set (fuzzyfication), designated by initial or current situation, are generated. After that the currect situation is compared with situations from knowledge data base and the biggest situation equiality degree is defined. Once maximum equiality degree is defined, the second value from the data base is selected, which are input data for fuzzy system.
- fuzzy system implements classical algorithm Mamdani except for fuzzyfication block, because fuzzy set enters the input.

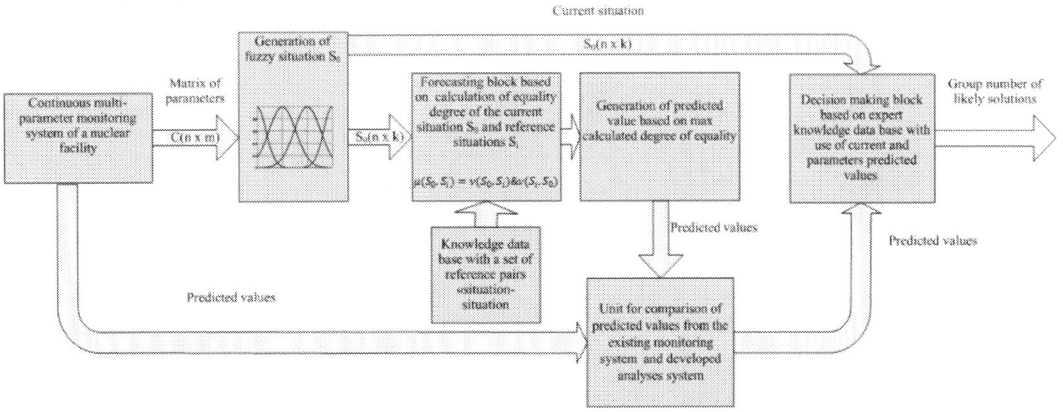

Fig. 2. Functional diagram of fuzzy situational model.

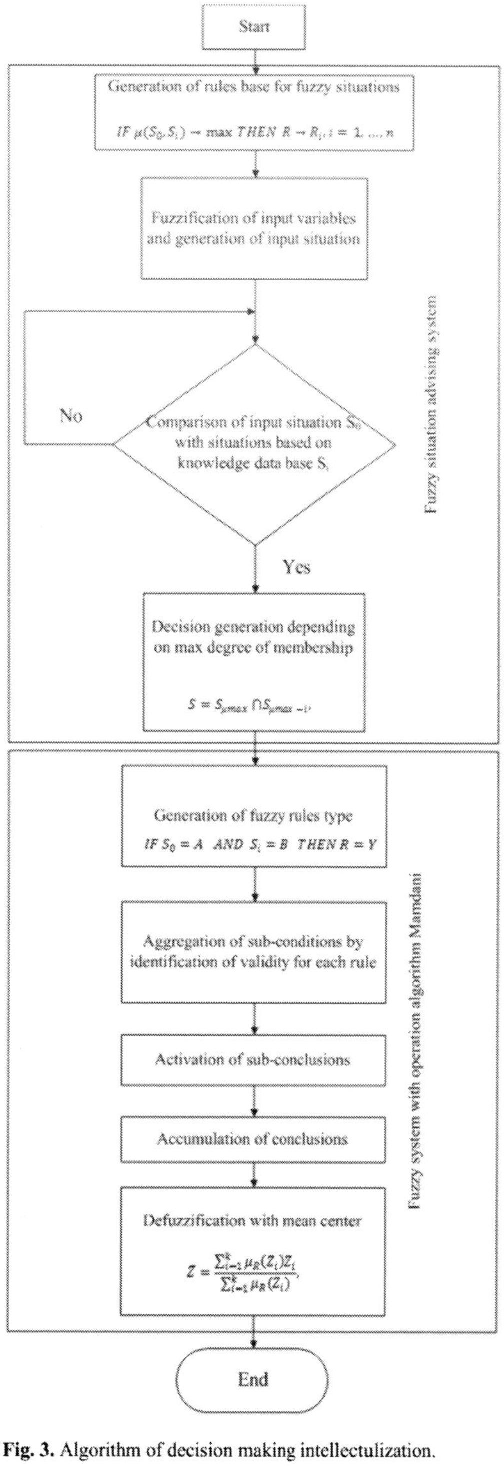

Fig. 3. Algorithm of decision making intellectulization.

4 Conclusion

The developed intellectual decision making system is a state-of-the-art tool for analyzing and forecasting status dynamics of hazardous technological facility. The main characteristic of the proposed integrated system is in implementation in the common format of two fuzzy systems together, serially communicating data without necessity of additional fuzzyfication-defuzzyfication.

References

1. V. Povarov, M. Bakirov, A. Danilov, *Automatic system of multi-parameter monitoring of nuclear power installations status* (Voronezh, Scientific book, 2017)
2. V. Burkovsky, K. Gusev, Voronezh technical university reporter, **3-1**, 13–17 (2014)
3. D. Rutkovskaya, L. Rutkovsky, M. Pilinsky, *Neural networks, genetic algorisms and fuzzy systems* (Moscow, Hot line-telecom, 2003)
4. S. Ayvazyan, I. Eniukov, L. Meshalkin, *Applied statistics and research dependencies* (Moscow, Finance and statistics, 1985)
5. K. Tanaka, H.O. Wang, *Fuzzy control systems design and analysis: a linear matrix inequality approach* (Wiley-Interscience, John Wiley and Sons, 2001)
6. N. Yarushkina *Fundamentals of the theory of fuzzy and hybrid systems* (M., Finance and statistics, 2004)
7. K. Jacek, *Analytical Methods in Fuzzy Modeling and Control* (Springer, 2009)

Maintenance problems of PWM-inverters in power networks with distributed generation

Pavel Kuznetsov[1,*], *Sergej* Solyonyj[2], *Yuriy* Sychev [1] and *Boris* Abramovich[1]

[1]Saint-Petersburg Mining University, Department of Power Industry and Electromechanics, 199106, Saint Petersburg, Russia
[2]Saint-Petersburg State University of Aerospace Instrumentation, Department of Electromechanics and robotics, 190000 Saint Petersburg, Russia

Abstract. This article observes maintenance problems of PWM-Electric Drive in Power Networks with distributed generation (DG). Compared to conventional centralized networks harmonics disturbances and non-sinusoidal waveforms happen faster in DG networks due to bigger amount of non-linear loads (Frequency and Power AC and DC Converters, PWM-Electric Drive, etc.). As a result, general stability, power quality and uninterruptible power supply can be disturbed. In order to prevent these factors a complex of power conditioning measures is proposed in this article. It demonstrates the possible influence of Active Power Filters with selective harmonics compensation function implementation together with developed controller and preventive algorithm. Almost all modelling was performed in Matlab/Simulink software.

1 Introduction

Recent statistics analysis demonstrates that power supply networks with DG are being implemented and created fast [1,2]. This happens due to some obvious benefits that DG networks have compare to conventional centralized grids:

- convenient location near end user;
- lower costs for energy transmission due to shorter distribution lines or their complete absence [3];
- excellent possibility to use low and mid power generators of renewable energy [4];
- integration of microturbines and creation of islanded electromechanical complexes;
- energy consumption from main network during peak odes and energy transfer into main network during standstill periods [5].

However, during real operation of DG networks many disturbances happen that lead to power quality worsening, emergency modes, and possible total blackouts [6]. One of these disturbances are current and voltage harmonics, which are injected into the grid by non-linear loads. In addition, they are closely connected with reactive power that implies the power factor and leads to overheating of power electronics, electromecanical equipment and total network overload.

2 Analysis of existing solutions

Classical reactive power compensation theory was very popular and widely implemented in conventional centralized power grids. Its idea is based on the fact when inductive or capacitive reactive power is compensated by opposite reactive element. For example, if industrial enterprise (petroleum or refinery plant) has a lot of induction motors, which produce inductive current and inductive reactive power it was enough to install capacitive batteries that would compensate this dangerous currents and correct total network's power factor.

However, situation changed with the fast development of semiconductor electronics. Integration of frequency converters, household power electronics, invertors and DG led to the change of the reactive power and harmonics nature, where convenient reactive power compensation theory was unsuitable. This happened due to the reason when passive reactive compensating elements can lead to resonances, overheating and even explosion trying to compensate reactive currents from non-linear loads. That is why new power conditioners were invented recently. The most perspective invention are Active Power Filters (APFs) that were created using Instantaneous Power Theory [7].

This article observes and describes protective complex of power conditioners that for implementation in islanded DG power networks on the example of petroleum plant. It also proposes protective algorithm against occurrence of emergency modes. Last part of the article demonstrates modelling results of protective algorithm in Matlab-Simulink libraries and proposed prototype of protective controller.

3 Protective algorithm for DG networks

DG power networks are considered to be more robust and efficient compare to conventional centralized power grids. This happens due to their more flexible structure

* Corresponding author: kuznetsovpavel@inbox.ru

and presence of bigger generators' amount. As a result, emergency modes consequences can vary and differ from their analogues in centralized grids (for example sudden power loss due to wave form worsening because of harmonics presence). This also has to be considered during their severity analysis. In order to make it complex and correct there was developed a modernized algorithm, first proposed in [8] where development of cascade outages was corrected according to DG power networks. It is shown on Figure 1.

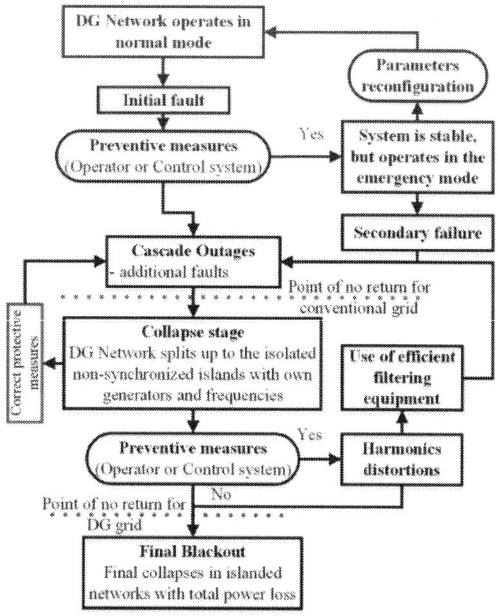

Fig. 1. Modernized algorithm of cascade outage's development in power network with DG.

As it was established during algorithm development DG power networks are more robust and can stand emergency modes and cascade outages. Even more so, after cascade outages total blackout has a low possibility to happen because power network will be split up to several islanded regions. However, the main problem is hidden in their future synchronization in one stable grid. In addition, presence of PWM-converters, electric dive, inverters, etc. leads to worsening of power quality and grid's destabilization. Next chapter studies possible solutions to minimize this influence.

4 Harmonics and existing protective devices

Non-linear loads create harmonics, and initially they are current waves. The spectrum frequencies (harmonic order) and the magnitudes of those current waves depend only on the nature of the load. For example, most DC drives have 6-pulse rectifiers, which generate mostly 5th and 7th order harmonics. The 5th harmonic has a frequency five times the fundamental and the 7th is seven times the fundamental. The harmonic is actually the power source at the high frequency. Since the harmonics exist in the current drawn, the source is treated as a current source. Unlike the voltage source, the ideal current source has infinity internal impedance and prefers zero load impedance to maintain the current.

The nonlinear load can be represented as linear load part and the high frequency current sources. For simplicity consider the situation where the only one of the connected loads is nonlinear and is produces only 5th order harmonic current. This scheme is shown on Figure 2.

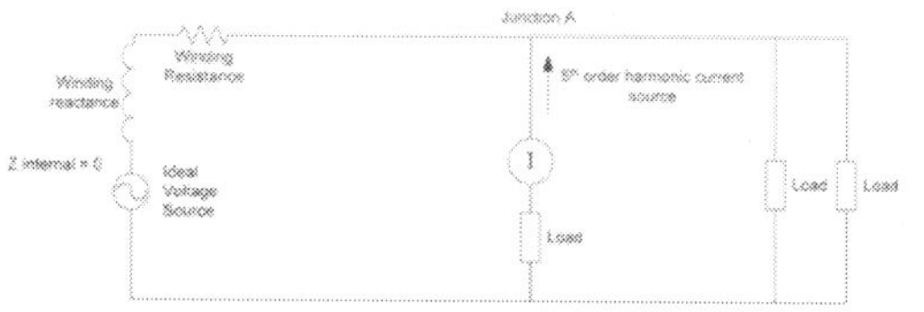

Fig. 2. Ideal current source scheme.

The 5th harmonic current is produced by the current source which is part of the load operation, and is fed to the network through the junction A. From the junction A the current will choose the lowest impedance path to flow. The left side impedance is the internal impedance of the distribution transformer in additional to the line impedance. The total impedance of the left side is very much lower than the fully loaded right side. Figure 3 shows how most of the harmonic current will flow towards the power transformer.

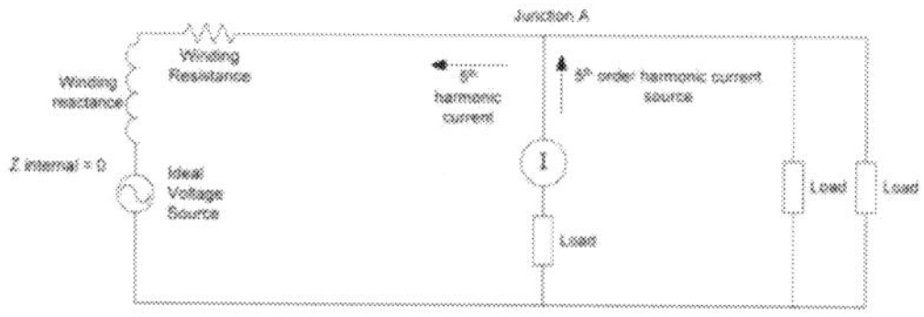

Fig. 3. Scheme of harmonic source that operates as a current source connected across the load.

Following the line and internal transformer impedances, the harmonic current creates the high frequency voltage drop. This is how voltage harmonics are created. In addition, the current, which passes through transformer secondary, will create the harmonics on the primary transformer side as well.

Now assume that power factor correction is required due to a low power factor at the fundamental frequency. Power factor correction capacitors are added to correct the power factor. Figure 4 reveals this situation.

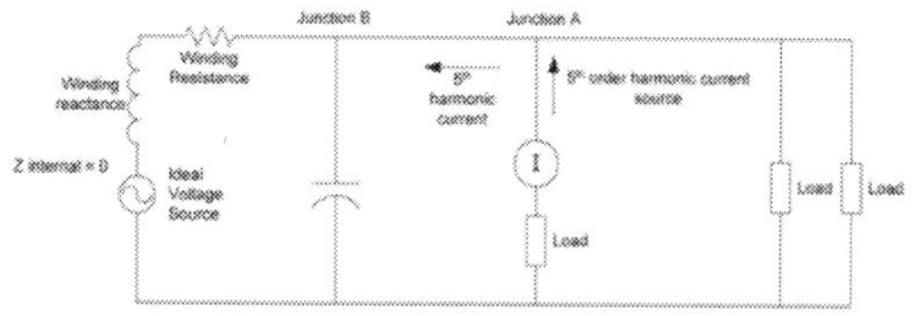

Fig. 4. PFC capacitors can create a resonant circuit at harmonic frequencies.

The harmonic current coming to the junction B will encounter the capacitor and connected in parallel to it resistance and reactance. The resistance part is negligible in most cases, so the above can be represented as parallel resonance circuit in the path of the 5th harmonic current. The case where *Xl* will be equal to *Xc* and parallel resonance will take place is shown on Figure 5.

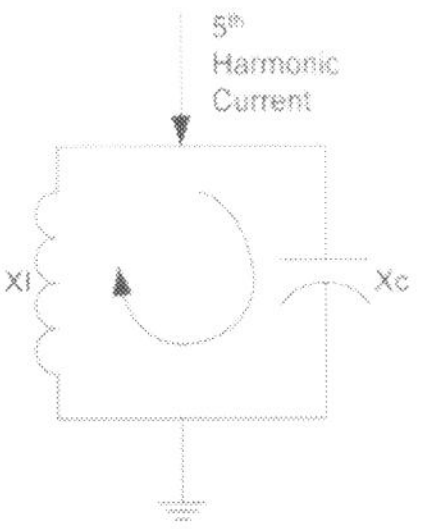

Fig. 5. Resonant circuit scheme at 5th harmonic.

Parallel resonance raises the circuit impedance. Current circulates between the capacitor and the inductance without being passed to the grounded terminal. In the case of parallel resonance or even close to resonance condition, the path, which includes distributing transformer and PF capacitors have a very high impedance at the harmonic current frequency. At parallel resonance condition the harmonic current is forced to go to the load. Since the path impedance is increased, the voltage harmonic is increased as well too. This situation may cause a significant damage to the electrical infrastructure. Normally, the weakest part, which fails first in such a condition, is the PF capacitor. The PF capacitor most likely cannot withstand the high harmonic current, which is circulating between the capacitors and the distribution transformer [9].

The parallel resonance frequency always exists in cases where the capacitors are used for PF correction. The way to prevent parallel resonance occurring is to insure that the frequency of resonance is located outside the harmonic range. This is the main idea behind the "detuned" system configuration (see Figure 6). The 7% or 5,67% inductors move the resonance point below the 4th harmonic which insures safe operation for the three phase delta capacitors in cases where the third harmonic is balanced.

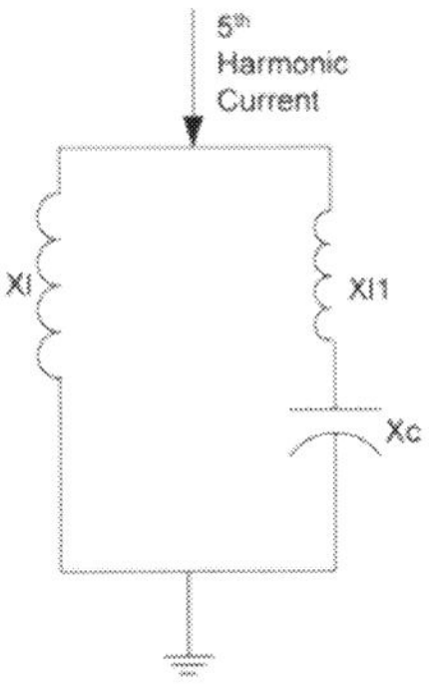

Fig. 6. Resonant circuit at 5th harmonic.

In the recent time, different types of APFs have obtained a huge support in conventional power networks due to their increased performance and harmonics compensation. Some of these devices were proposed in [10-12]. Figure 7 demonstrates possible implementation of APF in power network with non-linear load. These units have significant benefits: reduced switching-voltage ratings, smaller switching stresses dv/dt and an improved output voltage at lower switching frequencies. These characteristics have led to the development of many applications.

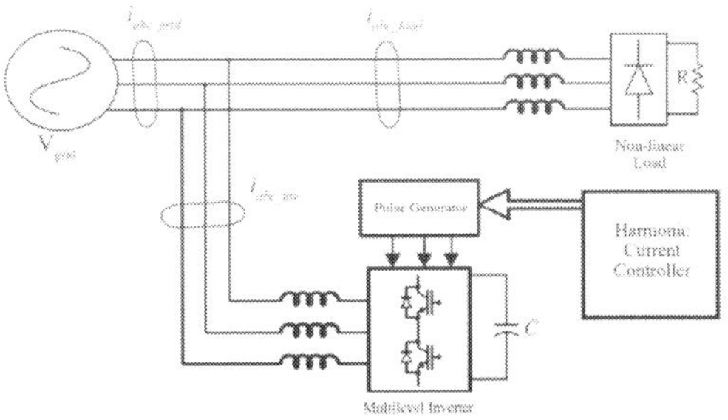

Fig. 7. APF's integration scheme in the DG power network.

The presented APF's modification can compensate harmonics selectively. This feature plays an important role in DG power networks because of different non-linear load types. Inverters, converters, electric dive, lighting, etc. produce different harmonics types in different amounts. Last researches [11,13-14] demonstrate that according to IEEE 1592-2013 requirements only two-three main harmonics bring 80% of disturbances. APF allow to compensate only them, not exceeding limited DG network's power limit.

During normal operation, generators and other electromecanical equipment create parasitical oscillations that can destabilize their stable rotational speed. Taking into account harmonics from non-linear loads in DG power networks a hazardous situation can occur when rotational mechanical oscillation summarize with harmonics distortion. In order to prevent this APFs are used together with Power system Stabilizers (PSS).

5 Modelling and discussion

In order to show positive effect of APFs application and demonstrate their positive effects in DG networks modelling was provided in Matlab-Simulink software.

Model developed in [15] was used as the main DG power network. Its part is shown on Figure 8. Power network consists of 39 buses, 19 loads, and 10 generators with generator one, connected to bus 39.

Generally, model has the following possibilities in order to research DG power networks properties:

- Three-phase PI section line blocks are used to represent transmission lines. Positive- and zerosequence resistances, inductances, capacitances, and line length can be set.
- All loads are represented by three-phase parallel RLC load in PQ type (Y-grounded configuration) since

corresponding references provided load data in active and reactive demands. Different types of connection configuration and types are available.

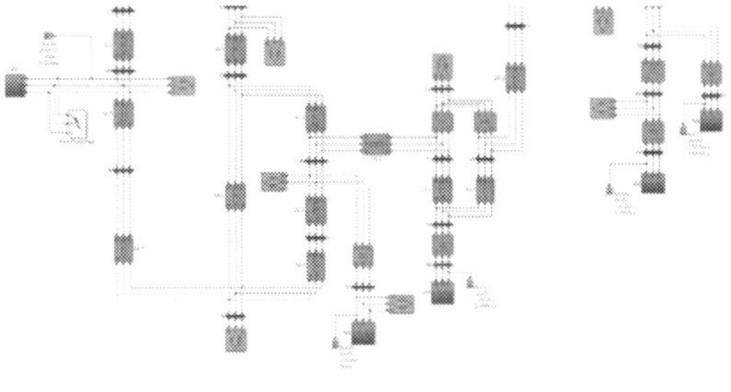

Fig. 8. Part of the test model in Matlab-Simulink.

• Three-phase transformer block are used to model all the transformers of the test systems. Y-Y connection for interconnecting transformer and Δ-Yg connection for generator transformer can be set.
• Generators are modelled by three-phase synchronous machine in dq rotor reference frame while d-axis and q-axis time constants vary to best fit with given data.
• If applied, single mass tandem compound steam prime mover including speed regulator, steam turbine and a shaft is set to model steam turbine and governor.
• Each generator equipped with three types of Power System Stabilizers (PSSs), Multi-Band PSS (MBPSS), Delta omega generic PSS, and Delta Pa generic PSS.

Non-linear loads with harmonics sources were added to the system in order to improve it functionality for power quality research. A simulation was done when a cascade outage starts close to bus 15 due to loss of synchronism because of harmonics presence. Also, a 3-pahse short-circuit occur at a time 2 s and lasts for 4 seconds.

First simulation (Figure 9) shows that without PSS and APF all generators loose synchronism after 6 seconds and cascade outage completely destroys system, leaving even no operating islanded parts.

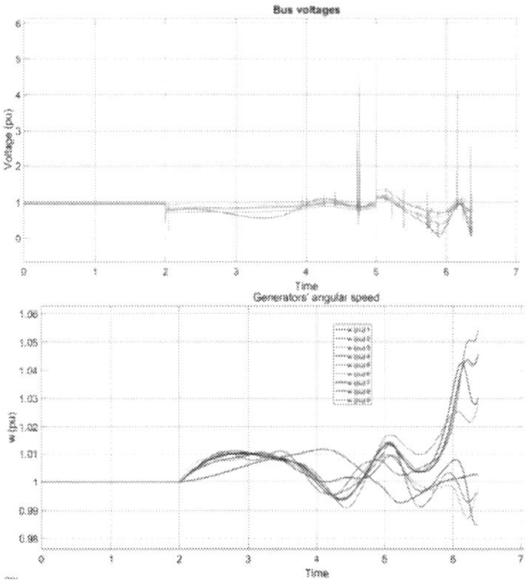

Fig. 9. Destruction of the system during cascade outage due to harmonics presence.

Second part of the modeling consisted in connecting APF fore 5th and 7th harmonics compensation together with PSS for stability improvement during emergency mode. Simulation plots are shown on Figure 10.

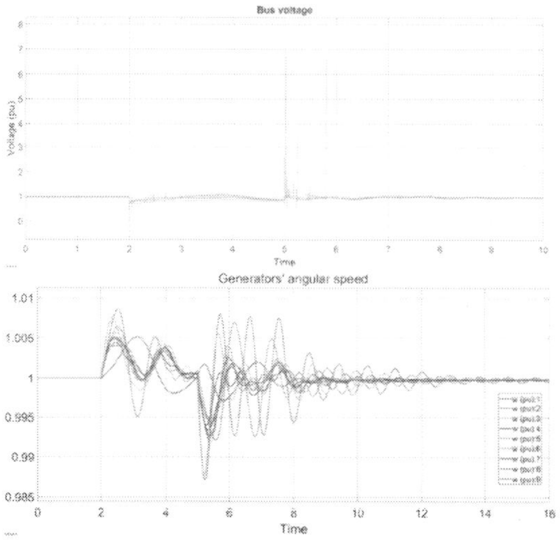

Fig. 10. Results of PSS and APF implementation.

Protective devices successfully overrode the same emergency mode. Bus voltages have completely recovered and got rid from pulsations after 7 seconds and generators' angular speed was restored after 13.5 seconds. This proves effectiveness of the proposed protective measures.

6 Conclusion

The research was conducted in order to demonstrate key benefits and incompleteness of power networks with DG. The developed protective algorithm that demonstrated development of cascade outage it DG power network can be used an implemented in real power networks in order to increase their stability.

Conducted research revealed that harmonics from non-linear loads can damage operating equipment and destabilize DG grid. APFs were proposed to be used in order to compensate disturbances selectively. Simulation of DG power network with APF demonstrated improved grid's stability and increased robustness. Simulation results and algorithm proposed to be realized in protective controller against cascade outages.

The presented results were obtained as a part of scientific research according to the contract №11036GU/2016 "The development of power network condition controller with blackout prevention possibility" provided by Foundation for Assistance to Small Innovative Enterprises in Science and Technology (FASIE).

The presented results were obtained as a part of scientific researches according to the contract № 13.3746.2017.8.9 "The designing on the basis of systematic and logic probability evaluations of rational and economically proved structure of centralized, autonomous and combined power supply systems with high reliability and stability level with usage of alternative and renewable power sources for uninterrupted power supply of enterprises with continuous technological cycle".

References

1. H. Martinez-Garcia, Jornades de Recerca EUETIB 2014, **12**, 100–109 (2014)
2. J.O. Rezende, L.M. Peres, G.C. Guimarães, A.J. de Moraes, M.A. Tamashiro, Renewable Energy and Quality Journal (RE&PQJ), **12** (2014)
3. A. Arjun M, B.K Uma Rao, C.A B Raju, Renewable Energy and Quality Journal (RE&PQJ), **12**, (2014)
4. B.N. Abramovich, Y.A. Sychev, International Journal of Applied Engineering Research, **11** (4), 2640–2645 (2016)
5. B.N. Abramovich, Yu.A. Sychev, A.S. Mingazov, V.V. Polishuk, **10**, 126–127 (2013)
6. P.A. Kuznetsov, O.A. Stepanov, Aerospace MAI Journal, **4**, 157–163 (2017)
7. H. Akagi, E.H. Watanabe, M. Aredes, *Instantaneous power theory and applications for power conditioning* (IEEE Press, 2007)
8. P. Pourbeik, P. Kundur, C. Taylor, IEEE Power and Energy Magazine **4(5)**, 22–29 (2006)
9. G. Whittle, Energize, **11** (2010)
10. V. Khadkikar, M. Singh, A. Chandra, B. Singh, 2010 Joint International Conference on Power Electronics, Drives and Energy Systems (PEDES) & 2010 Power India, 1 (2010)
11. M.B. Said-Romdhane, M.W. Naouar, I.S. Belkhodja, E. Monmasson, Energies, **10**, 1–19, (2017)
12. J.L. Monroy-Morales, D. Campos-Gaona, M. Hernandez-Angeles, R. Pena-Alzola, J.L. Guardado-Zavala, Energies, **10**, 1–23, (2017)
13. S. Golestan, M. Monfared, J. Guerrero, Power Electronics, Drive Systems and Technologies Conference (PEDSTC), 2013 4th, 510–517 (2013)
14. Yu. Huang, P.M. Pardalos, Q.P. Zheng, *Electrical Power Unit Commitment* (Springer, 2017)
15. A. Moeini1, I. Kamwa, P. Brunelle, G. Sybille, IEEE Power Engineering Conference (UPEC), 1–19, (2015)

Gaz Diffusion Layers from functional carbon materials for fuel cells used in energy installations

Elena Kiseleva* and *Anna* Vasilenko

Joint Institute for High Temperatures of the Russian Academy of Sciences, 125412, Izhorskaya st. 13 Bd.2, Moscow, Russia

Abstract. Gas Diffusion Layers (GDL) for fuel cells are two-dimensional carbon materials designed to obtain a smaller thickness, greater bending stiffness, lower residual deformation and lower compressibility compared to analogs. To produce it, functional carbon materials were used: activated carbon of own production, from vegetable raw materials, acetylene black and single-walled Tuball nanotubes. Dependence on the amount of binder is shown, as are the properties of electrodes made by calendering. The resistance of the rolled electrodes is much lower than that of the sputtered electrodes. There are proposed 4 variants of electrodes, which can be used in fuel cells and are used in power energy installations.

1 Introduction

The use of electricity as the most convenient, effective and widely used in various sectors of the economy and in the life of the energy carrier, in contrast to other intermediate energy carriers, whose energy is stored in chemical or other forms, requires at every moment a strict equality of the electric power supplied to the consumer and consumed by it. At present, research in the field of new materials and technologies for sources and accumulators of electrical energy is being carried out quite intensively all over the world. This is due to the hopes for solving a number of problems of energy, ecology, transport and some industries through the use of cheap and safe sources and accumulators of electrical energy with a long service life. The role of the development of new materials, technologies for the formation of active layers and electrodes, electrolytes and other components is crucial in achieving the targets for prices, service life, specific characteristics of energy sources.

At present, fuel cells (FCs) and batteries are widely used for power generation in transport, utilities, portable electronic devices and other applications. The most common types of fuel cells are phosphoric acid, with polymeric proton-exchange membranes, alkaline, methanol, direct conversion, regressive, zinc-air, solid oxide, etc.

One of the necessary components of the FC is a structural element in the form of sheet material with a high porosity, a good gas permeation, involved in electrochemical reactions, having a high electrical conductivity. According to the existing terminology, such an element is called a gas diffusion layer (GDL).

GDL is an important component of FC and perform the following functions: deliver gases and liquid reagents (fuel and oxidant) to the desired electrodes and distribute the reagents in a necessary and controlled manner; provide mechanical support for membranes and electrocatalytic catalyst beds; lead and withdraw electrons to / from the electrodes; help to remove heat; ensure the transfer of loads of the seals of the fuel cell assembly; participate in the interfacing of the internal parts of the FC battery.

It is not surprising that the world's leading manufacturers of fuel cell systems pay close attention to both the technical and operational characteristics of GDS, and their cost. The most important components for the fuel cell battery market are bipolar plates and membrane-electrode blocks (OIE). The OIE is a structure consisting of a catalyst-coated membrane between two layers of HDPE, which are sintered by hot pressing. GDL and catalytic inks are the most expensive subcomponents of the OIE in the FC battery. To create cheaper and more affordable GDL from functional carbon materials, our work is directed.

The proposed GDL has a two-layer construction in which a conductive microporous layer is applied over a macroporous carbon substrate. The sublayer of the basic material consists of hydrophobized macroporous activated carbon and acetylene black. Microporous sublayer is thinner and in turn consists of microporous activated carbon with nanocarbon additives (nanotubes, soot, etc.).

The porous structure and properties of the functional carbon material from which it is made play a decisive role in the efficiency of the electrode (GDL with the catalyst on its surface) functioning. To obtain the maximum specific electrochemical characteristics of MEAs, a certain type of carbon material is selected for each GDL sublayer, so for the main layer – macroporous carbon, with wider pores, and for the upper, active layer – microporous coal with a developed surface. The most suitable material is activated carbon

* Corresponding author: kanna787@mail.rug

(AU). It has a relatively low cost, is widely available, stable in many environments, has high conductivity [1].

The main surface area of the porous coal is in the micropores. However, it is necessary to select the optimal porous structure of coal, achieving the maximum specific electrochemical characteristics of the MEAs and the minimum internal resistance.

In addition to the effect of pore size distribution on activated carbons, an important factor is the porous structure of the layer itself, namely its formation method.

In our case, we offer a method of cold calendaring (rolling), in comparison with conventional spraying. The internal resistance of electrodes formed by the sputtering method is higher than when using electrodes made by calendering. Most likely, this is due to the peculiarities of the formation of the electrode structure for various methods of their manufacture. Sputtered electrodes have a more friable structure [2]. Their density is lower than that of electrodes made by calendering. Consequently, the contact of the coal particles with each other and, mainly, with the conductive substrate in the case of sputtered electrodes is much worse than in the case of the rolled up electrodes. Nevertheless, thin sprayed electrodes show the best performance even at high currents.

The method of calendering electrodes is more technologically advanced than the method of sputtering. It allows to manufacture electrodes in a wide range of thicknesses. The resistance of the rolled electrodes is much lower than that of the sputtered electrodes.

2 Technique of the experiment

For the production of GDL, functional carbon materials were used: activated carbon of own production, from vegetable raw materials and single-walled nanotubes Tuball. When mixing a mixture of coal and fluoroplastic, as well as in the manufacture of electrodes, the temperature of the mixture was always above 19°C. This is important in terms of fibrillation of fluoroplastic particles. Fluoroplast F-4 has two phase transitions: at 19°C and 30°C [3, 4]. At a temperature below 19°C, the formation of fibers (ie, "unwinding" of PTFE particles) does not practically occur. At temperatures above 19°C and 30°C, fluororesin is increasingly fibrillated. When kneading a creamy mixture of coal and a fluoroplastic slurry, it manifests itself in the adherence of the whole mixture to one lump (kneading at low temperatures does not lead to the formation of a lump). However, such a mixture does not yet have strength (like plasticine). This is due to the low percentage of "unwound" PTFE particles. During calendering, the number of unwound PTFE particles sharply increases and the electrode, in addition to becoming the desired shape, also acquires the necessary strength. However, bringing the electrode to the ready state has a number of subtleties. Even a small feature during calendering can affect the final characteristics of the electrode.

3 Results and discussion

With an increase in the proportion of the binder, the density of the GDL increases, which, firstly, is due to the replacement of porous carbon by a continuous fluoroplastic, and, secondly, to the denser packing of coal particles in the electrode (Fig. 1). For the rolled GDL, their hydrophobicity strongly increases with increasing the proportion of PTFE in the electrode. With a large proportion of the binder, hydrophobic formations from fluoroplastic filaments cover a substantial portion of the coal particles, blocking the access of the electrolyte to the micropores.

Thus, the choice of the amount of binder based on the aqueous suspension of PTFE, in the first place, can be determined by the requirements for manufacturability of the manufacture and strength of the electrodes and has little effect on the electrochemical characteristics of the electrodes.

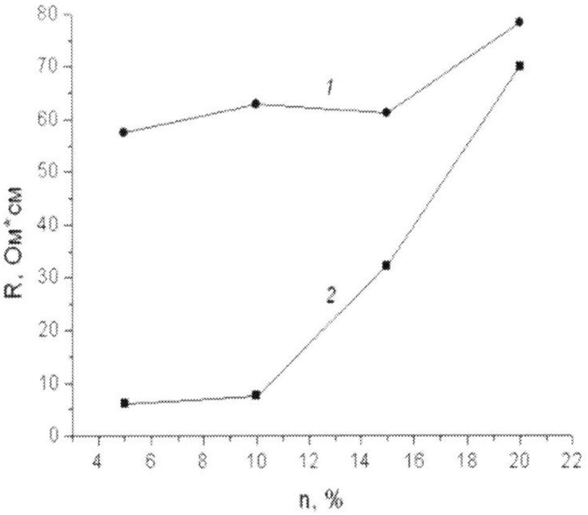

Fig. 1 Dependence of the internal resistance of the fuel cell on the content of fluoroplastic in the rolled GDL 1-thickness GDL 200 mkm, 2- electrode thickness 100 mkm.

Depending on the amount of binder, the properties of the electrodes vary, such as strength, specific and contact resistance, hydrophobicity and capacity. From the point of view of further introduction of these technologies into mass production of HDS, calendering technology is more preferable, since it is easier to implement. With its help, it is possible to produce electrodes of large thickness, which have low resistance and sufficiently high specific electrochemical characteristics. Sputtering technology can find application in the formation of a catalytic layer on an already prepared GDL.

The gas diffusion layer is a porous layer that ensures effective diffusion of the reagents to the catalytic layer. In addition, the GDL must have good electrical conductivity, since it is a conductor through which electrons are transferred to and from the catalytic layer. Usually GDL is made of a porous carbon material with a thickness of 100 ... 300 mkm. GDL also participates in water management, allowing some water to reach the membrane and moisten it.

One of the important components of the electrode is the insignificant (less than 1%) addition of carbon nanotubes increasing the strength and electrical conductivity of the GDS. Figure 2 shows a micrograph of the layer with the addition of Tuball nanotubes from OCSiAL, which shows the highly developed active surface of the GDL.

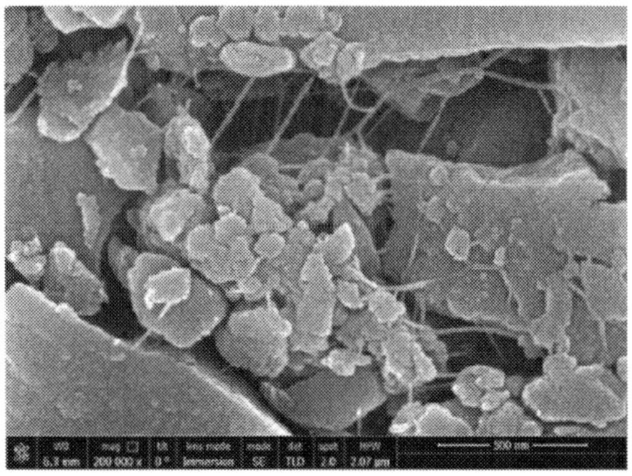

Fig. 2 SEM microphotographs for GDL electrodes with conductive additive – Tuball single-walled carbon nanotubes.

Gas diffusion layers for fuel cells are two-dimensional carbon materials designed to obtain a smaller thickness, greater bending stiffness, lower residual deformation and lower compressibility compared to analogs. Table 1 lists 4 types of GDL developed: GDL1A, GDL1B, GDL2A, GDL2B, with the corresponding characteristics.

Table 1. Properties of GDL.

Characteristic	Units	GDL1A	GDL1B	GDL2A	GDL2B
Thickness	mkm	190	235	190	235
Volume weight	g/m2	54	100	40	86
Porosity	%	84	76	88	80
Breathability	cm3/(cm2·s)	60	0,6	210	1,0
Electrical resistance (perpendicular to the surface)	mOm ·cm2	< 10	< 12	< 10	< 12

4 Conclusion

The GDL is proposed as a two-layer construction in which a conducting microporous layer is deposited over a macroporous carbon substrate. To produce it, functional carbon materials were used: activated carbon of own production, from vegetable raw materials, acetylene black and single-walled Tuball nanotubes.

The results of the research showed that the internal resistance of electrodes formed by the sputtering method is higher than when using electrodes made by the method of cold calendering (rolling). The method of calendering electrodes is more technologically advanced than the method of sputtering. It allows to manufacture electrodes in a wide range of thicknesses. The dependence on the amount of binder is shown, how the properties of the electrodes change, such as strength, specific and contact resistance, hydrophobicity and capacity.

Thus, four variants of electrodes have been developed, which can be used in fuel cells used in power energy installations.

References

1 M.A. Zhurilova et al., Izv. Vyssh. Uchebn. Zaved., Khim. Khim. Tekhnol., **60** (4), 82–87 (2017)
2 E.A. Kiseleva, F.V. Lelin, M.A. Zhurilova, E.I. Shkolnikov, RUSS J APPL CHEM, **90** (5), 712–715 (2017)
3 A.B. Ariawan, S. Ebnesajjad, S.G., Polymer engineering and science, **42**(6), 1247–1259 (2002)
4 M. Kontopoulou ed, *Applied polymer rheology. Polymeric fluids with industrial applications* (John Wiley&Sons, Inc. 2012)

Data support system for controlling decentralised nuclear power industry facilities through uninterruptible condition monitoring

Vladimir Povarov[1], *Aleksandr* Danilov[2], *Viktor* Burkovsky[2] and *Konstantin* Gusev[2,*]

[1]Novovoronezh Nuclear Power Plant, 396072, Novovoronezh, Russia,
[2]Voronezh State Technical University, 394026, Voronezh, Russia

Abstract. The article describes the automated uninterruptible multi-parameter system for monitoring operational vulnerability of critical NPP components, which differs from existing ones by being universally applicable for analysing mechanical damage of nuclear power unit components. The system allows for performing routine assessment of metal structures. The assessment of strained condition of a deteriorating component is based on three-dimensional finite element simulation with calculations adjusted with reference to in-situ measurements. A program for calculation and experimental analysis of maximum load and durability of critical area forms the core of uninterruptible monitoring system. The knowledge base on performance of the monitored components in different operating conditions and the corresponding comprehensive analysis of strained condition and deterioration rates compose the basis of control system data support, both for operating nuclear power units and robotic maintenance and repair systems.

1 Introduction

Rapid development of Information Technology extends its applicability and promotes intellectualisation of process management in every field of human activity. One of the most serious challenges is to provide information support by processing and analysing the information in the decision making process in managing decentralised nuclear power industry facilities. Nuclear Power Plant is a notable example of such facilities. Every power unit of a NPP is a dynamic facility and is under a high risk of emergency. Difficulties in controlling, monitoring and managing these power generation facilities are linked to the distance barriers between system elements and impossibility to describe their operation using standard modelling techniques. When managing nuclear facilities one will always face the issue of aggregating the results of assessing the current condition of NPP as a whole, as well as each of its subsystems (critical safety functions, parameters, readiness, etc.); aggregating or detailing the current information on any parameter group of elements and/or systems; detailing the information on any separate parameter, including current values, history and trends; diagnostic evaluation of power unit or equipment condition (operating mode); predicting power unit or equipment condition (operating mode); advising operating personnel where necessary. The uninterruptible on-line monitoring is essential due to the extensive use of automated repair systems in hardly accessible and highly radioactive elements of nuclear power units.

The main setback for this task was that the standard NPP design lacked systems for non-destructive damage control and actual thermal power load assessment of the most problematic equipment. During the operation of the nuclear power plants equipped with the VVER-type pressurised water reactors, uncompensated damage was identified, i.e. up to full depth cracks at welding joints of primary coolant header to steam generator connecting leg. Starting from 1980, inspections revealed dozens of in-service defects and even through-thickness cracks in pressuriser pipes, appeared due to stratification. It was discovered that a number of steam generators suffered similar damage; therefore the defects were recognised as a fundamental issue with a common cause.

Solving the problem required developing and introducing a new approach to calculation and experimental assessment of vulnerability of the most heavy-loaded elements of critical NPP equipment.

The present article describes the software and hardware complex for uninterruptible multi-parameter monitoring of condition of critical elements, susceptible to service damage by thermal distortion, as an integral part of the nuclear power unit operation control data support system.

2 Problem formulation

The analysis of accumulated experience in NPP operation revealed existence of critical elements (nodes, zones, junctions), susceptible to damage under heavy loads, including those beyond the design basis, as well as harmful environmental effects. It results in high risk of equipment integrity damage during the power operation due to a serious defect, such as coolant leak, with the possibility of further defect growth to the critical

* Corresponding author: gussev_konstantin@mail.ru

dimension, i.e. resulting in the immediate threat to the power unit operation. In case of such safety-related and dynamically developing service defects, the top-priority task is finding root causes of defect formation and making management decisions in order to restore damaged element and prevent the recurrence of similar events in the future. To adopt a correct decision, the management needs to conduct the analysis of both retrospective and current data with visualisation of real element condition.

2 Monitoring system

To resolve the aforementioned problem, we propose the technical solution, based on the uninterruptible multi-parameter monitoring and analysis of operational vulnerability of critical pieces of equipment, as the most effective way to visualise and control the formation and spread of defects in critical area. The main advantage of this system is its versatility, which allows for its' easy adaptation for monitoring almost any vulnerable zone of any component, including pipelines [1-4].

The practical result of the monitoring system implementation is the possibility for routine assessment of metal structures. Moreover, the use of the finite element simulation increases the accuracy of strain assessment, as calculations based on three-dimensional finite element simulation based on in-situ measurements are more precise, than approximation-based calculations.

All sensors are connected with the data collection and pre-processing unit, which in turn is connected to the data storage and transmission unit. The monitoring system is bundled with the software package that allows for automated remote system control, including data collection, storage and transmission, as well as real-time visualisation and express analysis of monitored parameters. The calculating module manages the three-dimensional finite element simulation and allows for comparing data on formation and spread of operational defects with the current condition of the equipment and its' operational modes. You can find the structural diagram of the proposed system at Figure 1.

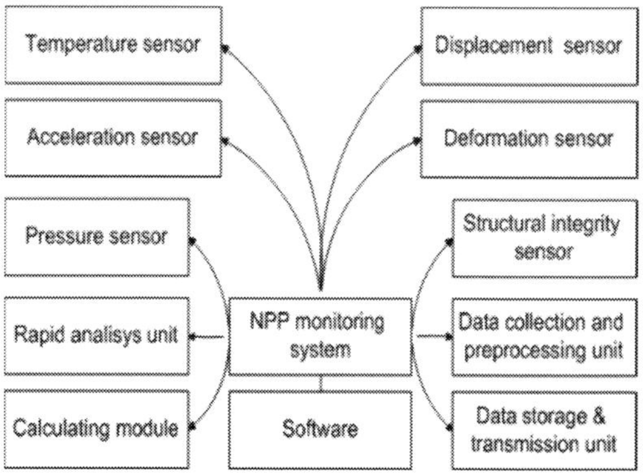

Fig. 1. Structural diagram of the uninterruptible multi-parameter monitoring system.

Due to the multitude of tasks carried out by the uninterruptible multi-parameter monitoring complex and in order to unify separate units, the system can be built on multi-channel microprocessor measurement modules, including National Instruments-based, and assembled in one complex. The data storage and transmission unit is a high-performance industrial server for data storage and backup. The monitoring system remote control, including monitoring the environment, regular performance tests of the subsystems and measurement channels, as well as data collection, storage and transmission, is performed via software means through the data storage and transmission unit. All received data is automatically recorded and transferred to the remote analytical centre via the Internet for further processing and detailed analysis (Fig. 2).

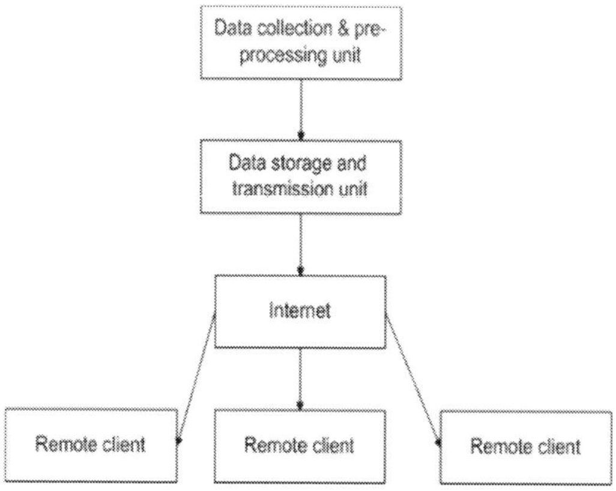

Fig. 2. Uninterruptible multi-parameter monitoring data transfer chart.

3 Algorithm for calibrating the simulation

Alongside the diagnostic data collecting, the uninterruptible monitoring system performs computational-experimental analysis of maximum load and durability of critical area based on the three-dimensional finite element simulation. The software performing the procedure is the core of the system. The simulation has nodes duplicating actual sensor placement, which allows for the comparison of calculated and measured parameters. The key feature of the three-dimensional finite element simulation is possibility to use data collected through monitoring (temperature, deformation, displacement) as the actual source data, as well as test data for adjusting and calibrating the three-dimensional finite element simulation. The algorithm for calibrating the simulation is described in Figure 3 [3, 5-11]. The calibrating process is performed by comparing calculated and measured parameters (deformation, displacement, temperature in different operational modes and load levels).

The actual data on pressure and temperature, registered by regular reactor unit sensors, as well as uninterruptible monitoring system sensors, are used as the load parameters. The failure to achieve satisfactory correspondence of calculated and measured parameters is followed by adjusting the calculating programme, aligning the geometry of monitored object, condensing the finite element array in certain areas, correcting the boundary conditions, etc. Usually, correspondence of calculated and measured parameters is achieved after few such iterations, which indicates operational integrity of the simulation model and adequate simulation of loading during various operational modes.

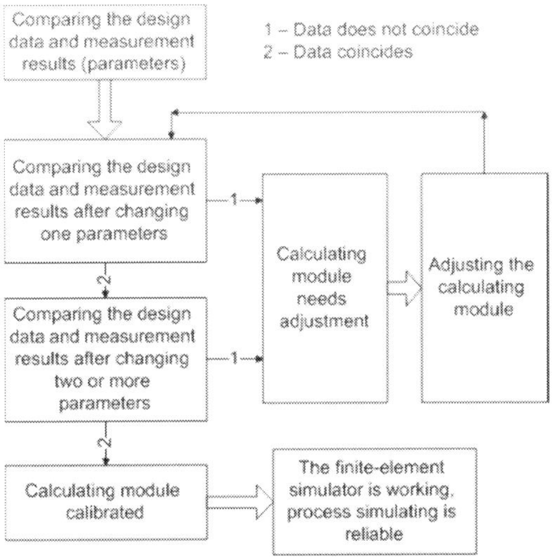

Fig. 3. The algorithm for calibrating the three-dimensional finite element simulation.

After the finite element simulation adjustment and calibration the calculation kernel is put into operation alongside the monitoring data collection process. The monitoring system enables real-time assessment of operational defect formation and advance kinetics, while calculation analysis allows for receiving stress distribution prospect at any time, based on actual load parameters. As a result, it helps to assess he actual accumulated damage rate with the fixed number of operational load cycles, as well as provide justified forecast for durability and service life of the monitored object.

With that said, all load history is collected during monitoring. Also, all events, noted by the monitoring system are compared to current operational modes, their characteristics and technological features. This allows identifying the link between the defect rate and certain factors and parameters.

4 Monitoring algorithm

On Figure 4 you can find a general algorithm for the multi-parameter calculation and experimental monitoring. The main stages of the algorithm work are as follows. First, a three-dimensional finite element simulation model is developed and calculations for displacements and deformations in key nodes for main operational modes and various loading values are conducted. It is followed by the comparative analysis of calculated and experimental data with fixed coincidence error. In case of the successful result, the system is considered ready for monitoring, otherwise the necessary corrections are implemented for the finite element model (see Fig. 3). When the system is ready for operation, calculations for static strength and cyclic durability are performed consistent with the detected deficiencies and thermal power impact beyond the design basis. After that the causes of the latter are identified and compensatory actions are developed based on the set of condition-action rules.

5 Conclusion

The described system for uninterruptible monitoring is a multipurpose software and hardware complex for assessing the vulnerability of critical NPP equipment and predicting damage development at any physical impact parameters (i.e. pressure, temperature, vibration, movement, etc.) and is applicable in case of a beyond the design basis events.

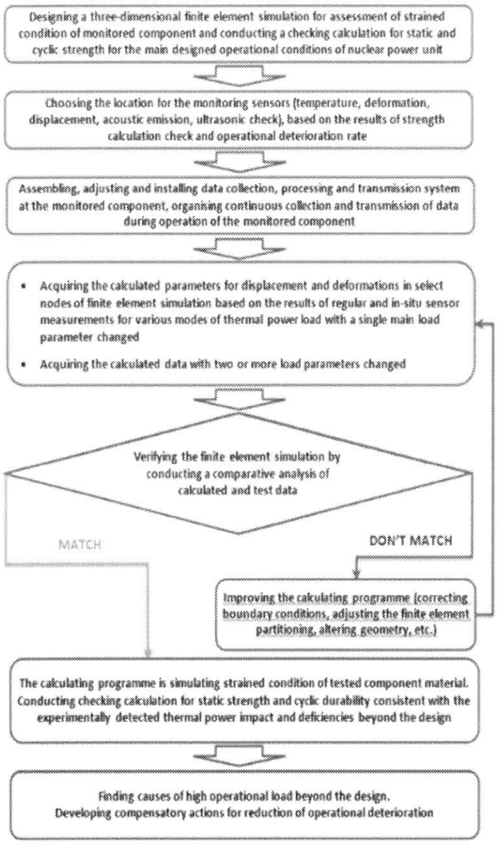

Fig. 4. Algorithm for the multi-parameter monitoring of condition and calculation analysis of operational deterioration rate.

The knowledge on operation of monitored equipment in various operation modes and corresponding analysis of the deflected mode and deficiency allow for the development of effective compensatory actions, aimed at reduction of main damaging factors' impact and improving durability of the monitored object. Moreover, it serves as the basis of control system data support for operation of both nuclear power units and automated repair systems. The use of set automated software and hardware complex in 2011 allowed to extend the service life of Novovoronezh Nuclear Power Plant Unit 5 on 30 years, as well as to assist the second service life extension of Unit 4 of the same power plant, which it currently undergoes.

References

1. V. Povarov, M. Bakirov, A. Danilov, *Automatic system of multi-parameter monitoring of nuclear power installations status* (Voronezh, Scientific book, 2017)

2. M. Bakirov, V. Povarov, A. Gromov, V. Levchuk, Izvestiya Wysshikh Uchebnykh Zawedeniy, Yadernaya Energetika, **3**, 15–24 (2014)

3. M. Bakirov, V. Povarov, D. Nikolaev, A. Gromov, V. Levchuk, S. Gorokhov, Izvestiya Wysshikh Uchebnykh Zawedeniy, Yadernaya Energetika, **4**, 5–10 (2014)

4. V. Povarov, M. Bakirov, A. Danilov, Voronezh technical university reporter, **6**, 58–63 (2017)

5. A. Danilov, V. Golovnev, *Digital control system* (Voronezh, VGLTA, 2007)

6. A. Danilov, *Microprocessor elements and devices of local automatic* (Voronezh, VGLTA, 2005)

7. A. Danilov, *Technical means of automation* (Voronezh, VGLTA, 2007)

8. A. Danilov, A. Pileich, *Software distributed computing heterogeneous dynamic system parameters in real time* (Voronezh, VGLTA, 2015)

9. V. Povarov, Nuclear Energy and Technology, **1**, 68–73 (2015)

10. M. Bakirov, V. Povarov, A. Gromov, V. Levchuk, Nuclear Energy and Technology, **1** (**1**), 32–36 (2015)

11. V. Povarov, M. Bakirov, Atomic Energy, **3**, 153–162 (2016)

A fuzzy adaptive sliding mode controller for uncertain nonlinear multi motor systems

Tran Xuan Tinh[1], *Pham Tuan* Thanh[1], *Tran Van* Tuyen[1], *Nguyen Van* Tien[1] and *Dao Phuong* Nam[2,*]

[1]Le Quy Don University (Military Technical Academy), Vietnam
[2]School of Electrical Engineering, Hanoi University of Science and Technology, Vietnam

Abstract. Multi-motor drive systems are nonlinear, multi-input multi-output (MIMO) and strong-coupling complicated system, including the effect of friction and elastic, backlash. They have been widely used in many modern industries. The control law for this dive system much depend on the determining of the tension being hard to obtain this tension in practice based on a load cell or a pressure meter due to the accuracy of sensors or external disturbance. An emerging proposed technique in the control law is the use of adaptive sliding mode control scheme to stabilize closed system. However, the control system would be affected by chattering phenomenon. In order to eliminate this term, fuzzy technique is proposed by adjusting equivalent coefficients. The theory analysis and simulation results point out the good performance of the proposed fuzzy adaptive sliding mode control for the drive system.

1 Introduction

Multi-motor drive systems have been investigated by many researchers in the recent times. The neural network technique based control law have been proposed by Yaoji Me et al. (2013) (see [1-9]). However, it is hard to find the equivalent networks as well as corresponding learning rules. Besides, the model of this system is approximately described as a linear system to use the transfer function to design the control law. Furthermore, the tracking ability or the stabilization of the whole system are not still solved under the effects of neural network based observer. In the multi-motor drive control systems, it is necessary to obtain the belt tension to design the suitable state feedback control law. However, it is hard to measure this belt tension based on sensors, ... and the high gain technique based observer is proposed in our work. Besides, the state feedback control design based on sliding mode control technique ensure that it is easy to remove efficient of disturbance and uncertainties. Therefore, an adaptive sliding mode controller is proposed to obtain tracking effectiveness. Moreover, fuzzy technique is considered to eliminate the chattering phenomenon disappearing by sliding mode control. The stability of closed system is obtained and verified by theory analysis, simulations.

2 Problem statements

Due to the effects by backlash and elastic (Fig. 1) and parameters (Table 1), we extend the model in [1] to obtain the following dynamic equation (2, 3) and the corresponding transfer function diagram (Fig. 2):

$$\begin{cases} \dot{\omega}_{r1} = \dfrac{1}{J_{L1}}\left[c_1.f_{11}(\Delta\varphi_1) + b_1\Delta\omega_1 f_{12}(\Delta\varphi_1) - (\mathrm{T}_{L1} + \mathrm{r}_1\mathrm{F})\right] \\ \dot{\omega}_{r2} = \dfrac{1}{J_{L2}}\left[c_2.f_{21}(\Delta\varphi_2) + b_2.\Delta\omega_2 f_{22}(\Delta\varphi_2) - (\mathrm{T}_{L2} + \mathrm{r}_2\mathrm{F})\right]. \\ \dot{F} = C_{12}\left[r_1\omega_{r1} - r_2\omega_{r2}(1 + \dfrac{1}{C_{12}J}\mathrm{F})\right] \end{cases} \quad (1)$$

Table 1. Parameters of a Multi-Motor System.

$K = E/V$	Transfer function
E	Young's Modulus of belt
V	Expected line velocity
$T = \dfrac{L_0}{AV}$	Time constant of tension variation
L_0, A	Distance between racks, Section area (m^2)
n_{pr}	Number of pole-pairs in the ith Motor
J_1, J_2, J_{L1}, J_{L2}	Inertia moment of Motors and Loads (kgm^2)
T, T_L, φ_r	Motor, Load torque (Nm), Flux of rotor (Wb)
L_r	Self-induction of rotor (H)
c_1, c_2, b_1, b_2	Stiffness and friction coefficient
$\Delta\omega_1, \Delta\omega_2$	The errors of angle speed in presence of backlash, elastic

We denote:

* Corresponding author: nam.daophuong@hust.edu.vn

$$\begin{aligned} u_{d1} &= \frac{1}{J_{L1}}\left[c_1 f_{11}(\Delta\varphi_1) + b_1\Delta\omega_1 f_{12}(\Delta\varphi_1)\right]; \\ u_{d2} &= \frac{1}{J_{L2}}\left[c_2 f_{21}(\Delta\varphi_2) + b_2\Delta\omega_2 f_{22}(\Delta\varphi_2)\right] \end{aligned} \quad (2)$$

are components including backlash and elastic to obtain the following equations:

$$\begin{cases} \dot\omega_{r1} = -c_1\omega_{r1} - \dfrac{1}{J_{L1}}\mathrm{T}_{L1} - \dfrac{1}{J_{L1}}\mathrm{r}_1\mathrm{F} + u_{d1} \\ \dot\omega_{r2} = -c_2\omega_{r2} - \dfrac{1}{J_{L2}}\mathrm{T}_{L2} - \dfrac{1}{J_{L2}}\mathrm{r}_2\mathrm{F} + u_{d2}. \\ \dot F = C_{12} r_1\omega_{r1} - C_{12} r_2\omega_{r2} - \dfrac{1}{l}\omega_{r2}\mathrm{F} \end{cases} \quad (3)$$

The model is described by eq. (3) belongs to the class of nonlinear systems as follows:

$$\begin{cases} \dot x(\mathrm{t}) = A(\mathrm{x},\mathrm{t})x + B(\mathrm{x},\mathrm{t})u + D(\mathrm{x},\mathrm{t}) \\ \mathrm{y} = x \end{cases}, \quad (4)$$

where

$$A = \begin{bmatrix} k_1 & 0 & k_2 \\ 0 & k_4 & k_5 \\ 0 & 0 & k_7 x_2 \end{bmatrix}; B = \begin{bmatrix} k_3 & 0 & 0 \\ 0 & k_6 & 0 \\ 0 & 0 & k_8 x_1 + k_9 x_2 \end{bmatrix};$$

$$D = \begin{bmatrix} u_{d1} \\ u_{d2} \\ 0 \end{bmatrix}$$

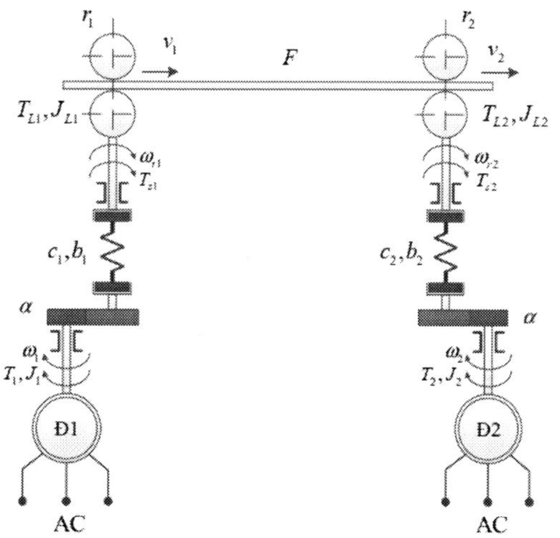

Fig. 1. The Two-Motor Drive System.

Remark 1:

The dynamic equations (1,2, 3) and Figures 1, 2 are described by the effect of friction, backlash, elastic and pointed out the nonlinear property of multi-motor systems.

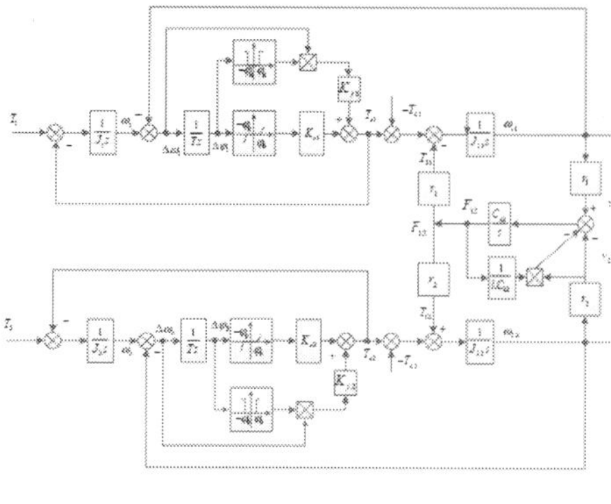

Fig. 2. The corresponding transfer-function diagram of the Two-Motor Drive System.

The control objective is to find the control input vector $u^T = [\mathrm{T}_{L1} \quad \mathrm{T}_{L2} \quad C_{12}]$ to obtain that the desired value are tracked by $x^T = [\omega_{r1} \quad \omega_{r2} \quad F]$ in presence of friction and elastic.

The following assumptions must be satisfied in order to design the control law:

Assumption 1: $x^T = [\omega_{r1} \quad \omega_{r2} \quad F]$ is measurable;

Assumption 2: There are real positive numbers $\omega_{r1\max}, \omega_{r2\max}, T_{L1\max}, T_{L2\max}, u_{d1\max}, u_{d2\max}$ such that $\omega_{r1}, \omega_{r2}, T_{L1}, T_{L2}, u_{d1}, u_{d2}$ are bounded by these values;

3 Fuzzy adaptive sliding mode control design

In this section, the main work is to find a state feedback control law based on the adaptive sliding mode control technique for the class of multi motor systems.

The proposed control law based on the following theorems as follows:

Theorem 1: The adjusting mechanism (5):

$$\begin{cases} \dot\theta_a^t = \eta_a\alpha\psi_a(\mathrm{x})\mathrm{s}, \\ \dot\theta_b^t = \eta_b\alpha\psi_b(\mathrm{x})\mathrm{s}u_{eq}, \end{cases} \quad (5)$$

with $s = \alpha e + \xi$,

$$\begin{aligned} u_{eq}(\mathrm{t}) &= \hat B^{-1}(\mathrm{x})\left(-\hat A(\mathrm{x}) + \dot x_d - \beta\,\mathrm{sgn}(\mathrm{s})\right) = \\ u_{eq} &= \hat B^T(\mathrm{x},\theta_b^t)\left[\varepsilon_0 I_m + \hat B(\mathrm{x},\theta_b^t)\hat B^T(\mathrm{x},\theta_b^t)\right]^{-1}; \\ &\left(-\hat A(\mathrm{x},\theta_a^t) + \dot{\mathrm{x}}_d - \beta\,\mathrm{sgn}(\mathrm{s})\right) \end{aligned} \quad (6)$$

$$\begin{aligned} A(x) &\approx \hat A(x,\theta_a^t) = \theta_a\psi_a(\mathrm{x}) = \sum_{i=1}^{p_a}\theta_{ai}\psi_{ai}(\mathrm{x}); \\ \mathrm{B}(x) &\approx \hat{\mathrm{B}}(x,\theta_b^t) = \theta_b\psi_b(\mathrm{x}) = \sum_{i=1}^{p_b}\theta_{bi}\psi_{bi}(\mathrm{x}). \end{aligned} \quad (7)$$

ensure some results as follows:

1. All signals of closed loop will be bounded and θ_a, θ_b converge to

$$\theta_a^* = \arg\min_{\theta_a}\left\{\sup_{x\in\Omega}\left|A(x,t)+D(x,t)-\hat{A}(x,\theta_{at})\right|\right\}$$

$$\theta_b^* = \arg\min_{\theta_b}\left\{\sup_{x\in\Omega}\left|B(x,t)-\hat{A}(x,\theta_{at})\right|\right\}$$

as $t \to \infty$

2. If $A^*(x,\theta_a^*) = \hat{A}(x,\theta_a^t); B^*(x,\theta_b^*) = \hat{B}(x,\theta_b^t)$ then the errors converge to zero in finite time;

3. If $A^*(x,\theta_a^*) \neq \hat{A}(x,\theta_a^t); B^*(x,\theta_b^*) \neq \hat{B}(x,\theta_b^t)$ then closed loop system converges to the neighborhood of sliding surface in finite time;

Proof:

The Lyapunov candidate function is selected as follows:

$$V = \frac{1}{2}s^2 + \frac{1}{2}\frac{1}{\eta_a}\tilde{\theta}_a^2 + \frac{1}{2}\frac{1}{\eta_b}\tilde{\theta}_b^2 \tag{8}$$

where $\tilde{\theta}_a = \theta_a - \theta^*_a; \tilde{\theta}_b = \theta_b - \theta^*_b$, we have:

$$\dot{V} = s\dot{s} + \tilde{\theta}_a\dot{\tilde{\theta}}_a + \tilde{\theta}_b\dot{\tilde{\theta}}_b \tag{9}$$

and:

$$\dot{s} = \alpha\dot{e} = \alpha(\dot{x}-\dot{x}_d)$$

$$= \alpha\begin{pmatrix} A(x,t)+B(x,t)u(t)+D(x,t) \\ -\hat{A}(x,t)-\hat{B}(x,t)u_{eq}(t)-\beta\,\mathrm{sgn(s)} \end{pmatrix}$$

$$= \alpha\begin{pmatrix} \left(A(x,t)+D(x,t)-\hat{A}(x,t)\right) \\ +\left(B(x,t)-\hat{B}(x,t)\right)u_{eq}(t)-B(x,t)u_c(t)-\beta\,\mathrm{sgn(s)} \end{pmatrix}$$

However, we have the relation:

$$\varepsilon_a(x,t)+A^*(x,\theta_a^*)-\hat{A}(x,\theta_a) = A(x,t)+D(t)-\hat{A}(x,\theta_a)$$

$$\varepsilon_b(x,t)+B^*(x,\theta_b^*)-\hat{B}(x,\theta_b) = B(x,t)-\hat{B}(x,\theta_b)$$

Therefore, we obtain:

$$\dot{s} = \alpha\begin{bmatrix} \left(\varepsilon_a(x,t)+A^*(x,\theta_a^*)-\hat{A}(x,\theta_a)\right) \\ +\left(\varepsilon_b(x,t)+B^*(x,\theta_b^*)-\hat{B}(x,\theta_b)\right)u_{eq}(t) \\ -B(x,t)u_c(t)-\beta\,\mathrm{sgn(s)} \end{bmatrix} \tag{10}$$

On the other side,

$$A^*(x,\theta_a^*)-\hat{A}(x,\theta_a) = \psi_a(\theta_a^*-\theta_a) = -\psi_a\tilde{\theta}_a$$
$$B^*(x,\theta_b^*)-\hat{B}(x,\theta_b) = \psi_b(\theta_b^*-\theta_b) = -\psi_b\tilde{\theta}_b$$

and we obtain:

$$\dot{s} = \alpha\begin{bmatrix} \left(\varepsilon_a(x,t)-\psi_a\tilde{\theta}_a\right)+\left(\varepsilon_b(x,t)-\psi_b\tilde{\theta}_b\right)u_{eq}(t) \\ -B(x,t)u_c(t)-\beta\,\mathrm{sgn(s)} \end{bmatrix}$$

and:

$$\dot{V} = s\alpha\begin{bmatrix} \left(\varepsilon_a(x,t)-\psi_a\tilde{\theta}_a\right)+\left(\varepsilon_b(x,t)-\psi_b\tilde{\theta}_b\right)u_{eq}(t) \\ -B(x,t)u_c(t)-\beta\,\mathrm{sgn(s)} \end{bmatrix} +$$

$$+\tilde{\theta}_a\eta_a\alpha\psi_a(\mathrm{x})\mathrm{s}+\tilde{\theta}_b\eta_b\alpha\psi_b(\mathrm{x})\mathrm{su}_{eq}$$

$$\dot{V} = s\alpha\left(\varepsilon_a(x,t)+\varepsilon_b(x,t)u_{eq}(t)-B(x,t)u_c(t)-\beta\,\mathrm{sgn(s)}\right) =$$

$$= s\alpha\left(\varepsilon_a(x,t)+\varepsilon_b(x,t)u_{eq}(t)-\bar{\varepsilon}_a-\bar{\varepsilon}_b\left|u_{eq}\right|-\beta\,\mathrm{sgn(s)}\right) \leq$$

$$\leq -\alpha s\beta\,\mathrm{sgn(s)}$$

where $\alpha > 0; \mathrm{s.sgn(s)} \geq 0$ and we obtain $\dot{V} \leq 0 \quad \forall x \in \Omega$.

Theorem 2: The control input $u(t) = u_{eq}(t) - u_c(t)$;

$$u_{eq} = \hat{B}^T(\mathrm{x},\theta_b^t)\left[\varepsilon_0 I_m + \hat{B}\;(\mathrm{x},\theta_b^t)\hat{B}^T(\mathrm{x},\theta_b^t)\right]^{-1}$$

$$\left(-\hat{A}(\mathrm{x},\theta_a^t)+\dot{\mathrm{x}}_d-\beta\,\mathrm{sgn(s)}\right);$$ with

$$u_c(t) = B^{-1}s\left(\bar{\varepsilon}_b\left|u_{eq}\right|+\bar{\varepsilon}_a\right)$$

$s = \alpha e + \xi$ ensure the nonlinear system (6) stability in finite time.

Proof:

We obtain the result:

$$s^T\dot{s} = s^T\alpha\begin{bmatrix} \left(\varepsilon_a(x,t)-\psi_a\tilde{\theta}_a\right)+\left(\varepsilon_b(x,t)-\psi_b\tilde{\theta}_b\right)u_{eq}(t) \\ -B(x,t)u_c(t)-\beta\,\mathrm{sgn(s)} \end{bmatrix}$$

$$\approx s^T\alpha\begin{pmatrix} \varepsilon_a(x,t)+\varepsilon_b(x,t)u_{eq}(t) \\ -\bar{\varepsilon}_a-\bar{\varepsilon}_b\left|u_{eq}\right|-\beta\,\mathrm{sgn(s)} \end{pmatrix}$$

$$\leq -s^T\alpha\beta\,\mathrm{sgn(s)}$$

$$\leq -\|s\|\alpha\beta\,\mathrm{sgn(s)}$$

Therefore, $s = \alpha e + \xi \to 0$ in finite time

Remark 1. it is necessary to ensure that the time of convergence of sliding surface is finite. The fact is described based on the following example:

We consider the system as follows:

$$\begin{cases} \dfrac{dx}{dt} = Ax + Bs \\ \dfrac{ds}{dt} = Cx + Ds \end{cases}$$

where: $A \in \mathbb{R}^{n\times n}, B \in \mathbb{R}^{n\times r}, C \in \mathbb{R}^{r\times n}, D \in \mathbb{R}^{r\times r}$, s is the sliding surface. Selecting A is Hurwitz matrix and $\begin{bmatrix} A & B \\ C & D \end{bmatrix}$ is not Hurwitz matrix. We obtain that although s converges to 0 in infinite time, x does not converge to 0.

Remark 2. The proposed control law ensure that the system trajectory converge to sliding surface in finite time.

Remark 3. We almost utilized theorem 1 to design adaptive sliding mode control technique for multi motor systems.

In order to eliminate the effect of chattering phenomenon, fuzzy technique (by Tagaki – Sugeno – Kang) would be proposed to adjust the coefficient α depending on the sliding surfaces s and $\dot{s}$, table 2, 3 and figure 3:

Table 2. Rule Matrix of control.

		s		
		N	Z	P
$\dot{s}$	N	B	M	B
	Z	B	S	B
	P	B	M	B

Table 3. Properties of controller.

AND method	MIN
OR method	MAX
Implication	MIN
Aggregation	MAX
Defuzification	Weighted average

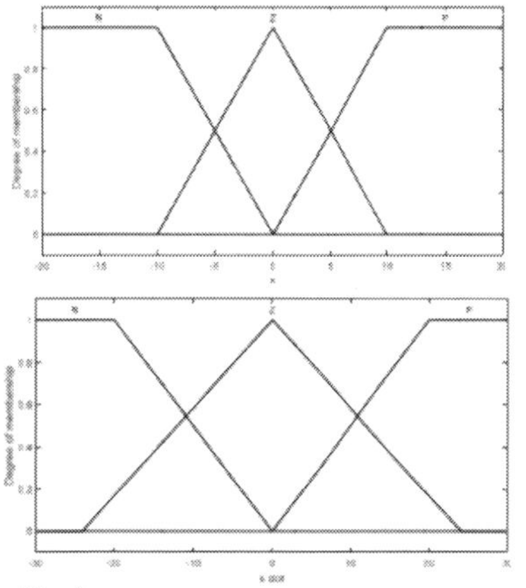

Fig. 3. Fuzzification.

4 Simulation results

In this section, we consider several simulation results to demonstrate the effectiveness of the proposed sliding mode control law based on the two-motor system having parameters as follows:

$$n_{p1} = n_{p2} = 4, J_1 = J_2 = 500Kgm^2, L_{r1} = 0.2H,$$
$$L_{r2} = 0.3H, \omega_{r1d} = \omega_{r2d} = 700v / p, F_d = 250N.$$

Figures 3, 4 show the tracking performance behaviour of velocity based on fuzzy adaptive sliding mode control law in presence of disturbance (figures 4, 5, 8). Figures 6, 7 show the high tracking performance behaviour of velocity based on adaptive sliding mode control law without disturbance.

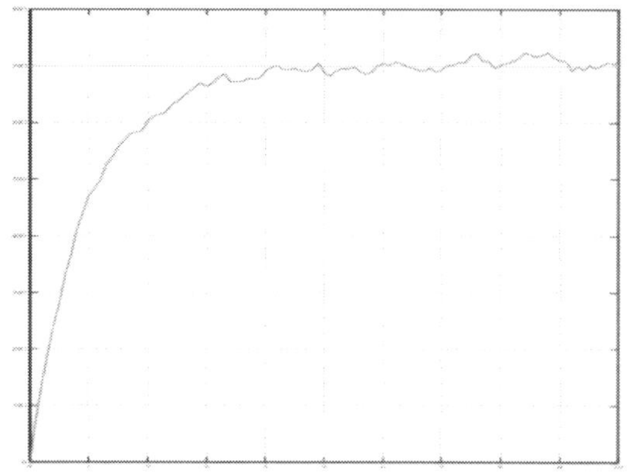

Fig. 4. The behaviour of the first motor's speed in presence of disturbance.

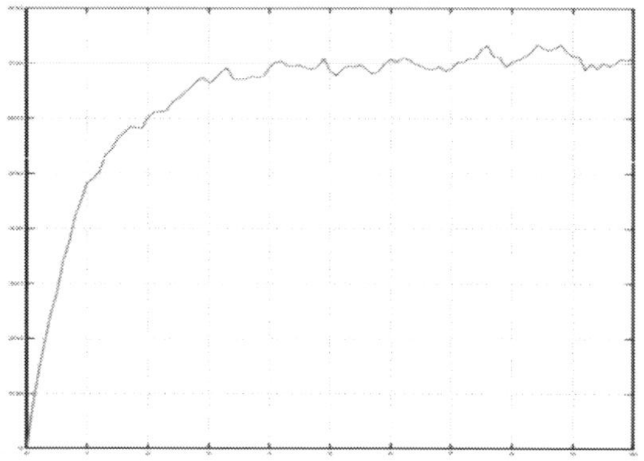

Fig. 5. The behaviour of the second motor's speed in presence of disturbance.

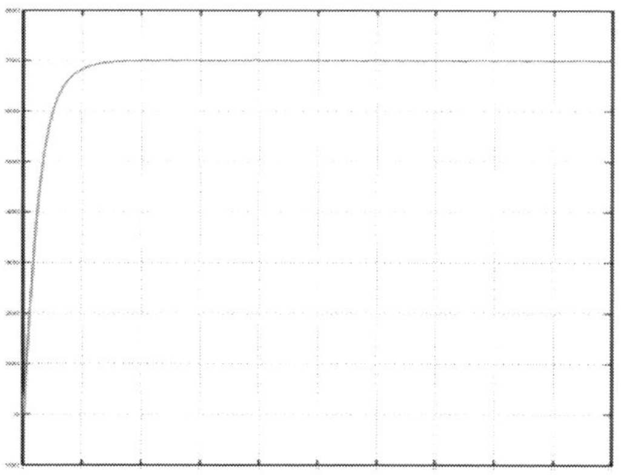

Fig. 6. The behaviour of the first motor's speed without disturbance.

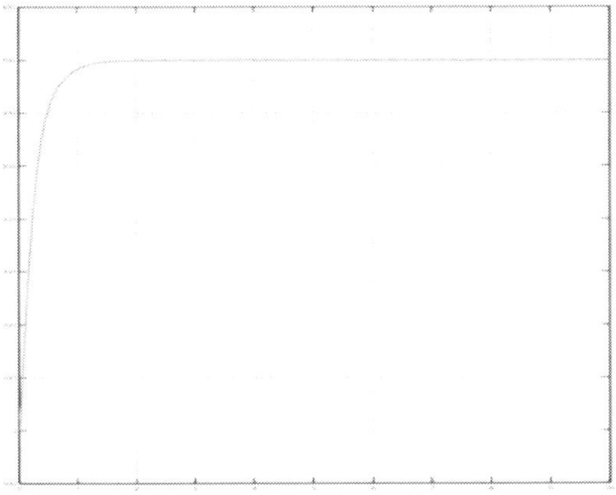

Fig. 7. The behaviour of the second motor's speed without disturbance.

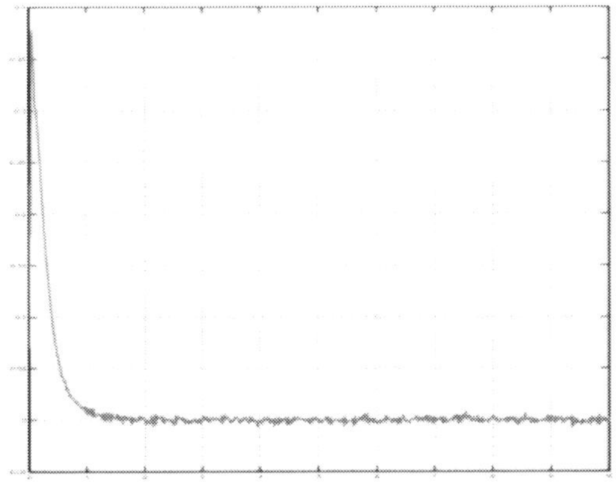

Fig. 8. The speed error between 2 motors.

5 Conclusion

This paper described a fuzzy adaptive sliding mode control law the two-motor system in presence of elastic and backlash, friction. The effectiveness of the proposed control scheme was pointed out by theoretical analysis and simulation results.

References

1. Y Mi et al., Proc. The 2013 International Conference on Electrical Machines and Systems, 2282–2285 (2013)
2. J Zhang et al, *Proc.* The IEEE International Conference on Intelligent Computing and Intelligent Systems, 178–182 (2009)
3. G Liu et al., Proc. The IEEE International Conference on Networking, Sensing and Control, 1476–1479 (2008)
4. L Jinmei et al, Proc. The IEEE International Conference on Industrial Technology, 1–6 (2008)
5. N.T.T. Vu, D. Yu, H.H. Choi, J.-W. Jung, IEEE Trans. Industrial Electronics, **60** (10), 4281–4291, (2013)
6. M. Zhihong, A.P. Paplinski, H.R. Wu, IEEE Trans. Autom. Control, **39**, 2464–2469 (1994)
7. J.J.E. Slotine, W. Li, *Applied Nonlinear Control* (Prentice Hall, New Jersey, 1991)
8. S. Labiod, M.S. Boucherit, T.M. Guerra, Fuzzy Sets Syst., **151**, 59–77 (2005)
9. V. Nekoukar, A. Erfanian, Fuzzy Sets Syst., **179**, 34–49 (2011)

Impact of the face angle to traveling trajectory during the riding standing-type personal mobility device

Jeyeon Kim[1], *Kenta* Sato[1], *Naohisa* Hashimoto[2], *Alexey* Kashevnik[3,4,*], *Kohji* Tomita[2], *Seiichi* Miyakoshi[2], *Yusuke* Takinami[2], *Osamu* Matsumoto[2] and *Ali* Boyali[2]

[1]Department of Creative Engineering, National institute of Technology, Tsuruoka College, 997-8511 Tsuruoka Yamagata, Japan
[2]Intelligent System Research Institute, National institute of Advanced Industrial Science and Technology, 305-8560 Tsukuba Ibaraki, Japan
[3]SPIIRAS, 39, 14th Line, St. Petersburg 199178, Russia
[4]ITMO University, 49 Kronverksky Pr, St. Petersburg, Russia

Abstract. In this paper, we investigate the impact of face direction during traveling by Standing-Type Personal Mobility Device (PMD). The use of PMD devices has been a popular choice for recreational activities in the developed countries such as in the USA and the countries in Europe. These devices are not completely risk free and various accidents have been reported. Since that, the risk factors leading to accidents have to be investigated. Unfortunately, the research studies on the risk factors on riding PMD devices have not been matured as much as the studies on driving cars. In this paper, we evaluate the impacts of face angle on travelling trajectory during travelling in a PMD. We showed by experiments that, the face direction is an important factor in risk assessment for traveling by a PMD.

1 Introduction

In the recent years, researches on personal mobility devices (PMD) have been actively conducted, as these devices are promising alternatives in solving the first and last mile. The PMDs can be classified into standing type PMDs, two-seater ultra-small devices and electric bicycles. Among them, the standing type PMDs have gained a great deal reputation due to their unique motion capabilities. However, the standing type PMDs have not been allowed in the public places due to the lack of regulations, insufficient research studies for assessment of their safety for policy makers. In Japan, traveling by a PMD is prohibited as well.

To enable the research on PMDs and their interaction with pedestrian and other entities in the public places, Japan has designated a special zone in which the PMDs are allowed. Various studies such as social acceptability, attitude towards them, safety, usability evaluation of the standing-type PMD have been published [1, 2]. Despite the fact that there have been frequent accidents caused by rider's behaviors reported due to the trends in increasing use of PMDs in the USA, and Europe [3, 4], inattentive operation and distracted ride profiles [5] have been less studied in the literature up to date.

Distracted driving is any activity that diverts attention from driving, including talking or texting on smartphone, eating, drinking, talking to other people, fiddling with the vehicle infotainment or navigation system [5]. That is, it indicates a driving operation in a state in which the rider does not concentrate on the operation of the device or concentrate on other activities than the required for the ride. From the research [8], it was found that the safety education contributes to reduce distracted driving [8]. Various studies have been also made on stick driving, dozing driving, etc. to prevent distracted driving.

The rest of the paper is organized as follows. Section 2 describes the mobile application developed for rider face tracking. Section 3 presents experiments methodology. Section 4 describes the results of the experiments. Finally, the conclusion summarize the paper.

2 Related work

The section presents research and development in the area of mobile applications utilisation for riders and drivers assistance.

The paper [9] describes the CarSafe driver safety application for Android phones that detects and alerts drivers to dangerous driving conditions and behaviour. It uses computer vision and machine learning algorithms on the phone to monitor and detect whether the driver is tired or distracted using the front-facing camera, while at the same time tracking road conditions using the rear-facing camera. CarSafe also tries to solve the problem of processing video streams from both the front and rear cameras simultaneously by using a context-aware algorithm. This app has two main disadvantages. The first problem is a lack of emotions, gestures and speech recognition. The second one is that it is not available in any application store.

* Corresponding author: alexey@iias.spb.su

The paper [10] presents a DriveSafe driver safety application for iPhones that detects inattentive driving behaviors and gives corresponding feedback to drivers, scoring their driving and alerting them in case their behaviors are unsafe. It uses computer vision and pattern recognition techniques on the iPhone to assess whether the driver is drowsy or distracted using the rear-camera, the microphone, the inertial sensors and the GPS.

The paper [11] describes WalkSafe application for Android-based smartphone that aids people that walk and talk, improving the safety of pedestrian mobile phone users. WalkSafe uses the back camera of the mobile phone to detect vehicles approaching the user, alerting the user of a potentially unsafe situation; more specifically This application uses machine learning algorithms implemented on the phone to detect the front views and back views of moving vehicles and exploits phone APIs to save energy by running the vehicle detection algorithm only during active calls. WalkSafe alerts the user of unsafe conditions using sound and vibration from the phone.

The paper [12] provides comparative analysis of the behaviors of non-professional and professional drivers on the basis of the undertaken individual studies. The results showed the correlation between non-professional drivers and ordinary and aggressive violations and errors, while professional drivers were associated with positive behaviors.

Authors of the paper [13] examines older people driving behavior which includes road selection, left/right turn and driving speed. There are 108 participants has been participated in the study. Authors confirmed that older drivers are more reluctant to drive on expressway than other participants for long trips. Considering that driving in non-expressway with long time may cause frequently facing complex and unsafe conditions, more efforts should be carried out to promote and train expressway driving among older drivers.

The paper [14] describes a tracking system that monitors facial feature and create 3D geometric reasoning. The system has three main components: robust facial feature tracking, head pose and gaze estimation, and 3D geometric reasoning to detect Eyes Off the Road state.

The system proposed authors in [15] is the one that determines dangerous states by sound. Authors present silence removal approach using short term energy and zero crossing rate prior to extracting the relevant features in order to reduce the computational time in a vehicular environment.

3 Rider face tracking

Modern smartphones are equipped with front-facing camera and variety of built-in sensors such as an accelerometer, a gyroscope, an ambient light sensor, a proximity sensor, a magnetic field sensor, and GPS, that are capable to measure parameters that are needed determine dangerous states.

The authors proposed the rider assistance system using a personal smartphone for PMDs [16]. The evaluation of this application was reported in [17] using a traveling simulator. In the paper [16] the reference model of a two-wheeled self-balancing vehicles rider assistant system is presented. The model consists of five main modules: mobile application, cameras, sensors, local database and cloud service. These modules allow the system to recognize rider and vehicle behaviour and producing alerts and warnings when dangerous situations are detected. The authors also detailed appropriate ontologies for the rider and vehicle behaviour recognized as well as the Segway motion detection algorithms using an Android-based smartphone. The cloud service is used for keeping statistics of the system utilisation.

At the moment the rider assistance system determines two dangerous states: drowsiness and distraction. The smartphone's front-facing camera monitors the head movements, facial expressions, and the prolonged and frequent eye blinks that indicates the micro sleep. Visual cues relevant to the drowsiness state are percentage of closure of eyelid (PERCLOS), eye blink time, eye-blinking rate, eye gaze, pupil movement and eyelid movement. Distraction dangerous state is related to maintaining the eye contact with the road. Distraction occurs when drivers divert their attention away from the driving task to focus on another activity instead.

Using the smartphone the rider assistant system generates alerts for the driver using vibration, audible signals, and visual information in according with the proposed recommendation schemes. The rider assistant system provides statistics about the driver's behaviour on his/her route. It tracks such parameters as recognized dangerous states, time, location, speed, acceleration, driver's face parameters, and other information from smartphone sensors. During the experiments presented in the paper the face angle has been measured and saved in the cloud service. Table 1 shows the example statistics for distraction dangerous state. The statistics is stored in the cloud systems that provides possibility to use the powerful resources to analyses it. The use cases for rider assistant mobile application evaluation using a ride simulator is introduced in reference [17]. The ride simulator has been proposed and briefly explained the concept, system configuration, employed mobility, and preliminary experiments.

In these studies, the impact of the face direction angle on ride trajectory during a ride has not been taken into account, therefore, we contribute these studies by investigating the effects of face direction angle based on the developed by authors [16] mobile application that has been adapted to us for PMD. Although there are rider assistance systems being studied, there has been no risk assessment from rider's perspective for PMD use. In this paper, we examine the risks associated by the riders by analysing the influence of face direction using a smartphone application.

Table 1. Statistics example accessible after the rider assistance system.

Date/Time	Latitude	Longitude	Acceleration (m/s^2)	Head angle	Speed (km/h)	City/Country	Light level (Lux)
27/04/17 18:38:08:048	59,94228254	30,262599	11,16	-22,51	12	City	116
13/05/17 16:06:32:875	56,18414201	36,9833062	2,1	20,43	26	City	630
13/05/17 16:08:35:733	56,19890477	36,95948511	2,09	-26,25	64	City	565
13/05/17 16:08:54:948	56,20137014	36,95550604	4,08	35,21	80	Country	465
13/05/17 16:36:43:879	56,38791051	36,66605464	9	32,82	67	Country	731

4 Methodology of experiments

In the experiments a Segway PMD [11] and collected position information by Total Station range sensors by Lacia Geosystems (TM, R) is used. The face direction angle of the subjects during the rides using the smartphone application presented in [9] is recorded. An Android-based smartphone was attached on the Segway as shown in Figure 1 and was synchronized with Total Station. The time synchronization error was measured around one second. The experiments were carried out in a gymnasium of Tsuruoka College to eliminate the effects of bumpy road and irregular road surfaces on the trajectories.

The experimental method is shown in Figure 2. The subjects gaze at the target and travel straight from the start point to the goal. Four subjects participated in the experiments who are between 19 and 22 years old. Each of the subjects repeated the experiment seven times with a speed of 4 mile/h. The starting position is chosen as the origin in the experiments.

Fig. 1. Installation of Smartphone.

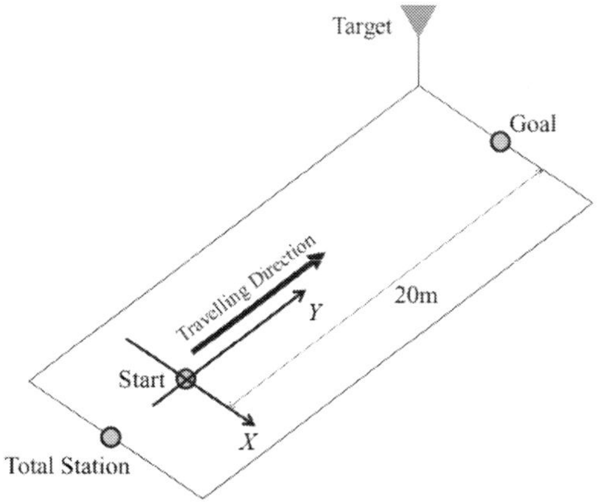

Fig. 2. Method of experiments.

Fig. 3. Scene of experiments.

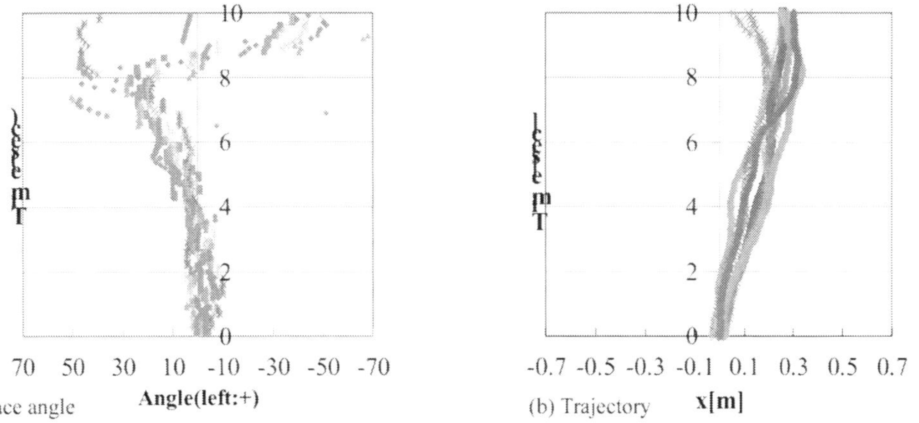

Fig. 4. A result of travelling experiment (Subject 1).

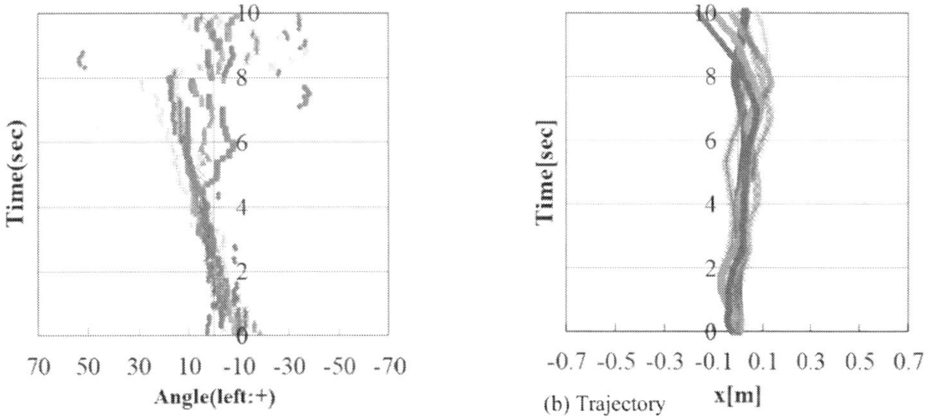

Fig. 5. A result of travelling experiment (Subject 4).

The subjects travel along the y axis. The x-axis shows the lateral direction (Fig. 2.). The positive face direction angle (clockwise) is towards to the right side of the subjects.

5 Results

The inputs to the assessments are the face direction angle and trip trajectory. Figures 4 and 5 show the face angle and the traveling trajectory in the experiments. In Figure 4 (a), the horizontal axis is the face angle and the vertical axis is time. In Figure 4 (b), the horizontal axis is the position in the x direction and the vertical axis is the time.

As shown in Figure 4(a), it can be seen that the face angle increases as the user gazes while watching the target. Then, it shows that the deviation (error) in the x direction increases as the face angle increases (Fig. 4 (b)). That is, it can be seen that a deviation occurs in the opposite direction to the gaze direction. Three of the subjects showed this tendency. Figure 5 shows an example in which the deviation in the x direction is not as much large as that of other participants while the face direction angle is increasing.

The experiments show that, there is a strong correlation between the face direction angle and the direction of the traveling trajectory. The deviation of the PMDs from the intended trajectory while the face direction changing create suspicion that the riders change the body center of gravity without realizing it while gazing at the target.

6 Conclusions

In this paper, we investigated the influence on the traveling trajectory by the angle of the rider face during travelling for risk assessment in PMD. Specifically, we conducted a running test in the indoor, and examined the relationship between the face angle during running and the traveling locus. As the experimental results show, the larger the face angle, the more the deviation shows that the deviation of the traveling locus becomes larger. In addition, it showed that the deviation of the running locus occurred in the direction opposite to the direction the subject was closely watching. These results showed

* Corresponding author: alexey@iias.spb.su

how distracted driving is dangerous and causes the risks of undeliberate trajectory changing.

There are outdoor experiments as future tasks. Indoor experiments were carried out without the influence of the slope of the road surface in this paper, but it is necessary to evaluate the risk caused by distracted driving outdoors.

This work was supported by JSPS KAKENHI Grant Number 16K18057, project funded by grant # 17-29-03284 of the Russian Foundation for Basic Research. The work has been partially financially supported by the Russian State Research # 0073-2018-0002, and by Government of Russian Federation, Grant 074-U01.

References

1. N. Hashimoto, K. Tomita, A. Kamimura, Y. Takinami, O. Matsumoto, International Internet of Things Summit, 157–162 (2014)
2. N. Sato, J. Kim, N. Hashimoto, K. Tomita, O. Matsumoto, A. Boyali, The First IEEE International Smart Cities Conference (2015)
3. J. Ashurst, B. Wagner, Western Journal of Emergency Medicine, **16** (5), 693 (2015)
4. K. Boniface, M. McKay, R. Lucas, A. Shaffer, N. Sikka, Annals of Emergency Medicine, **57** (4), 370–374 (2011)
5. Department of Transportation. What is Distracted Driving. URL: http://www.distraction.gov/content/get-the-facts/ facts-and-statistics.html (2012)
6. National Center for Statistics and Analysis, Distracted Driving: 2014 Data, Traffic Safety Research Notes (2015)
7. Research Bulletin of National Solid Wastes Management Association, Distracted While Driving, https://wasterecycling.org/images/documents/resources/Research-Bulletin-Distracted-Driving.pdf
8. Traffic Safety Facts, Training Novice Drivers to Shorten Distraction Time, https://ntl.bts.gov/lib/61000/61600/61653/tt387.pdf
9. C. You et al, Proceedings of MobiSys'13, 13–26 (2013)
10. L. Bergasa, D. Almería, J. Almazán, J. Yebes, R. Arroyo, IEEE Intelligent Vehicles Symposium, 240–245 (2014)
11. T. Wang, G. Cardone, A. Corradi, L. Torresani, A. Campbell, Proceedings of the twelfth workshop on mobile computing systems & application, 5 (2012)
12. M. Maslac, B. Antic, K. Lipovac, D. Pešic, N. Milutinovic, *Transportation Research Part F*, **52**, 101–111 (2018)
13. Y. Zhao, T. Yamamoto, T. Morikawa, Traffic and Transportation Engineering (to be published)
14. F. Vicente, Z. Huang, X. Xiong, F. De La Torre, W. Zhang, D. Levi, IEEE Transactions on Intelligent Transportation Systems, **16** (4), 2014–2027 (2015)
15. N. Kamarud, A. Wahab A. Rahman, K. Ikhwan Mohamad Halim, International Conference on Informatics and Computing (ICIC), 186–191 (2016)
16. A. Smirnov, A. Kashevnik, I. Lashkov, N. Hashimoto, A. Boyali, Proceedings of the 17th conference of FRUCT Association, 201–209 (2015)
17. N. Hashimoto et al, 2016 19th Conference of Open Innovations Association (FRUCT) , 47–53 (2016)
18. Segway Japan, http://www.segway-japan.net.

A dataset of head and eye gaze during dyadic interaction task for modeling robot gaze behavior

Mirko Raković[1,2,*], *Nuno* Duarte[1], *Jovica* Tasevski[2], *José* Santos-Victor[1] and *Branislav* Borovac[2]

[1] University of Lisbon, Instituto Superior Tecnico, Institute for Systems and Robotics, Av. Rovisco Pais, 1049-001, Lisboa, Portugal
[2] University of Novi Sad, Faculty of Technical Sciences, Trg Dositeja Obradovića 6, Novi Sad, Serbia

Abstract. In this work is presented a dataset of humans' head and eye gaze acquired with Pupil Labs gaze-tracking glasses and Optitrack motion capture system. The dataset contains recordings of adult subjects in dyadic interaction task. During the experiment, the subjects are asked to pick up an object and, based on the randomly defined instructions, to place it on the table in front of her/him or to give the object to a person sitting across the table. If the object is handed over, the second person takes the object and places it on the table it in front of her/him. The dataset is intended to be used to model the behavior of the human's gaze while interacting with another human and implement the model in a controller of a robot for dyadic interaction with a humans.

1 Introduction

Understanding human movements, actions, and intentions are important when two actors (human or robot) share a common workspace. In a human-human interaction (HHI) this process is called nonverbal communication in a dyadic scenario. This form of nonverbal communication is partially enabled by action observation. One of the main cues in action understanding is gaze which can influence observer's selection of motor activities. This explains why the attention should be paid to proper modeling of the robot's gaze behavior that needs to be easily understandable by humans.

Nonverbal signals are an integral part of all our communicative endeavors. In some cases, they are the most significant part of our message. In HHI, communication through nonverbal channels also influences the coordination of joint activity. We believe this is the same in the case of human-robot interaction (HRI). So the design of readable human-like behavior will support efficient and robust teamwork. Motivated by a desire to develop effective robot teammates for people, one of our goals is to model human-like gaze behavior as an integral part of nonverbal communication in HRI.

In [1] is shown that implicit nonverbal communication positively impacts human-robot task performance with respect to efficiency, and robustness to errors that arise from miscommunication. Authors in [2, 3] validated the use of gaze in addition to body pose cues as means of predicting human action. Thus understanding and effectively using nonverbal cues is of great importance to success for robots in dyadic tasks.

The initial step to model gaze behavior is to acquire quantitative data of humans' movement during dyadic interaction task. For this purpose, we prepared the experiment with two humans that have to interact in order to accomplish their tasks (Fig. 1.). In this experiment, we measured head and eye gaze.

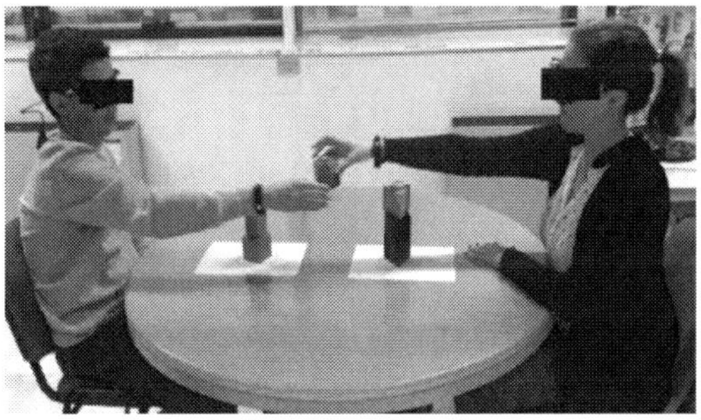

Fig. 1. Interaction in dyadic scenario

Paper [4] presents a data set that captures the gaze patterns of humans solving a recognition task. Eye movements were recorded using static gaze tracker, and the participants were looking at the content displayed on the LCD display. In [5] is presented the MPIIGaze dataset that contains 213,659 images collected from 15 participants during natural everyday laptop use over more than three months. Another publicly available dataset, called Eyediap [6], contains 94 video sequences of 16 participants looking at three different targets (discrete and continuous markers displayed on a monitor, and floating physical targets) under both static and free head motion. All the existing dataset are missing the gaze information during the interaction scenario. With this work, we tend to bridge this gap and

* Corresponding author: rakovicm@ist.tecnico.ulisboa.pt

to use it as a base for modeling the behavior of a humanoid robot.

Mutual gaze awareness is important in communication and collaboration in group activities. Prior works were mainly showing different ways of how to analyze gaze cues visually. Some recent studies are analyzing the behavior of two users in different scenarios by measuring the gaze using head mounted gaze tracker devices. Paper [7] gives early work-in-progress that explores the effects of gaze awareness on gameplay, in particular when the gaze of one or more players is augmented over the game and revealed to others. Authors in the paper [8] presented an exploratory study to understand how gaze cues can enhance collaboration between two users in front of a large shared display.

This paper presents the dataset of the head and eye gaze together with image sequences of eyes and scene view collected during the HHI tasks. In Section II is explained the task humans have to accomplish during which the movements are recorded. Section III describes hardware and software setup used to collect the data, while Section IV explains the details of the collected dataset. The conclusion and the future work is given in Section V.

2 Dyadic interaction task

Findings in neuroscience [9] suggest that the human motor control combines state estimation, feed-forward, and multiple feedback loops operating at different speeds. Its structure is highly modular and is believed to rely on the combination of motor primitives to generate complex movements.

Authors in [11] used coupled dynamical systems for the realization of coordinated complex movements of the robot's upper body when performing reaching and grasping motion in the presence of obstacles. The computational model of the eye–arm–hand coupling is based on human motion data collected with subjects performing a prehensile motion with obstacle avoidance. The coupling between dynamical systems is learned from the experiment with a single human performing reaching and grasping tasks. Paper [11] shows the experiment with a human performing bimanual movements and with two humans each performing reaching movements at the same time. In the case of dyadic interactions, the results indicated that co-actors synchronize the timing of their movements, although the task in itself is discrete and non-rhythmic.

With this experiment, we want to create a basis for research on how to integrate this coupling in robot's motor control system, in scenarios where both human and robot, share the same space and objects during task execution. For that purpose, the participants are asked to assemble a pair of towers inside a circle on the paper in front of them. Both towers are assembled from 3D printed objects of different shape or color as shown in Figure 2. The objects are marked with numbers 1-3. The numbers are used to define the position of the object in the tower.

Fig. 2. Objects for assembling the tower.

In the beginning, two stacks of three objects are placed next to each participant. A stack of objects is positioned below the table top in order to occlude them from the other person. Next to the stack of objects is given a paper with the desired order of the objects to build the tower. (Fig. 3.).

When the assembly of towers starts, the participants are asked, one at a time, to pick the first object from the stack. If the number of the object matches the number in their next level of the tower, they should use the object for their tower. Otherwise, they are instructed to give, i.e. handover the object to a teammate. Thus, there are two types of actions the participant can execute: (i) intra-personal action (pick and place an object on its tower, i.e. placing action) or (ii) inter-personal action (pick and handover an object, i.e. giving action).

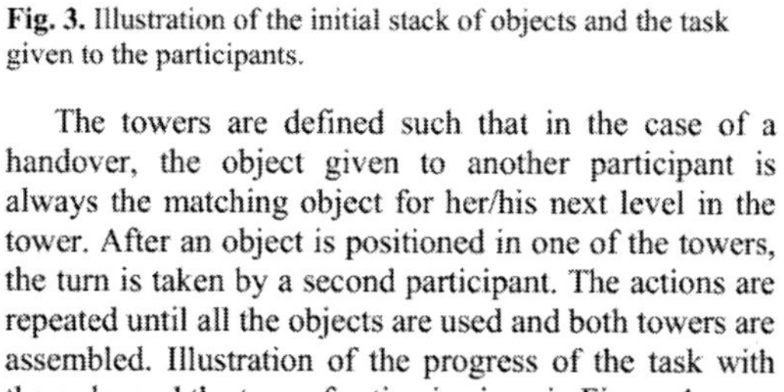

Fig. 3. Illustration of the initial stack of objects and the task given to the participants.

The towers are defined such that in the case of a handover, the object given to another participant is always the matching object for her/his next level in the tower. After an object is positioned in one of the towers, the turn is taken by a second participant. The actions are repeated until all the objects are used and both towers are assembled. Illustration of the progress of the task with the order and the type of action is given in Figure 4.

Once the assembly is finished, the new task and the new initial stack of objects is prepared and given to the participants. Each pair of participants had to repeat the task four times, i.e. to assemble four different pairs of towers.

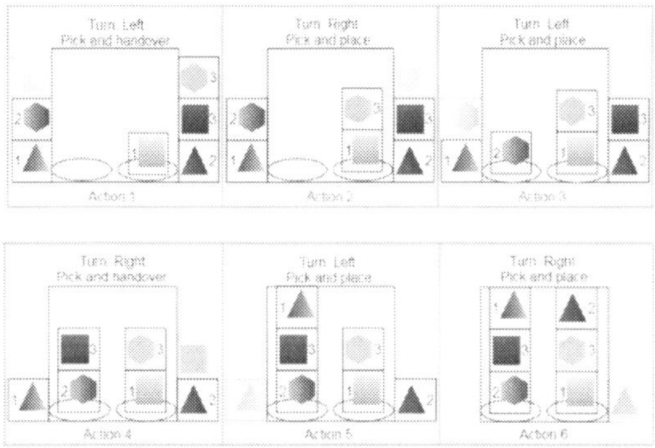

Fig. 4. Example of turn-taking order (left and right participant) and type of actions (pick and place or pick and handover) for assembling two towers.

The four tasks are defined in a way that there is always a different number of giving actions. This is to prevent the subjects to predict the action ahead of time. The goal here is to record a natural, unbiased human gaze behavior. The first task has two giving and four placing actions, the second task has six giving and no placing actions, the third task has no giving and six placing actions and the fourth task has four giving and two placing actions. Thus, during the experiment two participants performed together twelve giving and twelve placing actions.

3 The Data Acquisition Setup

When observing or scanning immediate surroundings, human eyes make jerky saccadic movements and stop several times, moving very quickly between each stop. The speed of movement during each saccade cannot be controlled, and the eyes move as fast as they are able [12]. To capture such eye movements, in this experiment both participants were wearing Pupil-Labs binocular gaze trackers [13].

During the performed actions, participants' head gaze was recorded using Optitrack motion tracking system [14]. Hardware and software setup used to acquire dataset is illustrated in Figure 5.

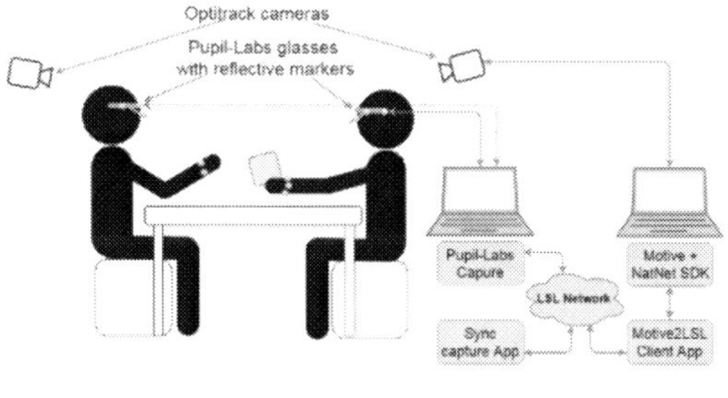

Fig. 5. Experiment hardware and software setup.

The Pupil Labs binocular gaze tracker is in the form of glasses equipped with three cameras. Two cameras are recording eyes at ~120Hz. A video stream of the egocentric view is recorded at 60Hz. The pupil detection algorithm does not depend on corneal reflection technique [15] and as reported in [13] the gaze tracker should work with users who wear contact lenses and eyeglasses. However, we experienced difficulties in calibrating the glasses with such participants, and we had to choose the participants not wearing glasses and contact lenses. Before the recording starts, each participant first calibrates his/her gaze tracker using the screen calibration method.

Optitrack motion capture system captures passive opto-reflective spherical markers at 120Hz. To record head gaze we fixate five opto-reflective markers on each glasses. Each group of five markers represented one rigid body whose position and orientation in the reference frame is being recorded.

Software setup is composed of following applications. For gaze data recording we used Pupil Labs Capture. For recording the body movements the Motive software platform is used. Since we want to capture synchronous data of head and eye gaze it was necessary to merge the input from two sensory systems. For that purpose is used Lab streaming layer (LSL) library [16]. LSL is designed to be a system for unified collection of measurement time series of various sensing equipment that handles both the networking, time-synchronization, (near-) real-time access and optionally the centralized collection.

In order to use LSL, we developed a Motive2LSL application that captures the broadcasted position of the markers and rigid bodies tracked within Motive software platform. Another application we developed is Sync capture application that receives the data measurements from two Pupil-Labs glasses and Optitrack cameras and records those data together with timestamps of the measurements into a file with synchronization timestamps.

4 The Dataset

We have acquired the data of three pairs, i.e. six participants. Participants were adults between 25 and 40 years of age. Acquisitions were performed at the Institute for Systems and Robotics, IST, University of Lisbon, during January 2018. The dataset contains:

- a video stream of the egocentric view with associated timestamps for both glasses,
- pupil data with gaze positions, pupil positions and its timestamps,
- position and orientation of rigid bodies representing head gaze with its timestamps and
- synchronization file with timestamps of gaze tracking and and motion tracking data.

In each recording, subjects had to perform 6 actions in one task, and each pair of subjects had 4 tasks. Thus, we collected the 24 actions in dyad scenario, i.e. 48 gaze and head motion for each pair (24 observer's movements and 24 performer's movements). The recordings are repeated for three different pair of humans, and thus we collected 72 actions and 144 gaze and head movements.

Figure 6 illustrates the recorded gaze movement of the eyes during the experiment for two different actions (placing and giving) and for two different roles (performer and observer) of the subject.

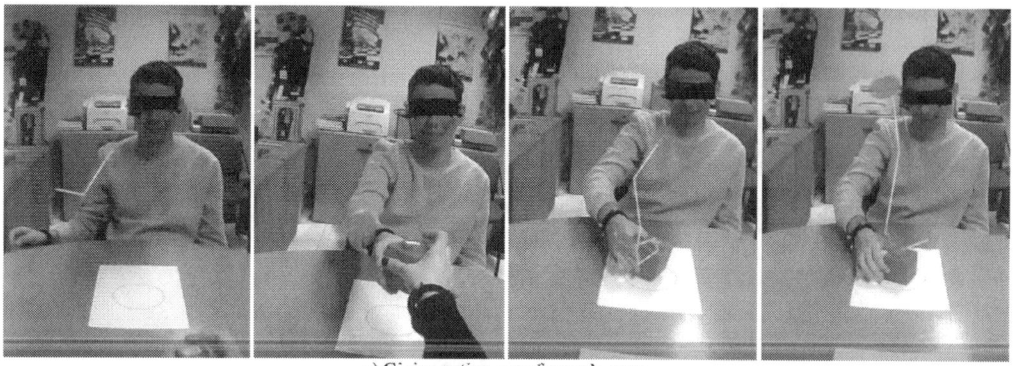

a) Giving action – performer's gaze

b) Giving action – observer's gaze

c) Placing action – performer's gaze

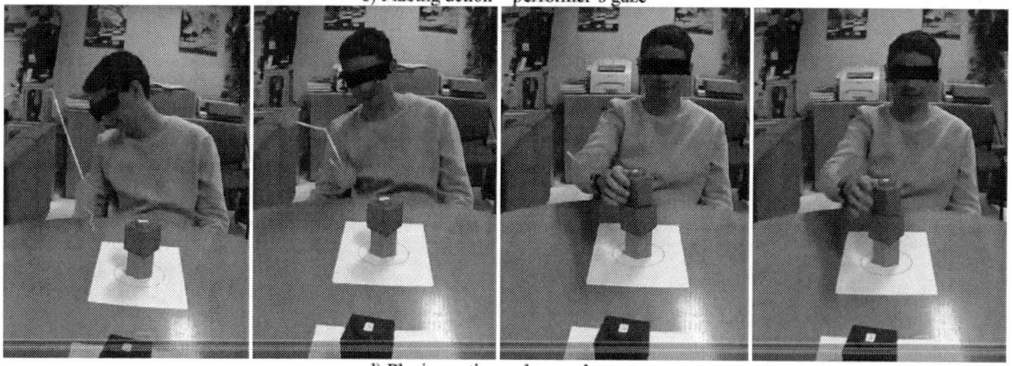

d) Placing action - observer's gaze

Fig. 6. Illustration of data set with an example given by an image sequences showing the gaze of a performer/observer during placing and giving actions (green circle represent the recorded gaze points, yellow line represent interpolation between recorded gaze points).

5 Conclusion

A multimodal dataset containing eye gaze and head gaze movement during placing and giving actions has been acquired using gaze tracking glasses integrated with motion capture system. Six adult subjects have participated in the recording sessions and made their data available for research. The acquired data has sufficient quality to investigate the behavior of gaze during dyadic interaction task. Our next steps will consist in annotating specific timestamps in the recorded movement so temporal correlations between important events can be established. After annotating the data, we will focus on modeling of the gaze behavior in interaction scenarios, and implement the model on a humanoid robot platform.

This work was partially supported by EU H2020 project 752611-ACTICIPATE, UID/EEA/50009/2013 FCT project and MNTR project III44008. The authors would like to thank all the volunteers that participated in the experiment.

References

1. C. Breazeal, C.D. Kidd, A.L. Thomaz, G. Hoffman, M. Berlin, IEEE/RSJ International Conference on Intelligent Robots and Systems, (IROS 2005), 708–713 (2005)
2. P. Schydlo, M. Raković, J. Santos-Victor, IEEE International Conference on Robotics and Automation, (ICRA 2018) (2018)
3. N. Duarte, J. Tasevski, M. Coco, M. Raković, J. Santos-Victor, arXiv preprint arXiv:1802.02788 (2018)
4. S. Mathe, C. Sminchisescu, IEEE Transactions on Pattern Analysis and Machine Intelligence (PAMI), **37** (2015)
5. X. Zhang, Y. Sugano, M. Fritz, A. Bulling, Proc. of the IEEE Conference on Computer Vision and Pattern Recognition (CVPR), 4511–4520 (2015)
6. K.A. Funes Mora, F. Monay, J.M. Odobez, Proceedings of the Symposium on Eye Tracking Research and Applications, 255–258 (2014)
7. J. Newn, E. Velloso, M. Carter, F. Vetere, Proceedings of the 2016 Annual Symposium on Computer-Human Interaction in Play Companion Extended Abstracts, 239–244 (2016)
8. Y. Zhang et al, Personal and Ubiquitous Computing, **21** (1), 173–186 (2017)
9. F.A. Mussa-Ivaldi, S.F. Gisztler, E. Bizzi, Proceedings of the National Academy of Sciences, **91** (16), 7534–7538 (1994)
10. L. Lukic, J. Santos-Victor, A. Billard, Biological cybernetics, **108** (2), 223–248 (2014)
11. C. Jung, A. Holländer, K. Müller, W. Prinz, Experimental brain research, **211** (3-4), 471–482 (2011)
12. R.C. Neil, *Psychology: the Science of Behaviour* (Toronto: Pearson, 2010)
13. M. Kassner, W. Patera, A. Bulling, Proceedings of the 2014 ACM International Joint Conference On Pervasive and Ubiquitous Computing: Adjunct publication, 1151–1160 (2014)
14. Point Natural. "Optitrack." Natural Point, Inc., Available at: http://www. naturalpoint. com/optitrack/
15. C. Jian-Nan et al, WRI Global Congress on Intelligent Systems, (GCIS'09), 133–138 (2009)
16. Kothe C., Lab streaming layer (LSL), Available at: https://github.com/sccn/labstreaminglayer

Mobile robot control based on noninvasive brain-computer interface using hierarchical classifier of imagined motor commands

Filipp Gundelakh[1], *Lev* Stankevich[1,*] and *Konstantin* Sonkin[2]

[1] Peter the Great Saint-Petersburg Polytechnic University, Russia
[2] Sagol School of Neuroscience, Tel Aviv University, Israel

Abstract. The study describes approaches of direct and supervisor control of a mobile robot based on a non-invasive brain-computer interface. An interface performs electroencephalographic signal decoding, which includes several steps: filtering, artefact detection, feature extraction, and classification. In this study, a classifier with hierarchical structure was developed and applied. Description of a committee of classifiers based on neural networks and support vector machines is given. The developed classifier demonstrated accuracy 50 ± 5% of single trial decoding of four classes of imaginary fine movements. Prospects of using non-invasive brain-computer interface for control of mobile robots was described. Key applications of the system are maintenance of immobilized patients and rehabilitation procedures both in clinic and at home.

1 Introduction

Nowadays various types of mobile robots are developed and widely used in many areas. *Mobile robots* designed to work alone (in extreme areas, security systems, service, etc.), are already produced in small series and have relatively simple intelligent control systems. Mobile robots capable of working in a group (exploring, providing communication, combat, security and gaming robots, etc.) are still under development; their control system is complex and intelligent.

Additionally, mobile *anthropomorphic* robots have been developed. They are named so because of their resemblance to a man. According to forecasts, in the near future anthropomorphic robots should be widely used as assistants to people, especially in cases of work in extreme areas and services sector including help to immobilized people in hospitals and at home.

Remote control of mobile robots can be implemented in various ways, but in order to work with immobilized people, contactless means of communication between a person and a robot are required.

A promising way to realize communication between a person with reduced motor abilities and a robot is the use of *"brain-computer interfaces"* (BCI). BCI is a modern technology and tool that can provide human interaction with external electronic and electromechanical devices based on decoding of signals of electrical brain activity [1]. It's shown that for immobilized people who need to control service robotic devices themselves, a BCI based on imaginary movements recognition can be applicable and useful [2].

As a mean of obtaining information about bioelectric activity of the brain, corresponding to motor commands, electroencephalography (EEG) is often used. Recent studies show that EEG as a non-invasive and relatively inexpensive technology has significant potential for BCI development. However, a factor limiting the practical application of BCI based on EEG signals is the complexity of reliable and reproducible interpretation (decoding) of brain signals. Another factor is the difficulty of classification of EEG patterns of imaginary movements in real time.

Currently, some Russian scientific groups and organizations carry out research related to the development of BCI, many of which are known internationally [3,4]. Thus, there are attempts to create a human-computer interface using BCI based on EEG signals. Commonly, such tools allow detecting some mental commands of a person using the P300 component of event related potential elicited in the process of decision-making [4]. However, such interfaces are limited by the number of degrees of freedom and can be used for simple control of a robot.

The aim of this work is to develop a mobile robot control system based on BCI, which performs classification of imaginary movements using EEG signals. Mobile robots with such control systems can be used, for example, to help people with limited motor abilities, as well as for rehabilitation in clinics and at home. In this study, BCI application for control of two types of mobile robots (robotised cart and the anthropomorphic robot) is discussed.

2 Mobile robots

Mobile robots appeared due to the need to expand the working area of manipulators and in order to perform transport operations.

* Corresponding author: stankevich_lev@inbox.ru

In modern mobile robots, many modern vehicles are used, including ones based on bionic ways of movement, borrowed from nature. First of all they include walking.

Depending on the environment, in which robots move and operate, they can be divided into terrestrial, water, air and space types of robots. Terrestrial robots include wheeled, crawlered, running on rails and walking robots.

Walking type of movement has higher maneuverability in rugged terrain than wheeled or tracked types. Additionally, this type of movement is more economical in these conditions because interaction with ground occurs only in places of contact with a foot, while wheeled and crawlered types of movement leave a continuous track behind machines. Since the main purpose of walking machines is movement on rough terrain, their control systems must be adaptive.

Anthropomorphic robotics engaged in research of humanoid robots has been actively developed in recent years. Such robots are designed to work in a weakly determined environment and are positioned as universal human assistants. Synthesis of control system of an anthropomorphic robot is a complex problem. Rapid development of this area is associated with the recent successes in the field of artificial intelligence.

3 Remote control

Remote control of a mobile robot can be implemented on different levels of control system, including direct control of individual drives of a robot (direct control) and supervisory control using a set of high-level commands.

Direct control provides maximum freedom but it is laborious and requires a lot of skill. Time lag in the communication channel makes it even more difficult to work in this mode.

In turn, supervisory control provides a limited set of commands but these commands can run complex behavior algorithms.

4 Brain-computer interfaces

At present, advances in neurophysiology, psychology and artificial intelligence, as well as in development of devices for bio signal recording, have led to the possibility to create new means of human-machine communications based on direct decoding of nervous system signals.

Currently, non-invasive BCIs based on EEG are implemented to recognize imaginary and real movements of large parts of the body, for example, hands and feet. However, such BCIs are not effective for decoding fine motor skills, for example, movements of fingers of one hand. This problem is difficult to solve because of the anatomical proximity of the brain structures participating in the realization of imaginary movements and because of subtle differences in corresponding EEG signals. Solution of this problem requires large computational resources for data analysis. On the other hand, BCI should operate in real time, which means to generate control signals with minimum delay determined by the speed of operation of external devices.

4.1 BCI structure

While controlling robotic devices, such as a wheelchair, robotic manipulator, prosthesis, exoskeleton, etc., a paralyzed person should be able to give commands that allow these devices to perform specific tasks for human movement, object manipulation or prosthesis control. In this paper, it is proposed to do this using a non-invasive BCI, which decodes brain activity signals and forms supervisory commands for external devices. For the practical implementation of such a system, it is necessary to use biological feedback through vision, which allows a user to learn how to control the system by amplifying EEG signals corresponding to imaginary commands [3]. Delays in calculating the response must be small enough so that in the user's mind mental command and response are clearly associated.

It is possible to formulate principles for development of systems for control of robotic devices through a BCI:
(1) aiming to achieve level of accuracy and speed of classification of EEG patterns, acceptable for use of BCI for control of robotic devices in real time;
(2) ensuring sufficient degrees of freedom of BCI, that is, the number of recognized imaginary commands (at least 4);
(3) optimization of computational resources to reduce run-time calculations including the classification of imaginary commands by single trial of imagination.

Data flow in EEG-based BCIs includes several main stages: recording and preprocessing of EEG signals, extracting features in spectral or temporal domains, classification of EEG patterns of imaginary commands, and forming a final decision (Fig. 1).

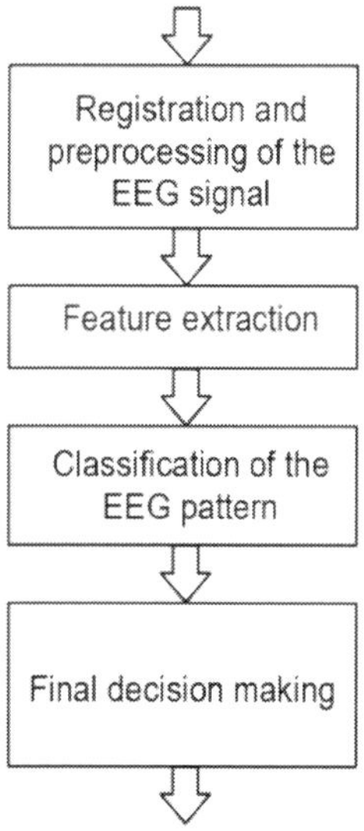

Fig. 1. Structure of EEG-based non-invasive BCI.

To study the possibility of implementation of a human-computer interaction system based on a non-invasive BCI, the classifier of imaginary finger movements of one hand, developed by the authors earlier [5, 6] was used.

The adaptation of this version of the BCI to the classification of imaginary commands was carried out using a specially developed program and methodology. According to the method, subjects were asked to consistently imagine 4 types of imaginary movements of fingers of one hand (thumb, index, middle and little fingers). These imaginary movements corresponded to control commands for the robotic device. During training subjects first had to press a computer mouse button with a finger assigned to control the necessary behavior of the device, in the rhythm defined by beeps. After a few real movements, sound disappeared and subjects had to *imagine* movements with the same finger.

Training was focused on the kinesthetic sensations of subjects. The test series of real and imaginary movements were repeated several times. As a result, the subject had to perform at least one hundred real and one hundred imaginary movements in a given rhythm in one block of the test. The number of test blocks should correspond to the number of types of performed movements (real / imaginary finger movements). Thus, the control of the robotic devices with the command buttons is simulated.

The EEG signals were recorded using a 32-channel digital electroencephalograph of the company «Mitsar». We used 19 electrodes, which were located using the 10-20 system. EEG signals with a duration of 600 ms were recorded in the frequency band of 0.53 Hz - 30 Hz. The sampling rate was 500 Hz. Artifacts such as eye movements, slow and fast waves, fragments of EEG signals with an amplitude of more than 100 μV were excluded. For analysis and subsequent classification, EEG signal registered from the sensorimotor region of the cerebral cortex were used: C3, Cz, F3 sites.

Analysis of EEG signals was carried out both in time and frequency domains. To ensure the necessary speed of the system, analysis of each individual trial (without accumulation) was made. An algorithm utilizing a combination of two feature spaces was used: length of the curve and the cepstral coefficients of segments of the signal. These features were extracted in a time sliding window, which allowed selecting the most informative segments of the signal.

Choosing the optimal size of the analysis window is important for increasing classification accuracy. Previously, it was shown that the selection of optimal values could significantly improve accuracy of classification of imaginary movements. On the basis of earlier studies [5], analysis window width was set 100 ms with a shift of 50%.

Classification of EEG patterns corresponding to imaginary movements was carried out using the calculated features as the inputs for two types of classifiers: artificial neural network and support vector machine.

4.2 Classification of imaginary movements

The developed committee of classifiers of EEG patterns is based on artificial neural networks (ANN) and support vector machines (SVM). Studies have shown that they are effective means of classification, including classification of imaginary movements.

In this work, neural networks were implemented in form of multilayer perceptron with three layers: two hidden and an output layer. As activating functions in hidden layers and in an output layer, sigmoid and linear functions were used, respectively.

SVM method is based on linear classification approach. The essence of the method is to divide the sample into classes using the optimal separating plane. Data conversion is determined by the kernel function. SVM with radial-basis kernel function was chosen for classification of EEG signal. This choice was based on the results of studies on the selection of the most suitable SVM type for this purpose.

A two-level committee of classifiers was used in the study. The first layer consisted of two artificial neural networks (ANN (C) and ANN (L)) and two classifiers based on support vector machines (SVM (C) and SVM (L)), and the second layer consisted of one artificial neural network (ANN (T)) combining results of the first level classifiers. The first-level classifiers analyzed two feature vectors (the length of curve L and the cepstral coefficients C) separately. For every trial, features were obtained for 20 segments of EEG signal (20 features for L and 20 features for C). Four classifiers of the first level gave their responses assigning a trial to one of classes. Every classifier's output consisted of 4 numbers – 4 probabilities of belonging to each of possible classes. On the basis of these outputs, a feature vector (16 components) for the second-level ANN (T) was formed. This ANN made the final classification decision for every trial (Fig. 2).

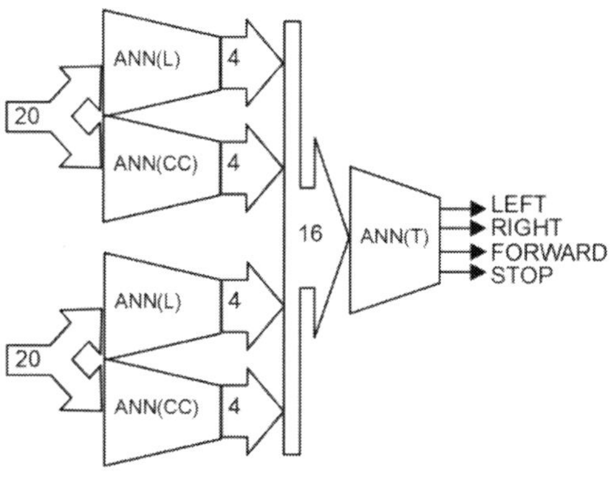

Fig. 2. Block diagram of the two-level classifier.

Real time control strategy based on BCI requires continuous EEG data recording. To ensure the continuity of input data and its parallel processing, multithreaded programming method was used. The data flow is as follows: as soon as the trial N ends and the trial N + 1 begins, thread 2 reads data related to the trial N, pre-processes it, extracts features, classifies, and returns the result of the classification (Fig. 3).

The conducted studies of the BCI prototype have shown that its characteristics correspond to real-time requirements. The time delays required to obtain a response from the classifier are within 150 ms, which is acceptable for BCI applications with biological feedback. Additionally, according to subjective self-reports of subjects, they really connected imaginary movements with received responses of the system.

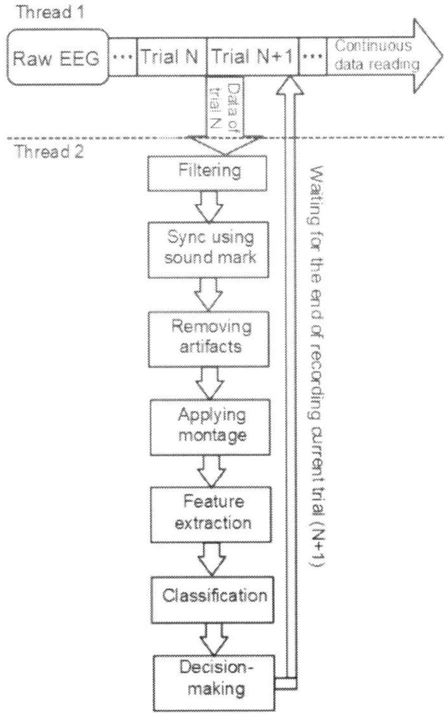

Fig. 3. Scheme of the classification system operating in real time.

The following results of classification were obtained in the study. In autonomous sessions, classification accuracy of single-trial EEG patterns of four imaginary commands was 50 ± 5% at average for channels F3, C3. Thus, the probability of successful classification of four types of imaginary commands using only two EEG channels significantly exceeded the guessing level of 25%. During testing of the system using feedback, classification accuracy of 58% was archived. At the same time, studies have shown and proved the fact that some subjects are not able to learn how to work with BCI on the basis of EEG signals.

5 Examples of application of the BCI-based control system

An operator should be able to give commands to change the behavior of robotic devices. As part of the study, it was proposed to remotely control medical mobile robots by means of BCI. The control system based on BCI can allow immobilized people both to self-service using robot and to train impaired motor functions.

Experiments were conducted to control a robotic cart and an anthropomorphic robot using noninvasive BCI based on classification system of imaginary movements.

The robotic cart (Fig. 4) has 4 controlled drives (2 for moving the platform and 2 for orientation of stereo vision cameras) and can be used to service immobilized patients.

An anthropomorphic robot NAO with 23 degrees of mobility can be used as a mobile device for interacting with a patient and performing simple actions (Fig. 5).

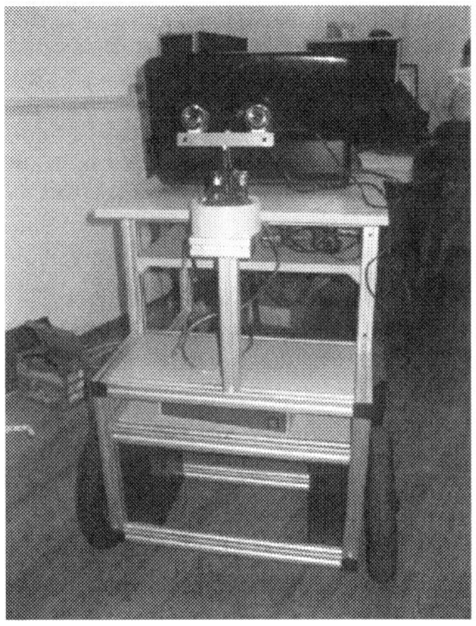

Fig. 4. The robotic cart controlled with noninvasive BCI

Fig. 5. An anthropomorphic robot NAO.

These mobile robots were controlled using BCI by issuing high-level commands (supervisory control). Supervisory control was performed by simulating pressing of control keys with different fingers: four classes of commands were formed: «Forward», «Stop», «Right», «Left». The experiments demonstrated possibility of training the BCI classifier to classify these commands and to control the cart and the robot. Since both the cart and the robot had vision systems and their own systems for recognizing and avoidance of obstacles, such supervisory control was sufficient for the practical application of these robotic devices.

Thus, a robotic cart is capable of delivering medicines and water and transporting away the used flatware. In the future we plan to equip the platform with an artificial hand for loading and unloading the required items.

Anthropomorphic robot NAO has intellectual control system, which allows it to perform a wide range of actions, such as hand and foot movements, steps, turns, etc. Such man-like robot equipped with BCI based control system can become a novel tool for rehabilitation of patients with severe movement disorders caused by stroke and brain injuries. There might be two use cases:

1. user imagines a movement of immobilized limb, BCI decodes this intention and robot performs this movement (imitating human movement);
2. user controls robot with supervisory commands («Forward», «Stop», «Right», «Left»), and robot, performing them, passes through a complex path.

In the first case user receives a feedback from robot, which significantly increases the user's motivation and enables binding of mental command and real movement. These results are the necessary conditions for facilitation of rehabilitation process [7].

Second scenario aims at objective tracking of rehabilitation process in absence of real movements, in a period when brain training based on brain plasticity has started but ability to control user's own movements has not yet appeared. In this case, path of the robot and required time might be used as objective markers to demonstrate day-to-day dynamics.

6 Conclusion

Thy study demonstrated that developed control system based on BCI meets the requirements of practical application in real time. Thus, maximum delay required for generation of control signals does not exceed 150 ms, which makes it possible to use bio feedback approach. The multithreading implemented in the system allows BCI to record and process EEG signals simultaneously without data losses.

BCI with a combined hierarchal EEG pattern classifier was given as an example of development of effective BCIs based on processing of EEG signals. This classification system learns in order to recognize imaginary motor commands not only by hands or feet, but also by small parts of the body (fingers of one hand). It's shown that such an EEG pattern classification system can provide acceptable accuracy and classification speed for BCI. Thus, it might be used to control various robotic devices, such as a robotic cart with computer vision system, an artificial hand with a five-finger wrist or an anthropomorphic robot

Further development of effective BCI-based systems for remote control of a robot requires new solutions for increasing the accuracy and the speed of classification of EEG patterns, as well as new approaches to increase degree of freedom of BCI and to optimize the computational resources required for BCI implementation. Solving these problems will create effective BCIs that will provide people with limited motor functions ability to control robotic devices using brain signals that can significantly improve their quality of lives in the world of healthy people.

This research is supported by Russian Found of Basic Researches (project #16-29-08296).

References

1. J.R. Wolpaw, E.W. Wolpaw, *Brain-computer interfaces: principles and practice* (Oxford University Press., 2012)
2. I. Daly et al. Clin. Neurophysiol, **124**, 1787–1797 (2013)
3. A.A. Frolov, V.U. Roshin. Brain computer interface. Available at: http://neurolectures.narod.ru/2008/Frolov-2008.pdf
4. A.Ya. Kaplan, A.G. Kochetkov, S.L. Shishkin, et al., Sibir. Med. Bull., **12**(2), 21–29 (2013)
5. K.M. Sonkin et. al. Art. Intell. Med., **63**(2), 107–117 (2015)
6. L.A. Stankevich et. al. Human Phisiology, **42**(1), 32–42 (2016)
7. C. Jeunet, F. Lotte, et al., Neuroscience (2017)

Mathematical support and software for data processing in robotic neurocomputer systems

Vitaliy Romanchuk *

Ryazan State University named for S.A. Esenin, Department of Informatics, CS and MTI, 390000, Svobody, 46, Ryazan, Russia

Abstract. The paper addresses classification and formal definition of neurocomputer systems for robotic complexes, based on the types of associations among their elements. We suggest analytical expressions for performance evaluation in neural computer information processing, aimed at development of methods, algorithms and software that optimize such systems.

1 Introduction

It is obvious now that productivity of computing devices and systems is insufficient for solving a number of tasks, such as processing of video and audio data, real-time management robotics, recognition of images, forecasting, optimization, and artificial intellect tasks for robotics. The problem is due to several causes: it is impossible to increase the frequencies of computing devices because of the "technological restrictions" we currently face; we lack effective methods, algorithms and software solutions for parallelization of operations when multiprocessor and multinuclear systems are used; we have a limited number of generic methods, algorithms and software means of parallelization, when employing specialized computing devices (neuroprocessors, DSP-processors, etc.).

One of the ways to deal with insufficient productivity in solving a number of tasks is to employ high-performance computing systems with massive parallelism. Such parallelism in numerous applications that will perform complex processing of video, images, communication, security, and other types of signals is mainly present at the data level (Data Level Parallelism – DLP). In the majority of modern processors, DLP is implemented through a single command stream and a multiple data stream, in «single instruction, multiple data» (SIMD). The above applications also show considerable Instruction Level Parallelism (ILP). In such cases, parallel assignment of multiple scalar operations and/or SIMD-operations provides a basis for their parallel performance at the instruction level. However, a SIMD-architecture may not be the best solution for applications with variable DLP, which will decrease productivity and increase power consumption. Modern computing systems use «very long instruction word» (VLIW) architectures that implement ILP and DLP. Nevertheless, popular von Neumann-type processors cannot implement parallelization effectively, because of complex-parallelized constructs in their algorithms, such as cycles, conditional branching, and dependence of data). This paper proposes a method to accelerate the development of methods other than frequency increase, to improve high-speed response: using high-parallel specialized computing hardware (e.g., neurocomputers), and development of innovative easily parallelized algorithms with elements of intellectual compilation. The idea behind this is to simplify the software and extract as much "implicit parallelism" as possible at the command level, using Wide Issue Width (WIW) in command outputs and long pipelines with Deep Pipeline Latency (DPL). Neurocomputers are next generation computing devices consisting of many concurrently running simple computing elements (neurons). These elements are connected in a neuron network. They perform uniform computing operations and do not require external control. The great number of concurrently running computing elements ensures high-response performance and low energy consumption. Besides, neurocomputing applications do not contain any elements that are hard to parallelize, and all computing they do is multiple, parallel, independent, and neuro-based.

The following research teams can be listed as active in similar fields of study:

- S. Pande, F. Morgan, S. Cawley, T. Bruintjes, G. Smit, D. McGinley, S. Carrillo, J. Harkin and L. McDaid [1], D. S. Modha, R. Singh [9], Preisslet [7] who study implementations of certain specific neuro-networks on embedded architectures of neurocomputers, i.e., research of specific architectures.
- N. Izeboudjen, C. Larbes. and A. Farah [2], M. Shulakeret [8], Eliasmithet [4] are active in cataloguing neurocomputing devices, which will further contribute to description of generic properties and principles of functioning for implementation of parallel programming methodology.
- P. Kolinummo, P. PASI Pulkkinen, T. Hämäläinen and J. Saarinen [3], E. Stromatias, F. Galluppi, C. Patterson, S. Furber [5], K. Esseret [6] have chosen a narrow field of research: parallel execution of self-organizing maps (Kohonen maps) on a neurocomputer (emulator).

* Corresponding author: v.romanchuk@rsu.edu.ru

2 Mathematical formalization of data processing in robotic neurocomputer systems

Thus, our intention to develop the first universal model of programming for neurocomputing devices and systems.

The goal of the current research is to develop mathematical tools for optimizing parallel, distributed and cloud computing systems based on neuroprocessors.

Our objectives are as follows:

- classification of computing systems built on neuroprocessors, according to associations among their elements;
- formalization of parallel, distributed and cloud computing systems – neuroprocessor computing systems (NPCS);
- defining analytical expressions of performance evaluation in neurocomputer systems of data processing;
- development of optimization methods for parallel, distributed and cloud systems of neurocomputer data processing.
- partitioning of software tools of optimization in parallel, distributed and cloud systems of neurocomputer data processing.

To solve these tasks, we employ the set-theoretic approach, methods of system analysis, methods of optimization and of computing process planning.

We introduce a classification of computing systems based on neuroprocessors, according to types of associations among their elements [4-7].

1. A parallel neuroprocessor computing system is one whose elements are connected on a bus level. The element of a computing system is its computing nucleus, the neuroprocessor. Computing nuclei may refer to specific processors (1 nucleus for each) or to several multinuclear processors. For purposes of simplifying the model, they are regarded as an integrated set of neuroprocessor computing modules (NPCMs) $P = \{P_1, P_2, ..., P_q\}$. Processes that are executed on computing nuclei can directly read and write information in the RAM and in the disk storage of the node, through the interface of a system bus or commutator.

A parallel system can be defined in this way:

$$RS = \{P, E, \Delta, S_w, Ttr, Tl\}, \quad (1)$$

where $P = \{P_1, P_2, ..., P_q\}$ is a set of computing nodes of the NCPS, in which each computing node (neuroprocessor) can be described with the following technical specifications:

$$P_i \rightarrow H_{ji}, H_{ki}, H_{li}, H_{mi}, \quad (2)$$

where $H_{ji}, H_{ki}, H_{li}, H_{mi}$ are the technical parameters that define the processor speed, number of its nuclei, RAM and disk storage capacity (if applicable) for the ith NPCM, accordingly;

S_w is the NCPS structure;

E is a set of directional associations between the nodes of the NCPS, whose quantity is defined by the structure type, i.e. $E = F(S_w)$;

Δ - controller (governing node) of the NPCS;

$Ttr : CL \cup \{\Delta\} \times CL \cup \{\Delta\}$ – defines the average throughput between nodes of the NPCS;

$Tl : CL \cup \{\Delta\} \times CL \cup \{\Delta\}$ – defines the latency, i.e., time required to initialize messages, send and receive data, etc.

We can single out the following properties:

$$\forall u, v \in CL \cup \{\Delta\} : Ttr(u, v) = Ttr(v, u)$$

$$\forall u, v \in CL \cup \{\Delta\} : Tl(u, v) = Tl(v, u)$$

As the processors are interconnected through a bus, delay times between data transfer Ttr and latency time Tl are negligible: $Ttr \approx 0$, $Tl = 0$.

We can also describe the system's dynamic parameters at any moment of time $t \in [0, +\infty)$:

$U_i(t)$ is the workload on the ith computing nucleus, $0 \le U_i(t) \le 1$;

$H_{li}(t)$ is the available RAM of the ith NPCM;

$H_{mi}(t)$ is the available disk storage capacity of the ith NPCM.

An important feature of neuroprocessors is their continuous performance and training (in each command), which consists in data exchange between the RAM and the computing nucleus. This makes it important to correctly choose the external bus configuration for interaction with RAM in a multiprocessor mode. Neuroprocessors, like most DSP processors, support both single-processor and multi-processor work modes along a single or multiple external bus. Let us analyze the most common option, with two buses. If two processors are connected to a shared memory, its access arbitration occurs without an external controller. It should be remembered that only oppositely charged processor buses can unite: a local bus of one processor with the global bus of another, as following a system reset, only one processor will be authorized to access the shared memory.

2. Distributed neuroprocessor computing systems (DNPCSs) are complexes of NPCMs or autonomous computer systems (CSs) that are situated at a distance and interact via programmable commutators and system devices. It should be noted that the rules of parallel and distributed processing are not the same. A GRID system is an example of distributed systems.

One of the features of a distributed system is absence of shared memory, except for a case when a unified address space [5] is provided (we do not discuss it here). Thus, processes that are executed on computing nuclei can directly read and write information in the RAM and disk storage of the node, through the interface of a system bus or commutator.

The notion of cluster is essential in a distributed system. We refer to a neuroprocessor cluster when we mean a group of computers, whose computing nodes are neuroprocessors; the computers have broadband connection and are perceived by the user as a single hardware unit. In other words, a cluster is a loosely connected combination of multiple computing systems that perform jointly in executing shared applications, and are viewed by users as a unified system. A cluster based on neuroprocessors has a number of advantages arising from high-level parallelism of the neurocomputer's paradigm.

Each cluster CL_i can be described in this way:

$$CL_i = \{P_i, K_i, E_i\}, \quad (3)$$

where $P_i = \{P_{i1}, P_{i2}, ..., P_{ig}\}$ is a set of the cluster's computing nodes CL_i;

K_i is a set of commutators;

E_i is a set of associations among them.

The number of computing nodes in set P_i can also equal 1. In cluster CL_i, all computing nodes of a P_i set are connected with a dedicated governing node P_{i0} through a separate governing network that is employed for transfer of signals of government, delivery of program files and input data, and also for output files. Node P_{i0} does not participate in computation. If it is necessary to transfer a message to another computing node within a cluster, a network subsystem is employed; it partitions the message into packages and sends them via a high-speed communication network.

Considering the nature of a neuroprocessor cluster, a distributed system can be described like this:

$$RS = \{CL, E, \Delta, S_w, a, h, q, Ttr, Tl\}, \quad (4)$$

where $CL = \{CL_1, CL_1 ..., CL_r\}$ is a set of computing clusters of the DNPCS;

E is a set of directional associations between clusters of the DNPCS, where some associations can be high-speed (strong), and others slow (loose).

Δ is the controller (governing node) of the NPCS;

S_w is the structure of the DNPCS;

a is the hardware architecture of each DNPCS node;

h is the software architecture of each DNPCS node;

$Ttr : CL \cup \{\Delta\} \times CL \cup \{\Delta\}$ defines the average throughput between each pair of computing clusters (bytes/sec);

$Tl : CL \cup \{\Delta\} \times CL \cup \{\Delta\}$ defines the latency, i.e., time required to initialize messages, send and receive data, etc. between each pair of computing cluster (bytes/sec).

3. Cloud neuroprocessor computing systems (CNPCSs) require ubiquitous access to the network infrastructure of configurable computing resources, such as a neuroprocessor system (we only analyze the *IaaS* service model, where users are given access to the neuroprocessor infrastructure). A significant feature of cloud systems is that there occur time delays due to the employment of relatively slow communication channels between a user and a cloud.

In cloud computing, these two options are possible:

a) The cloud's computing system is not distributed.

b) The cloud's computing system is distributed.

A cloud system can be described like this:

$$RS = \{CL, E, \Delta, S_w, a, h, q, Ttr, Tl, Ttrc, Tlc\}, \quad (5)$$

where $CL = \{CL_1, CL_1 ..., CL_r\}$ is a set of CNPCS computing clusters;

E is a set of directional associations among clusters of the CNPCS, where some associations can be high-speed (strong), and others slow (loose).

Δ is the controller (governing node) of the CPCS;

S_w is the structure of the CNPCS;

a is the hardware architecture of each CNPCS node;

h is the software architecture of each CNPCS node;

$Ttr : CL \cup \{\Delta\} \times CL \cup \{\Delta\}$ defines the average throughput between each pair of computing clusters (bytes/sec);

$Tl : CL \cup \{\Delta\} \times CL \cup \{\Delta\}$ defines the latency, i.e., time required to initialize messages, send and receive data, etc. between each pair of computing clusters (bytes/sec);

$Ttrc$ is the time delay in data transfer to the CNPCS governing device and back to the user;

Tlc is the latency time in data transfer from the CNPCS governing device and back.

If $TO_l^{(J)}$ is the execution time for subprogram $RO_l^{(J)}; l = \overline{1, L}$, defined as the total execution time for type 1 commands,

$$TO_l^{(J)} = TO_{1l}^{(J)} + TO_{2l}^{(J)} \quad (6)$$

then, due to the principles and specific features of neurocomputer functioning, the execution time for auxiliary commands tends to zero: $TO_{2l}^{(J)} \to 0$.

Then $TO_{1l}^{(J)}$ is defined as the total time of mathematical modeling of all neurons in the neuron network of subprogram $RO_l^{(J)}$. If $MK^{(1)}$ is a microcommand for neuron emulation in an artificial neuron network, M_l is the number of commands $MK^{(1)}$ in subprogram $RO_l^{(J)}$.

$$TO_{1l}^{(J)} = M_l * H_k * H_m, \quad (7)$$

where $H_k \in H$ is a technical specification defining the quantity of neurons modeled within a tact;

$H_m \in H$ is a technical specification defining the execution time for one tact of the neuroprocessor.

$$TO_l^{(J)} = M_l * H_k * H_m \quad (8)$$

If Ttr is the total time of data transfer delays between the NPCM in the system; Tl is the total latency time in data transfer.

1. Let us look into the sets that define delay and latency times in data transfer Ttr and Tl for the DNPCS,

$Ttr = \{Ttr_{ij}\}$ where

Ttr_{ij} is the transfer time from i to j NPCS.

The data transfer time can be:

$$\begin{cases} 0 \\ Ttr_{ij} = Ttr_{i0} + Ttr_{j0} \end{cases} \quad (2.72) \qquad (9)$$

where Ttr_{i0} is the data transfer time from node i to the Δ governing node.

$Ttr_{ij} = PS * N$,

where PS is the data throughput for the channel (bit/sec);

N is the bit count for transfer.

$$Ttr = \sum_{i=1}^{|E|} Ttr_{ij}, \qquad (10)$$

$Tl = \{Tl_{ij}\}$, where

Tl_{ij} is the latency time in data transfer between i and j the NPCM.

The latency time can be:

$$\begin{cases} 0 \\ Tl_{ij} = Tl_{i0} + Tl_{0j} \end{cases} \quad (2.72), \qquad (11)$$

where Tl_{i0} is the latency time in data transfer from node i to the Δ governing node;

Tl_{0j} is the latency time in data transfer from the governing node Δ to the j th node.

$$Tl = \sum_{i=1}^{|E|} Tl_{ij}. \qquad (12)$$

$$Ts_{ij} = Ttr_{ij} + Tl_{ij}. \qquad (13)$$

Thus, in distributed systems we obtain a transfer time delay in our NPCS:

$Ts = Ttr + Tl \neq 0$.

In such cases, the initial data transfer to the governing device (commutator) does not cause delays, i.e. $Ttrc \approx 0$.

2. Let us look into the Ttr and Tl sets for the CNPCSs that do not have a distributed structure.

In this case, value $Ttr \approx 0$, and the delay time $Ttrc$ in the system equals:

$$Ttrc = Ttr_{po} + Ttr_{op}, \qquad (14)$$

where Ttr_{po} is the transfer time from the user to the Δ node;

Ttr_{op} is the transfer time from the Δ node to the user.

$$Tlc = Tl_{po} + Tl_{op}, \qquad (15)$$

where Tl_{po} is the latency time in data transfer between the user and Δ;

Tl_{op} is the latency time in data transfer between Δ and the user.

Then we obtain the following:

$$Ts = Ttrc + Tlc \neq 0;\ Ttr = Tl = 0.$$

3. Let us look into the Ttr and Tl sets for CNPCSs with distributed structures.

In such cases, the total data transfer times are computed as:

$$Ts = \sum_{i=1}^{|E|} Tl_{ij} + Ttr_{po} + Ttr_{op} + \sum_{i=1}^{|E|} Tl_{ij} + Tl_{po} + Tl_{op} \qquad (16)$$

$$Ts = Ttrc + Tlc + Ttr + Tl \neq 0$$

That is, in this case we obtain $Ttrc \neq 0$; $Ttr \neq 0$.

For further consideration, we assume the total time loss in data transfer to be the following:

$$Tsc = Ttr + Tl + Ttrc + Tlc. \qquad (17)$$

Using the above ratios, we can define the analytical expressions for evaluating the chief criterion of effectiveness: performance of neurocomputer systems for various structures: pipeline, vector, pipeline-vector, and vector-pipeline.

Cycle $T_c^{(j)}$ of the pipeline structure is defined as the duration of the maximum processing time, i.e.

$$T_c^{(j)} = \max_{l \in L} TO_l^{(j)}, \forall l = \overline{1, L}. \qquad (18)$$

The first result of the information processing after implementation of algorithm $A^{(j)}$ will be obtained as the pipeline output after the $t^{(j)} = T_c^{(j)} * q + Tsc$ time.

This value will be the execution time $PR^{(j)}$, i.e.

$$T_o^{(j)} = \max_{l \in L} TO_l^{(j)} * q + Tsc, \forall l = \overline{1, L}. \qquad (19)$$

Each subsequent processing output will be obtained after $T_c^{(j)}$, and then the times of obtaining outputs can be calculated in this way:

$$T_o^{(j)} = \max_{l \in L} TO_l^{(j)} * q + (N-1) * \max_{l \in L} TO_l^{(j)},$$

where N is the ordinal number of the required processing output.

The time loss will be the difference in times between execution time $T_o^{(j)}$ and the total execution time for all subprograms:

$$T_n^{(J)} = q * \max_{l \in L} TO_l^{(J)} - \sum_{l=1}^{q} TO_l^{(J)} \cdot \quad (20)$$

On the other hand, processing each next word will yield a time gain as compared to a single-processor option of the system's implementation $T_l^{(J)} = \sum_{l=1}^{q} TO_l^{(J)}$. Then the gain time will equal the difference between the total operate time of all NPCMs and the total execution time for all subprograms:

$$T_k^{(J)} = \sum_{k=1}^{q} ((k-1) * TO_{q-k+1}^{(J)}) \cdot \quad (21)$$

The downtime can be computed as the sum of differences $T_c^{(J)} - TO_i^{(J)}$ for each ith subprogram in each NPCM:

$$T_{np}^{(J)} = \sum_{k=1}^{q-1} (k * (\max_{l \in L} TO_l^{(J)} - TO_{q-k+1}^{(J)})) \cdot \quad (22)$$

The total processing time is defined as the execution time for all q subprograms on all NPCMs:

$$T_p^{(J)} = \sum_{k=1}^{q} (k * TO_{q-k+1}^{(J)}) \cdot \quad (23)$$

Let us look at the time characteristics for distributed and cloud systems that have a pipeline structure.

For distributed or cloud systems at each pipeline run, data transfer time delays will be added, totaling the delay times of all $q-1$ associations:

$$Ttr = \sum_{i=1}^{q-1} Ts_{i(i+1)}, \quad (24)$$

$$Ts_{ij} = Ts_{i0} + Ts_{(i+1)0}, \quad (25)$$

Then, the DNCPS's pipeline cycle will equal:

$$T_c^{(J)} = \max_{l \in L} TO_l^{(J)} + \max_{Ts} Ts_{ij} \cdot \quad (26)$$

The first result of data processing during the $A^{(J)}$ algorithm implementation will be obtained at the pipeline output after $t^{(J)} = T_c^{(J)} * q + Ts$ time.

Then, considering the delay times, the $PR^{(J)}$ execution time will be:

$$T_o^{(J)} = (\max_{l \in L} TO_l^{(J)} + \max_{Ts} Ts_{ij}) * q \cdot \quad (27)$$

Each subsequent processing result will be the output in $T_c^{(J)}$, so the time when the next processed output is obtained can be computed in this way:

$$T_o^{(J)} = (\max_{l \in L} TO_l^{(J)} + \max_{Ts} Ts_{ij}) * q + + (N-1) * (\max_{l \in L} TO_l^{(J)} + \max_{Ts} Ts_{ij}) \quad (28)$$

The time lag is defined as the difference in execution time $T_o^{(J)}$ with consideration for time delays in data transfer and the total time of all subprograms' execution:

$$T_n^{(J)} = t^{(J)} - \sum_{l=1}^{q} TO_l^{(J)}, \forall l = \overline{1, L} \cdot \quad (29)$$

The time gain in a distributed or cloud system does not differ from the time gain in a parallel system.

The downtime in a distributed or cloud system:

$$T_{np}^{(J)} = \sum_{k=1}^{q-1} (k * (T_c^{(J)} - TO_{q-k+1}^{(J)})) \cdot \quad (30)$$

The processing time $T_p^{(J)}$ in a distributed or cloud system does not differ from the processing time $T_p^{(J)}$ in a parallel system.

3 Software for data processing in robotic neurocomputer systems

Basing on the obtained analytical expressions, we have developed optimization methods for CSs of neurocomputer data processing that relate to the task of achieving similar processing times for all subprograms and all segments of the program code [7].

For practical study, we chose the *NP Studio* software platform to analyze and optimize the CSs for neurocomputer data processing.

In keeping with the developed classification of structures, we have implemented a choice of five options: vector, pipeline, vector-pipeline, pipeline-vector, and arbitrary architecture.

We have developed a module for modulation, analysis and optimization. The results of this analysis are the following:

1. For each subprogram: the number of commands; number of nul commands; subprogram execution time; processor gain time; processor lag time; processor downtime; processing time; processing time as part of the NPCS; data transfer time; load quotient; average tacts per command; delay percentage; nul command percentage; command left-part nul percentage; delays of decoding, execution, transfer, loading; signals at the input bus, local bus, global bus, weight bus.
2. For NPCS: volume of transferred data; productivity factor, load quotient; system runtime; single-processor runtime; pipeline cycle; NPCS loss time; NPCS gain time; NPCS downtime; NPCS processing time; total NPCS lag time; total NPCS downtime; total NPCS processing time; total lag time; total downtime; total processing time; overall evaluation of NPCS without NPCM; overall evaluation of NPCS.

For work convenience with various source data a task model was implemented, to automatically control electric and mechanical systems in the developed Visual Programming subsystem of the NP Studio software platform. The chief item in the subsystem is the notion of a functional unit, connectable to other functional units to perform certain functions.

The workspace of the software application is divided into functional parts:

1. The area of functional units, instance of which may be transferred to the modeling workspace with the drag-and-drop function. The same area contains control items for visualization and deletion of connections between items. When added to the workspace, each item gets a unique ID, which is a concatenation of the item category, the item number within the category and the index number of the item, e.g., "O2.9" defines an item from category Output Signals, Number, with "9" as its unique ID.

2. The workspace (which, in its turn, can be represented as visual items and connections between them, or as a matrix of connections between items.

The theoretical and practical results of the study have been used for development (in cooperation with the Institute of Machinery Studies, Russian Academy of Science) of a specialized neurocomputing robotic device based on neuroprocessors for automated control of modules in electric-mechanical systems (specifically, hexapods) in a near real-time mode [10-13].

4 Conclusion

The paper proposes theoretical and practical conclusions that represent mathematical tools, as well as algorithms and software that are necessary for optimization of computing systems based on conceptually new computing hardware: neurocomputers. The results obtained can be employed in analysis and optimization of tasks that use multiple neural-based computing, e.g., implementations of neurocomputer systems with automated administration of SEMS modules (Smart electromechanical systems).

The research was carried out within the framework of the assignment for the performance of public works in the sphere of scientific activity within the framework of the initiative scientific project of the state task of the Ministry of Education and Science of the Russian Federation No. 2.9519.2017 / BC on the topic "Technologies for parallel processing of data in neurocomputer devices and systems".

References

1. S. Pande et al., Neural Process Lett., **38**, 131–153 (2013)
2. N. Izeboudjen, C. Larbes, A. Farah, Artif. Intell. Rev., **41**, 491–534 (2014)
3. P. Kolinummo, P. PASI Pulkkinen, T. Hämäläinen, J. Saarinen, Neural Processing Letters, **12**, 171–82 (2000)
4. C. Eliasmith et al., Science, **338**, 1202–1205 (2012)
5. E. Stromatias, F. Galluppi, C. Patterson, S. Furber, The International Joint Conference on Neural Networks (IJCNN), 1–8 (2013).
6. S.K. Esser et al., The International Joint Conference on Neural Networks (IJCNN) (2013)
7. R. Preisslet al., International Conference for High Performance Computing, Networking, Storage and Analysis (SC) (2012)
8. M.M. Shulakeret al., Nature **501**, 526–530 (2013)
9. D.S. Modha, R. Singh, Proc. Natl. Acad. Sci. U.S.A. **107**, 13485–13490 (2010)
10. V. Ruchkin, V. Romanchuk, R. Sulitsa, 2nd mediterranean conference on embedded computing (MECO), 58–61 (2013)
11. V. Ruchkin, V. Fulin, B. Kostrov, V.Romanchuk, E. Ruchkina, 4nd mediterranean conference on embedded computing (MECO), 45–50 (2015)
12. E. Diken et al., Microprocessors and Microsystems, **38**, 947–959 (2014)
13. R. Jordans, L. Jozwiak, H. Corporaal, 3rd mediterranean conference on embedded computing (MECO), 32–35 (2014)

A comparison of multi-style DNN-based TTS approaches using small datasets

Siniša Suzić[1,*], *Tijana* Delić[1], *Vladimir* Jovanović[2], *Milan* Sečujski[1], *Darko* Pekar[2] and *Vlado* Delić[1]

[1]Faculty of Technical Sciences, University of Novi Sad, 21000 Novi Sad, Serbia
[2]AlfaNum Speech Technologies, 21000 Novi Sad, Serbia

Abstract. Studies have shown that people already perceive the interaction with computers, robots and media in the same way as they perceive social communication with other people. For that reason it is critical for a high-quality text-to-speech system (TTS) to sound as human-like as possible. However, a major obstacle in creating expressive TTS voices is that the amount of style-specific speech needed for training such a system is often not sufficient. This paper presents a comparison between different approaches to multi-style TTS, with focus on cases when only a small dataset per style is available. The described approaches have been originally proposed for efficient modelling of multiple speakers with a limited amount of data per speaker. Among the suggested approaches the approach based on style codes has emerged as the best, regardless of the target speech style.

1 Introduction

Text-to-speech (TTS), the branch of speech synthesis based on producing artificial speech from an input text, has been a popular research subject for decades, owing to its far-reaching applications in the field of human-computer interaction (HCI). TTS-based screen readers help people with visual and reading impairments in their everyday lives. Driver assistants allow drivers to focus on the road while performing various other tasks. Automated call centers offer quicker response, since waiting time for human operator to become available is eliminated. Speaking robots help children and adults with autism to cope with their conditions [1, 2].

An important goal of speech synthesis is to make synthesized voice as human-like as possible, based on the assumption that such a voice is more pleasant for a human listener [3]. Synthesized speech should not just communicate information, it should do so in a natural manner, by conveying linguistic cues such as prosodic stress as well as paralinguistic information such as emotional state. Furthermore, there is a need to support task-specific speech styles, such as news, commercials, storytelling and warnings [4].

It has been shown that emotion, mood and sentiment affect attention, memory, performance, judgement and decision making in humans. Emotions are contagious, in a sense that a perceived emotional cue in an interlocutor can affect a person's emotions. Excitement for a product shown by an artificial speaker can make the human listener excited. A humorous voice and positive mood can affect the listener's mood [5], which has therapeutic implications.

Studies have shown that people already perceive the interaction with computers, robots and media in the same way as they perceive social communication with other people in real life. People are polite while communicating with computers, and react to male and female synthesized voices differently [6]. The importance of expressing emotions and attitudes in communication is shown in [7] since expressive robots are proven to be preferable over efficient ones. It is more comfortable to perceive an artificial agent as a real person than to think about the complexity and implications of a communicating machine. This innate human tendency to behave naturally in the interaction with computers should only be encouraged by improving the expressiveness of the synthesized voice.

These examples illustrate why speaker style adaptation is an area of very active research. Although earlier TTS techniques such as concatenative synthesis produce synthesized speech of good quality, it is impossible to adapt them to new voices or styles without recording and annotating large amounts of speech data for each new voice/style combination. For that reason, the concatenative approach has given way to parametric approaches, particularly deep neural networks (DNNs), which have been the focus of research of the majority of the scientific community for some time. Deep neural networks have benefited the most from the development of faster hardware and larger storage and bandwidth capabilities, since their performance does not reach a plateau as the amount of data increases, as is the case with other machine learning techniques [8]. They are also the first parametric method which gives results comparable with concatenative synthesis.

* Corresponding author: sinisa.suzic@uns.ac.rs

Clearly, an ideal TTS system should not only synthesize speech in a certain style or emotion but should also be able to deduce the target style or emotion based on the analysis of the input text. However, since the latter functionality is outside the scope of this text, the problems associated with it will be abstracted away and it will be assumed that the target style and emotion are known.

As is the case with neutral-style synthesis, emotional concatenative synthesis requires a fairly large amount of speech data for each target emotion [9], although there have been attempts to alleviate this problem by modifying F0 and duration curves using rules learned on small emotional databases [10]. Multiple speech styles and speaker emotions are much better handled by parametric approaches, owing to their flexibility [11]. A detailed analysis of concatenative and parametric approaches to emotional speech synthesis is given in [12]. However, the synthesis methods compared in that paper rely on a large emotional speech database consisting of approximately 100 minutes of speech per emotion. On the other hand, the main idea of this paper is to investigate the performance of different methods applied to multi-style DNN based synthesis in case when only on a small amount of speech data per style is available. The experiments were conducted on a speech database of American English, although it can be expected that the results are largely language independent.

The paper is organised as follows. The introduction is followed by an overview of single-style TTS based on DNN. Section 3 describes the methods used for speaking style modification in detail. Section 4 gives the details concerning the algorithms and models used, and presents the results of our experiments. The concluding section discusses the results and outlines the directions of further research.

2 DNN based TTS

The input text is firstly processed to extract linguistic features relevant for synthesis, which is referred to as *front end*. After that, the speech signal is produced from the phonetic transcription of the text augmented with the extracted linguistic features. The module charged with this task, referred to as *back end*, is based on two neural networks implemented using the *Merlin* toolkit [8].

The first network is used for modelling the durations of phonetic segments. The inputs and outputs of this network are phoneme aligned. The input for each hidden HMM state of the current phoneme is a binary vector containing linguistic information such as the identity of current phone and adjacent phones, number of words in the sentence, phones and syllables in a word, prosodic events indicated by ToBI tags [13], etc. The output is the vector of durations of the corresponding HMM state. State level durations are obtained using forced alignment proposed in *Merlin*. For all experiments in this paper, 5 HMM states per phoneme are used.

The second network is used for acoustic modelling. Its inputs and outputs are frame aligned. The input feature vector is the same as for the first network but extended with a number of frame and phone features such as the duration of the current phone, index of the current state, etc. In the training phase the target output is the vector of acoustic features extracted from wave files by the WORLD vocoder [14]. Acoustic features include MGCs (mel-generalised cepstral coefficients), BAP (band aperiodicities) and log F0, and they are further extended with their first and second derivatives as well as an additional flag indicating whether the current frame is voiced or unvoiced (V/UV). In all experiments in this paper, 40 MGCs, 1 BAP, 1 log F0 and 1 V/UV feature are used, yielding output feature vectors of length 127.

Both networks consist of 4 hidden layers with 1024 neurons each. The first three have tangent hyperbolic as the activation function, while the fourth layer is recursive and uses LSTM neurons. The output layer is linear. Objective function used is mean squared error. The input features are normalized to the interval [0.01, 0.99], while the output features are z-normalized. Networks are separately trained by backpropagation and stochastic gradient descent optimization.

In the synthesis phase, the outputs of the first network are provided to the second one, and the outputs of the second one represent target acoustic features. Smoothness of static features is achieved by using the maximum likelihood parameter generation algorithm [15], taking into account the predicted dynamic features. After the formants are further enhanced by post filtering, the acoustic features are fed to the WORLD vocoder in order to generate speech waveforms.

3 Methods for style conversion

In this chapter, three different approaches for the generation of synthesized speech in different speaking styles will be investigated. All of them have been originally proposed for efficient modelling of multiple speakers with a limited amount of data per speaker. The approach presented in 3.1 belongs to the group of methods for multi-speaker modelling, while the approaches presented in 3.2 and 3.3 can be considered as TTS adaptation methods.

3.1 Style code

The speaker code approach is presented in [16]. It is based on extending the input feature vector with information about speaker identity. Originally, this information is represented using a one-hot vector, and this concept is further extended in [17], by adding information about gender, age, etc. In [18], a similar approach is implemented, but instead of defining speaker codes, i-vectors are used. In the synthesis phase, one speaker voice is chosen by supplying the code of the desired speaker.

In case of multi-style modelling, styles are coded just as one-hot vectors. The input feature vector is extended with additional N binary features, one for each available style (angry, happy, etc.). Only one of the N features is set to 1, while all the others are set to zero (Fig. 1). In case of neutral style, all N features are set to zero.

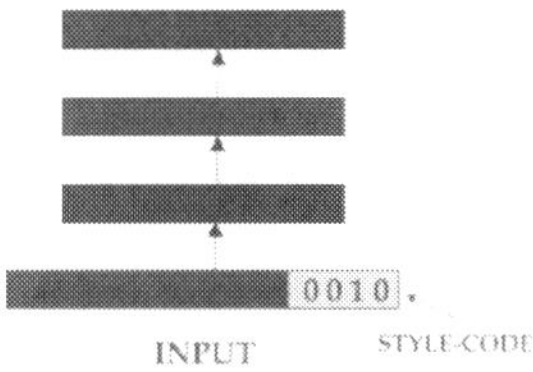

Fig. 1. Multi-style model with style codes.

The choice of a certain style is defined at word level since in natural speech a certain emotion does not need to extend over the entire utterance. A conventional DNN training is performed with input thus extended, both for the duration model and the acoustic model. At synthesis stage, the desired value is set for each word, the input is extended with a corresponding one-hot vector and the synthesized speech will have the characteristics of the desired style.

In multi-speaker modelling it is recommended to use equal amounts of data of each speaker in order to prevent the over-fitting of the final model to a specific speaker's voice. On the other hand, in multi-style modelling it is reasonable to assume the existence of much more speech data corresponding to the neutral style than to any other style. This is not necessarily a drawback and it can even be beneficial, since the speaker is unique for all styles, and many contexts not covered by a certain style can be compensated by the neutral style.

3.2 Separate output layer

Another approach of constructing a multi-speaker model is proposed in [19]. The idea is to have separate output layer(s) for each of the speakers in the database, while the hidden layers are shared between all speakers. This approach is based on the assumption that the shared layers are generally speaker independent and encode linguistic knowledge, while the output layer(s) encode the knowledge related to a particular speaker. This assumption is realistic having in mind the task of the NN in speech synthesis. Namely, NN transforms linguistic features to acoustic features. Linguistic features are extracted from text and are not dependent on any of the voice characteristics. For that reason, the lower layers can be considered as a speaker independent transformation of linguistic features. On the other hand, output acoustic features are speaker dependent and thus the output layer(s) should represent the acoustic space of each individual speaker.

Starting from this hypothesis in [19] is suggested that the shared layers should be trained on all available training data, while speaker dependant parts should be trained only on speaker-specific data. This can be achieved by propagating the error only through appropriate layers. The information about the speaker identity can be provided in different ways and it will be used for the selection of the output layer(s) through which the error will propagate during training, as well as for the activation of desired output layers at synthesis stage. Such an approach still requires a moderate quantity of data per speaker for the corresponding output layer to be trained adequately. Unfortunately, this requirement is rarely met in real-life scenarios.

However, Fan et al. also propose a method for the adaptation on a new speaker when very little data is available. In that case, they consider the speaker independent part of the network to be a global linguistic feature transformation, and one of the speaker dependent parts is used for adaptation to the new speaker. In the training phase only that part is updated using the limited speaker-specific data.

In case of multi-style modelling, the last layers are separated and there is one for each desired style, including neutral (Fig. 2). Since the output of the network is a vector of acoustic features, it is supposed that the last layer is sufficient to differentiate between styles. The network is trained in the same way, by back-propagating the error through the corresponding style-dependent output layer as well as shared layers.

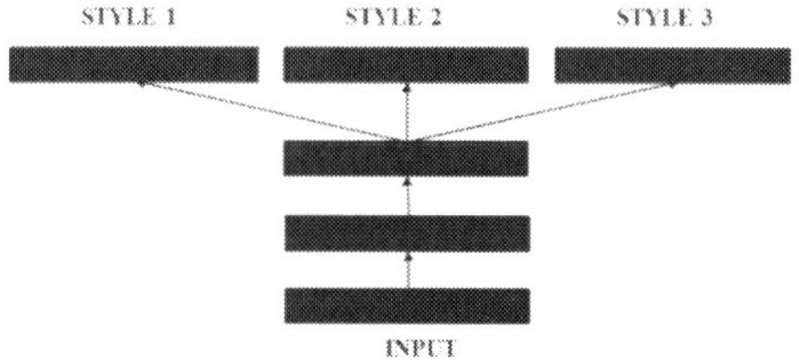

Fig. 2. Multi-style model with separate output layers.

One of the assumptions in the paper is limited amount of data per each style, insufficient to train the output layers from scratch, which would result in synthesized speech of poor quality. For this reason all outputs were firstly trained using neutral-style data. Afterwards only the output layers were trained using style-specific data by back-propagating the error through corresponding output layers only, keeping the weights in shared layers fixed.

3.3 Re-trained model

Although very intuitive and simple, this idea, to the best knowledge of the authors, has not been reported in literature yet, at least not with the aim of changing the speaker identity or speaking style. In [20] we firstly explained the idea in detail, but applied it to the adaptation to different speakers. If starting from model trained on a big database of speaker A, it has been shown that much less data of speaker B is needed to adapt the model to produce speech with characteristics of speaker B than when starting form randomly initialized model. The idea is based on the assumption that the model trained to any speaker has to be closer to the model for any other speaker of the same language than a randomly initialized model. Contrary to the idea presented in 3.2, the adaptation was applied to the entire network, not just the output layer(s).

The case of adapting the model trained on a big neutral-style database to another speaking style seems even easier than adapting it to a new speaker, since the model does not have to capture the characteristics of the speaker. This is in accordance with the fact that the voice of the same person in different emotional states exhibits far less variety than can be found in voices of different persons. The adaptation is completely straightforward. The initial model is trained on a big neutral-style database, and a small style-specific database is used for additional training of the model thus obtained. Both the duration model and the acoustic model are trained using conventional DNN training.

4 Results

For the synthesizers described in sections 3.2 and 3.3, 555 binary linguistic features were used, while for the synthesizer described in section 3.1 there were 3 more features related to the style (angry, happy and sarcastic). All synthesizers had the same architecture, except that in case of the synthesizer from 3.2 the single output layer was replaced with 4 style-dependent output layers. All other parameter values were as specified in Section 2.

For all experiments the same database of American English was used. It consists of 3 h of neutral speech and 5 minutes of speech for each of the 3 styles (3.5 h and ~7 min per style, if silences are included). The speaker is male, a native voice talent. For training the duration model, leading and trailing silences from each utterance were excluded and silences longer than 500 ms were shortened to 500 ms, while any remaining silent phonetic segment was used for training as is. Prior to training the acoustic model, all silences were shortened to 60 ms.

For the purposes of testing and comparing all synthesizers, two different listening tests were conducted among 20 amateur listeners, who are not native English speakers but are proficient in English. For each style, 5 sentences previously not seen during the training were synthesized to be used in the listening test.

4.1 MUSHRA test

The first test was a MUSHRA test [21] aimed at measuring the naturalness of synthesized utterances. It consisted of 15 questions (5 per style). In each question there was one reference utterance – an original recording of natural human speech, clearly indicated to the listener. Besides the reference utterance, there were 4 more utterances – one the same as the reference, and others synthesized by the synthesizers described in sections 3.1, 3.2 and 3.3 respectively – in a random order. Each subject had to rate the naturalness of each utterance, i.e. similarity to the reference utterance on a scale from 0 to 100. In this test the listeners were instructed to focus on the naturalness of style and intonation, and to disregard synthesis artefacts at segmental level. The results of this test are given in Figure 3.

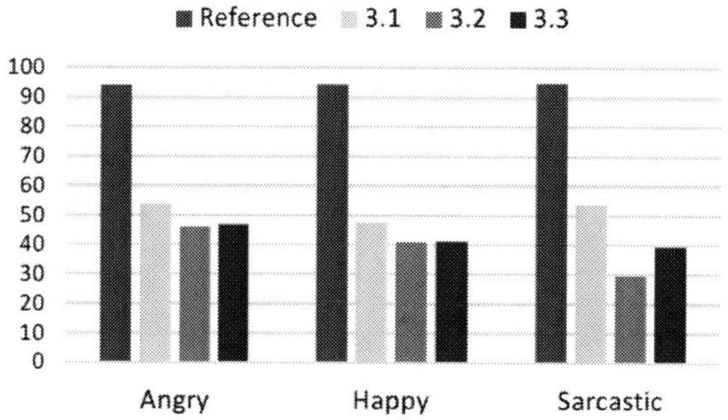

Fig. 3. Grades from MUSHRA test for angry, happy and sarcastic style, respectively.

The average grade for natural speech is 94, while the synthesizers described in sections 3.1, 3.2 and 3.3 were scored 52, 39 and 42, respectively. Based on these scores we can conclude that synthesizer 3.1 is significantly better than the other two, which achieved similar results. The angry style was consistently scored as the most natural-sounding one, regardless of the synthesis method used.

We have also investigated variations across utterances in order to check the consistency of each of the described methods. These results are given in Figure 4. It can be concluded that synthesizer 3.1 is not only the one with the highest grade, but also the most stable one. The range of the average grade is three times wider than the range for natural speech, but around 40% narrower than for other two synthesizers, and its value is 18, while the interquartile range is 10. The least stable synthesizer appears to be the synthesizer 3.2.

4.2 MOS test

The purpose of the second listening test was to measure the quality of synthesized utterances, with particular focus on the intelligibility and absence of synthesis artefacts. The test consisted of 3 groups of sentences. Each group represented one style and it consisted of 4 sets of utterances. One set consisted of original recordings, and other three were utterances synthesized by synthesizers described in 3.1, 3.2 and 3.3. In each of the 4 sets per style there were 5 sentences with identical textual content. The listeners were asked to rate the quality of each set independently on a scale from 1 (extremely bad) to 5 (extremely good). The results are presented in Figure 5.

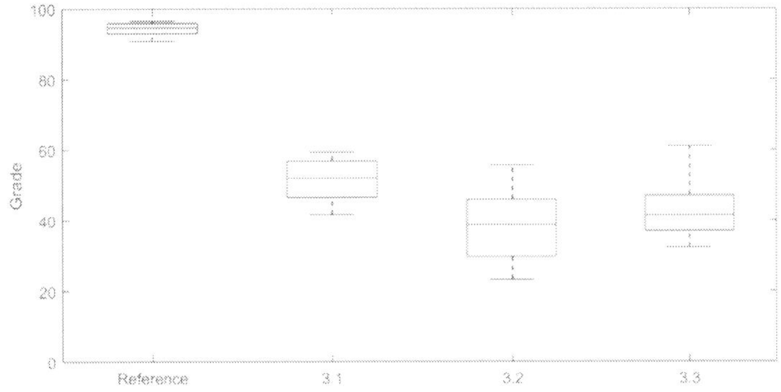

Fig. 4. Variations of grades across sentences.

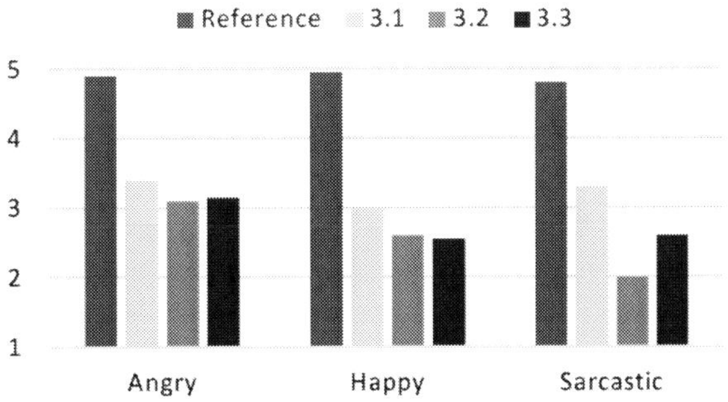

Fig. 5. Grades from MOS test for angry, happy and sarcastic style, respectively.

The average grade for natural speech was 4.9, while the sentences synthesized by the synthesizers 3.1, 3.2 and 3.3 were given grades respectively 3.2, 2.6 and 2.8 respectively. As in the previous test, the 3.1 synthesizer has shown to be the best. Further analysis reveals remarkable correlation with the results of MUSHRA test.

5 Conclusion

The paper presented methods that can make it possible to have machines which will be not only able to communicate, but also to express its virtual emotions and attitudes through speech. This will raise the level in HCI and make it a lot easier to communicate more naturally with call centres, appliances or robots.

Among the investigated methods the one based on style codes produced the speech of best quality. The re-training method and approach with separate output layers are comparable in terms of quality, although the latter is slightly more consistent but suffers from more audible artefacts. Among these three methods style-code approach is the only one in which the input feature vector is extended by additional information and which simultaneously updates all weights and biases regardless of whether expressive or neutral sentence features are used as inputs. In the other two approaches the nodes are updated with only a restricted amount of data for each style – retraining method updates all nodes in the network while in the approach with separate output layers, only the nodes from the output layer are updated. This suggests that it is beneficial to train the entire network with all databases simultaneously (with additional style-related information) than to tune the parameters of an already trained model with a small amount of new data. However, the main drawback of approach using style codes is that in case of adding a new voice, the entire model has to be re-trained from scratch, which can be extremely time-consuming due to the size of the database of neutral style speech. On the other hand, in case of other two approaches, the addition of a new style requires just the additional training on the small database representing the desired style, which is typically much faster, since style-specific databases are very small in comparison to the neutral style database.

As can be seen from Figures 4 and 5, the grades obtained in the two tests are quite similar. This is somewhat surprising, since in the first test the listeners were specifically instructed to disregard synthesis artefacts and focus on naturalness of the style and intonation, while in the second one they were asked to evaluate intelligibility, i.e. the absence of artefacts. This suggests that amateur listeners find it difficult to distinguish between different aspects of speech synthesis quality, and are very sensitive to synthesis artefacts. This conclusion has been confirmed by several TTS experts (with more than 10 years of experience in the field), who agreed that the degree to which the emotional states were conveyed was quite satisfactory, but objected that the synthesis is flawed by artefacts typical of systems trained with very small quantities of data. In order to prevent such situations, future experiments in this field should include a preliminary test intended to emphasize the difference between the intelligibility and the naturalness of synthesized speech to the listeners.

Our future research will investigate the possibility of separating multiple output layers as well as certain combinations of explained approaches, e.g. providing the style code in the training of shared layers within the approach described in Section 3.2 in order to get more stable and natural multi-style speech.

The research was conducted within the project "Development of Dialogue Systems for Serbian and Other South Slavic Languages" (TR32035) and "Interdisciplinary Research into the Quality of Verbal Communication" (OI178027), financed by the Ministry of Education, Science and Technological Development of the Republic of Serbia as well as the EUREKA project DANSPLAT (E!9944).

References

1. L.I. Ismail, S. Shamsuddin, H. Yussof, F.A. Hanapiah, N.I. Zahari, Proc. Eng, Robot-based Intervention Program for Autistic Children with Humanoid Robot NAO: Initial Response in Stereotyped Behavior, **41**, 1441–1447 (2012)
2. A. Csapo et al., Cognitive Infocommunications, IEEE 3rd International Conference, 667–672 (2012)
3. L. Gong, C. Nass, C. Simard, Y. Takhteyev, Usability Evaluation and Interface Design: Cognitive Engineering, Intelligent Agents, and Virtual Reality, **3**, 39–394 (2001)
4. M. Abe, Progress in speech synthesis, 495–510 (1997)
5. S. Brave, N. Clifford, The human-computer interaction handbook, 94–109. (2002)
6. B. Reeves, C. Nass, *The Media Equation: How People Treat Computers, Television, and New Media Like Real People and Places*, (Cambridge university press, 1996)
7. A. Hamacher, N. Bianchi-Berthouze, A.G. Pipe, K. Eder, 2016 25th IEEE International Symposium on Robot and Human Interactive Communication (RO-MAN), 493–500 (2016)
8. Z. Wu, O. Watts, S. King, Proc. 9th ISCA Speech Synthesis Workshop (2016)
9. M. Bulut, S. Narayan, A. Syrdal, Proc. ICSLP 2002, 1265–1268 (2002)
10. J. Pitrelli et al., IEEE Trans. Speech Audio Process, **14**, 1099–1108 (2006)
11. M. Tachibana, J. Yamagishi, T. Masuko, T. Kobayashi, IEICE Trans. Inf. Systems, **89**, 1092–1099 (2006)
12. R. Barra-Chicote, J. Yamagishi, S. King, J.M. Montero, J. Macias-Guarasa, Speech Communication, **52**, 394–404 (2010)
13. M. E. Beckman, J. Hirschberg, S. Shattuck-Hufnagel, Prosodic Typology – The Phonology of Intonation and Phrasing, 9–54 (2005)
14. M. Morise, F. Yokomori, K. Ozawa, IEICE TRANSACTIONS on Information and Systems, **99**, 1877–1884 (2016)
15. K. Tokuda, T. Masuko, T. Yamada, T. Kobayashi, S. Imai, Proc. IEEE International Conference on Acoustics, Speech, and Signal Processing, **3**, 1315–1318 (2000)
16. N. Hojo, Y. Ijima, H. Mizuno, INTERSPEECH, 2278–2282 (2016)
17. H.T. Luong, S. Takaki, G.E. Henter, J. Yamagishi, Acoustics, Speech and Signal Processing (ICASSP), 4905–4909 (2017)
18. S. Yang, Z. Wu, L. Xie, Proc. Signal and Information Processing Association Annual Summit and Conference (APSIPA), 1–6 (2016)
19. Y. Fan, Y. Qian, F.K. Soong, L. He, Acoustics, Speech and Signal Processing (ICASSP), 4475–4479 (2015)
20. T. Delić, S. Suzić, M. Sečujski, D. Pekar, 17th International Symposium INFOTEH-JAHORINA (to be published)
21. E. Vincent, M. Jafari, M. Plumbley, Proc. UK ICA Research Network Workshop (2006)

Processing models for conflicting user requests in ubiquitous corporate smart spaces

Dmitriy Levonevskiy*, *Irina* Vatamaniuk and *Anton* Saveliev

St. Petersburg Institute for Informatics and Automation of the Russian Academy of Sciences (SPIIRAS), 39, 14th Line V.I., 199178, St. Petersburg, Russia

Abstract. This paper considers processing of conflicting user requests in ubiquitous corporate smart spaces. The formulated problem consists in the contradiction between the limitation of available smart space resources to perform the conflicting user requests and necessity to provide the proper quality of service in corporate smart spaces. The principles of constructing the simulation model are described. The experiments were carried out basing on a model of the SPIIRAS digital signage service. Several task management strategies are discussed, an assessment of their effectiveness is given. The research is aimed at improving the quality of service and user experience in human-computer interaction within the corporate smart spaces.

1 Introduction

Smart spaces provide the most matter-of-course way of interaction between users and information environment of organizations [1]. The success of smart space implementation depends on three directions in science and technology: ubiquitous computing, ubiquitous communication and multimodal interfaces [2]. The easiness of access to information as well as the convenience of available user interfaces are the distinguishing features for such systems. Nevertheless, there are still a set of problems to solve while developing and deploying ubiquitous multimodal systems of human-computer interaction. These problems are related to information security, reliable and efficient node linking, managing heterogeneous devices, task priorities, assessing quality of service, etc. One of the significant aspects of task management is processing of multiple requests generated simultaneously by different users and securing the proper quality of information perception while interacting with multiple users. This paper discusses available approaches to clarify the problem mentioned above.

2 Related work

The main principles of providing the proper quality of multi-user service in multimodal distributed systems are considered in [3]. The authors propose a solution to handle multiple user requests for a multiscreen software-defined IoT digital signage system. The proposed method is based on the dynamic link allocation process which is derived due to the software-defined networking (SDN) architecture of the system. In such system, the functionality of the physical layer is decoupled from the control layer. So, the SDN controller of the digital signage cloud network estimates the network status and the number of user queries, and automatically updates the rules for query processing and reallocates system resources. Also, the authors propose the original technique for so called invisible communication which provides user with a tool for an easy access to the network services. Such interactivity between stationary terminals and user mobile devices substantially expands user experience.

As a rule, index of perceived quality is the most significant one for media streaming from the point of view of user experience. Its estimation and prediction based on the extended SSIM metric (SSIMplus index) are discussed in [4]. Recent approaches based on various techniques (machine learning, software-defined networking, flow dynamics et al.), can be found in works [5-7]. Estimation of the perceived quality has a great potential to provide the best possible service to each individual user in distributed systems, especially in multiuser ones.

Quality of visual service provided by a system of digital signage is discussed in [8]. This metric is determined based on the level of 2D visual sensitivity, and on the ability to perform 3D binocular fusion by users located at various viewing distances. Its use allows automatic control of visual information and radio resources for multiple users without additional interactions.

Systems of digital signage (DS) find their application in retail services, third-party advertisement, entertainment, wayfinding, education, and transportation. In [9], an architecture for providing quality of service (QoS) and mobility-aware telescreen services in vehicular networks is presented. It allows managing user registrations, service advertisements, network information, and service quality due to the adaptive

* Corresponding author: DLewonewski.8781@gmail.com

selection of QoS parameters and a handover decision mechanism for providing seamless imagery effect and enhance user's multimedia perception.

A multi-screen environment provides a new opportunity for digital signage applications as an effective means for advertisements and information [10]. A software-based synchronization is proposed in that work to address the frame-level synchronization between multiple screens. It provides better performance and extends possibilities of a DS system by allowing the use of non-standard screens and screens positioned unconventionally in space.

In [11] QoS and quality of experience (QoE) Aware N-Screen Multicast Service is proposed. In such multiscreen system, users are dynamically combined in multicast groups depending on their devices' capacity and internet traffic. The perceived quality of the multimedia content provided to each group is dynamically adjusted to meet better the QoE requirements according to available resources of user devices. This scheme helps to optimize the use of network resources for specified quality content requirements.

3 Model development

A typical application area of distributed multimodal systems is corporate smart spaces that provide ubiquitous human-computer interaction between users and components of the smart space. As an example, we consider the process of interaction between users and MINOS (Multimodal Information and Navigation Cloud System) information service [12]. The information and navigation services of MINOS are implemented via ensemble of webcams, monitors and information screens helping visitors to navigate within the corporate environment of SPIIRAS. The webcams are involved in processes of registration, authentication and recognition of users and allows automating work of checkpoint and reception. The monitors and touchscreens help users to get required information easily.

Users generate requests by means of different modalities (speech, gestures, etc.). They can also control the service using their mobile phones or personal computers (for administrators). Each request needs to be processed, i.e. some task has to be performed and requires allocation of some resource (display and sound channels, user session) during some time. In case when there is only one user working with the system at a certain moment of time, every new request implies suspending any other requests if they require the same resource, so there is no conflict between them. But in case of multiple users, the system may encounter the impossibility of allocating the needed resource for all tasks. Then it is necessary to determine the optimal strategy of request managing that results in the highest possible QoS metrics.

Each request is characterized by a set of properties that include the original user identifier, the session identifier, the active modality and the requested task.

The optimality of a certain strategy depends on a set of parameters (request intensity, resource requirements of tasks, number of users, etc.).

Let us consider the task classification in MINOS:
- by distribution: local (should be processed by only one channel), distributed (should be processed by several channels), global (use all available channels);
- by priority: critical (require immediate suspending all other tasks), regular (have a relative priority in a service discipline), background (are processed only if there are no other tasks and immediately suspended by any non-background tasks);
- by execution procedure: continuous and suspendable;
- by length: with constant or variable length;
- by bindings to other tasks: single, sessional.

The request structure can be described more detailed in JSON format:

```
request: {
  user: {
    id,
    role
   },
  session: {
    id,
    start_timestamp
  },
  modality,
  task: {
    id,
    options,
    resumable,
    resources: [ {
      id,
      requiredAmount,
      requiredTime
    } ]
  }
}
```

The task priority depends on these options. So, to obtain QoS values, we should define:
- source data;
- method of priority designation;
- task management strategy.

Let us consider these aspects in more detail.

Source data include information about the process of each request type generation. The conditions of task generation are not strictly defined and may change depending on the context. However, according to the central limit theorem of the probability theory for event streams, the process of single request entrance can be described using the Poisson distribution, and its parameter (λ) characterizes the intensity of request emergence. For priority designation, the following principles have to be taken into account.

Firstly, the probability of accurate perception and recognition of multimodal requests is very significant. Requests sent from mobile devices and PC's are recognized the most accurately, because a user has to start the service application, open a session, scan the QR code of the smart space component (for mobile devices) and send a command through the application. These actions must be performed purposefully, and the probability of their accidental execution is extremely small. In case when speech recognition is applied, the probability of a mistake is higher, for gesture-based control it is even higher [13].

Secondly, credibility and legitimacy of incoming requests are to be considered. Lack of credibility may take place, for example, if the request flow from a certain user does not look like a human-generated one (it can possess a very high frequency of events or be contradictory). This problem can be caused by perception inaccuracy as well as malicious activity (for example, when a bot connects to the service interface). In the latter case the requests are illegitimate. In both cases some requests are denied or assigned the lowest priority.

Thirdly, task processing technique should depend on the possibility of the user to receive the requested data. So, if there is a request from a user assuming some data output on a stationary display, but this user is out of the display's visibility zone, the task can be delayed or cancelled after a defined time-out.

Fourthly, the task itself has properties like urgency defined automatically by the task type (or the user priority) or manually by the administrator, suspendability, etc.

Task management strategy types are studied well in the theory of queuing systems [14]. Buffering and service disciplines differ in priority presence, rules of task selection, queue length, etc. Any new task can be processed in one of the following way:

- immediate processing (possibly replacing current task(s));
- immediate cancellation;
- adding to the queue.

In the latter case disciplines like FIFO (First In, First Out) and LIFO (Last In, First Out) may be used. To simulate the specified processes, we have implemented a model described in the next Section.

4 Experiments

The conducted experiments consist in the simulation of the incoming user request flow and the process of performing the requests. The output of the simulation reflects the efficiency of implemented strategies. While modeling, several assumptions listed below were considered.

At first, this model does not require the background tasks to be taken into account as their execution does not affect processing the tasks with higher priority at all. Critical tasks are also omitted due to their low probability and irregularity.

At this stage of work, we simulated only local tasks to simplify the experiments. In general, processing of a new task consists of the following steps:

1. If there is enough available resource to process the task, it is performed immediately.
2. If the resource is in use by tasks of the same user, these tasks are cancelled, and the new task is performed.
3. In other cases, the system uses one of the strategies described further.

Strategies without priority include:

1. Systems with denials: the task is cancelled if there is no enough resource.
2. Systems with limited waiting time/queue length.

Strategies with priority may imply:

1. Shift tasks with lower priorities.
2. Suspend tasks with lower priorities.
3. Cancel tasks with lower priorities.

To assess the quality of task processing several values can be measured:

- percentage of the fulfilled tasks (relative throughput);
- percentage of the fulfilled tasks weighed by their priority;
- average waiting time;
- percentage of resource/channel usage.

The system model used in the experiments is implemented in Python. It takes the following input data: N types of single requests from different users, request intensities λ_i, $i = 1 \ldots N$, task priorities P_i depending on the request modalities. The requests consist in user inquiries about help and reference information. The model is able to use strategies mentioned above as well as their combinations. Three priority levels that correspond to the modalities (speech, gestures, mobile application) were used. On appearance of a new request, some action is performed: A_H (if a new task is of a higher priority), A_E (equal priorities), A_L (lower priority). In case when $A_H = A_L = A_E = A$, the strategy without priorities is used.

The result of processing of every task is characterized by success and delay (waiting time). Strategies with queues show higher delays, but higher average success level at the same time. Strategies where denials prevail have lower delays and lower success level. To obtain a complex strategy estimate, an effect index for the task i may be composed. This index depends on both success and delay. In case of denial, we can assume that the task delay $D_i = \infty$, and the following estimate of the quality of service may be used to proceed to finite values:

$$E_i = \frac{D_0}{D_0 + D_i}, \quad (1)$$

where D_0 is the delay considered to reduce the effect estimate E_i twice.

If a task can be suspended with a partial effect, actual time of its processing should also be taken into account. In this case E_i may be represented as

$$E_i = \frac{T_{Fi} D_0}{T_{Pi}(D_0 + D_i)}, \quad (2)$$

where T_{Fi} is the actual time of the task i and T_{Pi} is the planned time. To get an integrated estimate of a certain strategy using task priorities, we can use the following formula:

$$E = \frac{\sum_{i=1}^{N} P_i E_i}{\sum_{i=1}^{N} P_i}, \quad (3)$$

where P_i is the priority of the task i and N is the number of tasks.

If the request intensity is low, the tasks have a relatively small number of conflicts, and the value E is near 1 and does not depend significantly on the strategy. With the growth of intensities, different strategies result in different effects. Figure 1 illustrates the effect dependencies on the request intensity if some common strategies without priorities are applied.

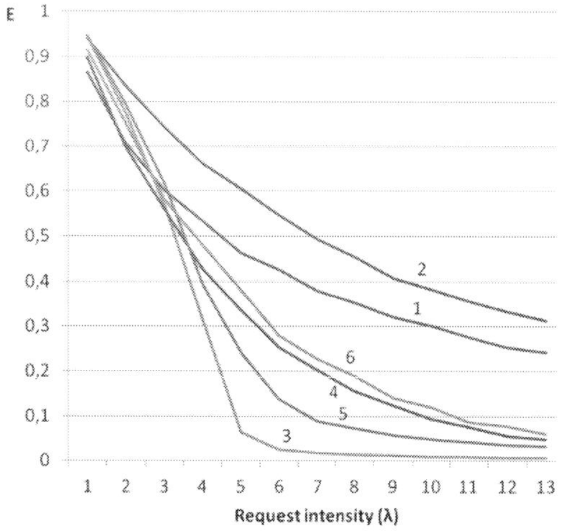

Fig. 1. Efficiency of the strategies without priority: 1 – denial, 2 – cancelling conflicting tasks, 3 – FIFO, 4 – LIFO, 5 – FIFO (with maximum delay 1.0), 6 – LIFO (with maximum delay 1.0).

The graph shows that strategies with denials are more efficient than ones with queues if the tasks are treated as if they were of an equal priority. In particular, this can be explained by the effect sensitivity to delays. In the experiments it was assumed that $D_0 = \overline{T}_P$, which means a considerably low sensitivity in the terms of human-computer interaction. The interface response time by which the system is perceived by the user as controllable does not exceed several seconds [15]. Thereby, it is evident that any stricter conditions for D_0 will only reduce the estimate for strategies with queues.

It is also noticeable that the limitation of queue length in strategies with queues affect positively on the strategy estimate because this condition leads to cancellations of tasks with high delays that hold the resource but do not make a perceptible impact on the effect estimate. Nevertheless, such strategies concede to the ones with denials.

Then, let us consider the strategies with priorities and compare them to the most efficient strategies without priorities. The conducted experiment included the set of strategies that have at least two different types of actions among A_H, A_L, A_E, and these actions do not contradict the task priorities. The graphs for the most efficient strategies revealed are shown on Figure 2.

Figure 2 illustrates the effect values for the following strategies:
1: denial (A = denial);
2: cancel conflicting tasks (A = cancellation of other tasks);
3: cancel conflicting tasks if the new task has an equal or a higher priority (A_L = denial, $A_H = A_E$ = cancellation of other tasks);
4: cancel conflicting tasks only if the new task has a higher priority ($A_L = A_E$ = denial, A_H = cancellation of other tasks);
5: add the new task to the FIFO queue only if it has a lower priority, otherwise cancel conflicting tasks (A_L = FIFO, $A_E = A_H$ = cancellation of other tasks).

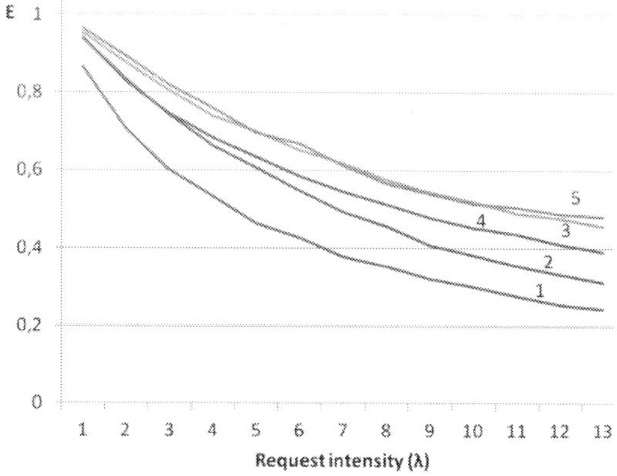

Fig. 2. Comparison of the strategies without priority (1, 2) and with priority (3, 4, 5).

For the majority of strategies, the simulation shows that adding tasks to the queue results in the lower value of the effect E. However, using the FIFO queue for processing low-priority tasks leads to a moderate rise of the effect value. This can be explained by the fact that denials for these tasks will cause $E_i = 0$, but using queues allows them to retain some probability of execution providing that there are no prevailing tasks. Thus, this strategy results in $E_i \geq 0$.

Experiments with the other strategies with priorities show that their efficiency is not so high (Figure 3).

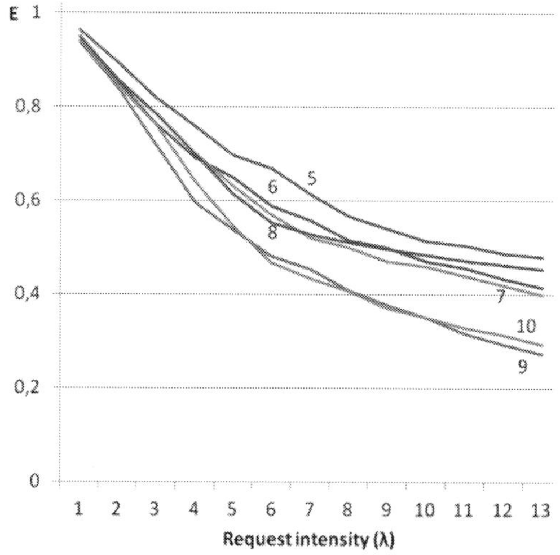

Fig. 3. Efficiency of the strategies with priority.

Figure 3 illustrates the strategies:
5: add the new task to the FIFO queue only if it has a lower priority, otherwise cancel conflicting tasks (A_L = FIFO, A_E = A_H = cancellation of other tasks).
6: A_L = denial, A_E = cancellation of other tasks, A_H = LIFO;
7: A_L = A_E = cancellation of other tasks, A_H = LIFO;
8: A_L = FIFO, A_E = cancellation of other tasks, A_H = LIFO;
9: A_L = FIFO, A_E = A_H = LIFO;
10: A_L = FIFO, A_E = LIFO, A_H = cancellation of other tasks.

So, it can be seen that under given conditions the 5th strategy is optimal. It uses the FIFO queue for low-priority tasks and cancels the conflicting tasks in other cases.

5 Conclusion

Choosing the right query strategy allows making the human-computer interaction friendlier, increasing the quality of service and improving user experience of interacting with corporate smart spaces.
The directions of further research include:

1. Analysis of sessional request features, working out the methods of determining their credibility and legitimacy and search of the optimal strategies of their processing. Sessional requests make the event flow not ordinary, but the processes of generating sessions and requests within a certain session may still be considered ordinary, so the models for Poisson event streams can be applied for these processes.

2. Analysis of interaction via several data output channels, homogeneous or using different modalities. These channels may include video and audio channels of stationary smart space nodes or portable user devices, as well as channels of human-robot interaction. The theory of multichannel queuing systems can be the theoretical basis for this study.

3. Analysis of how distinct types of tasks affect the efficiency of the smart space. In particular, this includes classifying tasks of various applied MINOS services besides Digital Signage (navigation, videoconferencing, etc.).

4. In the field of information security, there is a need to study the problem of trust to the user. In particular, the smart space has to be resistant to attacks from malicious users. This can be provided by means of integration with known methods of protection against attacks that are critical for smart spaces (denial of service, spoofing, etc.) [16]. There are also useful methods of protecting the smart space by distinguishing unreliable (deliberately false or mistaken) information [17]. Another approach to protect the smart space from incorrect information is the integrated simultaneous processing of data gathered using different modalities [18].

The presented work was supported by the Russian Science Foundation (grant No. 16-19-00044).

References

1. X. Zeng, H. Pei, International Conference on Information Computing and Applications 2012, **308**, 628–634 (2012)
2. R. Yusupov, A. Ronzhin, Herald of the Russian Academy of Sciences, **80**, 63–68 (2010)
3. M.A. Hossain et al., Int. J. of Distributed Sensor Networks, **7/12**, 1–14 (2016)
4. Z. Wang, K. Zeng, A. Rehman, H. Yeganeh, S. Wang, Applications of Digital Image Processing XXXVIII, **9599**, 95990Y (2015)
5. Y.T. Lin, E.M.R. Oliveira, S.B. Jemaa, S.E. Elayoubi, 2017 IEEE International Conference on Communications (ICC), 1–6 (2017)
6. H. Nam, K.H. Kim, J.Y. Kim, H. Schulzrinne, Global Communications Conference (GLOBECOM), 1317–1322 (2014)
7. Y. Xu, S. E. Elayoubi, E. Altman, R. El-Azouzi, Y. Yu, IEEE Transactions on Mobile Computing, **15** (11), 2762–2780 (2016)
8. H. Oh, H. Lee, I. Lee, S. Lee, (2015). Dig. Sig. Proc., **45**, 24-35 (2015)
9. G. Sarwar, F. Ullah, H.W. Lee, W. Ryu, S. Lee. Tehnički vjesnik, **24** (1), 177–185 (2017)
10. I. Shin, S. Lee, E. Lee, N.K. Lee, H. Lee, H. ETRI J., **38** (5), 868–878 (2016)
11. G. Sarwar, F. Ullah, S. Lee. J. of Sensors, **6**, 1–11 (2016)
12. I. Vatamaniuk, D. Levonevskiy, A. Saveliev, A. Denisov. 18th International Conference on Speech and Computer (SPECOM-2016), **9811**, 588–595 (2016)
13. H. Liu, L. Wang, Int. J. of Industrial Ergonomics, (to be published)
14. A. Penttinen. *Introduction to Teletraffic Theory* (2003)
15. S. Henty. UI Response Times. URL: https://medium.com/@slhenty/ui-response-times-acec744f3157
16. P. Eder-Neuhauser, T. Zseby, J. Fabini, G. Vormayr. Sust. Energy, Grids and Networks, **12**, 10–29 (2017)
17. V. Budkov, I. Vatamaniuk, V. Basov, D. Volf. 18th International Conference on Speech and Computer (SPECOM-2016), **9811**, 419–426 (2016)
18. K. Margiotoudi, S. Kelly, A. Vatakis. Procedia - Social and Behavioral Sciences, **126**, 154–155 (2014)

Implementation of impedance trajectory control on a 6-DoF manipulator

Igor Shardyko* and *Victor* Titov

Russian State Scientific Center for Robotics and Technical Cybernetics (RTC), 194064, 21 Tikhoretsky Prospect, Saint-Petersburg, Russia

Abstract. Aiming at enhancing the performance of a robotic system, i.e. of a robot manipulator, it is extremely important to ensure the safety of its functioning and the convenience of operating it. Special attention is required if there is a possibility of a collision, especially with costly equipment or with a person. Among the different means and approaches to prevent or react to such situations, implementation of force-torque control is considered in this paper. A trajectory task is analysed and an impedance control law is employed to achieve the desired robot behaviour. The effectiveness of the approach has been verified by using a 6-DoF elbow manipulator to perform two types of operations: a peg-in-hole task and path execution with obstacles on the way.

1 Introduction

There are a number of criteria to evaluate quality of robotic systems and their suitability to appropriate tasks. Among the whole variety of these criteria one can assess mechanical performance, functionality, reliability/safety and ergonomics. Mechanical performance comprises, for example, load capacity (and load to weight ratio), top and nominal speed, accuracy, and workspace dimensions. Functional criteria describe the range of tasks that a system is capable of due to its structure and overall design. However, even a fast and accurate manipulator with high load capacity and notable functionality does not value much if it cannot provide safe and reliable functioning, by which the ability to maintain its own operability is assumed along with preventing doing harm to workers that can be present in the workspace and the environment itself.

Easy operation and control is another important criterion, with different kinds of control under consideration, namely, manual control, copy control and automatic control. If it is too difficult for an operator to maintain control, it generally leads to the operator's fatigue and increases a chance of mistake. There are also situations when the operator physically cannot provide required motion, e.g. the motion of tool centre point (TCP) along a curvy surface keeping constant and limited contact with this surface. It is an extremely difficult task for the operator to keep the vector of TCP velocity in conformity with the surface curviness, not to mention that the available workspace view may not allow operator to estimate the surface shape. The other example is aligning the peg with the hole in a typical task illustrated on Figure 1. In this case it is possible to see the axes misalignment only in one projection, because of the static view of a single external camera.

Fig. 1. Aligning axes in peg-in-hole task.

Safe functioning and convenient control can be achieved through different approaches in robotics, most popular of which include machine vision [1, 2] and force-torque control [3-5]. An approach here is a general paradigm, first of all, a type of employed hardware, such as video cameras, 3D-scanners or force-torque sensors. A variety of specific methods implementing control laws and their modifications, and based on them concrete algorithms generally correspond to any particular approach. Applicability of methods depends primarily on present hardware and also on the description complexity of particular task in terms of selected method.

This article focuses on force-torque control that implies the presence of joint torque sensors and/or end-effector multi-axis force-torque sensor that should be situated between the manipulator output flange and the tool itself. Primary sensitive elements of such

* Corresponding author: i.shardyko@rtc.ru

sensors do not interact with the environment directly and their performance is not affected by the manipulator configuration unlike the performance of cameras and scanners, which is an issue with machine vision employment. Considering force-torque sensing in space robotic systems, the absence of interaction with the environment gives an advantage because of the lack of necessity for extra mechanical protection of additional weight.

There is a small chance that an external mobile vision system is available, whereas equipment of the manipulator with its own vision is limited by the overall mechanical design and the pursuit for workspace maximization. The location of the vision system, as a rule, cannot be changed during the manipulator lifetime. A survey manipulator, which is an additional manipulator with a camera as a tool, may be present; however, the area of possible camera location is limited within the workspace of this manipulator. Thus, machine vision is incapable of complete fulfilment of the specified criteria and hence force-torque control can be of interest in order to increase safety and operation convenience.

2 Control law

A topic of force-torque (FT) control represents a wide area within the scope of robot control. All developed FT methods can be divided into direct FT and indirect FT depending on a controlled variable that closes the outer control loop. The value of this variable (task value) should be specified either numerically or by the means of an input device, e.g. a joystick. Consequently, the direct FT control assumes active interaction (mechanical contact) with the environment, and a vector of TCP forces and torques (wrench) should be set to be performed by the manipulator. If a task is related to time-determined trajectory execution, then the direct FT is applicable only in special cases.

The indirect FT essentially is either position or velocity control constructed in such way that applied wrench is controlled or at least limited. Trajectory execution implies position control at each moment of time, thus we need an indirect FT control based on the position control, and trajectory impedance control is designed which perfectly suits. Consider a general form of impedance control law for a manipulator joint, i.e.

$$\tau_{req} = \tau_{obj} + S(q)(q_{req} - q_{cur}) + D(q,\ \dot{q})(\dot{q}_{req} - \dot{q}_{cur}),\quad (1)$$

where τ_{req} – control signal in the form of joint torque, τ_{obj} – objective joint torque, comprising generally gravity, centrifugal and Coriolis terms; S – virtual stiffness factor, D – virtual damping factor, q_{req} and $\dot{q}_{req}$ – required position and speed, q_{cur} and $\dot{q}_{cur}$ – actual position and speed. The last term may be absent, and then the law is often mentioned as stiffness control.

It is widespread to assume the TCP pose for the position variable. However, in the case when:

- Manipulator is non-redundant;
- Trajectory is smooth;
- Manipulator is rigid (rigid links and rigid joints);

every TCP pose in the workspace relates to a specific manipulator configuration, which is a vector of joint positions, or generalized coordinates. Typically, the last condition is not satisfied, but this fact effects either in complication of the model and computations or in deviation from the trajectory, the latter being often rather small and, thus, acceptable. Therefore, specifying trajectory for TCP pose results in obtaining joint-level trajectories.

In our implementation, on the user side, trajectory contains an arbitrary number of points, which are set by key point memorization, or by manual table input, or by teaching (record and repeat). The first two cases provide key points to be further linked with given geometric primitives (e.g. lines). Either way, the control system constructs a dense geometric trajectory, converts it into a joint space trajectory and then translates the latter trajectory into a time-dependent form [6].

Provided that τ_{obj} is computed with absolute accuracy and there is no mechanical interaction with external objects, the above law results in fine trajectory reproducing ($q_{cur} = q_{req}$). If for some reason an object appears on the path and a collision happens, the reaction force from the contact leads to the unbalanced torque

$$\tau_{un} = \tau_{ext} - \tau_{obj},\quad (2)$$

which in turn leads to joint position deflection until a new equilibrium is found, i.e. a point, where in the case of stiffness control,

$$\tau_{un} = S\cdot(q_{req} - q_{cur}),\quad (3)$$

so τ_{un} is a classical elastic torque, only of synthetic, or virtual, nature due to virtual stiffness S. Accordingly, the object itself (the joint) behaves as a spring, and changing the value of S, the controller actually changes the stiffness of this spring. Consequently, such a system allows deviation (deflection) from the desired path due to external forces action in a linear, smooth and safe way.

The general motion task formulation, though not necessary or unique, is to find a collision-free trajectory from the initial to goal position and perform it. However, obstacles still can exist on the manipulator path. It is possible that an obstacle gets on the path after the planning completion, e.g. a worker walks into the manipulator workspace. Another reason could be an error made in the determination of the manipulator position relative to external objects. Therefore, both the situations have a potential to employ impedance control. Superimposing an obstacle shape onto the given trajectory, position deviation can be computed for each point. If τ_{req} is no greater than the critical torque value for the considered joint, the transition to the goal position is possible, even though the trajectory itself cannot be reproduced. Otherwise, the motion is dangerous, and should be stopped when the critical torque is reached, until the next command comes from the operator. In the case of movable or elastic obstacles the analysis becomes more difficult, but as torque is being checked for limit exceeding, task execution is deemed to be safe. However, as this time the external objects change their location or/and shape, the operator should be sure the motion will not cause actual damage.

We implement the law (1) on the joint controller level, providing all the task values (required position, required speed, and objective torque) and factors (stiffness, damping) from the next-upper level, i.e. from the manipulator controller. For relatively slow manipulators functioning in earth conditions objective torque consists only of gravity torque. The overall control structure (data flow) is shown in Figure 2. To make the system work, the manipulator controller finds all the joint trajectories and loads them into the joint controllers before the motion begins, and provides joints with gravity torque values in the time of execution.

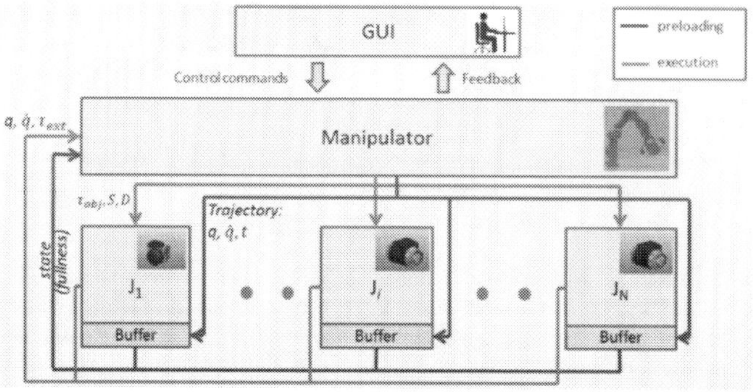

Fig. 2. Overall control structure.

3 Experiments

3.1 Test bed description

A series of experiments has been performed to tune the system and confirm the applicability of the proposed approach. Two tasks have been simulated:
a) peg-in-hole task;
b) execution of the path along the polygonal line intersecting a rigid object in the environment.

A general view of the experimental test bed is given in Figure 3 while the detailed description is provided in [7]. The controlled object is a 6 DoF manipulator devised for the verification of control system and design of a manipulation system for servicing spacecraft [8, 9] as a ground prototype. A photo and a simple kinematic model of the prototype are given in Figure 4. The manipulator comprises six torque sensors (one for each joint) and a three-axis force-torque output sensor. The sensors description is given in [10]. A software package has been written in Matlab Programmatic Workflow framework for the testing and control purposes, described again in [7]. Tasks (a) and (b) have been performed on the panels 1 and 2, respectively, of the test bed from Figure 3.

3.2 Peg-in-hole task

The task assumes motion in partly determined or erroneously determined environment. The initial position is shown in Figure 1. A peg is a probe rod that is mounted rigidly on the manipulator flange as a tool. A hole is a metal cylindrical glass with the inner diameter slightly larger than that of the tool. Task specification states: "perform a step of X mm in the tool axis direction with the speed of Y mm/s", X being a sufficient step size to touch the glass bottom. If the configuration is coaxial, task execution results in simple free motion until the bottom is reached. However, it is not so easy in the case when axes misalignment exists in the initial position. To secure the possibility of performing the task, the rod should be located so that initial entrance is possible, which is facilitated by the presence of a chamfer.

The task has been executed under trajectory impedance control. The limits were set on the TCP force (200 N) and torque (20 N·m) as well as on the joint critical torques to ensure the rod safety. There was no limit breaking during the motion until the bottom was touched, which was the purposeful end-of-motion condition. The task is illustrated by Figures 5 and 6.

Fig. 3. Experimental test bed.

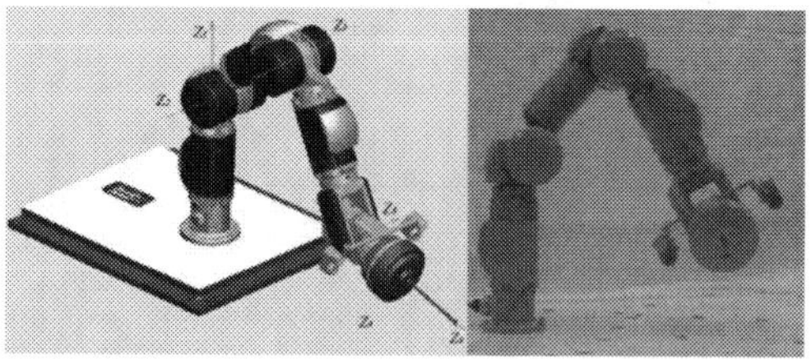

Fig. 4. Ground prototype.

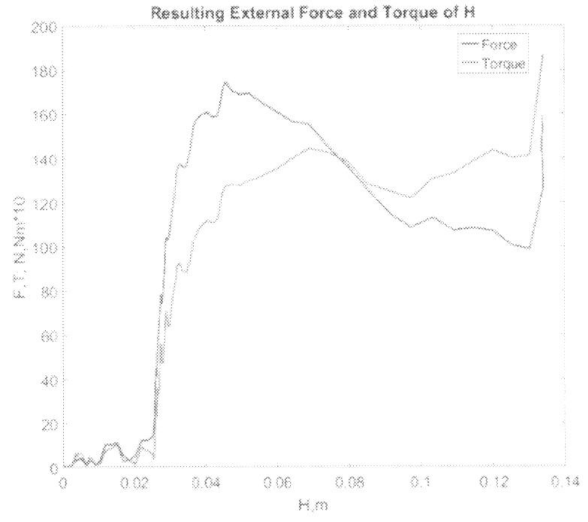

Fig. 5. TCP force and torque depending on the deepness.

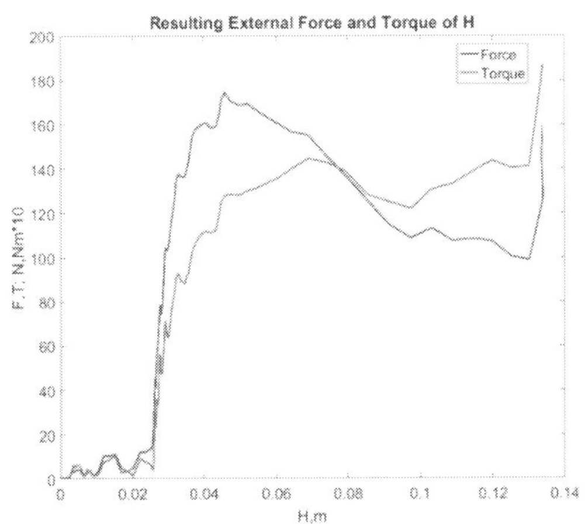

Fig. 6. TCP torque depending on deflection.

Figure 5 shows how the resulting force and torque on the TCP change depending on the tool depth. Despite the overall gradual force increase, a local drop can be seen at the TCP depth of 0.06 – 0.12 m. This part corresponds to passing over the local force maximum, after which the tool axis becomes close enough to the glass axis to allow continuing the motion into the glass. The orientation deviation is still non-zero due to the peg and hole diameter ratio. In the right part of the plot the force steeply increases and stabilizes, as the glass bottom is reached and the motion is halted when the force limit is triggered.

The relationship of the resulting torque and orientation, i.e. angle deviation, is shown in Figure 6. Phi here is the angle that the axes of peg and hole form, while Delta Phi is the difference between the initial value of Phi and transitional values, which is shown for the convenience of comparing angle deviation with torque (both start from zero). The relationship is close to be directly proportional according to the plot, thus it can be seen that the desired spring-like behaviour of the law (1) is actually achieved.

The important fact is that the angle deviation causes the TCP torque growth, consequently, increasing the initial deviation leads to higher maximal torques. The experiment has been performed for the maximal initial deviation, when the corresponding maximal torque values are still within the limits.

3.3 Trajectory task

This time the task objective is to test manipulator motion in the case of obstacles lying on its path. Task specification states: "perform the trajectory presenting a polygonal line connecting five given points that intersects solid metal", which is illustrated by Figure 7. The vertices of the task line ABCDE are set with the method of the key point memorization. The motion has been performed in the direction from A to E under the trajectory impedance control.

Fig. 7. Polygonal trajectory task.

Again the limits were set of the same size to secure safe motion, but this time they were not reached. The diagram of the motion is given in Figure 8.

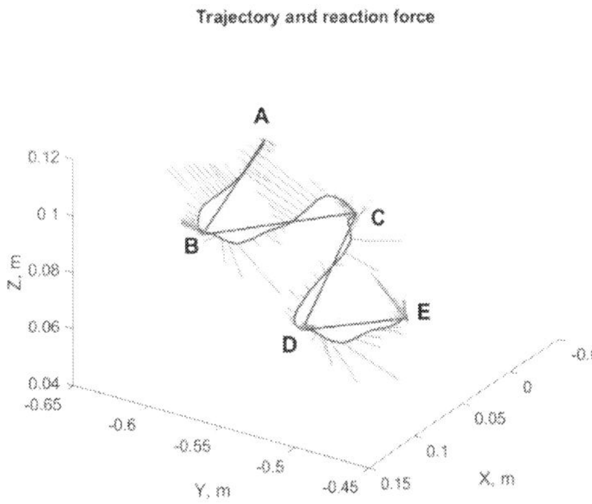

Fig. 8. Task path, actual path and reaction force.

The task path ABCDE is shown with straight (red) lines in Figure 8, while the actual path is shown with the (blue) curve. The contact with the metal profile has been not through TCP, but through some other point, though close to it. The exact position of the contact point is not identifiable with the available sensor set. Still, since the contact point is close to the TCP, the processing of FT information is done as though the contact actually is at the TCP, which is not a crucial assumption in the scope of the work. With short (green) lines the resulting force is shown at each observable point of the trajectory.

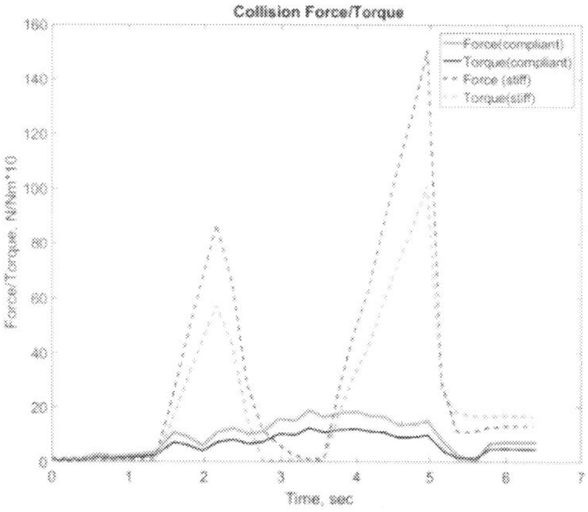

Fig. 9. Impedance and stiff position control comparison.

It follows from the plot in Figure 8 that the motion has substantial deviation from the task, but at the same time significant decrease (up to 5 times) of contact forces is seen in Figure 9 for the motion under impedance control ('compliant'). Thus, applying the impedance control law makes it possible to provide not only higher mechanical system compliance, but also higher functional compliance by allowing motions that are undesirable or even unacceptable in stiff position control. Intentionally permissible trajectory deviation substitutes abrupt contact force leaps when hitting obstacles with smooth force growth, lowering negative effects of the contact and increasing safety of manipulator performance.

As a main deduction, a basic criterion of impedance control applicability can be derived: this type of control is highly recommended for applications where safety prevails over trajectory execution accuracy. However, even for the tasks with accuracy in priority, impedance control can be of use, e.g. during a free space operation, impedance control has the minimal effect on the performed trajectory, but in the case of a worker located on the manipulator path, impedance control will mitigate the impact (due to smooth force changing) and possibly reduce the damage to the worker's health. The importance of accuracy in such situation concedes to person's health and the motion should be stopped.

There is one special observation on the last experiment. Provided the mutual position of the manipulator and the metal panel are known to the operator, who should transfer the TCP from A to E, he can deliberately make the movement through a sequence of lines instead of setting the complex curve. Despite of guaranteed contact with the metal, the goal will have been safely achieved while the technique itself is much easier for the operator and meets the criterion of easy operation and control.

4 Conclusion

In this paper, a method to enhance robot safety and operation convenience was presented, that implements trajectory impedance control. The conducted experiments prove the proposed benefits of this method, which has shown significant decrease of reaction force and confirmed improved ergonomic qualities. Since there are other methods and approaches (like machine vision), the issue of choosing a method becomes a design matter, both for a whole system design and for the planning of a specific mission. For example, a peg-n-hole task seems quite suitable for hybrid force-velocity control and the appropriate experiments are to be carried out soon. In any case, the choice of the method depends on the available hardware and on the particular task conditions. Furthermore, a task of trajectory impedance control on the operation level (directly for the tool) in a Cartesian coordinate system and a topic of force-torque control of redundant manipulators are to be considered.

Research are carried out (conducted) with the financial support of the state represented by the Ministry of Education and Science of the Russian Federation. Unique project Identifier: RFMEFI57517X0143.

References

1. K. Seong-Heum, Ch. Gyeongmin, A. Byungtae, K. In So, Proceedings of 2017 IEEE International Conference on Robotics and Automation (ICRA) (2017)
2. V. Gorbunov, F. Kulakov, Proceedings of the International Scientific and Technological Conference EXTREME ROBOTICS, 81–86 (2016)
3. E. Magrini, F. Flacco, A. De Luca, Proceedings of 2015 IEEE International Conference on Robotics and Automation (ICRA) (2015)
4. L. Le Tien, A. Albu-Schäffer, A. De Luca, G. Hirzinger, Proceedings of 2008 IEEE International Conference on Intelligent Robots and Systems (IROS), 3789–3795 (2008)
5. V. Titov, I. Shardyko, S. Isaenko, Procedia Engineering, **69**, 1232–1241 (2014)
6. I. Shardyko, V. Titov, Abstracts of the International Scientific-and-Technological Conference EXTREME ROBOTICS, 45–46 (2017)
7. I. Dalyaev, I. Shardyko, Proceedings of the International Scientific and Technological Conference EXTREME ROBOTICS, 411–415 (2016)
8. A. Lopota, I. Dalyaev, I. Shardyko, E. Kuznetcova, I. Belezyakov, Annals of DAAAM and Proceedings of the International DAAAM Symposium, 865–870 (2015)
9. I. Dalyaev, E. Kuznetsova, I. Shardyko, Robotics and Technical Cybernetics, **3** (8), 27–31 (2015)
10. I. Dalyaev, N. Zarutsky, Robotics and Technical Cybernetics, **4** (13), 64–70 (2016)

Investigation of the dynamical characteristics of the lower-limbs exoskeleton actuators

Andrey Yatsun[1,*], *Andrey* Karlov[1], *Andrey* Malchikov[1] and *Sergey* Jatsun[2]

[1]South-West State University, Robotics Department, 305040, 50 let Oktyabrya, 94, Kursk, Russia
[2] Mechanical Engineering Research Institute of the Russian Academy of Sciences, 101990, Moscow, Russia

Abstract. Authors present results of the theoretical modeling and experimental tests of the low-cost DC-motors, used in lower limb powered exoskeleton. Actuators work in difficult regime and it is important to achieve desired parameters, even for not robust motors. Results give us information and methods and means of experimental determination of the main characteristics of the robot's actuators. It gives possibility to tune control system and the whole system to achieve optimal walking regime.

1 Introduction

Nowadays robotic lower-limbs exoskeletons are commonly used for solving many problems, from reducing the heavy load on the human body during walking, up to the restoration of the locomotion abilities of injury patients [1, 2]. One of the key challenges in developing such devices is the synthesis and tuning of control system parameters, providing high precision movement of an exoskeleton parts in the presence of different disturbances, also taking into account the properties of the actuators [3, 4].

Non-linear characteristics of the actuator are one of the problems associated with accurate control of multi-link electromechanical systems. In particular, the actuator is characterized by the saturation of maximum torque and the presence of dead zone due to reducer. The saturation of the torque indicates that the drive is not able to produce a torque greater than some limit value, regardless of the strength of his signal. The effect of dead zones based on some range of values supplied to the drive signal at which the drive produces no torque.

In [7] indicates that the nonlinearity of the actuator has a greater impact on the quality of the control systems than the uncertainty in the model parameters of the mechanical system that was used in the design and configuration of the automatic control system. One of the most significant problems caused by the nonlinearity of the actuators as precision mechanical systems, is the controller windup. This effect is characteristic for a servo linear control systems feedback, especially for PID controllers. It based on the fact that the regulator requires from the drive more time than one can generate, therefore there is a growing error in the position tracking system. This leads to a change in the behavior of the control system, occurrence of overshoot and oscillations. For high-precision control systems, e.g. control systems of multi-link mobile robots, such effects are unacceptable.

To eliminate these problems, we tune controller in different approaches. Among the most common – increase the differential component of the PID controller, the scaling of the nominal exposure.

The aim of this work is to develop methods and means of experimental determination of the main characteristics of the actuator used in the creation of exoskeletons.

2 Experimental setup

To achieve this goal experimental setup have been developed and constructed for the study of the characteristics of the actuators. Figure 1 shows the model system for the study of the characteristics of the wiring.

The experimental setup consists of the following main parts: 1 – clamping chuck for motor's shafts with diameters from 3 to 16 mm; 2 – a shaft transmitting rotation from the output shaft of the electric drive; 3 – rotational disc with a large diameter; 4 – friction pair; 5 – rotation disk; 6 – pressing spring; 7 – rear carriage; 8 – the handle to move the carriage; 9 – basement; 10 – feedback optical sensor; 11 – gauge efforts; 12 – thrust bearing; 13 – front carriage; 14 – adjusting screw.

The device operates as follows. The power source control used to set the required rotation speed. After power on, motor with gearing, by the clutch drives to the measuring disc and the friction drive plate. Using the current meter and phototachometer, is determined by current consumption and the frequency of rotation of the drive shaft. Next, using the handle, the preload spring of the friction drive element pushes to the leading. On the contact surface of friction disks, the friction force leads to the displacement of the disk. Disk is connected via an adjustable rod, a spring-loaded force gauge, prevents the rotation of the driven disk. Thus, knowing the magnitude of the force F, acting from the drive element and knowing the rod length L,

* Corresponding author: ayatsun@ya.ru

we can estimate the friction torque *Mc* that occurs between the discs. Using the software signal is processed and plotted to graphs.

Fig. 1. General view of the experimental setup to study the characteristics of the actuator.

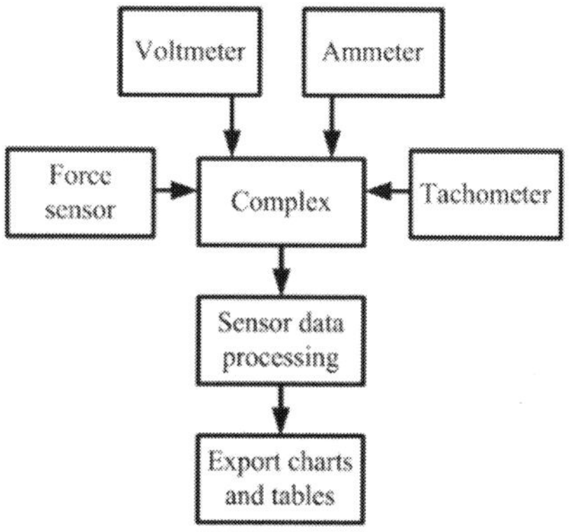

Fig. 2 Structural scheme of automated complex to study the characteristics of the actuator.

3. Comparative analysis of the actuators

On the basis of data obtained in an analytical way, the actuators of the lower limbs exoskeleton with a standing person should create torque al least 25 Nm and achieve rotation speed up to 1.8 rad/sec. Modern international companies offer brushed and brushless motors with similar characteristics but their products have high price (at least 150 $) [6]. So, we will investigate low-cost electric drive with similar characteristics for acquisition of the exoskeleton.

We offer brushed DC-motor DSH-12RSH3 with 3-speed planetary gearbox with price less than 20$ each. Nominal voltage is 12-14 Volts, and idle speed is 16000 rpm. The motor diameter is 38mm, the length is 74 mm, dimensions of gear is 50x43 mm, length is 97 mm. the total length of the assembly is 160 mm, weight is 475 grams (Fig.3).

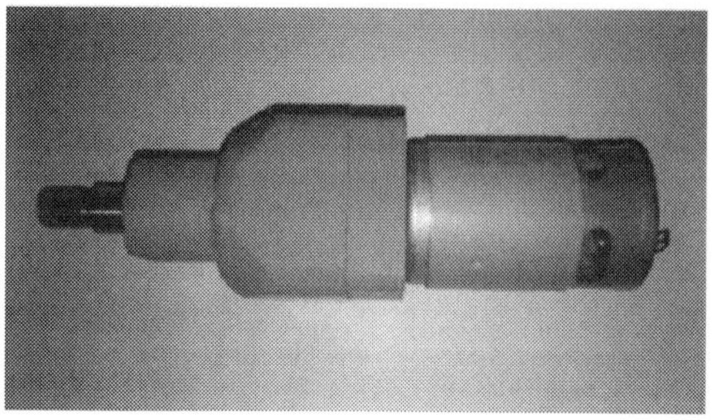

Fig. 3a Photo of the low-cost gear motor DSH-12RSH3.

Fig. 3b The 3D model of the 3 stages planetary gear box RSH3 own development.

The planetary gearbox has its own design, consists of 3 stages, gear ratio 6 each, so total reduction ratio is 206. Using CAD-system to study the characteristics of the actuator we have received a table of values for different supply voltage and has built a family of mechanical characteristics for this drive.

According to tests maximum torque is 16 Nm, and the maximum angular speed is equal to 7,98 rad/s at 12 Volts supply voltage. This drive does not allow to develop the effort to provide the speed necessary for the verticalization of the person in the exoskeleton (Fig.4).

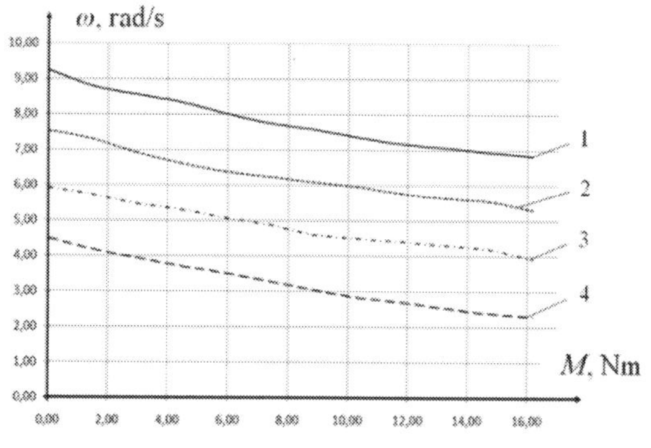

Fig. 4. The experimental family plots of mechanical characteristics DSH-12RSH3: 1 is for 12 Volts; 2 is for 10 Volts; 3 is for 8 Volts; 4 is for 6 Volts.

In connection with these experimental results it is necessary to add another gear stage that allows us to achieve maximum torque not less than 25 Nm on the output shaft with the angular velocity 1.8 (rad/s). So we developed brushed DC-motor DSH-12RSH4 with additional 4-th stage (Fig.5)

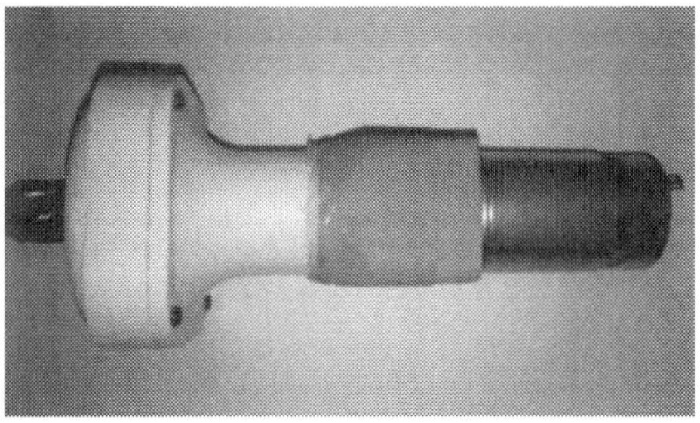

Fig. 5a. Photo of the low-cost gear motor DSH-12RSH4.

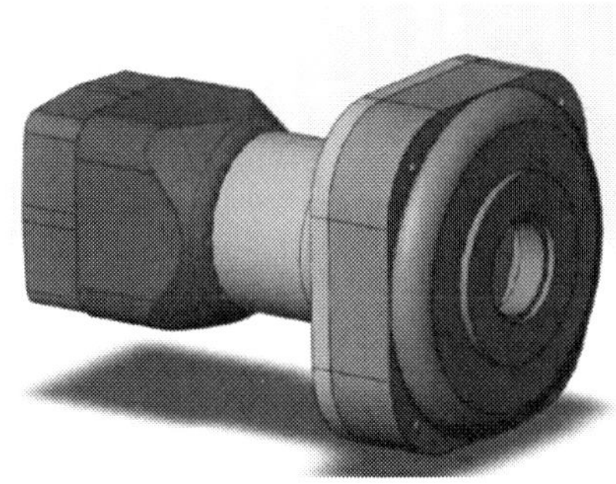

Fig. 5b. The 3D model of the 3 stages planetary gear box RSH4 own development.

Actuator called DSH-12RSH4 with a planetary 4-stages gearbox has such dimensions: gearbox size 72x72mm, gear length is 130 mm, total assembly length is equal to 200 mm, weight is 870 gr.

The planetary gear consists of 4 steps, based on the gearbox RSH3, and the fourth with a gear ratio of 4.15. Total reduction ratio is 855. Family of actuator mechanical characteristics are shown in Figure 6.

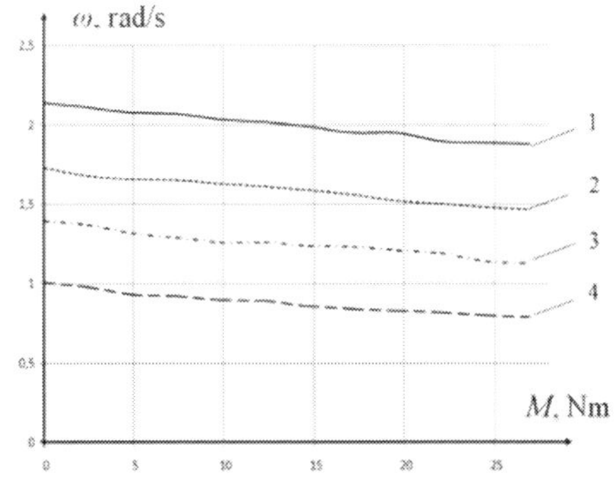

Fig. 6. The experimental family plots of mechanical characteristics DSH-12RSH2: 1 is for 12 Volts; 2 is for 10 Volts; 3 is for 8 Volts; 4 is for 6 Volts.

Maximum torque is 27 Nm, the maximum angular speed equal to 1.9 rad/s at 12 Volts supply voltage. This corresponds to the desired torque and speed characteristics of the actuator.

Further we investigate dynamical parameters of brushed DC-motor DSH-18, which nominal supply is 18 Volts and idle speed 19000 rpm with two-stage RSH2 and three-stage RSH3 planetary reducers (Fig. 7-9).

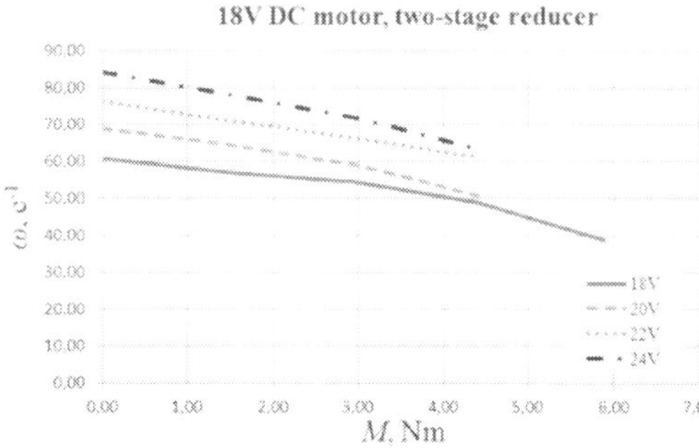

Fig. 7. The experimental family plots of mechanical characteristics DSH-18RSH2.

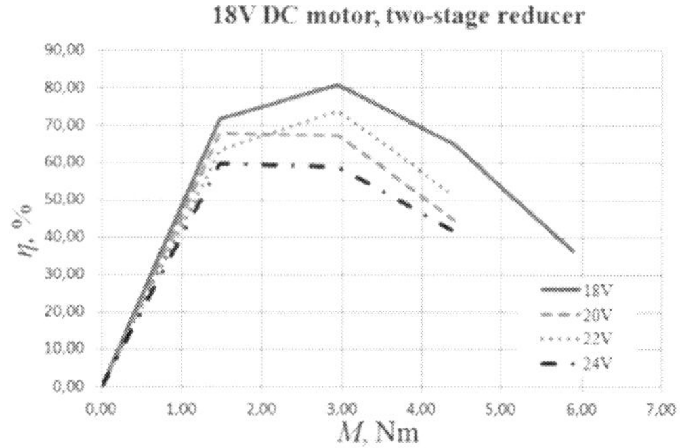

Fig. 8. The experimental efficiency for DSH-12RSH2 actuator.

The graphs show that the rotation speed and torque below the required parameters, but this can be compensated with the use of ball screw drive (linear actuator).

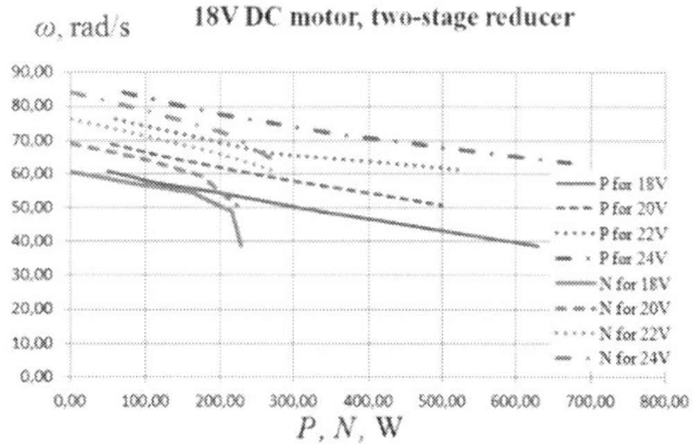

Fig. 9 Experimental graph of power in dependence on frequency of rotation: P – electrical power, N – mechanical power.

On the next experiment we get information about brushed DC-motor assembly DSH-18RSH3 (Fig. 10-12).

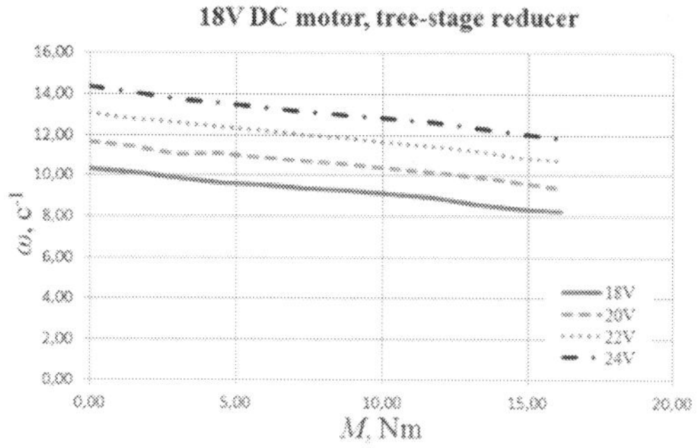

Fig. 10. The experimental family plots of mechanical characteristics DSH-18RSH3.

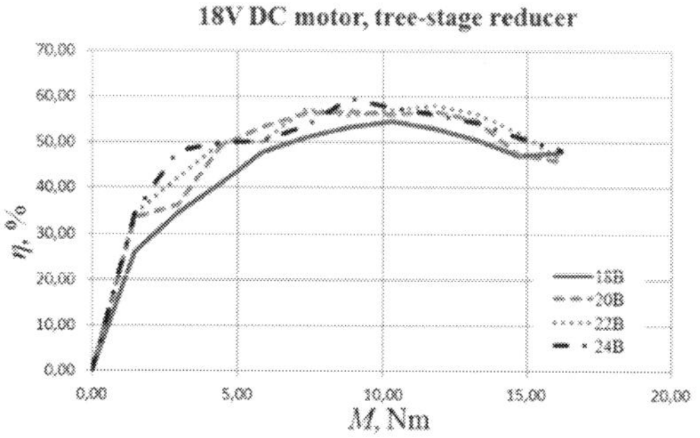

Fig. 11. The experimental efficiency for DSH-12RSH3 actuator

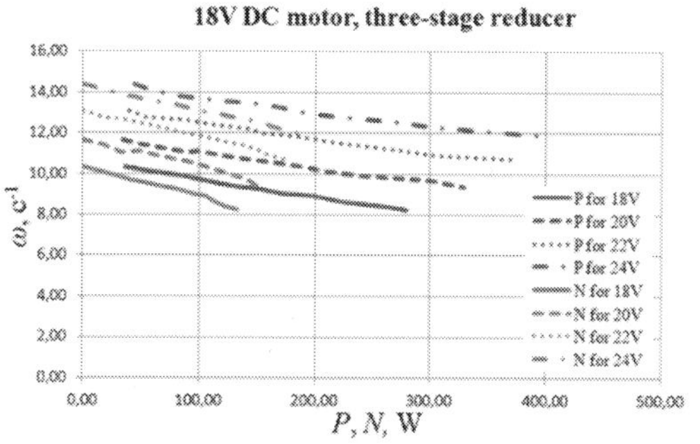

Fig. 12. Experimental graph of power in dependence on frequency of rotation: P – electrical power, N – mechanical power.

According to these results we can obtain optimal parameters of the assembly for the prototype of the lower-limb exoskeleton or bipedal walking robot. At figure 13 presents photo of the exoskeleton prototype with DSH-12RSH4 actuators, which was developed at the Robotics Laboratory of South-West State University (Kursk).

4 Analysis of the electric current in the windings

For the accurate control of the output torque it is important to get filtered feedback current into the controller. Due to real-time approximation, based on microcontroller we used such ways: a linear fit of the experimental data using the least squares method, linear approximation passing through the start and end points, logarithmic approximation passing through the start and end points.

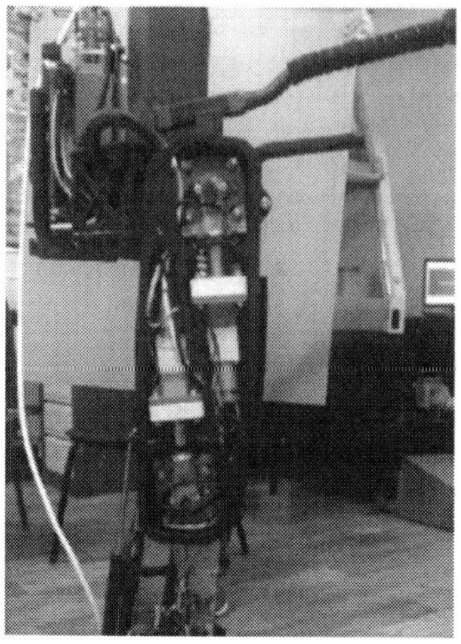

Fig. 13. Exoskeleton prototype with DSH-12RSH4 actuators.

The original data and the resulting linear approximation for the winding current of the DSH-12RSH3 at upper limit 22 Volts power supply is shown in Figure 14.

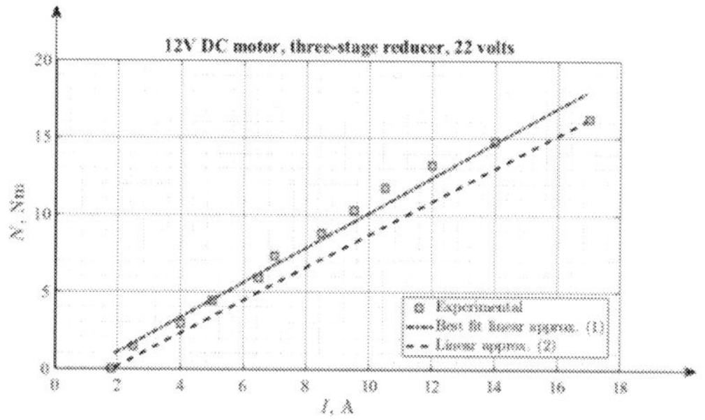

Fig. 14. Experimental data and linear fit for DSH-12RSH3.

The logarithmic approximation of the experimental data passing through the start and end points of the data is more curving. The original data and the resulting approximation for the winding current of the DSH-18RSH at 20 Volts power supply is shown in figure 15.

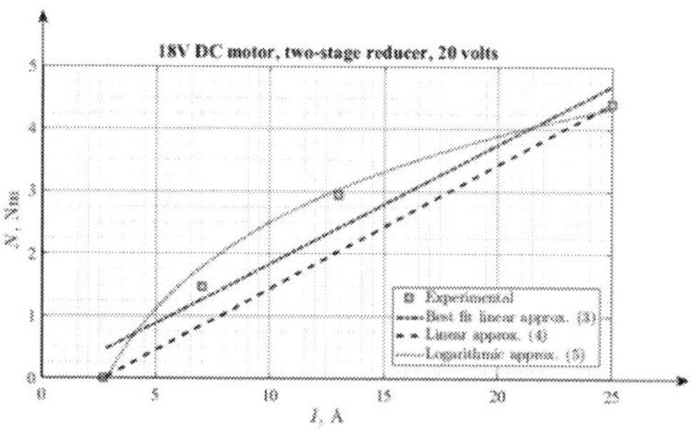

Fig. 15. A linear fit of the experimental data for DSH-18RSH2.

Thus, the torque of the actuator from current is described by the formulae:

$$\begin{cases} M = sign(I)\cdot(2\log(|I|)+2.5) \quad if \quad |I|>2.7 \\ M=0 \quad if \quad |I|\le 2.7 \end{cases}$$

We will consider the motor parameters L = 0.001 H, R = 0.5 Ohm, C_e =0.0033, C_T = 0.0033, N = 100 and will conduct numerical experiments for the following values of the controller parameters:

$$k_p = 500,\ k_d = 20,\ A_p = 20.$$

At figure 16 we obtain results of modelling : desired and real angular speed of the motor shaft depending on time, which corresponds to DSH-12RSH3 with torque feedback control.

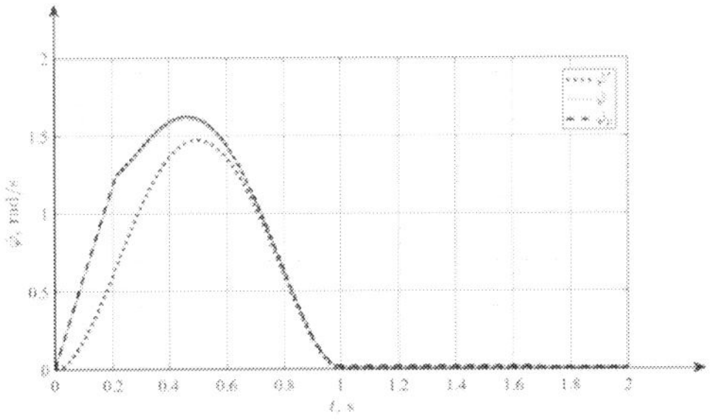

Fig. 16 Angular speed time dependence for actuator 12V with a three-stage reducer.

This graph is comparable with that obtained in experimental part, one can notice that the maximum torque developed by the actuator, which in this case was 5.5 Nm, which was not taken into account the experimental dependence of the torque of a current, this value was 8.5 Nm.

The obtained simulation results (without taking into account the experimental data on the dependence), show that the steady-state value of the current in the windings of the motor is 26 A. In this case, the steady-state current value was 44 A, i.e. 69% more. The steady-state value of the voltage was 12.5 , i.e. 80% more. Static error in this case has decreased significantly.

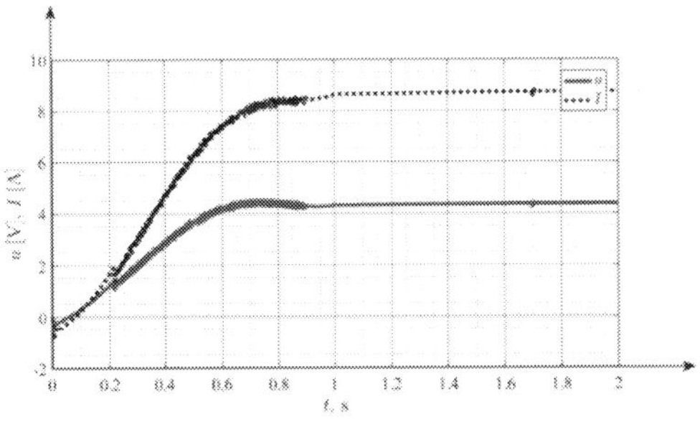

Fig. 17. Time dependence of supply voltage and current; used the model corresponding actuator 12V with a three-stage reducer.

In this case the steady-state current value amounted to 8.7 A, i.e. 66% less than in the experiments. The steady-state value of the voltage was 4.5 V, i.e. 64% less than in the data section

5 Conclusions

Description of the experimental stand, on which the various tests developed for electric drives is presented. The experiments show that the optimal design of an Assembly for use in walking robots and exoskeletons. The use of experimental data allows to confirm the simulation results and to determine the optimal coefficients of the controller for a feedback system in which the currents do not exceed the level of permissible values. According to the modeling results we obtained transfer characteristics for the control system.

This research is supported by grant of the RF President for young scientists MK-2701.2017.8.

References

1. J. Zhang, P. et. al., Science, **356** (6344), 1280–1284 (2017)
2. M. Sczesny-Kaiser et. al. Frontiers in Neuroscience, **11**, 449 p. (2017).
3. A.P.P.A. Majeed et al, Procedia Computer Science, **105**, 183–190 (2017)
4. G.Ya. Panovko, S.F. Jatsun, S.I. Savin, A.S. Jatsun, Machinery and engineering education, **2** (47), 2–10 (2016)
5. A.J. Young, J. Foss, H. Gannon, and D.P. Ferris, Frontiers in Bioengineering and Biotechnology, **5**, 4 (2017)
6. S. Jatsun, S. Savin, A. Yatsun, 24th Mediterranean Conference on Control and Automation (MED), 322–326 (2016)
7. F. Tyan, D.S. Bernstein, Journal of Robust and Nonlinear Control, **5** (5), 521–537 (1995)
8. S. Jatsun, 2015 10th International Symposium on Mechatronics and its Applications (ISMA), 1–6 (2015)
9. S. Jatsun, Proceedings of the International Conference on Pure Mathematics – Applied Mathematics, 83–87 (2015)
10. S. Jatsun, Proceedings of the 24th International Conference on Robotics in Alpe-Adria-Danube Region (RAAD), **371**, 165–172 (2015)

Synthesis of structural scheme of drive of adaptive multiple-link gripper

Aleksej Bogdanov[1], *Aleksandr* Permyakov[1] and *Yulija* Zhdanova[2,*]

[1]JSC 'SPA 'Android technics', Moscow, Russia
[2]Institute of Cybernetics, Moscow Technological University (MIREA), Moscow, Russia

Abstract. The problem of construction of drive of multiple-link gripper capable to operate with human infrastructure objects is considered in the article. Technical specifications of structural scheme of actuating link group of a gripper are determined with reference to analysis of human hand finger bones motion. Structural scheme of system of motion transmission to the output links that realizes group drive and does not have kinematic correspondence of links is recommended. Synthesis of structural scheme is accomplished on the basis of providing of adaptability of output links location to the grasped object surface contour. The key point of the construction is variability of output link according to the interaction with grasped object. This is provided by passive power connections which are initiated between output links and motion transmission system links. Sequence of functioning of actuating link group is represented.
Keywords: synthesis, structural scheme, drive, multiple-link gripper, actuating link group, passive power connections.

1. Introduction

Active works on designing of anthropomorphic robots (AR) are determined by factors of necessities and capabilities of designing of them as technical object capable to supply the place of human in performance of work (technological operations) in dangerous environment. Technical design specification of 'FEDOR' AR (as part of 'Spasatel' project financed by Advanced Research Foundation) covered 54 tests including wide range of activities specific to human: from going upstairs to using injection syringe. Technologies realized during AR designing are based on modern hardware components and original architecture of actuating complex and controlling system.

The whole range of the AR operations may be divided in several connected functions: analysis of outdoor environment, decision making, global moving in the outdoor environment, affecting the outdoor environment objects. Each function is essential and significant. At the same time the first two functions may be accomplished by operator in master-slave mode [1] and the third one may be accomplished by means of external subject.

Regional and local moving of gripper (AR working tool) and its interaction with outdoor environment objects are accomplished only by AR. Regional and local moving is provided by manipulator. Seven degrees of freedom manipulator makes it possible to remove the gripper to the localization point of required orientation angle and obstacle avoidance capability.

2. Task description

Analytical research of structural schemes of AR manipulators is accomplished considerably [2]. Analysis of gripping tool received far less consideration of researchers. On one side it is determined by a larger variety of potential constructional decisions, on the other side – by absence of universal methodology of its analysis. Interaction with undetermined objects of outdoor environment requires designing of cross functional gripper having both fine motor skills and power capacities. Multiple-link (anthropomorphic) gripper (MG) having kinematics similar to human hand [3] meets set requirements. MG has from three to five actuating link groups of the same structural scheme (Fig. 1, a). Each actuating link group (ALG) includes three links connected in sequence with each other by B and C rotating pairs with parallel axes (Fig. 1, b). Link 1 is connected with gripper base by two-degrees-of-freedom spherical pair A. Three from four degrees of freedom have parallel axes of rotation. Location of each i link is determined by α_i relative rotation angle (Fig. 1, b). The total number of degrees of freedom is no less than 14.

Usage of two versions of drive schemes construction that provide links motion is possible. In this case drive shall mean complex of power section (electro-, pneumo-, hydraulic motor) including gear and motion transmission system (MTS) if necessary. According to the first version the drive is located on set in motion link or the previous one. In this case MTS may be absent and actuating group link that is set in motion is connected

* Corresponding author: musrosjk@gmail.com

with output link of power part directly. In the second version power part is placed on the manipulator link and the motion is transmitted through the MTS.

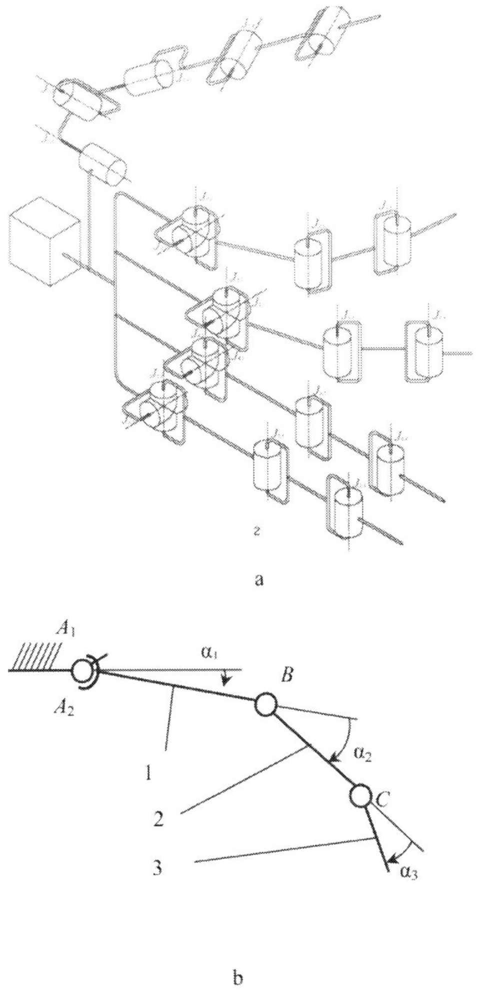

Fig. 1. Structural schemes: a – anthropomorphic gripper, b – actuating link group.

The first version is relatively easy in realization however it does not make it possible to use drives with sufficient force on the output link in virtue of restrictions placed on the links dimensions. The second version provides usage of high power motors. At the same time no more than 8-10 power parts may be placed on a manipulator link compactly. Usage of individual drive for each degree of freedom causes difficulties in its arrangement on a manipulator. The number of power parts used may be decreased by means of group drive. In this case the motion according to the degrees of freedom with parallel axes of rotation A_1, B, C will be provided from one power part. The number of power parts for each AGL decreases from four to two. Kinematic correspondence of output links motion is typical for classic scheme of group drive construction. The employment of it in ALG results in stoppage of all output links as soon as one of them reaches the object surface.

3. Formation of ALG drive terms of reference

Kinematics of ALG must be similar to human hand kinematics and must provide interaction with external objects according to one of two scenarios either grasp or plucked grab. In the first case three phalanges operate with the object. This provides significant stress on the object surface and its safe fixation. In this case the location of each phalange is determined by proper generalized coordinate γ_i. In the second case only one distal phalange operates with the object.

Analysis of fingers grasp of different objects determined that distinct sequence is maintained. Table 1 represents values of relative rotation angles of phalanges (except thumb) while grasping cylinder object of 5 mm in diameter. Numerical values of 20-25 years old males with the hand size of 9 are noted. Interval in angles values is dictated predominantly by incongruous values between finger bones of different fingers and in a smaller degree by individual characteristics of examined men.

Table 1. Initial position and limiting position of phalanges.

Phalange	Phalange position, deg	
	Initial $\gamma_{i.0}$	Limiting $\gamma_{i.Max}$
Proximal γ_1	$0^0 - 20^0$	$90^0 - 110^0$
Mesial γ_2	$0^0 - 35^0$	$100^0 - 135^0$
Distal γ_3	$0^0 - 15^0$	$45^0 - 75^0$

Realizing of plucked grab is provided by means of motion of only just proximal phalange.

4. Synthesis of structure of motion transmission system

Analysis of phalanges motion sequence in the process of grasping objects of different surface contour makes it possible to formulate main aspects of finger functioning:
- grasp is provided if only one generalized coordinate primary changes;
- absolute sequence of phalanges motion is maintained from proximal to distal;
- changing of generalized coordinate of each phalange is carried out until it reaches the object surface or the limiting value;
- each phalange has fixed value of limiting generalized coordinate.

5. Technical specifications of drive structural scheme

The results of the analysis make it possible to formulate technical specifications for development of ALG drive structure.

- The motion of links should be realized from one power part of a drive.
- All the links are output with capability to change proper generalized coordinate independently.
- Limiting position of links must be adaptive (posses adaptive feature) to the grasped surface contour.

The last requirement leads to the fact that the order of changing of generalized coordinates and their final values are determined by the object surface contour. The motion of adaptive ALG may be provided from one or several motors.

MTS that meets the set technical specifications may be designed by realization of one or two standard approaches:
- usage of differential devices;
- usage of changing structural scheme.

Opportunity to employ the first approach is determined by the function of differential device capable to divide the motion of input link in several motions of output links.

The initial basis of the second approach includes the following statements:
- Structural scheme of MTS must provide motion of all output links of actuating group;
- While i link is moving a part of structural scheme providing the transmission of motion to $i+1$ link must be blocked;
- The transmission of motion to i link is provided in case of stoppage of $i+1$ link.

6. Structural scheme of system of transmission of motion to the actuating group links

The general structural scheme including the main scheme (Fig. 2, b) and MTS scheme must have three degrees of freedom; in such case two of them must be blocked initially. Its unblocking must happen in case of stoppage of one of output ALG links. As a result the general structural scheme has one degree of freedom at every instant. Realization of unit of blocking/unblocking of the kinematic pair degree of freedom is the most problematic issue of the approach. This problem may be solved by means of including of active or passive power connections.

The general structural scheme of ALG which has three output links (1, 2 and 3) is represented in Figure 3. Link 4 is the input. Kinematic pair A_3 made of 4 and 1 links, and pair A_1 fit geometrically in a general structural scheme. Pair B_1 made of 8 and 2 links and pair B have the same position similarly.

The number of degrees of freedom is determined by known formula

$$W=3n-2p_5-p_4\ . \qquad (1)$$

Where n is the number of moving links, $n=9$,

p_5 is the number of kinematic pairs of the fifth type, $p_5=12$,

p_4 is the number of kinematic pairs of the fourth type, $p_4=0$.

$$W=3*9-2*12=3.$$

Two passive power connections a and b by way of springs are included in the general structural scheme. Spring a is fixed between links 4 and 1, spring b is fixed between links 8 and 2. If the links 1-3 do not contact with the object surface the connections a and b eliminate (block) relative motion in kinematic pairs A_3 and B_1. The number of kinematic pairs decreases to ten. In this case springs combine links 4, 1 and 8, 2 into one correspondingly. The number of links decreases to seven and the number of degrees of freedom is determined by correspondence (1) $W^1=1$. The power part of the drive provides the motion of link 1 only from the initial position.

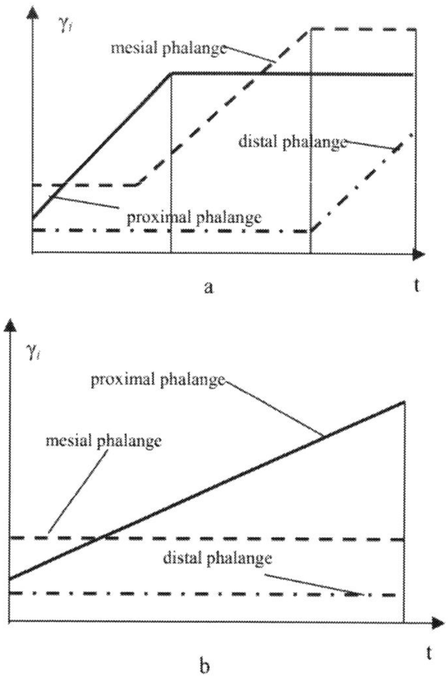

Fig. 2. Cyclogram of phalange motion: a – while grasping an object of minimum diameter (d = 5 mm), b – while plucked grabbing.

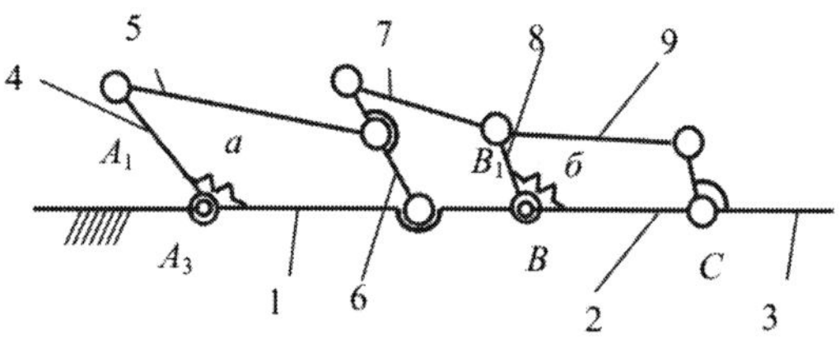

Fig. 3. The general scheme of ALG.

The process of grasping the object surface is carried out in three stages. The rotation of link 1 is realized from the initial position (Fig. 3). The generalized coordinate (Fig. 4, a) is changed. When link 1 reaches the object surface it stops (Fig. 4, b). Thus mobility of kinematic pair A_1 is not further realized. Generalized coordinate α_1 takes the fixed value α_{1K}, the first stage of grasping is finished.

The stoppage of the link 1 and continuing motion of output link 4 provide unblocking of kinematic pair A_3. In this case spring a is compressed. The number of mobile links is not changed as well as the number of kinematic pairs. Thus the number of degrees of freedom is $W^{AI}=W^{A}$. The motion of input link 4 provides the rotation of link 2 in a kinematic pair B. The position of links 2 and 3 of the actuating group is determined only by generalized coordinate α_2 (Fig. 4, c). The second stage of the object contour grasping is realized.

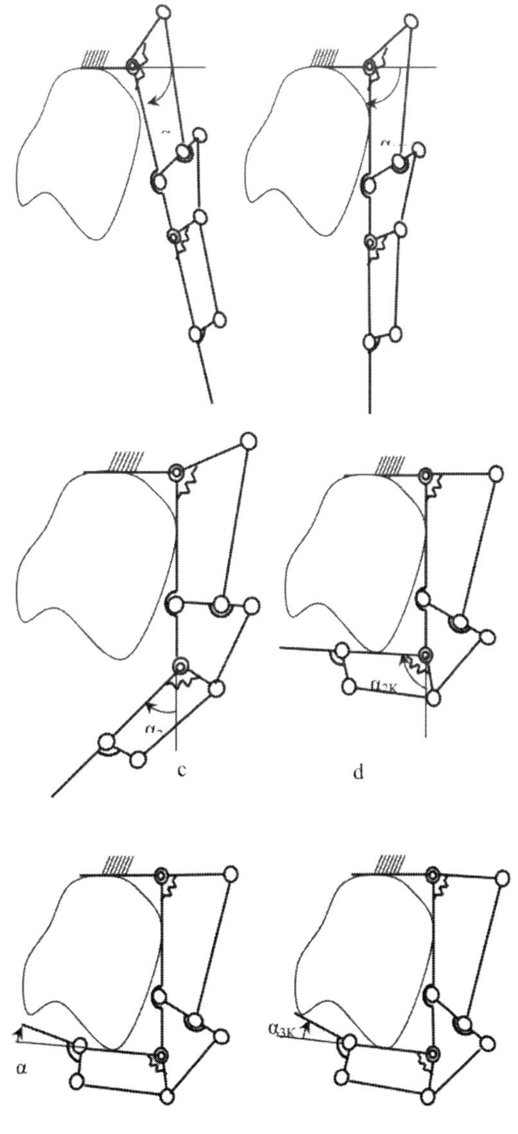

Fig. 4. Stages of the process of grasping the object contour by ALG links.

After the link 2 stops on the object surface (Fig. 4, d) the mobility in the kinematic pair B is eliminated. Generalized coordinate α_2 takes fixed value α_{2K}. The second stage of the object contour grasping is finished.

The stoppage of link 2 provides unblocking of pair B_1. Link 8 has the opportunity to rotate then compression of spring b (Fig. 4, e) is realized. The number of degrees of freedom stays the same $W^{AII}=W^{A}$. The position of link 3 is determined by generalized coordinate α_3. The third stage is finished when link 3 reaches the object contour (Fig. 4, f).

7. Conclusion

Group drive which structural scheme includes power connections should be used in design of grippers having kinematics similar to human hand at minimum number of motors. The location of them between links of actuating group and links of motion transmission system makes it possible to eliminate kinematic correspondence of motion. The found structural scheme of motion transmission system provides the grasp of the object contour by links of the gripper with adaptation of their position to the object shape.

References

1. A.A. Bogdanov, V.B. Sychkov, I.G. Zhydenko, I.M. Kutlubaev, Reshetnev Readings, **1** (16), 230–231 (2012)

2. I.G. Zhidenko, A.A. Bogdanov, I.M. Kutlubaev, V.B. Sychkov, Reshetnev Readings, **1**, 278–280 (2013)

3. A.A Bogdanov, A.S. Gorbaneva, I.M. Kutlubaev, Yu.I. Kutlubaeva, Reshetnev Readings, **1**, 273–274 (2014)

The control algorithm of the lower limb exoskeleton synchronous gait

Vladimir Antipov[1], *Alexey* Postolny[1], *Andrey* Yatsun[1,*] and *Sergey* Jatsun[2]

[1]South-West State University, Robotics Department, 305040, 50 let Oktyabrya, 94, Kursk, Russia
[2]Mechanical Engineering Research Institute of the Russian Academy of Sciences, 101990, Moscow, Russia

Abstract. In this article a study of algorithms for human movement in the lower limbs exoskeleton is presented. Human-machine system is considered, the classification of the existing exoskeletons by type of power distribution, the features of stable motion of the mechanism are presented. The law of the necessary trajectory of the center of mass of the exoskeleton is shown in the sagittal and frontal planes to maintain stability. The synchronous motion scheme of the centre of mass and the foot is described.

1 Introduction

A human's gait in the exoskeleton is one of the most difficult locomotion. That's why it is a very significant problem. There are a great amount of papers concerning bipedal gait [2, 3, 6, 7, 12-16, 22]. However, in the majority of cases the gait is considered in a single sagittal plane, thus limiting the application sphere of the results obtained for exoskeleton applications. The thing is that the locomotion in the exoskeleton is similar to the quasi-static gait. That's why the locomotion modeling should be considered in respect to the spatial kinematics, since the motion takes place both in the sagittal and frontal planes. This approach to the locomotion modeling appears only in some papers [1, 3, 7, 13, 15, 18, 21].

The locomotion research in the paper is done with relation to the quasi-static locomotion modeling with the help of kinematics approaches, while the inverse kinematics problem is solved by approximation methods. The obtained results, concerning the determination of the rotation angles of the exoskeleton links for different positions of the foot taking into account the position of the center-of-mass projections are used for the development of the control algorithms of a person's locomotion in the exoskeleton. The special attention is focused on experimental researches, which were carried out on the prototype of the exoskeleton.

One of the most important moments, that requires the adequate mathematical description is the locomotion of a person in the skeleton on uneven surface. The creation of the locomotion control algorithm, taking into account the recognition of a surface irregularities, and the formulation of the conditions for the gate on the basis of "intellectual foot", opens completely new possibilities for the creation of the lower-limb exoskeletons, which provide a confident, steady locomotion of a person across rough terrain. So that, this monograph is first of all devoted to the problems, concerning the modeling of a person's locomotion in the exoskeleton, taking into account the interaction of the feet with the supporting surface. Besides, much attention is paid to the energy distribution in the human-exoskeleton system. An energy criterion is offered, it allows the estimation of the power inputs from the actuators, taking into consideration the human muscles work.

2 Exoskeleton classification

A person in the exoskeleton can be described a human-machine system. The cooperation of this system elements is many-sided. Different criterion can be used for quantities analysis of the interference of a human and an exoskeleton. Let's consider in detail only one of them, that allows the estimation of the power inputs level from both the person and the machine, while doing one of the exercises, for instance at verticalisation, walking and so on.

Figure 1 shows the classification of exoskeletons by the criterion of power inputs in the human-machine system. Let's choose three different classes of exoskeleton systems.

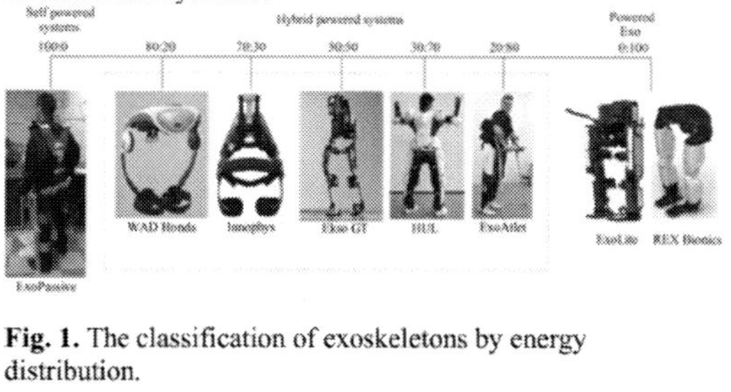

Fig. 1. The classification of exoskeletons by energy distribution.

The first class is self-powered systems (SPS), the second is hybrid systems, driven by both a human and the exoskeleton drives (hybrid powered systems (HP), and the third is the systems set in movement only by exoskeleton drives, the so-called Powered Exoskeleton (PE).

* Corresponding author: ayatsun@ya.ru

As noted earlier many researches are devoted to the bipedal gait analysis, and the results were published in many papers [5, 7, 8, 20-22]. Both theoretical and experimental methods for locomotion study were developed. The kinematic and dynamic aspects of the movement are thoroughly considered. Basically, these papers describe the motion of bipedal robots in the sagittal plane. This approach is valid only for a fairly fast walk, in which the inertia forces become significant.

At the same time, a human gait in the exoskeleton, especially at the very beginning, is extremely slow and has been studied insufficiently. This is due to the fact that such a movement should occur both in the sagittal and frontal planes. The kinematics of such a motion is quite complicated, so the mathematical modeling of the spatial motion of two-legged robots is presented only in several works due to considerable difficulties [1-8, 21]. The mathematical description of the slow (quasi-static) gait of a human in an exoskeleton, according to our analysis, is not present anywhere.

Besides, special attention should be paid to the formulating of conditions that ensure a stable vertical position of the patient during movement, and this becomes possible if the patterns of movement of the feet and the center of mass of the exoskeleton with the patient are studying.

3 The gait description

For mathematical modeling of a person's gait in an exoskeleton, it is necessary, on the one hand, to analyze and establish kinematic connections between the movement of exoskeleton links during walking, and on the other, to study the nature of the motion of the center of mass and to investigate the interaction of the feet of the exoskeleton with the supporting surface.

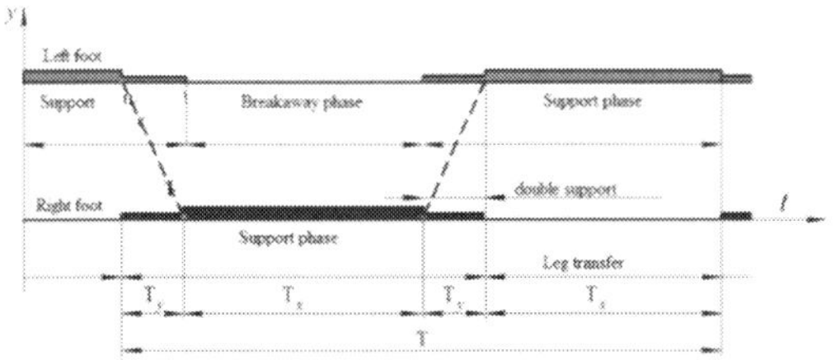

Fig. 2. The scheme of the interaction of the feet with the supporting surface in instants of time.

Let us consider the scheme of the interaction of the feet with the supporting surface at different instants of time, for this we use the diagram shown in Figure 2, such a diagram in the biomedical literature is called apodogram [20, 21]. A thin line indicates the the absence of the contact mode of one or another foot with a support. A wide line indicates the mode when the foot contacts the support. From the podogram you can see that the locomotor cycle (step) consists of two two-support and two single-support phases.

The two-support phase is the period of step Ty, during which both legs are in contact with the supporting surface and the center of mass is transferred from one foot to the other and a single-support phase is the period of step Tx, during which the transfer is carried out, starts from the moment of detachment of the foot from the support surface and finishes at the moment when the same foot contacts the support surface, at the same time the center of mass is transported along the supporting foot. The time of step T determines the rhythm of walking and is determined by the formula:

$$T = 2T_y + 2T_x \quad (1)$$

We will assume that walking begins with a two-support phase, when the feet are located symmetrically with respect to the longitudinal axis. Figure 3 shows the scheme for the exoskeleton feet moving along the longitudinal axis Ox along the support surface.

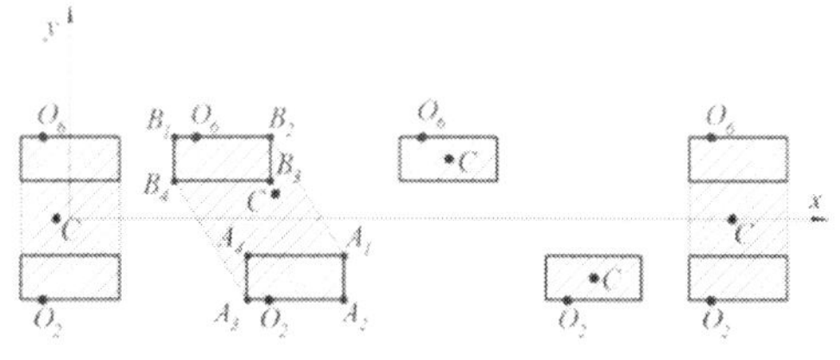

Fig. 3. The scheme for the exoskeleton feet moving along the longitudinal axis Ox along the support surface.

Here we use following annotation: $A_1A_2A_3A_4$ are the boundaries of the right foot, $B_1B_2B_3B_4$ are the boundaries of the left foot, O2, G6 are the ankle joints, C is the projection of the center of mass on the reference plane. In the two-support phase, the dimensions of the supporting surface are determined by the points A1A2A3B2B3B4, and in the single-support phase by the dimensions of the foot.

To visualize the position of the exoskeleton at different instants of time, we can consider both the spatial image of the exoskeleton in the 3D model format, and the projection of the exoskeleton on the frontal and sagittal planes, the so-called 2D models. Figures 4 and 5 show flat images of the exoskeleton for the movements sequence in the frontal and sagittal planes.

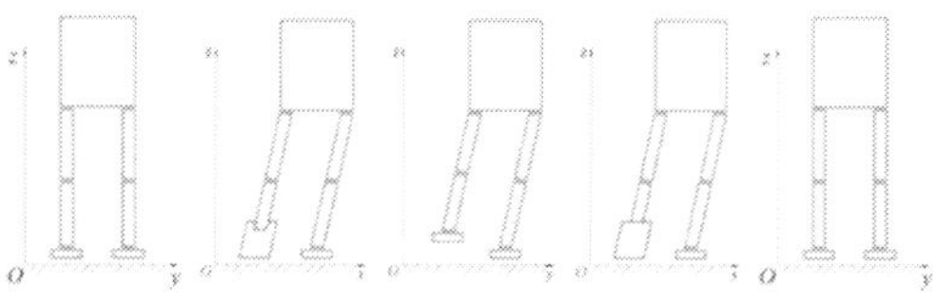

Fig. 4. The scheme of the exoskeleton weight transfer in the frontal plane.

The distinctive feature of such a gait is that the movement begins in the frontal plane and the weight is transferred to the left foot by the work of the ankle and hip joint drives.

After the center of mass is transferred to the left leg, a one-support phase begins, at which the transfer of the right foot is carried out through the movement of the links of both the right and left legs. This phase is completed after the right foot contacts the support surface. After this, two-support phase begins, at which the center of mass is transferred to the right foot. Then all the movements are repeated. The left foot is moved forward by the length of the step, the center of mass moves along the right foot and at the moment of the left foot contacts the support surface is transferred to the left foot.

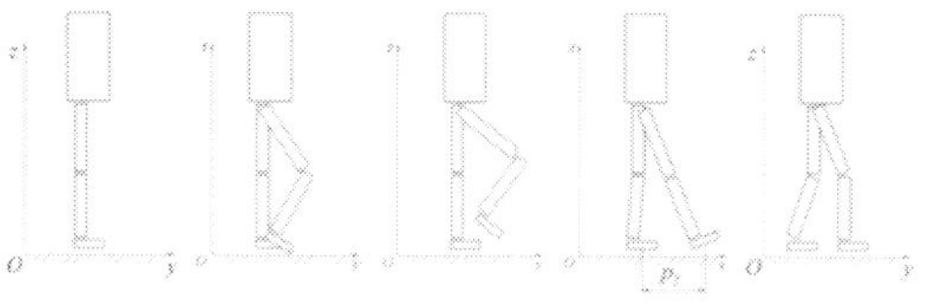

Fig. 5. The scheme of the transfer of the exoskeleton weight in the sagittal plane.

This step corresponds to the so-called "quasi-static gait". The name suggests that the exoskeleton is statically stable in every intermediate position of the mechanism.

In order to formulate conditions that ensure a stable gait, we choose three points of the exoskeleton. These points include the center of mass of the exoskeleton "C" (CME) and the "Ai" points of the feet, where i = 2 for the right foot and i = 6 for the left foot.

Let' s introduce the parameters that determine the gait and definine the motion of the foot: L is the length of the step, determined by the distance between the point of one foot at the previous and following steps; ym is the maximum deviation of the point "A" of the foot from the longitudinal axis of symmetry, T is the execution time of one step, H is the maximum height of the lift of the foot. The position of the points "Ai", at any time is determined by parametric equations:

$$\begin{cases} z_{Ai} = z_{Ai}(t) \\ x_{Ai} = x_{Ai}(t) \cdot \\ y_{Ai} = y_{Ai}(t) \end{cases} \tag{2}$$

The position of the center of mass projections (point "C") in a fixed coordinate system is given by the corresponding equations:

$$\begin{cases} z_C = z_C(t) \\ x_C = x_C(t) \\ y_C = y_C(t) \end{cases} \tag{3}$$

These equations contain parameters that determine the nature of the motion of the exoskeleton, including: L_C - distance passed by the center of mass at one step, x_{C0}, y_{C0}, z_{C0} - the initial location of the center of mass.

We will consider the motion along the supporting plane Oxy of the projections of the points "C" and "Ai". It is obvious that a stable vertical position can be ensured if the condition are fulfilled: the CME must move along a certain trajectory both along the supporting foot and when moving from one foot to the other, and the transferred foot must be adjusted to this movement and synchronously move to the point determined by the movement center of mass.

Now suppose that the exoskeleton moves along straightly along the Ox axis (the plane Oxz is the plane of symmetry of the exoskeleton in the initial position), and the trajectories of the points "Ai" are in the plane parallel to the plane Oxz and, respectively, perpendicular to the support plane, which is horizontal. This means that yA = const. Then, taking into account the symmetric character of the feet motion of the relatively to the axis Ox, we obtain

$$y_A = \pm y_m \ .$$

The "+" sign corresponds to the left foot, and "-" to the right, respectively.

Fig. 6 shows the various phases of carrying the weight of a person (CME) while walking, and the trajectory along which the center of mass moves through the foot in a single-support phase is clearly visible. There are four characteristic points $K_1K_2K_3K_4$ (for the right and $N_1N_2N_3N_4$ for the left), which determine the position of the projection of the point C at different instants of time.

The projection of the center of mass on the supporting surface coincides with the point of application of the vector of the reduced reaction of the support. The graphs shown in Figure 6 represent the movement of this point from the external part of the heel along the outer edge of the foot in the medial direction to the toes of the foot. The movement trajectory varies according to the pace and type of walking, the support surface relief, the type of footwear, namely the heel height and the rigidity of the exoskeleton foot. The pattern of support reaction is largely determined by the functional state of the muscles of the lower limb and the innervation structure of the walk [1, 2, 21].

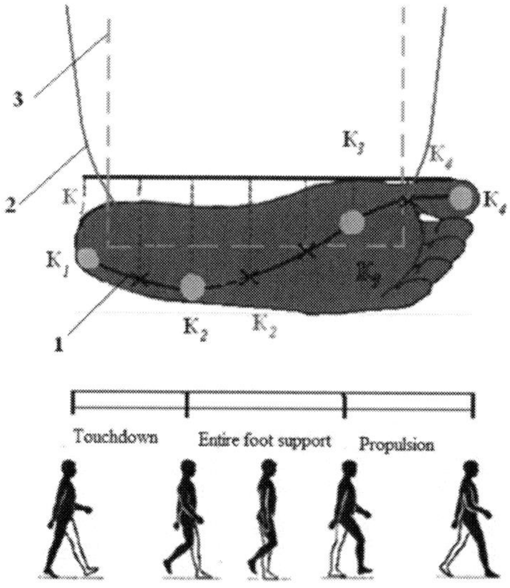

Fig. 6. The movement trajectory of the point "C" through the foot during walking: 1-fast walking, 2-slow walking, 3-mathematical model of the trajectory.

So the diagram 1 corresponds to the fast (dynamic) walking, the diagram 2 describes the displacement of the center of mass during the slow (quasistatic) walking. The diagram 3 shows some mathematical interpretation of the experimental dependences.

The movement of the point C throughout the foot is connected with a single-support walking phase, when the weight is transferred to this foot, and the second foot performs a predetermined displacement with a detachment from the surface. It is obvious that the exoskeleton in the single-support phase has a much smaller stability margin, because in this case the support area is several times smaller than in the two-support phase.

As can be seen from Figure 6, the real trajectory of the point C movement throughout the foot has a complicated shape and depends on many factors. Therefore, further let' s consider the conditions of stable walking for a number of simplified models of the trajectory, the motion of the point C both between the stop and throughout the supporting foot.

4 The stable quasistatic walking condition

To ensure a stable gait, it is necessary to synchronize the motion of points A i and C along the axis Ox, that is, the position of the feet Ai must be determined by a certain equation:

$$x_{Ai} = x_{Ai}(x_C)$$

The form of this equation depends on a number of factors. In a simplified version, for the sagittal plane, we can limit ourselves to a linear model, then it is necessary to fulfill the condition at each instant of time:

$$x_{Ai} = \lambda x_C + x_{C0}, \quad v_{A_i} = \lambda v_{Cx} \quad \text{or} \quad \dot{x}_{A_i} = \lambda \dot{x}_C$$

where v_{A_i}, v_{C_x} - the projections of the velocities of the points "Ai" and "C" on the Ox axis, is the coefficient that determines the movement synchronism of the points "Ai" and "C". This parameter depends on the height of the center of mass relative to the support surface. For the frontal plane, the synchronization condition looks similar.

Figure 7 shows the motion scheme of the center of mass and foot along the support plane in synchronous mode. The projection of the center of mass on the support plane Oxy moves with velocity V_C along the trajectory O-$N_1N_2N_3N_4K_1K_2K_3$-K_4, ensuring the transfer of the exoskeleton weight from one foot to the other. On the segment ON_1, the so-called two-support phase is realized, at which the feet are on the supporting surface and remain fixed, and the center of mass moves due to the movement of the exoskeleton links. On the segment $N_1N_2N_3N_4$, a single-support phase is realized, the weight of the exoskeleton is moved to the left foot and at this moment the right foot, with the point A, can move with the velocity V_A simultaneously with the displacement of the point C, performing the synchronization condition of the motions.

Further, the exoskeleton passes into a two-support phase, that is, the right foot places at the support surface and the point C moves first along the segment of the trajectory N_4K_1, then the weight moves to the right foot, and then the point C moves along the trajectory $K_1K_2K_3K_4$, ensuring weight transfer along the foot - this is already the single-support phase.

When the synchronization condition is fulfilled, the foot essentially follows the position of the exoskeleton center of mass, which provides stability to the quasi-static gait.

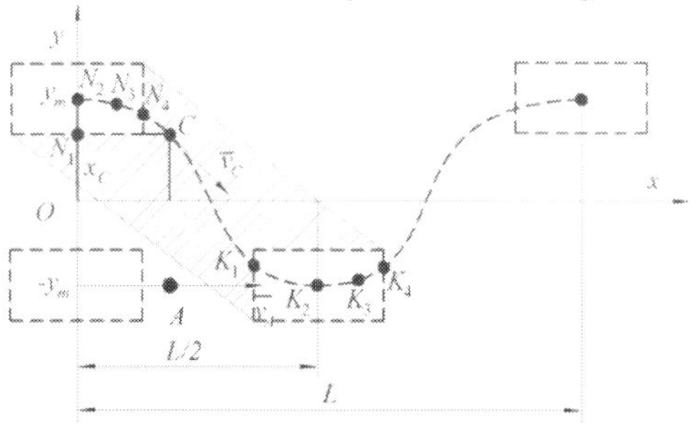

Fig.7 The synchronous motion scheme of the centre of mass and the foot.

Here a_i, b_i and c_i are the coefficients, determined by the limiting conditions. The feet motion can be expressed by the following equations:

$$z_{6,2} = 4\frac{H}{L}\cdot V\cdot t - 4\cdot\frac{H}{L^2}\cdot V^2\cdot t^2 \quad (4)$$

$$y_{6,2} = V\cdot t$$

$$V = \frac{L}{T_ц} \quad (5)$$

$$P = \frac{y_6 + y_2}{2} \quad (6)$$

$$y_m = y_{6,2} - \Delta y \quad (7)$$

$$y_C = P + \frac{P - y_m}{T_ц}\cdot t$$

$$z_C = z_{c0}$$

$$z_{c0} = \begin{cases} H & l_2 + l_3 < l_2^0 + l_3^0 \\ H - \Delta H & l_2 + l_3 \ge l_2^0 + l_3^0 \end{cases} \quad (8)$$

The formulas that describe the motion of the points A_i and C along to corresponding plane trajectories can be represented in the form:

$$\begin{cases} y_A = \pm y_m \\ x_A = \sum_{i=0}^{3} a_i\cdot t^i \\ x_C = \sum_{i=0}^{3} c_i\cdot t^i \\ y_C = \sum_{i=0}^{3} b_i\cdot t^i \end{cases} \quad (8)$$

The formulas (9) obtained allow us to take into account the begining of the motion and the approach of the point Ai of the foot and point C to the end point of the trajectory with zero speed. This allows to minimize the acceleration that occurs when the point C moves unevenly along the trajectory.

5 Conclusions

As a result of the analysis of the quasi-static gait of a person in the exoskeleton, three characteristic points determining the walking of the exoskeleton are chosen: this is the center of mass C and the points Ai of the feet. It is shown that to ensure stable walking it is necessary to move point A synchronously with the projection of the center of mass on the longitudinal axis. It is established that for the trajectories under consideration, when the exoskeleton moves along a straight line, the velocity of the foot along the longitudinal axis must be much higher than the velocity of the center of mass. Various mathematical models of the trajectory of the center of mass motion are proposed, and a kinematic analysis of the motion of the center of mass and the points of the feet on the horizontal plane is performed.

The investigation is supported by Russian Foundation of Basic Research grant 18-08-00773.

References

1. S.L. Francesca, Frontiers in human neuroscience, **8** (2014)
2. E. Grant, 2013 IEEE International Conference on Rehabilitation Robotics (ICORR), 1–6. (2013)
3. H. Gou, Journal of Electrical and Computer Engineering, **2** (2014)
4. K. Harada, IEEE/ASME Transactions on Mechatronics, **12** (1), 53–62 (2007)
5. M. Hassan, Sensors, **14** (1), 1705–1722 (2014)
6. Q. Huang, IEEE Transactions on robotics and automation, **17** (3), 280–289 (2001)
7. D. Jasińska-Choromańska et al., Bulletin of the Polish Academy of Sciences: Technical Sciences, **61** (2), 419–431 (2013)
8. S. Jatsun, A. Savin, A. Yatsun, Proceeding of the 19th International Conference on CLAWAR 2016, 797–805 (2016)
9. S. Jatsun, A. Savin, A. Yatsun, 24th Mediterranean Conference on Control and Automation (MED), 322–326 (2016)
10. S. Jatsun, A. Savin, A. Yatsun, 2016 IEEE International Conference on Automation, Control, and Intelligent Systems (CYBER), 239–244 (2016)
11. S. Jatsun, A. Savin, A. Yatsun, Proceedings of 20th International Conference on System Theory, Control and Computing Joint Conference SINTES 20, SACCS 16, SIMSIS 20, 107–112 (2016)
12. S.F. Jatsun, In Assistive Robotics: Proceedings of the 18th International Conference on CLAWAR 2015, 175–182 (2015)
13. S. Jatsunб S. Savin, B. Lushnikov, A. Yatsun, ITM Web of Conferences, **6**, 01001 (2016).
14. S. Jatsun, Advances in Robot Design and Intelligent Control, 165–172 (2016)
15. S. Jatsun, 2015 10th International Symposium on Mechatronics and its Applications (ISMA), 1–6 (2015)
16. B.I. Ergasheva, Scientific and technical journal of information technologies, mechanics and optics. **17** (6), 1153–1158 (2017)
17. N.N. Rukina et al., Modern technologies in medicine, **8** (4), 231–240 (2016)
18. N. Vitiello et. al. Robotics, IEEE Transactions on 29, **1**, 220–235 (2013)
19. M. Vukobratović, B. Borovac, M. Vukobratović, M. Raković, Research and Education in Robotics—EUROBOT 2008, 42–53 (2008)
20. Y. Wen, J. Rosen, Cybernetics, IEEE Transactions on cybernetics, **43** (2), 673–684 (2013)
21. K.A. Witte, J. Zhang, R.W. Jackson, S.H. Collins, IEEE International Conference on Robotics and Automation (ICRA), 1223–1228 (2015)
22. W. Yang, Y. Yang, W. Liu, R. Wang, International Conference on Mechatronics, Electronic, Industrial and Control Engineering (MEIC-15) (2015)

Map representation using hidden markov models for mobile robot localization

Jesus Savage*, *Oscar* Fuentes, *Luis* Contreras, and *Marco* Negrete
Bio-Robotics Laboratory, Universidad Nacional Autónoma de México (UNAM)

Abstract. This paper describes a map representation and localization system for a mobile robot based on Hidden Markov Models. These models are used not only to find a region where a mobile robot is, but also they find the orientation that it has. It is shown that an estimation of the region where the robot is located can be found using the Viterbi algorithm with quantized laser readings, i.e. symbol observations, of a Hidden Markov Model.

1 Introduction

It is the general case that Robots are conceived to perform a task, and the chosen world representation is highly correlated with the performance of the robot in such task [1]. The mapping process is restricted to the robot features, where a map is generated using a collection of sensor readings from a traversed trajectory in the scene given by the robot shape and its degrees of freedom, and this information might be useful to define the map [2] [3].

In the many years of research of robot navigation [4] and SLAM [5] [6] [7], several map representations have been proposed. From metric and topological to symbolic and probabilistic representations [8] [9].

In this work, we propose a probabilistic representation given by a Hidden Markov Model from a number of sensor readings and robot displacements; this approach has been proven useful for robot localization and navigation [10], [11]. We extend those works by incorporating the trajectory information in the Markov Model, allowing us to obtain the position and orientation related to a number of nodes on the trajectory.

The remaining of the paper is divided as follows: in Section 2 we introduce the VIRBOT system while in Section 3 we detail the cartographer module and develop the map representation using Hidden Markov Models based on the system characteristics. Then, in Section 4 we present a localization method using the proposed map representation. In Section 5 we detail the experiments and results and in Section 6 we conclude this work.

2 The VIRBOT system

There are several architectures to control mobile robots, in our approach, the VIRBOT system [12], the mobile robot's operation it is divided into four general layers: Inputs, Planning, Knowledge Management and Execution, having each of them several subsystems, see Figure 1. Each subsystem has a specific function that contributes to the final operation of the robot.

This system has similar features presented in the INTERRAP agent architecture [13]

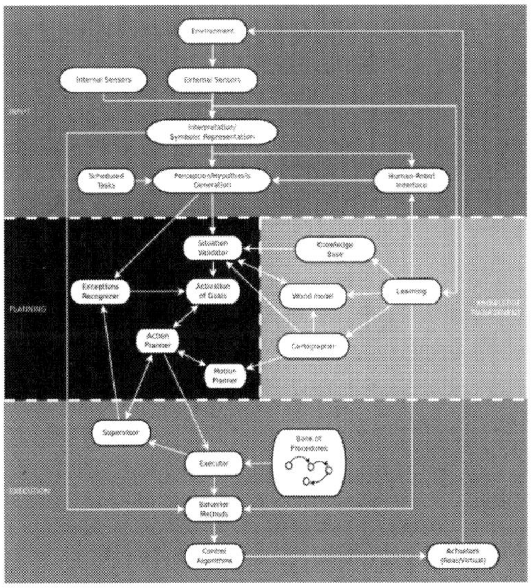

Fig. 1. The VIRBOT System architecture consisting of several subsystems that control the operation of a mobile robot.

2.1 Inputs layer

In this layer are the robot's internal and external sensors, as well as, the simulator that simulates these sensors when a simulated Robot is used.

*e-mail: robotssavage@gmail.com

2.2 Planning layer

Action planner: The objective of action planning is to find a sequence of physical operations to achieve the desired goals.

Motion planner: This module receives, from the action planner, a set of locations that the robot needs to visit and it finds paths to reach them using a topological map and the A* Algorithm.

2.3 Knowledge management layer

Cartographer: This module creates and contains different types of maps for the representation of the environment. This section is explained in more detail in section 3

Knowledge representation: A rule-based system, CLIPS, developed by NASA, is used to represent the robot's knowledge, in which each rule contains the encoded knowledge of an expert.

2.4 Execution layer

This module executes the actions and movements plans.

3 Cartographer

The cartographer module is responsible for estimating the region where the robot is located. One of the problems of using only dead reckoning to estimate the robots' position x, y, θ is that, due to errors in the movement of the robot and errors in the sensors that measure them, after several movements, the real position of the robot can not be estimated accurately. In some robotics architectures, the ones based only on reactive behaviors [14], it is not necessary to know the exact pose of the robot but only an estimation of the region where is located. There have been several approaches based on Hidden Markov Models and Bayesian techniques to solve this problem [15]. In our system this task is done by a localization module, using laser readings, Hidden Markov Models and the Viterbi algorithm.

3.1 Hidden Markov models

A Hidden Markov Model (HMM) is a two random variable process, in which X one of the random variables is hidden, and the other random variable O is observable [16]. The hidden random variable X represents the states, in a model that is a collection of states connected by transitions. In our system, the states represent the centroids of the regions in the free space where the robot navigates, see Figure 2.

Each state has two sets of probabilities: an output probability, that provides the conditional probability that given that the system is in a particular state, a symbol is generated from a finite set of symbols, and a transition probability from going from one state to another.

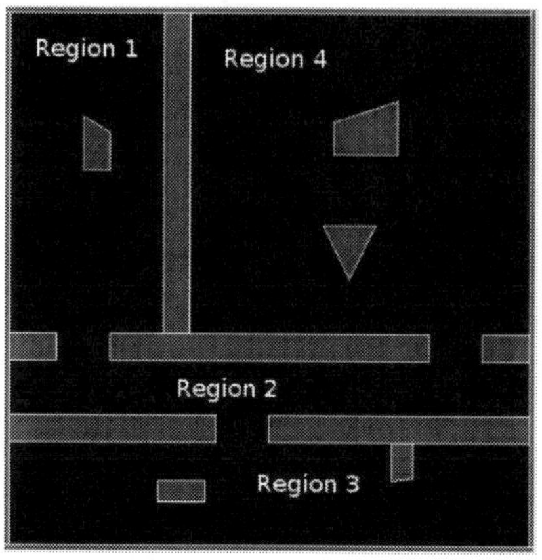

Fig. 2. Navigation regions.

3.1.1 Generation of the HMM symbols

The symbols are obtained using clustering techniques like K Means or Vector Quantization (VQ). VQ techniques are used for data compression, given a set of vectors, $S_j = [s_1, s_2, ..., s_m]$ that represent lasers readings, see Figure 3, a set of centroids are found which represent them.

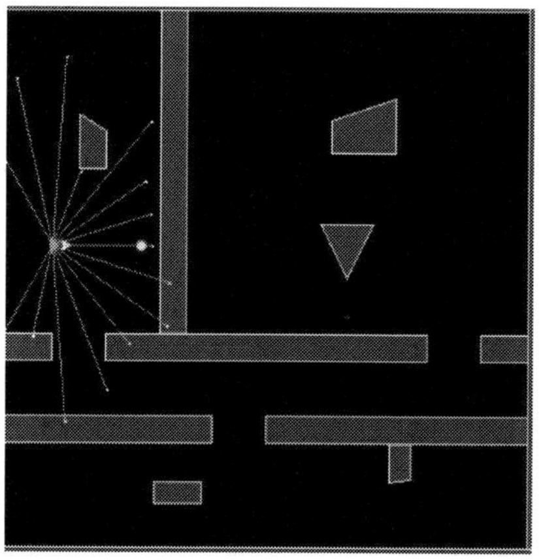

Fig. 3. Laser data $S_j = [s_1, s_2, ..., s_m]$.

The collection of centroids is called a codebook. This codebook is designed from a long training sequence that is representative of all data, laser readings, to be encoded by the system. To generate the codebook, the Lynde-Buzo-Gray algorithm is used [17].

After the codebook is created, each vector S_t of the laser readings to be encoded is compared with each of the stored vectors C_i, and the vector S_t is coded by identifying the vector C_k that best represents S_t according to some distance measurement d.

$$d(S_t, C_k) = min(d(S_t, C_i)), \quad i = 1, ..., L \quad (1)$$

where L is the size of the codebook. The distance measurement used was the Euclidean distance.

3.1.2 HMM model

The symbols are the indexes of a vector quantizer, for each laser reading we have L possible observations vectors that correspond to the L vectors of a quantizer. Then for a vector S_t the symbol O_t corresponds to the index of the centroid that best fits the laser readings, $O_t = k$.

And the probability of emitting symbol k being in state i, $b_i(k) = p(O_t = k|X_t = i)$, where $X_t = i$ means the Markov chain was in state i at time t, and $O_t = k$ means the output symbol at time t was k.

The transition probability defines the probability of taking the transition from a particular state to another state or itself. Where a_{ij} is the probability of taking a transition from state i to state j:

$$a_{ij} = p(X_{t+1} = j|X_t = i).$$

Also included is the initial state distribution probability:

$$\pi_i = P[x_1 = X_i]. \qquad 1 < i < N$$

In the first order HMM, it is assumed that whether the Markov model will be in a particular state X at time $t+1$ only depends on the present state at time t, and not on the past states.

$$P(X_{t+1} = x_{t+1}|x_i, ..., x_t) = P(X_{t+1} = x_{t+1}|x_t).$$

Another assumption is that if a symbol will be emitted it only depends on the present state at time t, and not in the past states or emitted observations.

$$P(o_t = k|o_i, ..., o_{t-1}, x_i, ..., x_t) = P(O_t = k|x_t).$$

3.2 Discrete hidden Markov models for laser readings

Each region R_j of the environment where the robot navigates is represented by a Markov model $\lambda_j = (A, B, \pi)$, where the matrices A and B are: $A = a_{ij}$ and $B = b_i(k)$ for all i, j and k, with N states and L observation symbols. The probabilities of each Markov model λ_j are found during training. For each region $j, 0 < j < N_{regions}$, the robot is set in region j with orientation of 0 degrees according to an global origin and then it follows the steps described in Algorithm 1.

HMM states from 1 to 8, corresponds to the rotation process and from 9 to 16 to the navigation one, see Figure 4. For each path that the robot takes in a closed region, see Figure 5, we get the observations $O^n(t_1), O^n(t_2), ..., O^n(t_i), ..., O^n(t_T)$, corresponding to the vectors of quantization that best fit each of the laser readings at t_i, where n corresponds to the nth-repetition.

For each laser reading, we have L possible observations that correspond to the L vectors of a quantizer. This quantizer was created using laser data during several paths that the robot followed. With the set of observations an estimation of the parameters of the HMM λ_j, that is A, B and π for each region j, was found using the BaumWelch algorithm.

Algorithm 1: Training Behavior for region R_j

Data:
Number of navigation steps NF
Training Vector Size TS
Result: Observations Vector $\vec{O}$
$n \leftarrow 0$
$aux \leftarrow 0$
for $n \leq TS$ **do**
  $\vec{O}_n \leftarrow$ Quantized(laser readings)
  if $aux > NF$ **then**
    The robot rotates 45°
    HMM state $\leftarrow$ possible states (1,2,...,8)
    if $aux > NF + 8$ **then**
      $aux \leftarrow 0$
    end
  end
  else
    Robot in reactive navigation mode
    HMM state $\leftarrow$ possible states (9,10,...,16)
  end
  $aux \leftarrow aux + 1$
  $n \leftarrow n + 1$
end

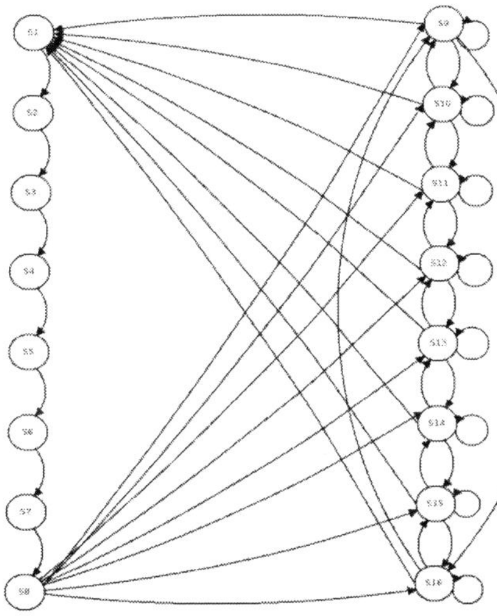

Fig. 4. HMM for each region.

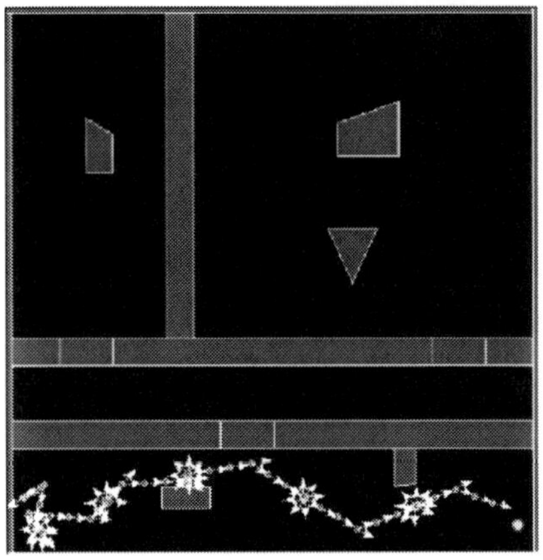

Fig. 5. Training procedure for one region of the environment.

In the Figure 6 we can see the complete HMM for the environment in Figure 2. This HMM represents the concatenation of each region in the environment, where each node is the particular HMM λ_j for each region.

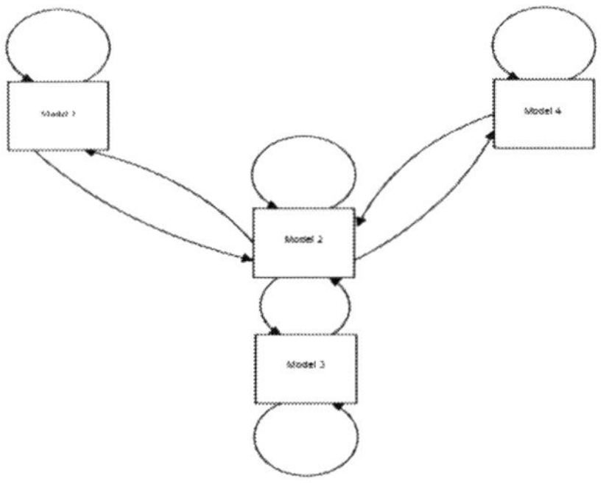

Fig. 6. HMM generated from the concatenation of each model HMM of the regions shown in Figure 2.

4 Robot localization using discrete hidden Markov models

To recognize the region where the robot is and its orientation, first, the robot start rotating every 45 degrees, until it rotates 360 degrees, for each rotation it makes a laser reading, generating a vector of laser readings that is encoded through a vector quantizer of L vectors. This quantizer gives a set of observations

$$O = O(t_1), O(t_2), ..., O(t_i), ..., O(t_T),$$

that correspond to the VQ vectors that best fit the laser vector at time t during its rotation.

Then using the observation vector with all the HMM_j for each region, the probability of each model λ_j is found using the Forward algorithm [16], see Figure 7. Then the robot is in $Region_j$:

$$Region_j = argmax_{1<i<N}[p(\lambda_i, O))]. \quad (2)$$

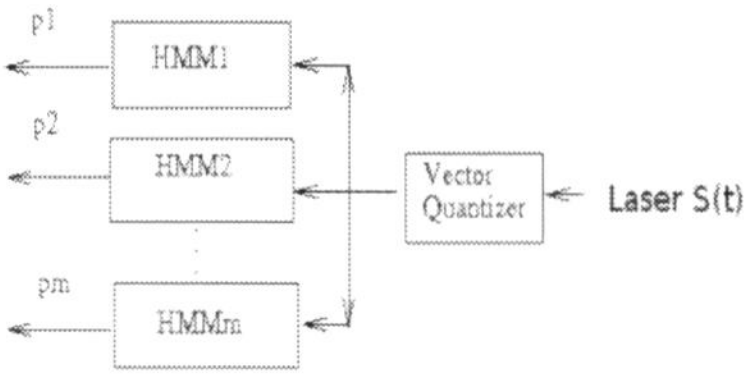

Fig. 7. Probability evaluation for each region.

After it is found in which region is the robot as well its orientation the robot starts navigating to a defined destination, collecting again another observation vector O, and using this vector, the hidden variables, the states, are found using the Viterbi algorithm. This algorithm finds the best sequence of states in the network, and thus the best sequence of regions that the robot follows during its path. Figure 8 shows the best path followed by the robot going from one region to another.

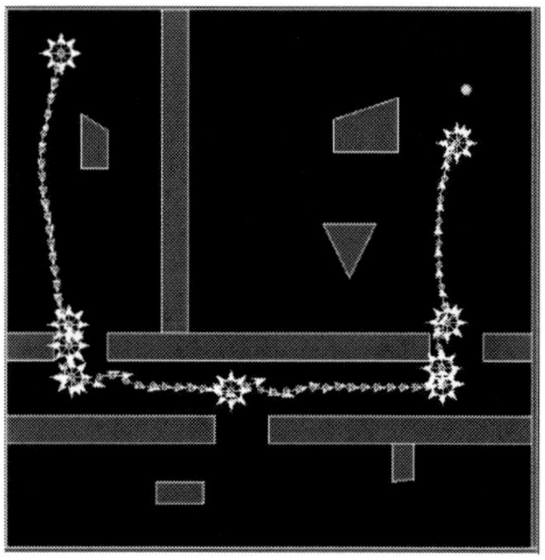

Fig. 8. Best path followed by the robot.

The Viterbi algorithm is as follows [16]:

Given a HHM model $\lambda = (A, B, \pi)$ with N states and L observation symbols, and an observation sequence $O = O_1, O_2, ..., O_T$. The following equation gives the highest probability to reach state i along a single path at time t, given that the system has been in states $x_1 ... x_{t-1}, x_t$ and with observations $O_1, ..., O_t$:

$$\delta_t(i) = maxP[x_1 ... x_{t-1}, x_t = i, O_1, ..., O_t|\lambda].$$

In order to retrieve the state sequence, it is necessary for each t and j to keep track of the argument which maximize the previous equation. Using the array $\psi_t(j)$, the procedure is as follows:

2.1) Initialization:

$$\delta_1(i) = \pi_i b_i(O_1), \qquad 1 < i < N. \tag{3}$$

$$\psi_t(i) = 0.$$

2.2) Recursion:

$$\delta_t(i) = max_{1<i<N}[\delta_{t-1}(i)a_{ij}]b_j(O_t), \tag{4}$$

$$2 < t < T, \qquad 1 < j < N,$$

$$\psi_t(j) = argmax_{1<i<N}[\delta_{t-1}(i)aij], \tag{5}$$

$$2 < t < T, \qquad 1 < j < N.$$

2.3) Termination:

$$P^* = max_{1<i<N}[\delta_T(i)], \tag{6}$$

$$x_T^* = argmax_{1<i<N}[\delta_T(i)]. \tag{7}$$

2.4) State sequence backtracking:

$$x_t^* = \psi_{t+1}(q_{t+1}^*), \tag{8}$$

$$t = T-1, T-2, ..., 1.$$

Figure 9 shows the overall system to find the best sequence of regions that the robot visited.

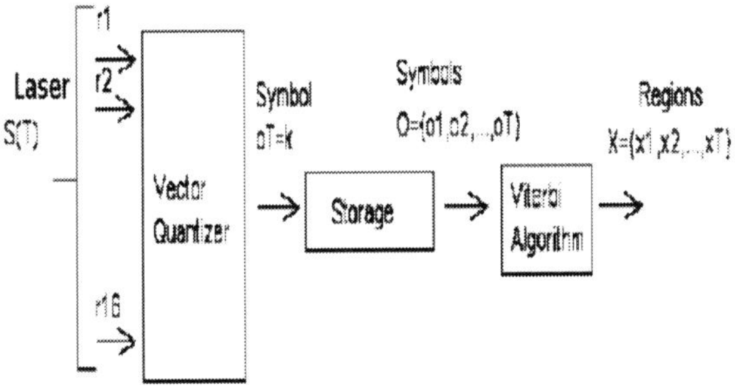

Fig. 9. Overall system.

5 Experiments and results

We tested the algorithm using the VIRBOT simulation module, in which laser readings can be simulated, see Figure 3, as well as, with a TurtleBot robot, see Figure 10. The simulator simulates laser readings by finding the distance of the objects in front of the robot, adding Gaussian noise to the values. The virtual mobile robot and the real

Fig. 10. TurtleBot robot used in the experiments.

one have a 180-degree laser, with a separation of 12 degrees each, making 15 laser readings in each sensing. The real robot uses a Hokuyo laser.

The algorithm was tested in an environment with 4 regions, shown in Figure 2. The robot was put in any of the regions of each environment, and from this original position, it tried to reach a set of destinations using reactive behaviors. In Figure 8, the robot starts in a particular region and to reach the goal destination, that is in another region, it needs to pass through the nodes that link the regions, when it reaches each of them it rotates to localize it shelf. With the symbols that were found using the vector quantizer, the start orientation of the robot, as well as, the best states were found on using the Viterbi algorithm. Table 1 shows the results with the robot's simulator, column $Tr.Size$ represents the number of vectors used to create the laser VQ; column $N.Symbols$ indicates the number L of centroids of the VQ, that is, the number of symbols of the HMM; column $N.Obs.$ indicates the number of observations T used in the Viterbi and Forward algorithms, and finally the column $Error$ shows the percentage of error for detecting in which region the robot is.

Table 1. System performance with the simulated robot.

Tr.Size	*N. Symbols L*	*N. Obs. T*	*% Error*
4000	32	36	56%
8000	32	36	32%
4000	256	36	35%
8000	256	36	16%
4000	512	36	31%
8000	512	36	13%
4000	32	64	45%
8000	32	64	26%
4000	256	64	30%
8000	256	64	15%
4000	512	64	17%
8000	512	64	11%

As we can see the best performance was obtained with a training set of 8000 vectors, 512 observation symbols and using 64 observations each time the Forward and Viterbi algorithm was performed. From the estimated states we found 89% of the times the region where the simulated robot was, using only the laser readings without dead reckoning.

Table 2 shows the results with the TurtleBot robot. As we can see the best performance was obtained, as with the simulator, with a training set of 8000 vectors, 512 observation symbols and using 64 observations each time the Forward and Viterbi algorithm was performed. From the estimated states we found 85% of the times the region where the real robot was, using only the laser readings without dead reckoning.

Table 2. System performance with the real robot.

Tr.Size	*N. Symbols L*	*N. Obs. T*	*% Error*
4000	256	36	36%
4000	512	36	29%
8000	256	64	24%
8000	512	64	15%

6 Conclusions

We proposed a method to estimate the region where a robot is by using laser readings combined with Hidden Markov Models and the Viterbi algorithm. Using only laser reading without using dead reckoning, this technique was able to find the region where the robot was located. This approach worked well with our simulator, as well as, with a real mobile robot. The next step is to increase the number of regions, that is, to increase the accuracy of the estimated position.

This work was supported by PAPIIT-DGAPA UNAM under Grant IG-100818.

References

[1] L. Contreras, W. Mayol-Cuevas, *O-POCO: Online POint cloud COmpression mapping for visual odometry and SLAM*, in *IEEE International Conference on Robotics and Automation* (2017)

[2] L.P. Berczi, T.D. Barfoot, *It's like deja vu all over again: Learning place dependent terrain assessment for visual teach and repeat*, in *IEEE/RSJ International Conference on Intelligent Robots and Systems* (2016)

[3] M. Paton, K. MacTavish, M. Warren, T.D. Barfoot, *Bridging the appearance gap: Multi-experience localization for long-term visual teach and repeat*, in *IEEE/RSJ International Conference on Intelligent Robots and Systems* (2016)

[4] D. Scaramuzza, F. Fraundorfer, IEEE Robotics and Automation Magazine 18(4) 80–92 (2011)

[5] H. Durrant-Whyte, T. Bailey, IEEE Robotics and Automation Magazine 13(2) 99–110 (2006)

[6] T. Bailey, H. Durrant-Whyte, IEEE Robotics and Automation Magazine 13(3) 108–117 (2006)

[7] Thrun, J.J. Leonard, *Simultaneous Localization and Mapping* (Springer Handbook of Robotics, 2008)

[8] J. Savage, L. Contreras, I. Figueroa, A. Pacheco, A. Bermudez, M. Negrete, M. Matamoros, C. Rivera, *Construction of Roadmaps Maps for Mobile Robots' Navigation Using RGB-D Cameras*, in *13th International Conference on Intelligent Autonomous Systems* (2014)

[9] L. Contreras, W. Mayol-Cuevas, *Towards CNN Map Compression for camera relocalisation*, in *arXiv* (2017)

[10] S. Shahriar, A. Zelinsky, *Mobile Robot Navigation based on localisation using Hidden Markov Models* (1999)

[11] S. Shahriar, A. Zelinsky, *Robot localization from minimalist inertial data using a Hidden Markov Model*, in *IEEE International Conference on Autonomous Robot Systems and Competitions* (2014)

[12] J. Savage, M. Negrete, M. Matamoros, J. Cruz, *The Role of Robotics Competitions for the Development of Service Robots*, in *IJCAI, Workshop on "Autonomous Mobile Service Robots"* (2016)

[13] J. Muller, *The Design of Intelligent Agents: a layered approach* (Springer-Verlag, 1996)

[14] R. Arkin, R. Murphy, IEEE Transaction on Robotics and Automation 6(4) 445–452 (1990)

[15] S. Thrun, Technical Report, CMU-CS-96-122, School of Computer Science, Carnegie Mellon University. (1996)

[16] L. Rabiner, Readings in Speech Recognition Morgan Kaufmann Publishers 267–296 (1990)

[17] Y. Linde, A. Buzo, R.M. Gray, IEEE Transactions on Communications, 84-95 (1980)

State observer design for a walking in-pipe robot

Sergei Savin *, *Sergey* Jatsun and *Ludmila* Vorochaeva

Southwest State University, department of mechanics, mechatronics and robotics, 305040 Kursk, Russia

Abstract. In this paper, a state observer design for a walking in-pipe robot is studied. The necessity of using a state observer is related to the fact that sensors have limited accuracy and are prone to producing noise. This is especially problematic for in-pipe walking robots, since they use model-based control and require accurate information of their current state. The paper shows that an iterative state observer based on solving Riccati equation provides significant improvements in the behaviour of the control system. It allows to smooth out the spikes in the control actions requested by controller and to minimize tremor of the robot links. In order to study the behaviour of the observer when different sensors are used, a performance function was introduced. It was shown that the observer allows to improve the performance of the control system for a wide range of sensor parameters. Additionally, it was shown that the introduction of the observer allows to choose higher feedback controller gains, enabling more precise control. Simulations on the full robot model, taking into account mechanical constraints and contact forces showed that the linear observer is capable of improving the behaviour of the control system of the walking robot, if measurements of the reaction forces are provided. The effects that the noise and quantization in the reaction forces measurements have on the behaviour of the state observer is studied.

1 Introduction

Pipeline inspection robots are complex electro-mechanical systems that are designed to move inside pipelines while carrying necessary equipment. Popular in-pipe robot designs feature parallel linkages with wheels or tracks [1-4]. Some other designs include flexible links, pneumatic or Bowden cable activation, serial linkages designs (such as snake-like robots), multi-module designs, etc. [5-7]. This diversity and complexity is the result of the diversity and complexity of the tasks posed for this type of robots and reflects the advances in the field of in-pipe robotics.

Currently, in-pipe robot designs have been able to solve a number of mobility tasks, however the problem of navigating pipelines has not been completely solved yet. In part, this is due to the diversity of the mobility tasks the in-pipe robots face. For example, pipelines can have significant changes in diameter, vertical sections, L-shaped and U-shaped turns, Y-shaped and T-shaped junctions and other features that make it difficult to navigate for most of the known robot designs. This motivates the use of more agile robot designs, such as walking in-pipe robots [8-10].

All types of in-pipe robots require solving a number of sensor-related challenges. These include estimation of the robot's position and configuration, navigation and mapping. Some of the sensor-related problems are common for all types of multilink robots, and arise due to the fact that real sensors are characterized by quantization, discretization, noise, time lag and other strongly nonlinear behavioural features. Estimating the configuration of walking in-pipe robots is an additionally challenging problem because of their structure, which features multiple serial linkages connected to the floating body (for the terminology see [11]). Since calculating the positions of links at the end of a sequential kinematic chain requires measurements of a number of joint parameters (angles or linear displacements), the nonlinear behaviour of the joint sensors will manifest significantly in these calculations. Incidentally, the contact elements of walking in-pipe robots, whose trajectories need to controlled [12-15], are located at the ends of sequential kinematic chains and therefore their positions are hard to infer exactly. This motivates the use of special configuration estimation algorithms.

There are a number of methods that can be used for configuration estimation (or state estimation). For linear systems, state observers can be used. In papers [16-18] state observers had been used to control DC motors, underwater robots and wheeled mobile robots. This shows that although these methods are developed for linear systems, they can be used for nonlinear mechanical systems as well. This is similar to how linear control methods, such as linear quadratic regulator, can be used to control robots with nonlinear dynamics (see [19-21] for examples). In this paper, we look at the use of linear state observer to provide predictions for joint parameters of an in-pipe robot.

The rest of the paper is organized as follows. The section 2 provides description on an in-pipe walking robot which is used in the numerical studies. The section 3 discusses the observer design, and sections 4 and 5 show the work of the observer in simulation.

* Corresponding author: savinswsu@mail.ru

2 Walking in-pipe robot description

In this paper, we consider an in-pipe walking robot with nine links. One of the links is called "body" and other links are organized into pairs and are connected to the body in a serial manner. The robot is shown in figure 1.

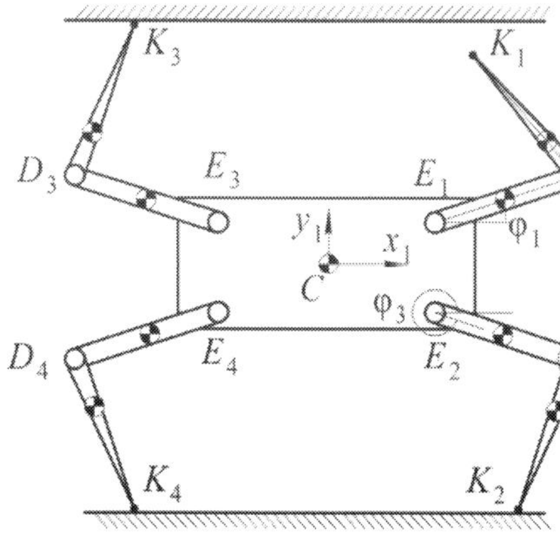

Fig. 1. Diagram of the in-pipe walking robot.

The diagram in fig. 1 uses the following notation. D_i and E_i are active joints, K_i are contact elements, φ_i are joint angles, x_1 and y_1 are axes of a local coordinate system, attached to the robot's body and C is the center of mass of the robot's body.

In the following sections, we will consider one of the robots legs separately from the rest of the device. Assuming that the robot's body remains stationary, we can introduce a vector of generalized coordinates that describes the configuration of one of the robot's legs:

$$\mathbf{q} = [\varphi_1 \quad \varphi_2]^{\mathrm{T}}. \tag{1}$$

Using this generalized coordinates we can describe robot's dynamics in the following way:

$$\mathbf{H}\ddot{\mathbf{q}} + \mathbf{c}_0 = \boldsymbol{\tau}, \tag{2}$$

where $\mathbf{H}$ is the generalized inertia matrix, $\mathbf{c}_0$ is a bias vector and $\boldsymbol{\tau}$ is a generalized motor torques (see [22] for terminology, derivations and practical algoriths for generating these equations). These equations can be linearized in the following way:

$$\dot{\mathbf{x}} = \mathbf{A}\mathbf{x} + \mathbf{B}\mathbf{u} + \mathbf{c}, \tag{3}$$

where $\mathbf{x} = [\mathbf{q}^{\mathrm{T}} \quad \dot{\mathbf{q}}^{\mathrm{T}}]^{\mathrm{T}}$ is a vector of state coordinates, $\mathbf{u} = \boldsymbol{\tau}$, $\mathbf{A} = \partial\mathbf{f}/\partial\mathbf{x}$, $\mathbf{B} = \partial\mathbf{f}/\partial\mathbf{u}$ and $\mathbf{c} = \mathbf{f} - \mathbf{A}\mathbf{x} - \mathbf{B}\mathbf{u}$ are constant linearization matrices and vectors calculated at a given point $\mathbf{x}_0$ and $\mathbf{f} = \mathbf{H}^{-1}(\boldsymbol{\tau} - \mathbf{c}_0)$ is the function describing the dynamics of the robot.

The robot is controlled by a computed torque controller (CTC), which is given by the following control law:

$$\boldsymbol{\tau} = \mathbf{H}(\ddot{\mathbf{q}}^* + \mathbf{K}_d\dot{\mathbf{e}} + \mathbf{K}_p\mathbf{e}) + \mathbf{c}_0, \tag{4}$$

where $\ddot{\mathbf{q}}^*$ is the desired value of the generalized coordinates, $\mathbf{e} = \mathbf{q}^* - \mathbf{q}$ is the control error and $\mathbf{K}_p$ and $\mathbf{K}_d$ are gain matrices. The use of this controller is discussed in [23] and the tuning of the gain matrices is studied in [24-25].

3 State observer design

The idea behind a state observer is to predict the actual evolution of the state coordinates of the robot given a model of the robot dynamics and possibly incorrect measurements of the state coordinates. Let us introduce a vector of state coordinate predictions $\mathbf{z}$. We define the observer dynamics as follows:

$$\dot{\mathbf{z}} = \mathbf{A}\mathbf{z} + \mathbf{B}\mathbf{u} + \mathbf{c} + \mathbf{L}(\mathbf{x} - \mathbf{z}), \tag{5}$$

where $\mathbf{L}$ is a gain matrix of the observer. In order to find the optimal gain matrix we can define $\mathbf{e} = \mathbf{x} - \mathbf{z}$, subtract (5) from (3) and simplify to get:

$$\begin{cases} \dot{\mathbf{e}} = \mathbf{A}\mathbf{e} + \mathbf{I}\bar{\mathbf{u}} \\ \bar{\mathbf{u}} = -\mathbf{L}\mathbf{e} \end{cases}, \tag{6}$$

where $\mathbf{I}$ is an identity matrix. This suggests that the optimal gain matrix $\mathbf{L}$ can be computed by solving the Riccati equation, same as it is done in linear quadratic regulator design. Examples of how this is done can be found in [20-21]. The weight matrices that appear in the Riccati equations are chosen to be identity matrices in all experiments shown in this paper.

If the mechanical system has explicit mechanical constraints imposed on it, the observer dynamics equation can be modified as follows:

$$\dot{\mathbf{z}} = \mathbf{A}\mathbf{z} + \mathbf{B}\mathbf{u} + \mathbf{F}^{\mathrm{T}}\boldsymbol{\lambda} + \mathbf{c} + \mathbf{L}(\mathbf{x} - \mathbf{z}), \tag{7}$$

where $\mathbf{F}$ is the constraint Jacobian taken with respect to the state coordinates $\mathbf{x}$ and $\boldsymbol{\lambda}$ is a vector of Lagrange multipliers that are equivalent to reaction forces. Values of $\boldsymbol{\lambda}$ can me read from force sensors, and they are subject to measurement errors.

Thus, the observer is given by the update law (5) or (7), where the gain matrix is obtained by solving the Riccati equation for the dynamical system (6). For nonlinear system this is done iteratively, by linearizing the system along the path that the system follows in the state space. In order to prevent the erroneous values of sensor measurements from appearing in the dynamics model during linearization, the model can be linearized along its desired (rather than actual) trajectory in the state space. This is what is done in all experiments shown in this paper.

4 Simulation results

In this section, we study the behaviour of a single leg of the walking in-pipe robot, taking into account the nonlinear effects of the joint sensors. In particular, we model sensor quantization and sensor noise. The sensor model is then given as follows:

$$\tilde{x} = x - (x \mod \sigma) + r(c,\delta), \quad (8)$$

where $\tilde{x}$ is the measured value of x, $(x \mod \sigma)$ is remainder after division of x by σ and $r(c,\delta)$ is a random number that has a flat distribution in the interval $[c-0.5\delta \quad c+0.5\delta]$. Figure 2 shows the sensor output during a single robot step for the case when $c=0$, $\delta=0.05$ and $\sigma=0.05$ for both position and velocity sensors.

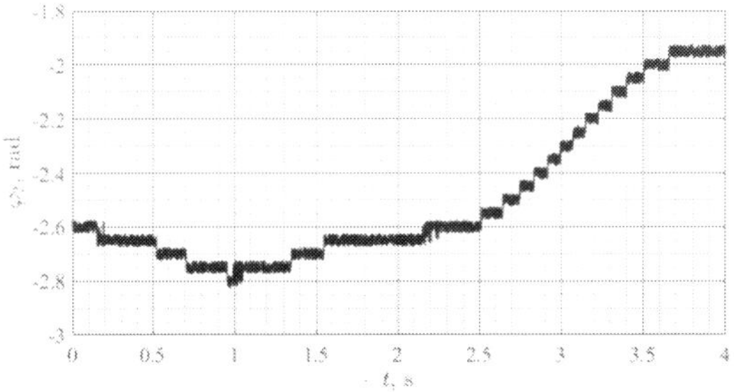

Fig. 2 The measured values of $\varphi_2(t)$.

As we can see, the measured value of joint angle increases by visible increments, and the noise levels are so high that they make it difficult to determine the time when the increments took place. Expression (4) suggests that the nonlinearities in $\varphi(t)$ translate to nonlinearities in $\boldsymbol{\tau}(t)$. Figure 3 shows the generalized torques $\boldsymbol{\tau}(t)$ obtained for the case discussed above. This and the following simulations are done with $\mathbf{K}_p = 1000\mathbf{I}$ and $\mathbf{K}_d = 100\mathbf{I}$, where $\mathbf{I}$ is a 2-by-2 identity matrix.

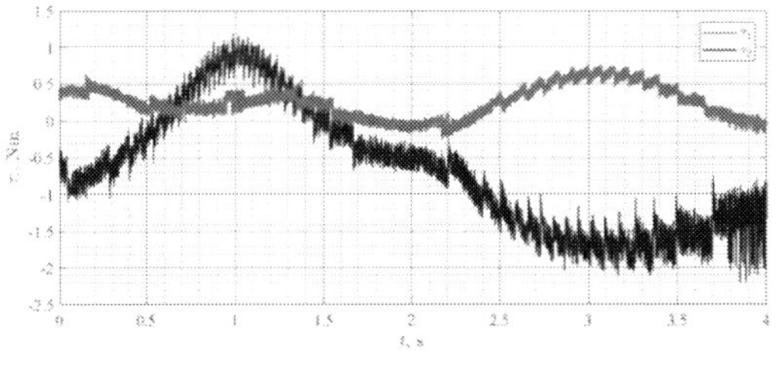

Fig. 3. The obtained values of $\tau_1(t)$ and $\tau_2(t)$ for the experiment with sensor noise and quantization.

Figure 3 shows that the torques change rapidly, possibly faster than is possible for some motor types. This illustrates another reason why state estimation is important, since it is needed to prevent the control system from requesting impossible behaviour from the robot motors.

Let us study how the behaviour of the controller changes when the state observer (5) is introduced. Figure 4 shows the time functions $\boldsymbol{\tau}(t)$ for this case.

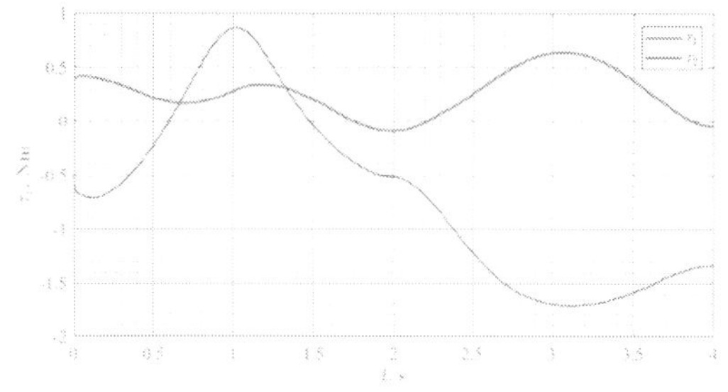

Fig. 4. The obtained values of $\tau_1(t)$ and $\tau_2(t)$ for the experiment with sensor noise and quantization, with the state observer used.

As we can see, for this experiment the time functions $\boldsymbol{\tau}(t)$ requested by the controller are smooth, which suggests that the state observer works well. To make this observation more quantitative, let us introduce a cost function J:

$$J = \int[(\mathbf{x}^* - \mathbf{x})^{\mathrm{T}}\mathbf{Q}(\mathbf{x}^* - \mathbf{x}) + (\mathbf{u}^* - \mathbf{u})^{\mathrm{T}}\mathbf{R}(\mathbf{u}^* - \mathbf{u})]dt, \quad (9)$$

where $\mathbf{x}^*$ is the desired value of $\mathbf{x}$ and $\mathbf{u}^*$ is the value of $\mathbf{u}$ found using inverse dynamics and $\mathbf{Q}$ and $\mathbf{R}$ are weight matrices of the additive quadratic cost J. In the following experiments we use the weights $\mathbf{Q} = 1$ and $\mathbf{R} = 0.1$.

Let us study the cost J as a function of the sensor resolution σ. Figure 5 shows this graph for the case when observer is not used.

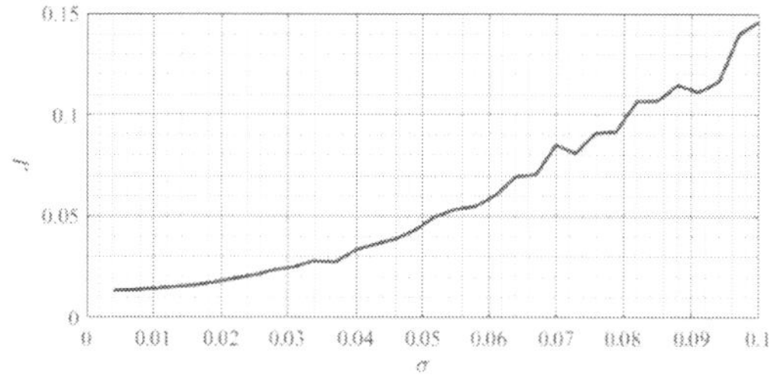

Fig. 5. Dependence of J on sensor resolution σ (without using the observer, c=0, δ=0.05)

As we can see, the cost function steadily grows as σ increases (the sensor resolution falls). This can be expected from the shape of the graphs shown in Figure 3. The same dependence for the case when the linear observer is used is shown in Figure 6.

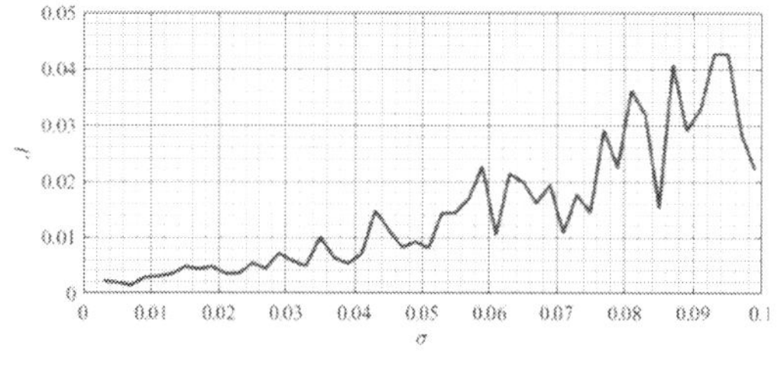

Fig. 6. Dependence of J on sensor resolution σ (with using the linear observer, c=0, δ=0.05).

Figure 6 shows that introducing the state observer improves the cost. When the state observer is used, the cost does not exceed 0.05, whereas previously the cost was reaching values of 0.15.

We should note that in all experiments shown here the parameter *c* was chosen to be equal to zero. Experiments with different values of *c*, such as 0.01, showed the same results, and thus were omitted from this paper for conciseness.

Let us study how the noise levels affect the performance of the system with and without observer. Figure 7 shows both dependencies.

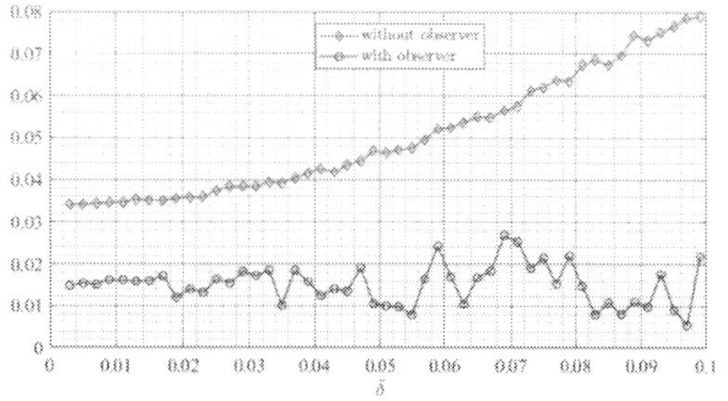

Fig. 7. Dependence of *J* on sensor noise δ (with the linear observer, *c*=0, σ =0.05).

Figure 7 demonstrates that the cost increases with the noise level when observer is not used, while introducing the observer makes the performance of the system seemingly independent from the noise.

It is well known, that high gain feedback control is more sensitive to sensor noise and measurement inaccuracies, compared with low gain feedback. In order to study this relation we fix the sensor parameters at $c=0$, $\delta=0.05$ and $\sigma=0.05$, and vary gain matrices $\mathbf{K}_p=k_p\mathbf{I}$ and $\mathbf{K}_d=k_d\mathbf{I}$, while changing scalar coefficients k_p and k_d. Figures 8 and 9 illustrate the dependence of *J* on the scalar coefficients k_p and k_d for cases with and without observer.

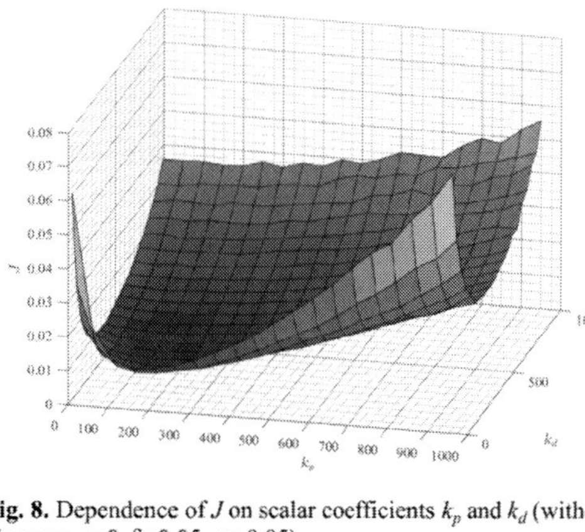

Fig. 8. Dependence of *J* on scalar coefficients k_p and k_d (without observer, *c*=0, δ=0.05, σ =0.05).

Figure 8 shows that there exists a minimum value of *J*, obtained at k_p=160 and k_d=150. The optimal value of *J* for that surface is $7.5\cdot10^{-3}$, while the mean value is 0.023. Increasing the values of the gain coefficients k_p and k_d worsens the performance of the system. This is an undesirable result, as higher controller gains allow higher motion precision.

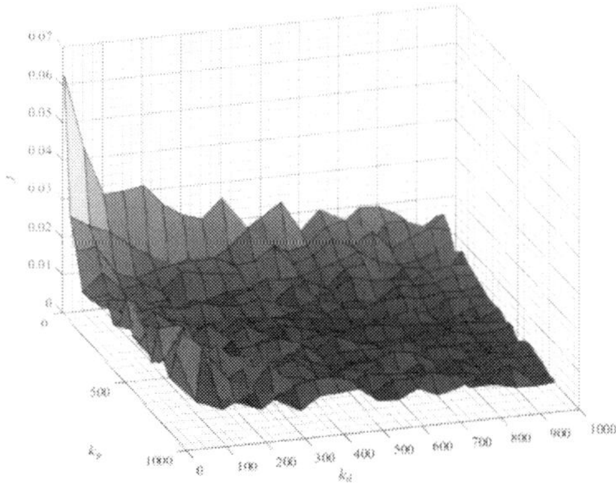

Fig. 9. Dependence of *J* on scalar coefficients k_p and k_d (with linear observer, *c*=0, δ=0.05, σ =0.05).

Figure 9 shows that introducing the linear state observer allows to negate this problem, improving the performance of the system when the controller gains are high. The mean value of *J* for that surface is 0.011.

5 Simulations on a full scale robot

In this section, we present simulation results for the full robot model. This model takes into account mechanical constraints and the reaction forces that arise from them. In these experiments LQR-based controller discussed in [26] is used. Figure 10 shows the generalised velocities for the front legs of the robot for the case when $c=0$, $\delta=0.05$ and $\sigma=0.05$.

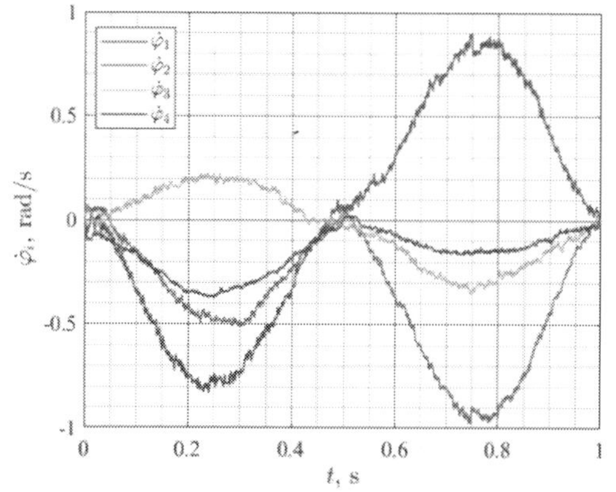

Fig. 10 Time functions for the generalized velocities for the full robot model (without observer, *c*=0, δ=0.05, σ =0.05).

As we can see, the robot exhibits high frequency oscilations. This reflects the fact that the chosen LQR-based controller was tuned to produce high feedback gains and thus the resulting control law is sensitive to the sensor noise. Figure 11 shows the same time functions obtained for the experimet with the linear state observer (7). The values of λ are quantized and sensor noise is added, same as it is done in (8). The corresponding constants c_λ, δ_λ and σ_λ have values $c_\lambda = 0$, $\delta_\lambda = 0.01$ and $\sigma_\lambda = 0.01$.

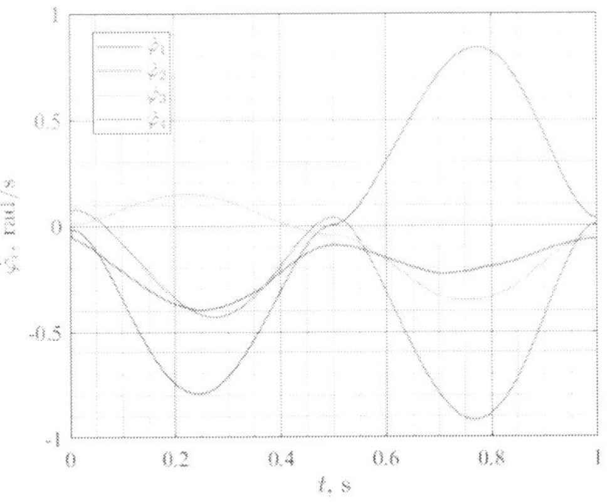

Fig. 11 Time functions for the generalized velocities for the full robot model (with the linear observer, c = 0, δ=0.05, σ =0.05).

As we can see, the observer improved the behaviour of the robot, removing the tremor. Let us study how the behaviour of the observer depends on the quality of the reaction forces measurements. Figure 12 shows the dependance of *J* on the sensor resolution σ_λ.

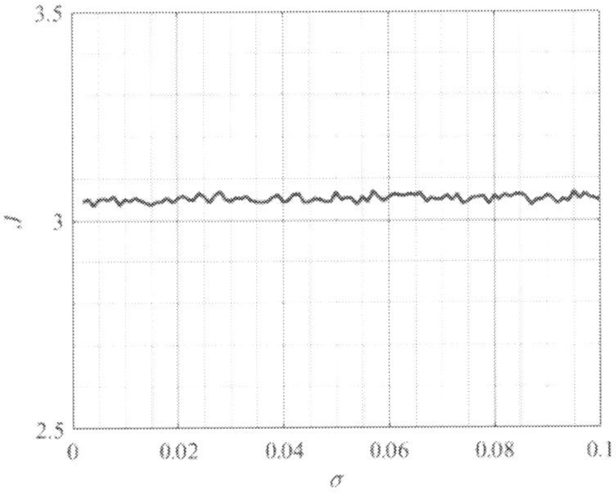

Fig. 12 Dependence of *J* on sensor resolution σ (without using the observer, $c_\lambda = 0$, δ_λ=0.01).

Figure 12 demonstrates that the performance function does now show dependence on the sensor resolution σ_λ in the chosen range. This is an important result, as it encourages the use of the state observer in systems with mechanical constraints, where the sensors measuring reaction forces have low accuracy.

We can also observe a significant increase in the values of the cost function. This is related to control errors, characteristic of the chosen controller. This is discussed in paper [14].

6 Conclusions

In this paper, the topic of state observation for in-pipe walking robots had been studied. These type of robots depend on accurate state information and accurate robot models, which should be derived from the erroneous sensor data, accessible to the robot. This signifies the question of choosing effective state observer. In this paper, an interactive linear state observer had been studied.

Simulations of motion of a single leg of the walking robot have showed that this type of observer significantly improves the behaviour of the control system in the presence of sensor noise and data quantization. It was demonstrated, that the introduction of the state observer allows the use of higher feedback controller gains for the case of a computed torque controller.

Simulations performed on the full robot model showed that the introduction of the state observer allows similar improvements. However, in order to use the state observer on the full robot mode, the information about reaction forces need to be included in the observer design. The observer demonstrated good performance even in the presence of noise and quantization of the reaction forces data.

The following research can be focused on designing a an observer for reaction forces and on using the information of contact points locations in the state observer design.

This work is supported by the Presidential grant MK-2577.2017.8.

References

1. N. Sh. Roslin et al, Procedia Engineering **41**, 1456–1462 (2012)
2. S.-g. Roh, and H. R. Choi, IEEE transactions on robotics, **21** (1), 1–17 (2005)
3. S.-g. Roh, J.-S. Lee, H. Moon, and H. R. Choi. IEEE/RSJ International Conference on Intelligent Robots and Systems (IROS 2008), 1724–1729 (2008)
4. S.M Ryew et al., Proceedings of the 2000 IEEE/RSJ International Conference on Intelligent Robots and Systems (IROS 2000), **3**, 1652–1657 (2000)
5. A.M. Bertetto, M. Ruggiu. Proceedings of the 2001 IEEE/ASME International Conference on Advanced Intelligent Mechatronics, **2**, 1226–1231 (2001)
6. C. Anthierens, A. Ciftci, M. Betemps. Proceedings of the IEEE International Symposium on Industrial Electronics (ISIE'99), **2**, 968–972 (1999)
7. S. Jatsun, A. Yatsun, S. Savin, Adaptive Mobile Robotics, 713–719 (2012)
8. F. Pfeiffer, J. Eltze, H-J. Weidemann, Intelligent Automation & Soft Computing, **1** (3), 307–323 (1995)

9. J.A. Gálvez, P.G. De Santos, F. Pfeiffer, IEEE/ASME Transactions on mechatronics, **6** (1) 26–35 (2001)
10. A. Zagler, F. Pfeiffer, Proceedings of the 2003 IEEE International Conference on Robotics and Automation (ICRA'03), **3**, 2954–2959 (2003)
11. R. Featherstone, *Rigid body dynamics algorithms* (Springer, 2014)
12. S. Savin, L. Vorochaeva, International Conference on Industrial Engineering, Applications and Manufacturing (ICIEAM), 1–6 (2017)
13. S. Savin, L. Vorochaeva, 2017 International Siberian Conference on,Control and Communications (SIBCON), 1–6 (2017)
14. S. Savin, S. Jatsun, L. Vorochaeva, Dynamics of Systems, Mechanisms and Machines (Dynamics), 1–6 (2017)
15. S. Savin, S. Jatsun, L. Vorochaeva, MATEC Web of Conferences, **113**, 02016. (2017)
16. B. Yodwong et. al., 19th International Conference on Electrical Machines and Systems (ICEMS-2016), 1–6 (2016)
17. R. Cui, L. Chen, C. Yang, M. Chen. IEEE Transactions on Industrial Electronics, **64** (8), 6785–6795 (2017)
18. H. Yang, , X. Fan, Y. Xia, C. Hua, Advanced Robotics, **30** (1), 68–78 (2016)
19. Y. Alothman, D. Gu, 2016 8th Computer Science and Electronic Engineering (CEEC), 168–173 (2016)
20. S. Jatsun, S. Savin, A. Yatsun, International Workshop on Medical and Service Robots, 305–317. (2016)
21. S. Jatsun, S. Savin, A. Yatsun, 2016 IEEE International Conference on Cyber Technology in Automation, Control, and Intelligent Systems (CYBER), 239–244 (2016)
22. R. Ortega, and M. W. Spong, Automatica, **25** (6), 877–888 (1989)
23. S. Jatsun, S. Savin, B. Lushnikov, A. Yatsun, ITM Web of Conferences, **6**, 01001. (2016)
24. S. Jatsun, S. Savin, A. Yatsun, ROMANSY 21-Robot Design, Dynamics and Control, 361–368 (2016)
25. S. Jatsun, S. Savin, A. Yatsun, 2016 20th International Conference on System Theory, Control and Computing (ICSTCC), 107–112 (2016)
26. S. Mason, N. Rotella, S. Schaal, L. Righetti. 2016 IEEE-RAS 16th International Conference on Humanoid Robots (Humanoids), 63–68 (2016)

Highly passable propulsive device for UGVs on rugged terrain

Valery Gradetsky[1], *Ivan* Ermolov[1,*], *Maxim* Knyazkov[1], *Boris* Lapin[2], *Eugeny* Semenov[1], *Sergey* Sobolnikov[2] and *Artem* Sukhanov[1]

[1]Institute for Problems in Mechanics RAS, Moscow, Russia
[2]MSTU STANKIN, Moscow, Russia

Abstract. One of the priority functional tasks of both industrial and mobile robotics is to perform operations for moving payloads in space. Typically, researchers pay attention to control the movement of the robot on different soils. It is necessary to underline the specificity of the movements of mobile robots, the main functional purpose of which is the movement of different objects. Unlike other robot applications there is the fact that transported cargo may have different mass-dimensional characteristics. The payload should be comparable to the mass of the mobile robot. This article addresses the issue of passability on rough terrain for a mobile robot performing the transport task and proposed a technical solution in the field of mechanics of propulsion to improve propelling of the traction wheel of the mobile robot with the ground.

1 The problem statement

A number of robotics research centers investigate transportation task performed by group of robots. This includes a number of scientific problems to be solved [1-3]. The interaction between the mobile robot's mover and the ground is a key factor to perform the mobile robot's passability.

In mobile robotics, the transportation operations are associated with the planning of the trajectory of movement for both the mobile robot and a payload. There are many factors that influence the formation of control, both at the stage of motion planning, and at the stage of maneuvering. Among these factors, we can point the main ones:

- the effect of distributed load on the propulsion of a mobile robot,
- special requirements for cargo transportation (orientation in space, vibration, speed and acceleration requirements),
- the influence of soil parameters on the passability of the mobile transport platform,
- force effects on the structural elements of the robot during movement, caused by external factors.

Among those for single robot-transporter the following tasks were solved: consideration of geometrical and inertia properties of the object to be transported, object's slippage effect referring to UGV, consideration of undulations of ground while keeping desired orientation of transported object, interaction of soil with wheels. These tasks were discussed throughout scientific community during last 15 years [4, 5]. However, we have not met such solutions for coordinated transportation task performed by a group of robots.

In order to effectively fulfill the transportation task mobile robot requires information about the type of soil, the irregularities of the surface, information about moving and stationary obstacles. Thus, the solution of this task should be comprehensive and take into account significant number of factors.

Coordinated transportation task implementation resembles communication coverage task by a group of robots. A potential fields approach for this task was proposed and discussed in [4].

In practice there are also a lot of cases of transporting deformable loads. This is a case of [5] which discusses transportation of deformable load by a group of wheeled UGVs equipped with manipulators. There they study centralized and decentralized control approaches. An algorithm presented there secures obstacle avoidance for both static and dynamic obstacles. A remarkable point is that during whole transportation task a desired shape of deformable load is secured.

The tire is the communication element between the robot and the ground. Through tires all the forces and torques influence on the system during acceleration and deceleration of the robot. The tire takes the action of lateral forces, keeping the robot on the operator selected trajectory. Therefore, the physical conditions of the tire friction with the road surface define the boundaries of the dynamic loads acting on the mobile robot.

The dependence of the pressure in the contact area q_t on the geometric parameters of the tire is shown in Figure 1. This dependence is given in accordance with variation of the width and height of a wheel's tire of the robot.

The pressure in the contact area q_t depends on a number of parameters. It is expressed as follows:

$$q_t=\pi h_g \frac{(p_0+p_w)}{2_b}\left(\frac{B}{H}+\frac{3H}{2B}\left(1-\frac{h}{B}\right)\right) \quad (1)$$

* Corresponding author: ermolov@ipmnet.ru

Here p_0 is the pressure caused by the stiffness of the tire carcass on the ground, 105 Pa; p_w is the internal air pressure in the tire, 105 Pa; b is width of the tire tread, m; B – width of the profile of tire, m; H is the height profile of the tire, m, h_g is the track depth, h is magnitude of wheel's bending.

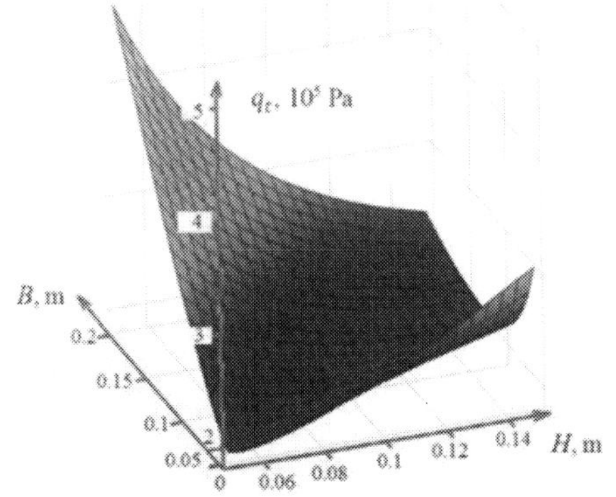

Fig. 1. The dependence of the pressure in the contact area from the geometric parameters of the tire.

Here one can see that the slight increase of the height of the tire profile with increase of the width of the tire profile leads to increase of the internal air pressure in the tire. Further increasing of the height of the tire profile with increase of the width of the tire profile decreases the internal air pressure in the tire.

Figure 2 shows the change of resistance coefficient f depending on the magnitude of wheel's bending h and on the load. The soil type on which the robot moves through influences on the resistance coefficient f.

After determining the soil type, the control system of the robot can generate the necessary control for drive system of the robot to overcome the possible obstacles, or to plan a new trajectory, avoiding the dangerous area. The resistance coefficient is determined as follows:

$$f = 3.5 p_0 y\, h^2 (B^2 + 1.5H^2) \frac{B - 0.3h}{PHB^2} 0.6qz \frac{b + b_{ki}}{p}. \quad (2)$$

Here y is the coefficient of free traction; P is payload, H; q is roughness vertical coordinate, m; z is RMS displacement value; b_{ki} is the width of the tire tread, m.

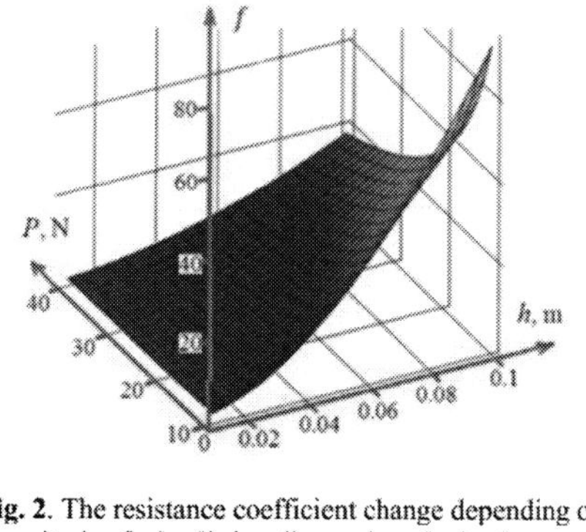

Fig. 2. The resistance coefficient change depending on the magnitude of wheel's bending and on the load.

Figure 2 presents increasing of the wheel's bending according to the quadratic law with the increasing of the resistance coefficient and payload. As it can be seen from (2), the rolling resistance coefficient depends more on the geometry of the tire in the contact zone than on the payload value.

The dependence of the load on the wheel from deformation of the tire and the soil type is shown in Figure 3. The payload also influences the parameters of the wheel and, as a consequence, its interaction with the soil.

$$p = \frac{Eh_g}{\frac{2Eh_g}{pp_{so}} arctg\left(\frac{p(H_g - h_g)}{2b_t}\right) + ab_t arctg \frac{H_g - h_g}{ab_t}}. \quad (3)$$

Here E is deformation modulus; h_g is the track depth; H_g is the depth of soft soil layer; p_{so} is inner pressure; b_t is tread width; a = $0{,}64(1 + b_t / H_g)$ is passability coefficient.

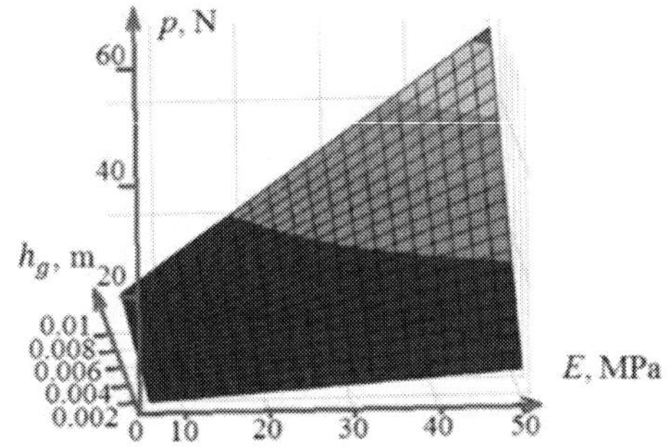

Fig. 3. The dependence of the axial load from deformation of the tire and the soil type.

When the load increases than the area of contact with the surface increases too, but in loose soil it can lead to slippage and penetration of the wheel into the soil with the formation of ruts.

The idea is to combine round wheels of mobile robot as the most effective movers on the plain areas and legged movers designed for overcoming obstacles and rough terrain.

Moving through cross-country mobile robot meets different types of the ground. That specifies the degree of the grousers' rotation. That changes different ground-wheel's interaction parameters such as width and height of a wheel's tire, pressure in the contact area, the resistance coefficient, the track depth and magnitude of wheel's bending.

Balanced action of the forces is controlled in the process of moving. Suppose that the wheels of the robot equipped with spikes. The change of direction of any object can only be achieved by the application of additional external forces. The robot is subjected on many forces when driving.

Thus tires perform important functions: every change of speed or movement direction of the robot causes the reactive forces.

In this paper we presented some simulation results that can be used in motion planning for a group of robots, considering target trajectory, type of load, load distribution among agents, type of terrain and specifics of soil (Fig. 1-3). During transportation different robots involved in the coordinated movement of a particular cargo may be located on different type of soil. Thus, in order to ensure the coordinated movement of cargo, it is necessary to take into account the interaction between each mover of each robot of the group with the ground. This is discussed in more detail in [4] and [5].

Research and development of wheel devices for various mobile robotic platforms are carried out by many research groups in different countries. Analysis we have done has shown that for motion planning and control for group-based transportation usually assumptions are done which are far from reality in cross-country conditions. E.g. wheel-soil interaction is not considered or soil considered being non-deformable, only slippage factors are considered, some works include only 2 robots into transportation. Some studies do not include practical experimentation.

Recent studies [6] for improving traction-coupling properties of a wheel, have led to the development of the anti-skid device that provides the mover to move on soils with low passability. Depending on the slippage factor, rotation of the segmented disc fixes its parts in a certain position relatively to the wheel axis; the device pushes the anti-skid elements out, which increases the adhesion to the ground. The disadvantage of this design is the lack of automation and adaptation to the actual soil parameters.

Another solution [7] is the construction of the wheel containing the soil hooks (grousers), as well as the mechanism of extension of the soil hooks with limiters (Fig. 4).

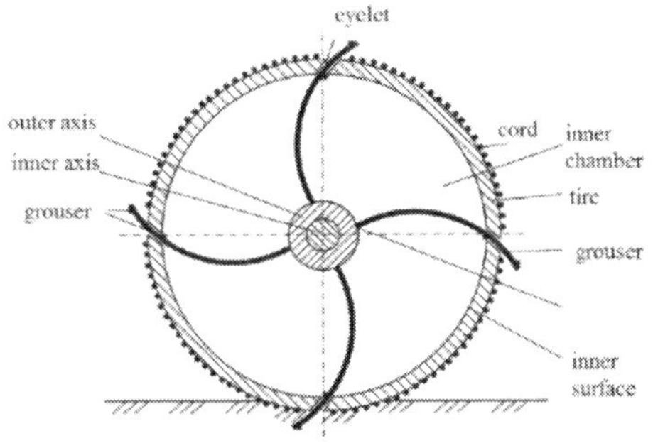

Fig. 4. Wheel with grousers.

However, the design of this device does not allow improving the adhesion to the ground during the reverse movement of the wheel, which is essential for using in mobile robotics.

Researchers are also interested in the development of wheel rotation algorithms in the slippage and towing. Work [9] proposes the method of controlling the thrust of a wheeled robot with a console suspension when moving on rough soils. The purpose of the proposed method is to reduce the slippage of wheels. The method is based on measuring the contact angle between the wheel and the ground using wheels of special design (tactile wheel), equipped with tactile sensors. The sensors are located on the wheel's rim, which allows fixing the point of contact with the ground wheels (Fig. 5).

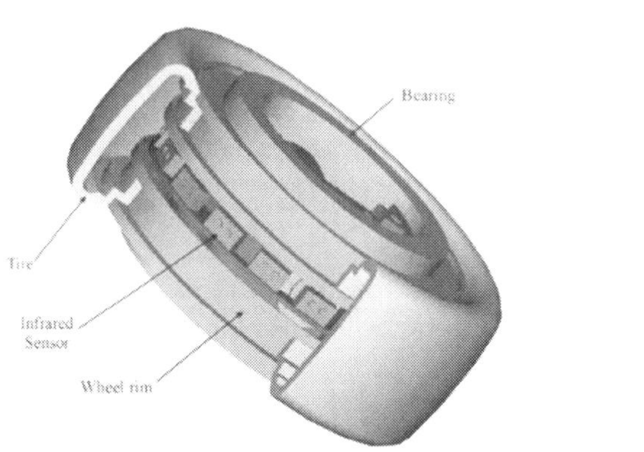

Fig. 5. Wheel equipped with tactile sensors.

The presented technique makes it possible to exclude complex models of interaction between the wheel and the surface, parameters of which are often unknown. The authors note the advantage of the presented method of steering the wheels torque in comparison with the traditional speed control. However, the work did not consider the interaction with deformable soils or grass.

Another way of solving the problem of increasing the robot's passability is the creation of a wheel-walking mover. The work [10] describes a transformable wheel shown in Figure 6. These wheels allow the robot to move steadily on a flat surface, using a conventional round wheel, and overcome obstacles as wheel-stepping machines.

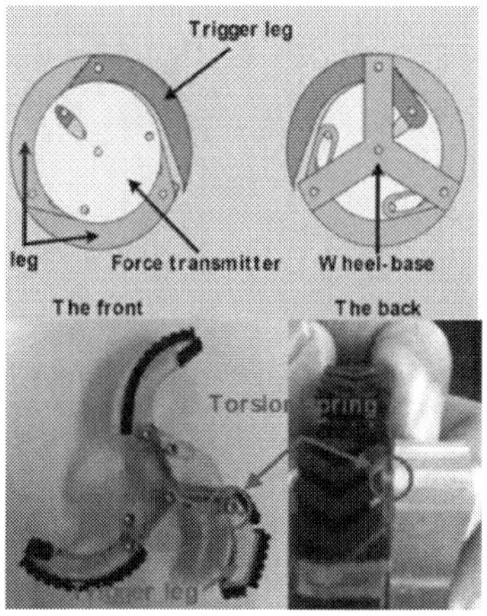

Fig. 6. Transformable wheel.

The advantage of this mover is the ability to overcome obstacles and to move in a loose environment (grass, swamp), but the disadvantages of the design can be attributed to the inability to set the amount of disclosure blades. As a result, this solution does not provide the ability to adapt the movement to different types of deformable soils.

In addition to devices that improve adhesion to the round, different methods andalgorithmsfor determining the parameters of the soil on which the robot moves are being developed. One of such methods is the Generalized Newton Raphson (GNR) method, which allows determining all the necessary parameters of the soil for the formation of the traction force. This method was considered in [8]. This method requires calibration and preliminary adjustment of the control system for a specific type of surface.

2 The proposed design of the mover for the mobile robot

In terms of this work a new propulsion system was developed. While moving on various soils this propulsion system may function in 2 modes: rotating and stepping. This allows combining high robot velocity with improved cross-country ability.

The proposed solution is to equip the wheels of the robot with additional blades that improve the coupling properties of the wheel with the ground. At the same time, unlike other similar solutions, the proposed scheme implies a mechanism for adapting the degree of extension of the blades depending on the nature of the actual interaction of the wheel with the ground. The general scheme of a wheel is shown in Figure 7. The mover consists of a disk 1 and connected to it retractable soil hooks (blades) 2. Each hook is provided with two rods 3 and 4 fixed on it at a distance from each other. One end of the first rod 3 is connected to disk 1, and the other end of the other rod 4 is connected to the plane mounted on the disk of the rotating element 5. All rotating elements 5 are connected by kinematic coupling 6, and one of them is equipped with a drive.

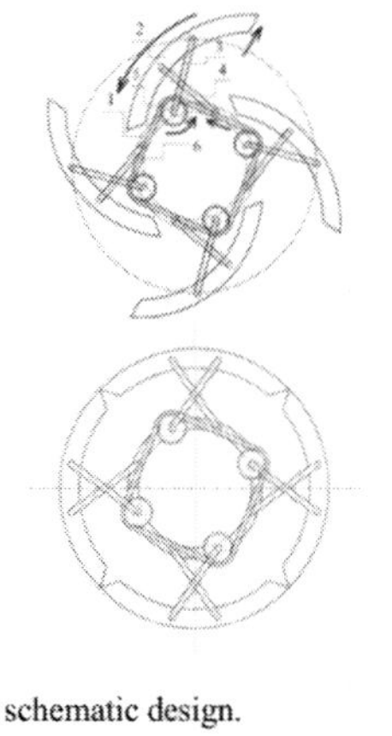

Fig. 7. Proposed schematic design.

A feature of the proposed wheel design is receiving and processing the information from the built-in bend-sensors (Fig. 8) located on the inner surface of the flexible blades. They measure the level of vertical deflection of the tire relative to the unloaded condition of the blade. The vertical deflection of the blade depends on the level of rigidity of the surface (soil), on which the robot moves and the rigidity of the blade itself.

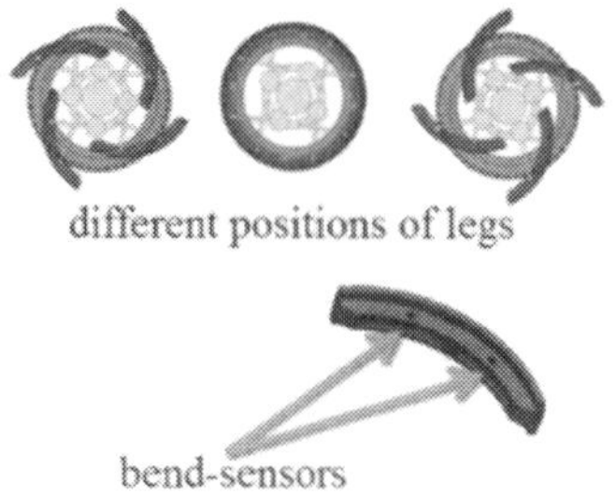

Fig. 8. Proposed design of the wheel.

When the robot moves over rough terrain, the control system is checking the soil type in database and, based on the vertical tire deflection value obtained from the bend-sensors, determines the most appropriate parameters. Depending on the type of soil, the speed of rotation and the applied torque of each wheel are adjusted, which reduces the probability of slipping. It is possible to make a map of the terrain for one mobile robot, which will be useful for other robots in the group, and contributes to the efficiency of movement and overcoming possible obstacles for all robots in the group (Fig. 9).

Fig. 9. Overview of the device.

Thus, the use of bend-sensors in the proposed configuration allows to calculate the reaction forces in the contact area and to adapt the mover to the type of soil with the extraction of the flexible blades outside from the wheel to the desired value.

Robot's movers are equipped with bend-sensors located on the inner surface of the flexible blade. They measure the change in the vertical deflection of the wheel giving the command to change the configuration of the wheel [11]. The smaller the value of the vertical deflection, the less the force of adhesion of the wheels with the surface movement. The second condition for changing the configuration of the wheel is to reduce the torque on the shaft of the developed mover, which may indicate its slippage or blocking due to external factors [12].

It was experimentally shown [11] that the coefficient $m(\lambda)$ of the longitudinal friction force is subject to Gauss law, i.e.

$$m(\lambda) = m_c e^{-(l^2)} / b = m_c \exp(-(\lambda^2)/b), \quad (4)$$

for b=0,26, if m_c J 0,9.

This coefficient is determined by the following ratio:

$$\lambda = \frac{j\ P_c - v_p}{\max(v_p, f\ R)}.$$

Here j is the angle of rotation of the wheel, v_p is the linear speed of the robot's wheel, R is the radius of the wheel, m_c is the static friction coefficient. Slippage is absent, if $\dot{v}_p = \dot{v}_c = \ddot{\varphi}R$. In the case of the movement with slippage $v_c = \dot{\varphi}R > v_p$. Linear velocity v_p of the robot equals the linear velocity v_c of the wheel under absence of slipping. The driving moment causing slippage can be found as follows:

$$\mu_K = \mu_c\, exp(-(\lambda^2)/b)\, GR \qquad (5)$$

For the wheel rolling without slipping requires that the adhesive force satisfies the condition $F_s \geq \left(R_x{}^2 + R_y{}^2\right)^{1/2}$, here R_x – the tangential reaction force from the ground, R_y – the transverse reaction force from the ground. Considering that $F_s = R_z{}^2 \varphi_x{}^2$, we may obtain $R_y \leq \left(R_z{}^2 \varphi_x{}^2 - R_x{}^2\right)^{1/2}$, that is, the transverse force applied to the wheel and not sliding, the greater the force of adhesion of the wheel to the ground than the less the tangential reaction of the road, φ_x – coefficient of longitudinal adhesion of the wheel to the surface.

The evaluation of robot passability in the group is necessary to identify impassable areas and further form the optimal route or restrictions for other mobile robots in the group. This parameter for the all-wheel-drive mobile robot can be obtained as follows:

$$\Pi = \frac{G_i V_1 q_1}{G V_2 q_2} \qquad (6)$$

here G_i – payload on the wheel, obtained by bend-sensors, G – mass of the $\frac{1}{4}$ robot, V_1 – the desired speed of the wheel, V_2 – the real speed of the wheel in the coordinate system associated with the object, and calculated through the instantaneous center speeds, q_1 - power consumption of the electric motor in ideal conditions, q_2 – power consumption of the electric motor in real conditions.

3 Conclusion

The proposed technical solution in the field of propulsion mechanics improves the adhesion of mobile robot wheel to the ground. It can significantly improve the technical characteristics of the mobile robot on the parameter of passability. A distinctive feature of the proposed solution is the ability to adapt the degree of extension of the soil hooks (blades) depending on the type of the soil.

The proposed design of the mover should keep the average speed for certain mobile robot in group under rough conditions of the cross-country territory. This leads to increasing of passability for the whole group and increasing efficiency for joint transport task solution.

This work is supported by RFBR grant №16-29-04199.

References

1. A.V. Vasev, V.P. Noskov, I.V Rubtsov., S.G. Zaichenko, Izvestiya YuFU. Technical science, **2**(175), 127–139 (2016)
2. N.C. Volskaya, *Evaluation of the permeability of wheeled vehicles when driving on uneven ground surface* (Moscow: MGIU, 2007)
3. T. Muro, J. O'brien. *Terramechanics Land Locomotion Mechanics* (A. A. Balkema Publishers, Tokyo, 2004).
4. V.G. Gradetsky, L.I.,Ermolov, M.M. Knyazkov, E.A. Semenov, A.N. Sukhanov, Mechatronics, automation, control, **12**, 819–824 (2017)
5. V.G. Gradetsky et al., Robotics and technical Cybernetics, **3**, 57–62 (2016)
6. M.M. Makhmutov, *Anti-Skid Device For Wheeled Vehicle* (Patent RU, B60B 15/18, 1996)
7. Yu.I. Jakimov, *Vehicle Drive Wheel* (Patent RU, B60B 15/26, 2014)
8. S. Hutangkabodee, Y. Zweiri, L. Seneviratne, K. Althoefer, International Journal of Advanced Robotic Systems, **5**(4) (2008).
9. Pierre Lamon and Roland Siegwart, Wheel Torque Control in Rough Terrain - Modeling and Simulation, Available at: https://infoscience.epfl.ch/record/97563/files/P782_final.pdf (11.01.2018)
10. Y-S. Kim et al., Proc. of ICRA-2013, 5605–5610 (2013)
11. T. Zielinska, A. Chmielniak, Proceed. of Twelfth Int. Conf. of Climbing and Working Robots (CLAWAR-2012), 1039–1046 (2012)
12. T. Okada, S. Tezuka, H. Wada, Procced. of CLAWAR-2015 Conf., Adaptive Mobile Robotics, 585–594 (2015)

Approaches to the robotization of agricultural mobile machines

Vladimir Serebrenny, *Madin* Shereuzhev and *Ivan* Metasov

Bauman Moscow State Technical University, Mechanical Engineering Department, Moscow, Russia

Abstract. Agriculture is the extremely important and developing economic movement in all times. Automation of agricultural machines occurs by different ways. One way is through the creation of specialized technical solutions for the required technological processes, another way is the construction of automatic agricultural machines, including mobile ones. The state of modern technology allows to create autonomous machines. The agriculture robotization trends are the high precision and unmanned farming. The article considers the issues of robotization of agricultural machinery. Stages of robotization of agricultural mobile machines were analyzed. The factors affecting the autonomous movement of mobile agrorobots were shown.

1 Introduction

Currently, the production efficiency increase and the replacement of people in the hard and dangerous works is gained through using automation and robotic systems [1].

Trends in agricultural production are attempts of introduction of precision agriculture technologies [2, 3] and unmanned production cycles of agricultural products.

In Russia there are researches of problems associated with the development of an agricultural robot for collect vegetable crops growing above the ground, such crops are cabbage, melons (watermelons), cucumbers, tomatoes, eggplants, peppers (e.g. [4]). There are attempts for creating robot for field state monitoring [5].

High productivity of agriculture machines in tasks (soil cultivation and sowing; fertilization; protection of plants from pests, diseases and weeds; harvesting of grain, legumes, oilseeds and groats; post-harvest processing of cereals, legumes, oilseeds and cereals, etc.) today is supported by:

- Increasing engine power,
- Increasing working width,
- Increased operating speed,
- Increased capacity,
- Automation of the process.

The only way to increase the efficiency of agricultural machinery without negatively affecting the soil is automation [6, 7].

Automation includes creation of control system for follow processes:

- Internal systems and processes control,
- Fleet control,
- Process chain control,
- Autonomous machines control.

Automation of agricultural machines occurs on the one hand through the creation of specialized technical solutions for the required technological processes, on the other hand, by construction of automatic agricultural machines [8,9], including mobile ones. Today's state of technics and technology allows to create not only automatic, but also autonomous machines. The trend is shifting from automation to the robotization of industries, which, within the agriculture, occurs by using high-precision farming [10, 11] and unmanned cycles of growing crops [12].

The efficiency of mobile agricultural machines is ensured by measuring, recording and controlling their parameters, protecting the main mechanisms, regulating and controlling the machine as a whole.

The automatic system of a mobile agricultural machine, as a rule, consists of the following main parts: an object (measurement, accounting, control, protection, regulation), a sensitive device, an amplifying converter, an actuator, a power source, an operator.

In the autonomous system, there is no human operator. The creation of autonomous mobile agricultural machines requires the addition of new elements in the design to carry the control signals from the control system to the actuators. In the robotic system, adaptability of the control system is assumed, which also causes the inclusion in the control system of the vision system, navigation system, etc.

2 Robotization of traditional mobile agricultural machinery

To analyze the stages of robotization of traditional agricultural equipment, we consider the basic stages of robotization of a tractor with an internal combustion engine (ICE).

2.1 Control levers

Based on the previous experience [13, 14], for the robotization of the tractor with ICE, tractor control devices should primarily be provided with electronic control interfaces. This problem should be solved by using industrial servo drives. For the robotization of mobile agricultural machines, servos must be installed on the following controls:
• Throttle control lever,
• Steering control lever,
• Brake control levers,
• Transmission control levers,
• Clutch control lever,
• Control levers for hydraulic drives of attachments,
• Control levers for auxiliary systems (lighting, heating, etc.).

2.2 Executive control system

This system is necessary for interfacing high-level commands of the top-level control system, with commands applied to the servo drives of the controls levers. In addition, this system should provide monitoring of current parameters from on-board sensors, such as temperature, pressure, voltage, speed sensors. The main stages of the development of this system consist of:
• Selecting a microcontroller model,
• Development of a system for matching signals from airborne sensors, with a microcontroller,
• PCB development,
• Development of the management program.

2.3 Upper level control system development

It is advisable to build such a system as computer based because of the large computational load [15]. The main goals of this system are:
- Receiving and processing of information from sensors of the system of technical vision,
- Receiving and processing of information from sensors of the global positioning system,
- Receiving and processing information from other sensors, such as accelerometers and gyroscopes,
- Receiving and execution of commands from the operator's panel,
- Construction of the external environment model,
- Determination of the patency of the external environment,
- Solving the problems of mapping and localization,
- Trajectory planning,
- Submission of commands to the executive management system for trajectory development,
- Transfer of control commands to attachments.

2.3.1 Selection of the software platform

The architecture of the hardware platform, the number of required libraries for a specific operating system and the cost of work are the parameters that determine the choice of the operating system.

2.3.2 Machine vision system sensors selection

To obtain the necessary amount of data about environment, it is advisable to use a combination of a LIDAR and cameras as a vision system [16]. To obtain a wider range of information about the environment, ultrasound and ultraviolet sensors can be added to the system.

2.4 Power system development

When developing a mobile robot power system, the following points should be considered:
• Voltage range of the on-board network,
• The range of permissible supply voltages for the control systems of the upper and lower levels,
• Power consumed by the upper and lower control systems,
• Excess of power in the on-board power supply network.

2.5 Development of communication interfaces

These interfaces are required for sending to the on-board upper level control system the tasks that the operator determines. In addition, these interfaces can be used for direct control.

2.6 Development of parts for mechanical connection of chassis with mounted systems

The main problem of this stage is the mechanical connection of the servo drives to the control levers.

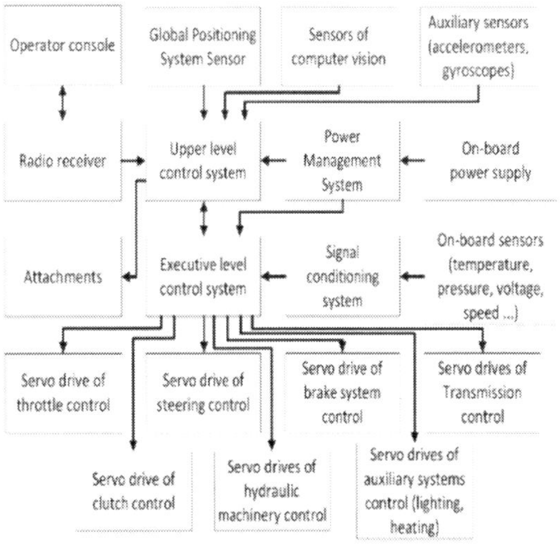

Fig. 1. Scheme of the robotized tractor with internal combustion engines.

The resulting system has a large number of elements, which make it difficult to build such a system with high reliability. In addition, due to the installation of servos on the controls, it becomes difficult to control the tractor from the driver's cab. In Figure 1 the developed structural scheme of a robotic tractor with ICE is shown.

3 Robotization of electric mobile agricultural machinery

Thanks to the latest world trends in the electrification of vehicles, the first electric tractors have begun to appear recently [17] (Fig.).

Fig. 2. Electric tractor Fendt e100 Vario.

For robotization of electric tractor control levers installation and executive control system for control levers are not required for electric tractor. On the electric tractor, a built-in control system with an electronic interface already exists. The main difference of this system for an electric tractor, from a system for a tractor with ICE, is a link with an existing on-board control system.

Stages of selection software platform, machine vision system sensors selection, power system and communication interfaces development are similar to those for the tractor with ICE.

Stage of development of parts for mechanical connection of chassis with mounted systems for electric tractor is much easier compared to a similar stage for a tractor with ICE, because there is no need for a mechanical connection of servos with control levers.

The amount of elements for the robotized electric tractor control system is much less, than the one of tractor with ICE. This simplifies autonomous control system creation for electric tractor, and provides higher reliability of this machinery.

In Figure 3 the developed structural scheme of a robotic tractor with an electric motor is shown.

For mobile agriculture robot with ICE it is necessary to solve the problem of automated refueling, this process can be fire-dangerous.

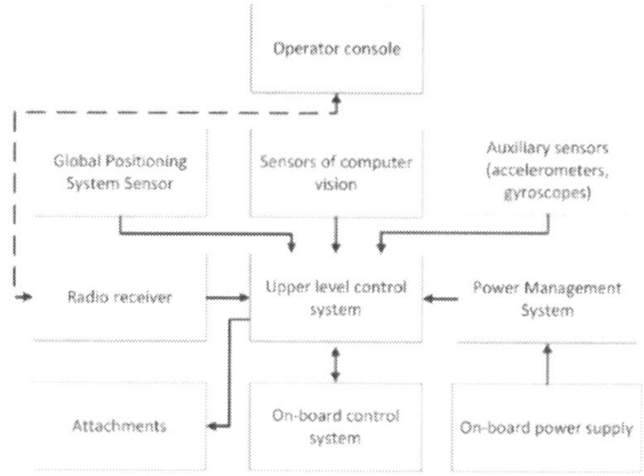

Fig. 3. Scheme of the robotized electric tractor.

In Figure 4 factors, affecting the straightness movements of the tractor is shown.

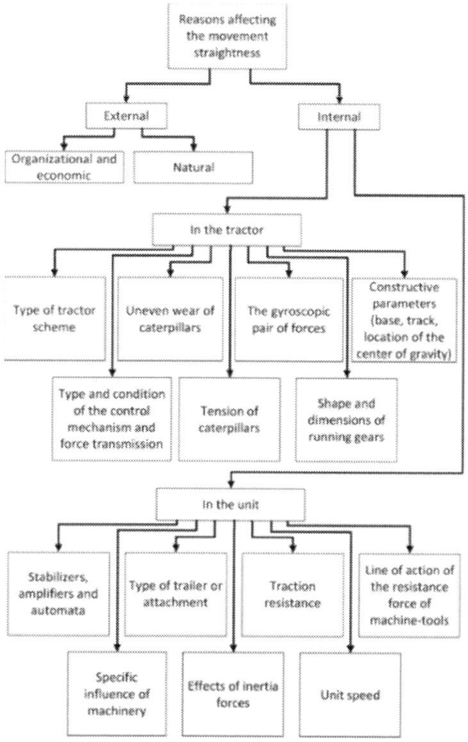

Fig. 4. Factors affecting the straightness movements of the tractor.

All reasons are divided into external and internal. Internal causes, such as "Type and condition of the control mechanism and force transmission", in the case of a tractor with ICE will have a stronger influence on the straightness of the movement. This is due to the complexities that will create control of the clutch and transmission, as well as the nonlinearity of the ICE.

4 Conclusion

In the work the analysis of the basic stages of robotization of mobile agricultural machinery is carried out. The result of the analysis is a functional diagram of the mobile robot control system for agricultural purposes. Based on experience in the field of robotization of military equipment, recommendations on the structural composition of the management system were formulated.

The comparative analysis of the complexity of the robotization of mobile agricultural equipment with ICE and electric drive was conducted, which showed a higher reliability of the robotic system with factory electronic controls and a lower cost of its robotization, and also showed the need to promote the electrification of agricultural machines to accelerate the robotization process.

References

1. T. Bock Construction Automation and Robotics, (InTech, 2008)
2. S. Blackmore From precision farming to phytotechnology. In I: Mechatronics & Robotics '04, Aachen, Germany, (2004)
3. Drenjanac, D., & Tomic, S. D. K. Middleware challenges in robotic fleets for precision agriculture.
Journal of Mechanics Engineering and Automation, 3(11), (2013).
4. Robotic system for harvesting vegetable crops growing above the ground, Internet Journal "Naukovedenie".3(40) vol. 9, (2017)
5. Development work on the creation of robotics farmland, Scientific Journal of KubSAU, 119(05), (2016)
6. P. Gonzalez-de-Santos, A. Ribeiro, C. Fernandez-Quintanilla, F. Lopez-Granados, M. Brandstoetter, S. Tomic, S. Pedrazzi, A. Peruzzi, G. Pajares, G. Kaplanis, M. Perez-Ruiz, C. Valero, J. del Cerro, M. Vieri, G. Rabatel, B. Debilde Fleets of robots for environmentally-safe pest control in agriculture, (Precision Agric, 2017)
7. J. F. Reid, Q. Zhang, N. Noguchi, and M. Dickson, "Agricultural automatic guidance research in North America," Computers and Electronics in Agriculture, vol. 25, (2000)
8. M. Perez-Ruız, D. C. Slaughter, F. A. Fathallah, C. J. Gliever, B. J. Miller Co-robotic intra-row weed control system, (Elsevier, 2014)
9. S. J.Moorehead,C.K.Wellington, B. J. Gilmore, and C.Vallespi, Automating orchards: a system of autonomous tractors for orchard maintenance, in Proceedings of the IEEE International Conference of Intelligent Robots and Systems, Workshop on Agricultural Robotics, (2012)
10. A. Peruzzi, C. Frasconi, L. Martelloni, M. Fontanelli, and M.Raffaelli, Application of precision flaming to maize and garlicin the RHEA project, in Proceedings of the 1st International Conference on Robotics and associated High-Technologies and Equipment for Agriculture (RHEA '12), Italy, (2012)
11. R. Eaton, J. Katupitiya, K. W. Siew, and K. S. Dang, Precision guidance of agricultural tractors for autonomous farming, in Proceedings of the 2nd Annual IEEE Systems Conference Proceedings (SysCon '08), pp. 314–321, April (2008)
12. A. T. Nieuwenhuizen, J. W. Hofstee, and E. J. van Henten, "Performance evaluation of an automated detection and control system for volunteer potatoes in sugar beet fields," Biosystems Engineering, vol. 107, no. 1, pp. 46–53, (2010)
13. A.N.Kutuzov, V.S.Lapshov V.P. Noskov, I.V. Rubcov Experience in developing and creating an autonomous intelligent robotic complex based on the T-72 serial tank, (Defense technology: scientific-technical, 2000)
14. A. V. Kalinin, V. P. Noskov, I. V. Rubcov The system for automative driving the tank T-72B, (Bulletin of BMSTU: Mechanical engineering, 2012)
15. V.V. Serebrenny, I.E. Metasov, M.A. Shereuzhev Development of the agricultural machinery robotization methods (News of KBNC RAN, 2017)
16. L. Zalud, P. Kocmanova, Fusion of Thermal Imaging and CCD Camera-based Data for Stereovision Visual Telepresence, Safety, Security, and Rescue Robotics (SSRR), (2013)
17. Electronic resource https://www.fendt.com//int/fendt-e100-vario.html

Hierarchical classification of robotic grippers applied for agricultural object manipulations

Quyen Vu[1*], *Mikhail* Kuzov[2] and *Andrey* Ronzhin[1,2]

[1]SUAI, Department of Electromechanics and Robotics, 190000, Saint-Petersburg, Bolshaya Morskaya Str., 67, Russia
[2]SPIIRAS, Laboratory of Autonomous Robotic Systems, 199178, Saint-Petersburg, 14-th line of V.I., 39, Russia

Abstract: An overview of agricultural grippers used to control weeds and harvesting is presented. The classification of the grippers, which are installed on robotic agricultural tools for manipulation of fruits, weeds and other objects, is compiled. There are 22 types of grippers depending on 6 selected criteria: drive type, the presence of the drive in the grip, the number of fingers, the type of gripper movement, the type of mechanism, the type of sensors. In this classification, we mainly consider the characteristics of the gripper, which is installed at the end of the manipulator and is responsible for physical contact with the object. Therefore, the main attention is paid to problems requiring direct capture of objects by the agrobots. The issue of joint interaction of a group of heterogeneous terrestrial and airborne robots in the performance of the target agrarian task in an autonomous mission will also be investigated.

1 Introduction

Precision technologies have been developed mostly for traditional crops since the advent of precision agriculture in the early 1990s. Taking into account these technologies, the productivity of experienced agricultural plots increased significantly. The concept of precision agriculture involves the assessment of in-field spatial variability of different factors such as fertility, soil type and characteristics, and water content in a field and the subsequent management of each crop production input in a more precise and site-specific manner according to the variability.

With the development of science and technology, various sensors and sensor systems are increasingly being used in precision farming to solve a number of important tasks, such as monitoring of land, detection of fruits, weeds, diseases, etc. For example, over the past decade, much effort has been made to develop sensing technologies for site-specific crop management in Florida, USA. Citrus production in USA has been seriously affected by the emergence of exotic diseases such as citrus greening or Huanglongbing (HLB), citrus canker, and citrus black spot. It has been estimated that citrus greening alone has cost about $3.63 billion in lost revenues. [1]. In managing plant disease, detection is one of the most important steps. Detecting disease at an early stage, especially at the asymptomatic stage, could be the most cost-effective method of disease control. Currently, human scouting is most commonly used technique for disease detection in the majority of crops. However, human scouting is costly, time-consuming, and limited to human senses for the detection of disease. In recent years, progress in the area of low-cost, low-altitude satellites and unmanned aerial systems have provided an opportunity for continuous monitoring of plants. Among the sensing systems developed, the following have great potential to be adopted by the growers: immature green citrus detection, debris detection and cleaning system (currently in the process of being patented), blueberry yield mapping system, the HLB detection systems developed using optical properties and starch accumulation, silage yield mapping, grain insect detection using NIR, soil P sensing systems, citrus debris detection system from mechanical harvesting, and detection of dropped citrus fruit on the ground.

The main phases of the agricultural cycle are soil preparation, planting, production, and harvesting. Soil preparation refers to the mechanical manipulation of soil that alters its structure and strength, in order to provide and maintain optimum soil conditions for germination, growth, and development of plants, thus showing their productive capacity. In outdoor conditions, the planting process consists of dropping the seeds or placing them into the soil. In the case of intensive crops, this process of planting has automated machines that perform this task. In outdoor conditions, the planting process consists of dropping the seeds or placing them into the soil. In the case of intensive crops, this process of planting has automated machines that perform this task. Production includes the period between the transplant or the appearance of the first true leaves, and the last harvesting in the case of annual crops. It is composed of tasks such as pesticide spraying, pruning, precision fertirrigation, plague recognition, weed removal, harvesting, and crop removal. Harvesting is the task that requires most inputs and resources [2].

The advent of agricultural robots has the potential to raise the quality of fresh produce, lower production costs, reduce the drudgery of manual labour, and, in some parts

* Corresponding author: vuquyennntk@gmail.com

of the world, to compensate for the lack of workers in some agricultural sectors. Agricultural robots are generally designed to execute a 'main task', which is usually a specific agricultural task such as planting, weeding, pruning, picking, harvesting, packing, handling, etc. [3]. The nature of these analyzed tasks requires the use of mobile robots combined with manipulator robots with different end-effectors (spraying nozzles, irrigation droppers, vacuums, harvesting cutters, suction pads, etc.). End-effectors may consist of a gripper or a tool. Grippers are subsystems of handling mechanisms which provide temporary contact with the object to be grasped [4].

In section 2, we will analyze the existing robotic solutions in the most complex and time-consuming stages of the agricultural cycle: weed control, disease monitoring and harvesting. In Section 3 we give a classification of the grippers used for manipulating agrarian objects.

2 Analysis of existing robotics in the field of agriculture

2.1 Weed control and disease monitoring

Different abiotic and biotic stresses affect the potential yield of crops. 40% of world food production is lost through diseases, insects and weeds. To achieve high yields in agricultural crop systems, the control of biotic stress is highly relevant. Monitoring diseases and pests during the growing and harvesting stages is essential to fulfilling the plant's production potential, detecting and preventing disease spread, and avoiding significant yield losses. Pest, disease and weed control is a frequently performed and often time-consuming task, which sometimes exposes the HO to the danger of contamination with hazardous chemicals. Development of systems for weed control, including weed detection and removal, has been one of the major fields of research in agricultural robotics in the last few decades. To date, some complete weed-control ARS have been tested under field conditions. The main stages of weed control include: guidance, weed detection and identification, precision in-row weed removal, and mapping. Four types of weed-removal mechanisms are suitable for selective in-row weed control by ARS: mechanical, thermal, chemical, and electrical means [5].

Knowledge of the precise position of crop plants is a prerequisite for effective mechanical weed control in robotic weeding application such as in crops like sugar beets which are sensitive to mechanical stress. In work [6] the potential of using knowledge about the crop seed pattern is investigated based on simulated output from a perception system. Cereals like barley and wheat are placed in rows with no clear structure, whilst maize, sugar beets and other high value crops are placed in rows with a clear defined intra-row spacing between crop plants. In robotic weeding applications plant recognition is often based on machine vision either using spectral properties or plant morphology/shape information. Plant classification based on spectral properties and plant morphology are vulnerable to variations in plant appearance. The results of previous studies have shown that a priori information about the location of plants can be used to recognize seed crops. An experimental model for evaluating context-based crop recognition based on simulating plant positions of both crop and weed plants, in an artificial field. A list of all the plant positions was then used as input to a context-based plant recognizer, which localized crop plants based on the known sowing geometry. The results of the tests showed that the recognition reliability can be described with the positive predictive value (PPV), which is limited by the seeding pattern uncertainty and the weed density according to the inequality: $PPV \leq \left(1 + 2\pi\rho\sigma_x\sigma_y\right)^{-1}$, where ρ - weed density, σ_x, σ_y - crop plant pattern position.

In organic crop management the use of conventional pesticides is prohibited, placing a major challenge and priority on most organic farms for mechanical weed control. Agricultural workers who perform manual weeding are exposed to several musculoskeletal disorder (MSD) risk factors, particularly prolonged trunk flexion angles. Hand weeding (and thinning for lettuce) operations in organic production of these crops represents ~95% of their total weed control costs. Currently, some commercial machines for intra-row weeding are available to farmers, such as: the finger weeder, the torsion weeder, the weed blower, flame weeding, current state-of-the-art intelligent systems. In work [7] is presented the design and assess the performance of an automatic intra-row mechanical weeding corobot with automatic hoe positioning based only on a low cost (below US$100) odometry sensing technique. In place of machine vision or RTK-GPS sensors, a human partner will provide visual crop detection capability. To reduce human exposure to musculoskeletal disorder risk factors, the human will ride in a seated position on the hoeing platform equipped with a pair of small (~7 cm wide) hoes (shown as red triangles) was used to control weeds in the intra-row zone and a standard cultivator was used to control inter-row zone weeds. The following results of tests of the collaborative robot on the tomato field are obtained: the average speed of movement is 1.2 km / h, the time for processing weeds in the ripening zone decreased by 57.5% compared to manual processing. Based on the cost of production of broccoli in 2012, an average farm labour cost of US$12.33 ha^{-1} for hand hoeing can be estimated. The estimated co-robot's labour savings of 13.8 h ha^{-1} suggests US$170.15 ha^{-1} in labour savings for weed control.

In recent decades, research into weed detection technologies, data integration from heterogeneous sensors and selective crop management through herbicides or other treatments has improved significantly. Of particular note is the improved navigation capabilities of unmanned vehicles and agricultural implements by combining information from the sensors of the global navigation satellite system and from local measurement devices. Nevertheless, many of these studies were focused exclusively on specialized tasks in which only individual components of robotic systems in a narrow context were studied, tested and evaluated. The idea of using groups of heterogeneous robotic systems for agricultural applications was discussed several years ago, but the first real attempts to use them were carried out only recently [8].

In tests for field resistance of potato to late blight, crop scientists rate the disease severity exclusively using visual examinations of infections on the leaves. However, this visual assessment is generally time-consuming and quite subjective. In work [9] is proposed to use a new a new estimation technique for disease severity in a field using RGB imagery from an unmanned aerial vehicle (UAV). Potato late blight is one of the most serious diseases affecting potato production in Japan. The disease rapidly destroys leaves, and it consequently leads to yield losses and tuber quality deterioration. The results of studies based on the analysis of images from UAVs confirmed the possibility of a high-throughput phenotyping system to assess disease resistance. When monitoring the field, UAVs are widely used for aerial photography due to low cost, high resolution images compared to satellite images due to low flight altitude, the ability to move in any direction, hover, maintain a stable position in flight, etc. In work [9] images of the potato field were taken by a UAV (HiSystems GmbH Mikrokopter, Germany) with four counter rotating propeller pairs and eight brushless motors. The test field consisted of 36 rows laid out along the long side of the field in an area that was 53.8 m x 27.0 m, and the space between rows was 75 cm. The area of 42.2 m x 22.5 m, located in the centre of the field, was partitioned into 360 experimental plots. Phased image processing included geometric image correction, parametric processing, threshold severity assessment of the disease using the developed algorithm. The results of the tests in 2012 showed: the Root mean square error of disease severity estimated by the image processing error calculated was 14.7%, the determination coefficient was 0.77. In 2013, the results remained approximately at the same level - 17.1% and 0.73%, respectively, which confirms the promise of using the image processing method derived from UAVs to assess plant diseases and resistance to late blight.

2.2. Harvesting

Currently labour for harvesting constitute more than 50% of total production costs and about 71% of the total human labour required for sweet cherry production. With the growth of competition, the possibility of reducing the cost of production through the mechanization of harvesting is becoming increasingly relevant. Variability of physical and geometrical characteristics of fruits and methods of their collection does not allow to create universal robotic systems, therefore at the current moment researches on designing mechanisms of processing of fruits of separate cultures are conducted. Fruit detection accuracy is critical for obtaining high harvesting efficiency because sweet cherry is characterised by many small fruit. Automated mechanical shakers may be more practical than robotic harvesting for crops like sweet cherry. One advantage of mechanical shaking method is that not every fruit needs detection as long as concentrated areas of fruit in branches are detected. For automatically harvesting cherries using mechanical shakers, a machine vision system needs to be capable of detecting and localising fruit as well as branches. In [10], machine-vision system was developed to segment and detect cherry tree branches with full foliage is given, when only intermittent segments of branches were visible. The specific objectives are to: 1) segment branch pixels in the cherry tree canopy images captured in the presence of leaves and fruit in orchard environment, and 2) detect individual branches using segmented branch regions and estimate their location and orientation in 2D images and assess the detection accuracy. The method of detecting cherry tree branches included 4 main stages: image acquisition, image preprocessing, pixel-based image classification and individual branch detection. Bayesian classifier was used to classify image pixels into four classes e branch, cherry, leaf and background. The algorithm achieved 89.6% accuracy in identifying branch pixels. The morphological properties of the segmented branch sections were used to filter out noises as well as to group together the segments of the same branch in a specified neighbourhood. A curve fitting method was then used to fit an equation through detected branch segments to connect them together. The overall accuracy in detecting individual branches was 89.2%.

In work [11] the method of detection of tomatoes is considered on the basis of the images obtained from UAVs in remote sensing. Tomatoes grow in clusters and are generally partially hidden by their leaves and stalks. At the stage of image acquisition, a quad-copter with a takeoff weight 2.1 kg including a payload of 0.1 kg was used. Based on the spectral-spatial segmentation and classification, tomatoes were searched on the general background of images obtained from different heights with different resolutions - 293 x 415 pixels and 1080 x 1920 pixels, with an accuracy of 92%.

Fruit harvesting is one of the most difficult agricultural tasks for autonomous robotic systems. This task consists of several different stages and involves physical contact with the fruit at high location and orientation accuracy, decision-making in real time, detachment of the fruit from the plant without damaging either, and temporarily storing the fruit under safe conditions. The need to use robotic harvesters is associated with a shortage of labor in agriculture and fierce competition, which requires lower production costs. For example, the costs of harvesting citrus fruits in US currently exceeding their cost of production, and approaches four times Brazilian harvesting cost. According to economic studies, harvesting costs must be reduced by 50% to maintain global competitiveness [12].

In work [13] a vision-based estimation and control system for robotic fruit harvesting was developed. In this work, a cooperative visual servo controller was developed to developed to regulate a robot end-effector to the target fruit location. The Lyapunov-based stability analysis is presented to guarantee uniform global exponential stability of the closed-loop system. The robotic system includes a manipulator, a fixed camera and a closed-loop camera-in-hand (CiH). The fixed camera provides a global view of a tree canopy, while the CiH, due to proximity, provides high resolution fruit images. For improved dexterity and accuracy, a seven DOF kinematically redundant, electric motor driven manipulator is selected for the task. In order to control the gripper of the robot during the harvesting process, a rotation controller for orientation of the robot end-effector and a translation controller for regulation the CiH to the target fruit position are developed. The results of the depth estimation show that the depth error estimation graph depends on the distance between the fruit and the camera. The depth estimation exhibits small error (< 2%) for camera distance greater than 450 mm. The accuracy of the controller was observed to be about 15 mm, thus making the system suitable for harvesting medium and large varieties of citrus fruit but may limit operation for small varieties such as page and blood oranges.

Planning the movement of the robot on the field is one of the important and not completely solved tasks for autonomous harvesting. An analysis of this problem was carried out in [14], taking into account the specificity of the production of sweet pepper. The manipulator and end-effector need to avoid these obstacles to successfully approach a target fruit and prevent damages to the plant or nearby fruit. Testing of the sweet-pepper harversting robot was made in a greenhouse with a pre-known number of fruits and stem location. During preliminary testing in the greenhouse, it was found that the end-effector could successfully detach a fruit if a collision-free azimuth angle was selected for the end-effector, whereas the skew and elevation angle of the fruit hardly influenced detachment success. The sensitivity analysis revealed that reducing end-effector dimensions and widening stem spacing significantly improve the goal configuration success.

3 Classification of grippers for robotic agricultural manipulators

Based on the results of the analysis of recent publications, a classification of grippers (Fig. 1), which are used in robotic agricultural vehicles for manipulation of the fruits, weeds, and other objects. Range, accuracy, speed and other capture characteristics also depend on the parameters of the manipulator. In this classification, we mainly consider the characteristics of the gripper, which is installed at the end of the manipulator and is responsible for physical contact with the object. It should also be noted that the figure shows tasks that require direct capture of objects by an agrobot, therefore, directional spraying of weeds or pruning of branches and leaves, in which manipulators also participate, but the objects of impact are most often not captured later, are also relevant, but not fully are reflected here. In addition, there are examples where, after trimming, objects fall into a pipe channel and are delivered to a container on a robotic platform. For example, in [14] the manipulator is equipped with a knife and a pipe channel, into which the cut fruits of sweet pepper fall under the force of their gravity.

In the right part of Figure 1 there are 22 types of capture depending on 6 selected criteria: the type of drive, the presence of the drive in the gripper, the number of fingers, the type of movement of the gripper, the type of mechanism, the type of sensors. Figure 2 shows examples of existing research agricultural robots equipped with combined grippers according to the proposed classification, referring to different types.

4. Conclusion

The performance of monotonous physically heavy operations in agricultural production leads to the risk of disorders of the musculoskeletal system of workers, and in some cases, contamination with chemical preparations. Therefore, the use of robotic means for physical contact and manipulation of objects in agricultural production is an urgent task, ensuring a reduction in the cost of production and improving the quality of operations performed.

Operations to remove weeds, pruning branches, harvesting fruits require an accurate three-dimensional definition of the position of all objects involved in the interaction. The task is complicated by the fact that image analysis is performed on a complex background with the presence of similar ones (garden trees, a number of plants) and overlapping objects (leaves, branches, fruits). To compile soil and crop cartograms necessary for the operation of ground-based robotic means, unmanned aerial vehicles are already being actively used.

Variability of physical and geometrical characteristics of fruits and methods of their collection does not allow to create universal robotic systems, therefore at the current moment researches on designing mechanisms of processing of fruits of separate cultures are conducted. The proposed classification lists the main types of seizures used to collect fruits and vegetables.

Further work will be devoted to the study of the problems of the physical interaction of agrobots with processed objects, differing in weight, density, geometry, surface roughness and other parameters. When developing the capture and design of the manipulator, experience will be used to develop software and hardware for anthropomorphic robots [20, 21]. The issue of joint interaction of a group of heterogeneous terrestrial and flying robots in the performance of the target agrarian task in an autonomous mission will also be investigated [22-23].

The present research was partially supported by state research № AAAA-A16-116033110095-0 and project № 16-08-00696 of the Russian Foundation for Basic Research.

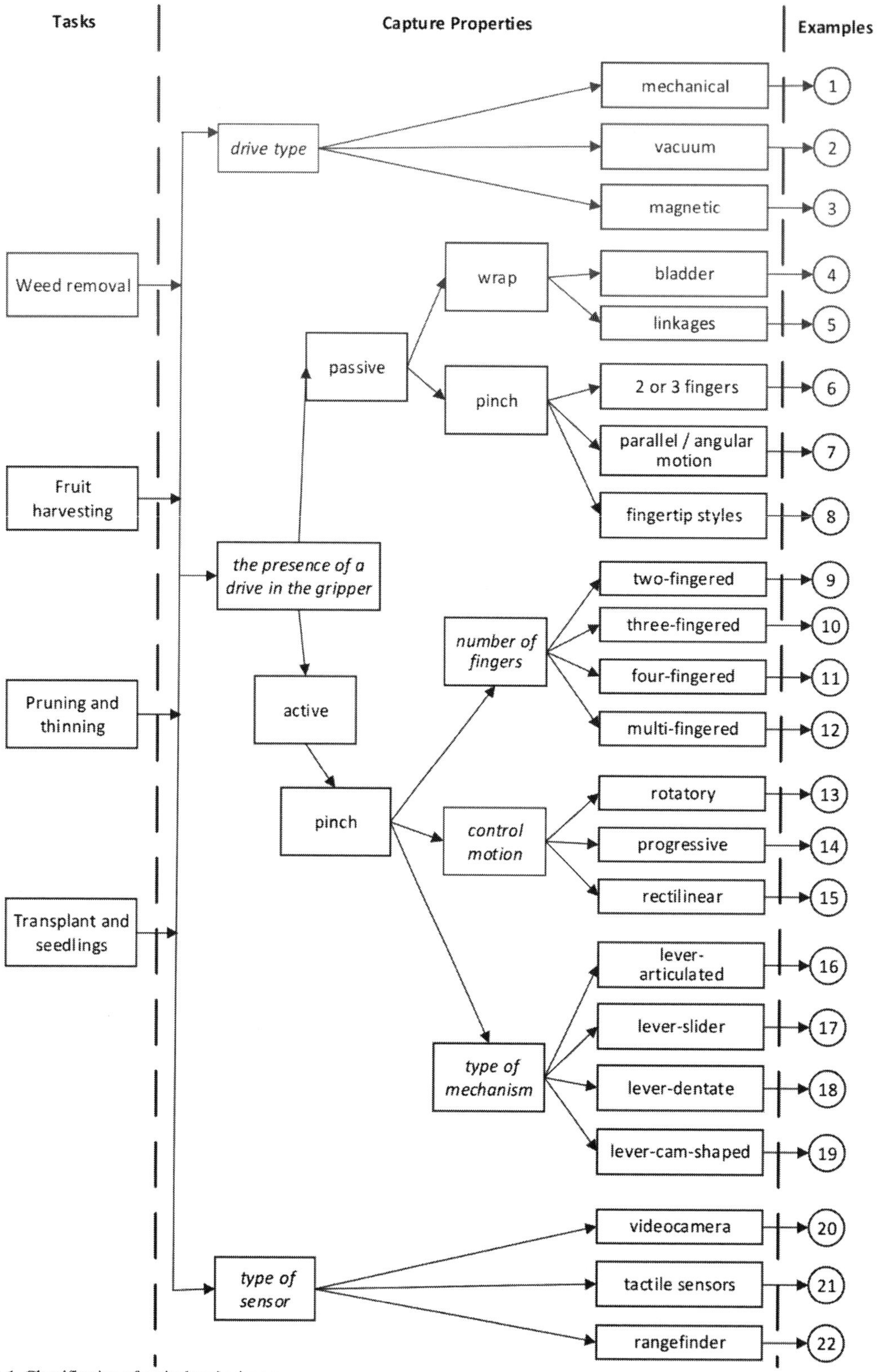

Fig. 1. Classification of agricultural grippers.

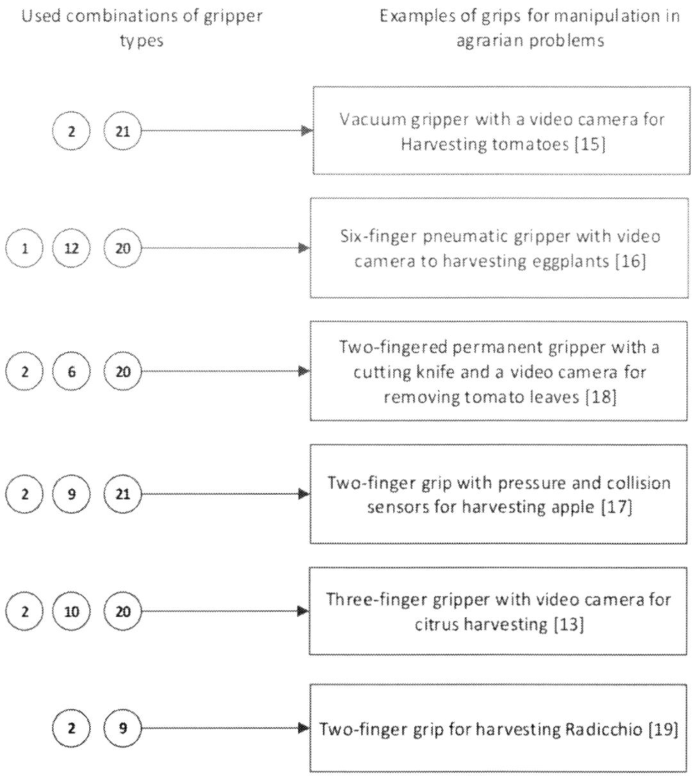

Fig. 2. Examples of grippers used in agriculture.

References

1. W.S Lee, R. Ehsani, Computers and Electronics in Agriculture, **112,** 2–9 (2015)
2. F. Rodriguez, J. C. Moreno, J. A. Sanchez, M. Berenguel, Grasping in Robotics, 385–409 (2012)
3. A. Bechar, J. Stewart, Postharvest Review, **6** (3), 1–11 (2010)
4. G.J Monkman, S. Hesse, R. Steinmann, H. Schunk *Robot grippers* (Wiley-VCH, Weinheim, 2007)
5. A. Bechar, C. Vigneault, Biosystem engineering, **153** 110–128 (2017)
6. H.S.Midtiby, B. Astrand, Biosystems engineering, **146**, 183–192 (2016)
7. M. Perez-Ruız, D.C Slaughter, Biosystems engineering, **126,** 45–55 (2014)
8. RHEA Project: A robot fleet for highly effective agriculture and forestry management. URL: http://www.rhea-project.eu
9. R. Sugiura, S. Tsuda, Biosystems engineering, **148,** 1–10 (2016)
10. S. Amatya, M. Karkee, Biosystems engineering, **146,** 3–15 (2016)
11. J. Senthilnath, A. Dokania, Biosystems engineering, **146**, 16–32 (2016)
12. G.K Brown, HortTechnology, **15**(1), (69–72) 2005.
13. S.S. Mehta, T.F. Burks, Biosystems engineering, **102**, 146–158 (2014).
14. C. W Bac, T. Roorda, Biosystems engineering, **146**, 86-97 (2016)
15. Q. Feng, X. Wang, G. Wang, Z. Li, ICIA, 949–952 (2015)
16. S. Hayashi, K. Ganno, Y. Ishii, K. Tanaka, Japan Agricultural Research Quarterly, **36**(3), 163–168 (2002)
17. Z. De-An, L. Jidong, J. Wei, Z. Ying, C. Yu, Biosystems engineering, **110**, 112–122 (2011).
18. P. Kompano https://www.priva.com/discover-priva/news-and-stories/priva-kompano-deleaf-line
19. M.M. Foglia, G. Reina, J. of Field Robotics, **23** (6/7), 363–377 (2006).
20. A. Denisov, R. Iakovlev, I. Mamaev, N. Pavliuk, MATEC Web of Conferences, **113** (2017)
21. V. Dashevskiy, V. Budkov, A. Ronzhin, ICR 2017, **10459**, 50–58 (2017)
22. V. Nguyen, Q. Vu, O. Solenaya, A. Ronzhin, MATEC Web of Conferences, **113** (2017)
23. Q. Vu, V. Nguyen, O. Solenaya, A. Ronzhin, ICR 2017, **10459**, 260–267 (2017)

Remote control library and GUI development for Russian crawler robot Servosila Engineer

Ilya Mavrin[1], *Roman* Lavrenov[1,*], *Mikhail* Svinin[2], *Sergey* Sorokin[3] and *Evgeni* Magid[1]

[1]Intelligent Robotics Department, Higher School of Information Technology and Information Systems, Kazan Federal University, Russia
[2]Robot Dynamics and Control Laboratory, College of Information Science and Engineering, Ritsumeikan University, Japan
[3]Mechanical Engineering Department, Chuvash State University, Russia

Abstract. Recently a large variety of robots are available on the global market for a reasonable price. Often it turns out that even though a robot has a high-quality hardware and architecture, an original software of the robot may have a number of drawbacks or lacks some important features, which are required for a particular application of the robot. In such cases a user may decide to implement new libraries and functionalities using a provided by a manufacturer application programming interfaces as in most cases code of a commercial robot system is not an open-source. A number of our ongoing research projects concentrate on applications of various robots in urban search and rescue operations, and in most of these projects utilize Russian crawler robot Servosila Engineer. This paper describes the development of a new ROS-based control software and graphical user interface for Servosila Engineer robot.

1 Introduction

Today robotic applications replace a human in numerous scenarios that range from social-oriented human-robot interaction [1] to dangerous for a human urban search and rescue scenarios [2]. Modern mobile robots serve in indoor and outdoor environments and require skillful autonomous navigation within unknown GPS-denied environments [3], efficiency in dealing with computa-tionnal complexity of simultaneous localization and mapping (SLAM) [4], aptitude of negotiation [5] and collaboration with other robots within a group [6] and other functionality. Crawler-type UGVs are typical for urban search and rescue (USAR) operations, especially in situations where such mission deals with dangerous environments and high risk for human rescuers.

In our research, we employ Russian crawler robot Servosila Engineer as a USAR-oriented platform [7]. Original Servosila Engineer robot does not contain any operator oriented software. Customers can use only a client-server system to control the robot in a simple teleoperation mode that implies usage of 3D vision goggles (or single camera view on an operator screen) and joystick. Our goal within this research project is to create a robot control software for an operator, which does not need a joystick and the control could be performed via Robot Operating System (ROS) framework. Since a large variety of algorithms for path planning, localization, and SLAM are implemented in ROS, in order to apply them with our robot our new library is allowing to receive control commands from ROS as well and not only from the user interface [8].

The rest of this paper is structured as follows. Section 2 describes robot Servosila Engineer and its original API. Section 3 presents the developed software, including a remote control library, a user interface program and a ROS node. We conclude in Section 4.

2 Servosila Engineer robot

2.1. Hardware and sensors

A crawler-type mobile robot Servosila Engineer (Fig. 1) is designed and manufactured by Russian company "Servosila" for operating in difficult terrain conditions (e.g. pipeline and tunnel inspection or inspection underneath a static vehicle). The robot is equipped with radiation-hardened electronics and a sensors pack, which includes an optical zoom camera, a pair of stereo vision cameras, back camera and IMU. The robot is tooled with a headlight for operations in low illumination conditions and a manipulator for grasping, pushing or pulling potentially dangerous objects or opening doors with its gripper in teleoperated mode. In addition to a standard sensor package, we equipped the robot with an extra laser range finder Hokuyo UTM-30LX-EW [8].

2.2 Original robot software for teleoperation and application programming interface

The software provided with the robot consists of a server and a client. The server starts immediately after booting the robot, receives commands from the client program and sends back the robot telemetry to the client. The

* Corresponding author: Roman Lavrenov, lavrenov@it.kfu.ru

client is an Operator Control Unit (OCU), which is installed on the user's computer and sends commands to the server. To control the robot, the client requires an XBox joystick that should be calibrated before each use. The client program displays an image from a single camera and the current schematic image of the robot in the corner of the window. The operator has an opportunity to switch between the four cameras of the robot and to select a particular camera to stream the data into client GUI window.

Fig. 1. Servosila Engineer robot (courtesy of the manufacturer).

Table 1. Remote Control Packet.

Field	Size	Type
Frame type ID	1 byte	uint8
Axis #0	2 bytes	int16
Axis #1	2 bytes	int16
...	...	...
Axis #15	2 bytes	int16
Button #0	1 byte	uint8
Button #1	1 byte	uint8
...	...	...
Button #15	1 byte	uint8
Video bit rate telemetry	8 bytes	double
Total size	57 bytes	

Five times per second the OCU sends Remote Control Packets (Table 1), which contain information about buttons and joysticks of Xbox 360 controller. On the other side once per second the robot sends to OCU Telemetry Packets (Table 2) with information about cameras and servo drives, where each motor provides information about its state, position, speed, electric current (in amperes), commands etc. (the details about motor data are presented in Table 3). This protocol has an opportunity to send video frames with delay from a single camera of the robot. While using the robot in teleoperated mode with the original software, we noted the following drawbacks of the protocol:

- Strong dependency of the protocol on game controller. Thus, the protocol allows performing only those commands (both with the original GUI or any new GUI of an external developer) that are available for the game controller.
- The protocol allows to send only speed values to the servo drives and control by position is implemented on OCU side.
- The protocol does not allow checking the battery charging level. For this reason, an operator has to constantly check battery charging level by connecting an additional display to the embedded computer of the robot.
- The protocol prevents a simultaneous rotation of the waist joint and chassis. Thus it is impossible to rotate the waist joint while robot is in motion and the robot has to stop first.
- The protocol does not allow checking the headlight state. Thus an operator has to check visually (e.g., through one of the robot front cameras) whether the headlight is on or off.
- The protocol does not allow gripper joint rotation.

Due to these restrictions the protocol, our library could only emulate an interaction with the game controller. In order to implement all required for a comfortable teleoperation features, the protocol should be reviewed and the server side should be re-implemented, which is a part of our future work.

Table 2. Telemetry Packet.

Field	Size	Type
Frame type ID	1 byte	uint8
Tick number	8 bytes	uint64
Number of motors	1 byte	uint8
Motor data #0	24 bytes	struct
Motor data #1	24 bytes	struct
...	...	...
Motor data #9	24 bytes	struct
Not used	25 bytes	-
Total size	275 bytes	

Table 3. Motor data.

Field	Size	Type
Device ID	1 byte	uint8
Device state	1 byte	uint8
Operation mode	1 byte	uint8
Position	4 bytes	uint32
Speed	2 bytes	int16
Electric current (in amperes)	2 bytes	int16
Status bits	2 bytes	int16
Position command	4 bytes	uint32
Speed command	2 bytes	int16
Electric current command (in amperes)	2 bytes	int16

3 Software development

The original GUI provides basic functionality and an operator could control robot in real time only using Xbox 360 game controller. For this reason, we decided to develop a new GUI and remote control library with an extended functionality, which should provide robot control from the ROS framework and assure a more comfortable teleoperation process.

Our new software consists of three parts: a remote control library, a graphical user interface (GUI), and a ROS node program. The structure of robot control environment with description of remote control library, GUI for an operator and a ROS node are illustrated with in Figure 2.

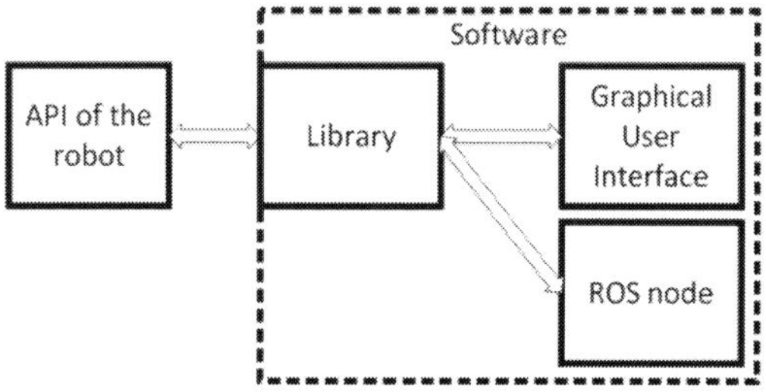

Fig. 2. The scheme of our software.

3.1 Remote control library

Remote control library is a mediator between GUI and ROS node on one side and the robot (through its API) on the other. It is used for sending commands to the robot and receiving information from the cameras and the servo drives. The library has a set of API functions, which are used for controlling the robot via Wi-Fi (or radio communication). The commands are encapsulated by the developed by Servosila company remote control protocol (Table 1).

The remote control library is implemented as a shared library (*.so file), which gives an opportunity to use the library with other applications. For example, pathfinding algorithms could be implemented for Servosila Engineer robot using API of the remote control library. The library consists of the following blocks (Fig. 3):

- Structure RobotConfiguration contains information about the configuration of the robot, including a set of maximal velocities of the joints;
- Class Robot is a wrap of the class Controller, which sends joint velocities from RobotConfiguration to Controller methods;
- Class RobotController starts a thread for UDPClient and controls the robot by updating the required values in remote control packet;
- Class RobotPositionController extends RobotController, so it can control particular joints by position control;
- Struct RobotSI converts commands to SI System (International System of Units) in order to handle velocity values in m/s and rad/s;
- Struct RobotPackets contains implementation of the packets from Servosila's API protocol;
- Class UDPClient provides data sending service from the client to the server (on the robot);

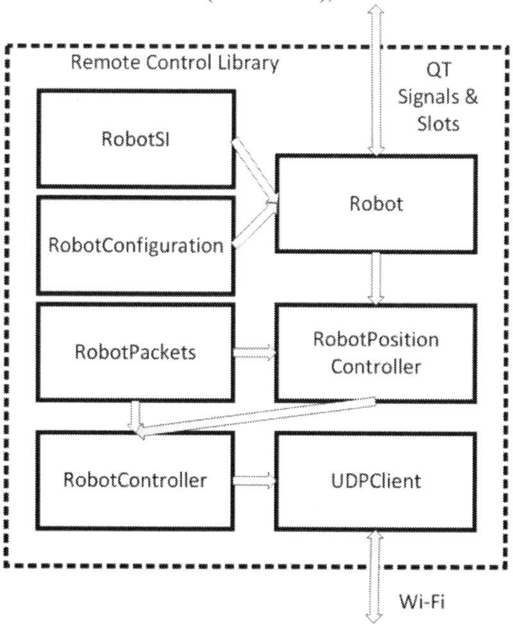

Fig. 3. The remote control library scheme.

The library is built with qmake and g++ using Qt framework [9]. The library works in its own thread and connects to GUI and our ROS node (that is currently in development) using Qt signals and slots mechanism. While this solution allowed a quick development of the robot control environment, is has a number of drawbacks.

First of all, every application, which plans to use the library should include Qt libraries to support signals and slots mechanism. Therefore, the application should include Qt libraries in its binary (which dramatically increases size of binary) or dynamically link Qt libraries (which sets constraints on OCU). For this reason, as a part of our future work we plan to rewrite library from scratch without Qt using pure C++, without any libraries, frameworks and other significant dependencies.

UDP client in library encapsulates static object with the structure of Remote Control Packet and Qt implementation of UDP socket. UDP client runs in its own thread and sends the packet five times a second. Each packet could be constructed via operations with GUI or commands from ROS node. Telemetry packets bring useful information to OCU. A packet contains information about each motor, including motor state, motor shaft position and speed, electric current, speed command etc. (Table 2). Motor shaft position information is used for position control, but the chassis motor drives do not have position telemetry information. This issue may be resulted by absence of encoders in these motors or by the original API. This way, the only available approach for travelled distance calculation with robot odometry turns out to be multiplying travel velocity in m/s by travel time is seconds, which is significantly less precise with regard to encoders usage.

The protocol stores values of each servo speed as integers in the range [-32768, 32767], which correlates the velocities with positions of control sticks of Xbox 360 game controller. The remote control library should be able to accept velocity values in m/s, which is necessary both for a user convenience and for ROS node program. In order to identify mapping between abstract API commands and real-world motion with reverse engineering approach, we performed a set of experiments for linear and rotational motions of the robot.

The main idea of the experiments is to convert API numbers to real-world measure units, which requires to explore tracks rotation velocity depending on API commands sent to robot. The correlation between these values should be expressed as a ratio between API commands numbers and real-world velocity in meters per second. The experiments were performed as follows:

1. Install the robot in way that nothing influences tracks rotation velocity (including friction, Fig.4);
2. Measure length of the tracks l;
3. Send constant velocity commands to the robot;
4. Pick a relatively large number N of full rotations of the tracks;
5. Measure time duration t which is required for full rotation of the tracks N times;
6. Compute real-world velocity in meters per second with the equation:

$$vr = l*N/t \quad (1)$$

In order to neglect any external influence on tracks rotation velocity, the robot was installed on two blocks in the way that its tracks do not touch the blocks or any other surface during their movement (Fig. 4). The robot base and a track were marked with external markers for rotations count. Every time the two markers match, the number of rotations is incremented by one. The track length l is 1.15 m. We set number N to 40, which is large enough to eliminate random friction influence between tracks and driving wheels.

Fig. 4. The robot fixed position during the experiments.

Based on the results of the experiments, we concluded that the integer values, which are sent to the robot, are linearly proportional to the velocity of linear movement and rotation. We calculated these two coefficients by converting integer values to the SI system using least squares method. The linear motion value was lv = 71553.837 and the rotation motion value was lr = 25171.216. In order to obtain the velocity in m/s and rad/s we divided the integer values by these constants, and the resulting values were incorporated into RobotSI class.

3.2 Graphical User Interface program

Graphical User Interface (GUI) is an application, which can be used to control the robot in real time. An operator could set speed (for all joints) and position (for several joints) values. These values are sent to the Remote control library using QT signals and slots, which in turn sends these values to the robot.

The GUI consists of two active windows. The velocity control window (Fig. 5) contains elements of the robot control, a connect button, a button for switching to the maximal velocity window and a checkbox to enable\disable the robot torch. In addition, there are switches for flippers and gripper control. In the maximal velocity window (Fig. 6) an operator could set a maximal velocity for each joint using an appropriate slider or a textbox, which will set the upper limit for a corresponding velocity slider and hotkeys.

GUI offers a variety of ways to control the robot:

- Using keyboard hotkeys;
- Using sliders or textboxes for velocity control of each joint;
- Using textboxes for position values (only for several joints);

An operator could use keyboard to control the robot in real time.

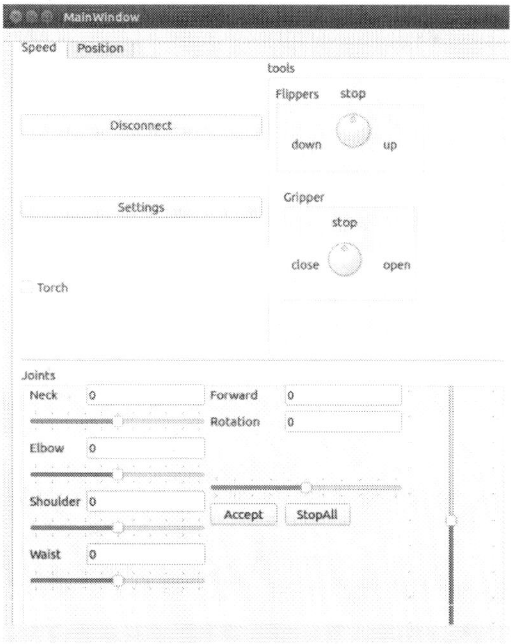

Fig. 5. Velocity control window.

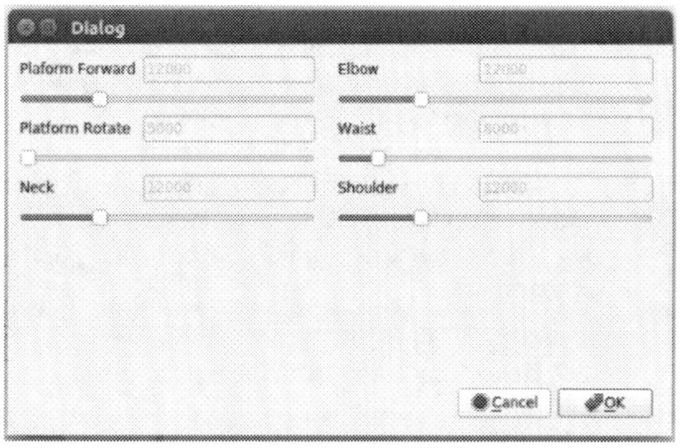

Fig. 6. Maximal velocity window.

3.2.1 Keyboard controls

In this mode an operator should hold the key to move the particular joint (Table 4). When he/she releases the key, the servo immediately stops. This mode allows to hold multiple keys simultaneously. The implementation uses Qt keypress events.

3.2.2 Sliders

An operator could use sliders for controlling all joints of the robot. Once he/she sets the slider, the joint starts moving. A user can set velocities in textboxes and joints will move after he/she presses the button «Accept». At any time, the operator could press an emergency stop button «StopAll» (or press hotkey «Space»), which immediately stops all servo drives.

Table 4. Keyboard layout.

Key	Movement
W	Drive forward
S	Reverse
A	Rotate left
D	Rotate right
Q	Waist rotate left
E	Waist rotate right
Y	Shoulder up
H	Shoulder down
F	Flippers up
R	Flippers down
G	Grippers close
T	Grippers open
U	Elbow up
J	Elbow down
Space	Emergency stop

3.2.3. Position Values

The joints, which support position control, include the elbow, the neck, the shoulder, the waist (these joints have the corresponding labels in GUI). An operator can use position values for these joints by setting values in the textboxes of the joint position control window (Fig. 7). The joints will automatically move into a required position after the operator presses button «Accept». The contents of telemetry packets that arrive from the robot are displayed in the same window. Unfortunately, due to drawbacks of the existing remote control protocol of the robot, it is impossible to send position values directly to the robot.

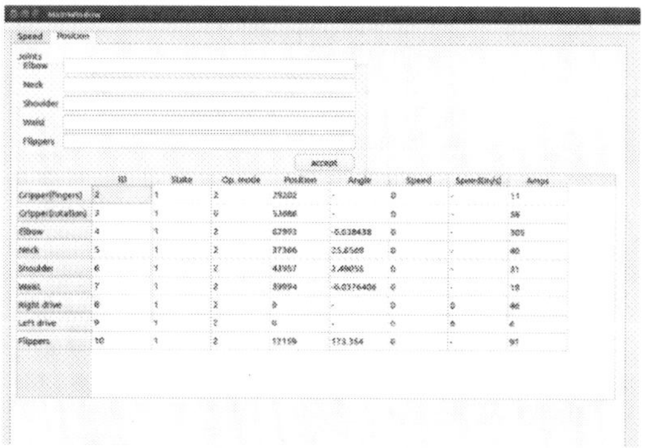

Fig. 7. Joint position control window.

We implemented position values handler on OCU side. The handler finds motor shaft's position values in telemetry packets for each required motor and compares these values with required position values for each motor respectively. It calculates the difference between the values and sets motor velocities, positive or negative, depending on current and required motor shaft positions. Currently its implementation is rather simple: motor speed doesn't depend on the difference between required and current values.

3.3 ROS node program

Since Servosila Engineer robot uses some of the existing in ROS algorithms for autonomous localization, mapping and path mapping, a ROS node is required. In ROS framework nodes generate motion commands in \geometry_msgs\Twist format, which contains linear (for all axes) and angular velocities (for all axes) being measured in meters and radians respectively. Therefore, a velocity conversion function, which is a part of our remote control library, is required.

Our remote control library is included into ROS node, which sends ROS commands from ROS node \move_base to the robot via Wi-Fi connection. In order to do that, library was statically compiled and its headers were included into ROS node source file.

ROS node is a console program. The node workflow, which runs in a loop, works as follows:

1) Wait for messages on the specified topic (e.g.\cmd_vel\).

2) When a new message arrives, it is split in two parts: linear and angular. If the linear (angular) part exceeds or equal to 0.2 m/s (0.2 rad/s) then, it is sent to the robot. If the linear (angular) part is positive, but does not exceed 0.2 m/s (0.2 rad/s) then, the value of 0.2 m/s (0.2 rad/s) is sent to the robot. The values of 0.2 m/s (0.2 rad/s) were selected experimentally as these are the lowest velocities that are required by robot servos to start moving.

Before the node shuts down, it sends zeros to angular and linear velocities in order to stop the robot and prevent any uncontrolled motion.

4 Conclusions and future work

Often robot original software lacks some important features, which are required by a user for a particular application. These forces a user to implement new functionalities using a provided by a manufacturer API.

A number of our ongoing research projects utilize Russian crawler robot Servosila Engineer, and this paper describes the development of a new ROS-based control software and graphical user interface for Servosila Engineer robot. With our new software and GUI a user can operate the robot without using a joystick, which was the only control possibility with the original GUI. Moreover, because our software is based on robot operating system ROS, it broadens practical applications functionality due to capability of incorporating different existing and open-source ROS-based standard useful algorithms and functions into the robot control system.

Currently, we further improve the GUI by allowing to simultaneously displaying data from all four robot cameras in real time instead of the original single camera data display that allows only a sequential switching between the cameras. As a part of our future work, we plan to add new features into the GUI, including position control for all joints and visual displaying of a current robot configuration with a 3D schematic model. Next, we will add several automatic functions including a pilot advisory system that considers static and dynamic balance of the robot while selecting a suitable path [10].

This work was partially supported by the Russian Foundation for Basic Research (RFBR) and Ministry of Science Technology & Space State of Israel (joint project ID 15-57-06010). Part of the work was performed according to the Russian Government Program of Competitive Growth of Kazan Federal University.

References

1. V.Y. Budkov, M.V. Prischepa, A.L. Ronzhin, A.A. Karpov, Int. Congress Ultra Modern Telecommunications, 485–488 (2010)
2. A. Birk, S. Schwertfeger, and K. Pathak, IEEE Wireless Communications, **16.1**, 6–13 (2009)
3. K. Yakovlev, V. Khithov, M. Loginov, A. Petrov, IEEE Int. Conf. on Intelligent Systems, Springer Int. Publishing, 49–56 (2015)
4. A. Buyval, I. Afanasyev, E. Magid, Int. Conf. on Machine Vision, 103411K (2016)
5. B. Li, S. Ma, T. Liu, J. Liu, IEEE Int. Symp. on Safety, Security and Rescue Robotics, 53–57 (2008).
6. N. Michael et al. Journal of Field Robotics, 832–841 (2012)
7. M. Sokolov, R. Lavrenov, A. Gabdullin, I. Afanasyev, E. Magid, Int. Conf. on Control, Mechatronics and Automation, 61–65 (2016)
8. N. Alishev, R. Lavrenov, Y. Gerasimov, Int. Conf. on Artificial Life and Robotics, 204–207 (2018)
9. Qt framework, Available at: https://www.qt.io/
10. E. Magid, T. Tsubouchi, E. Koyanagi, T. Yoshida, Int. Conf. on Intelligent Robots and Systems, 349–356 (2010)

Automated system of monitoring and positioning of functional units of mining technological machines for coal-mining enterprises

Yaroslav Meshcheryakov and *Roman* Meshcheryakov*

Tomsk State University of Control Systems and Radioelectronics, 634050, pr. Lenina, 40, Tomsk, Russia

Abstract. This article is show to the development of an automated monitoring and positioning system for functional nodes of mining technological machines. It describes the structure, element base, algorithms for identifying the operating states of a walking excavator; various types of errors in the functioning of microelectromechanical gyroscopes and accelerometers, as well as methods for their correction based on the Madgwick fusion filter. The results of industrial tests of an automated monitoring and positioning system for functional units on one of the opencast coal mines of Kuzbass are presented. This work is addressed to specialists working in the fields of the development of embedded systems and control systems, radio electronics, mechatronics, and robotics.

1 Introduction

At present time, in connection with the presentation of ever higher demands on the processes of managing technological machines in the coal mining industry, the problem of identifying objects, systems, phenomena and processes play an extremely important role [1]. It is impossible to ensure the effective and qualitative management of the system if its model is not known with a sufficient degree of accuracy. Therefore, when designing control and control systems, numerous studies are conducted to determine the mathematical models of these systems [2, 3].

The creation of control and management systems for complex technological facilities, such as coal mining enterprises, is connected with the tasks of selecting and setting up automated monitoring and positioning systems (AMPS) for the functional units of typical devices of mining technological machines (similar units in different stages of the production process) [4].

Currently, these parameters are monitored by mechanical plummets and do not meet modern requirements for accuracy and reliability. In addition, there are no devices on technological machines that allow not only performing continuous monitoring of the angular orientation of functional elements but also to register and analyze the information received in order to assess the quality of mining operations for the final time interval - hour, shift, week, month [5, 6].

The main existing manufacturers of position monitoring systems in the space of functional elements of mining machines are: one-dimensional (1D), two-dimensional (2D) and three-dimensional (3D) leveling systems for hydraulic excavators and drilling rigs firms: Topcon (Japan), Trimble (USA), Prolec (UK), Leica Geosystems (Switzerland), Moba (Germany), which are presented in Table 1.

These systems, using data from satellite navigation systems and signals from inertial sensors, determine the position of the bucket and the platform of the excavator.

The systems are relatively expensive (more than $ 30,000), are complicated to operate, have a multi-level interface, require careful debugging and long training for the operator-operator. In addition, information about the position of the platform excavator in foreign systems is secondary and not displayed on the monitor [7, 8].

This function can be successfully implemented using modern microelectromechanical systems (MEMS) - gyroscopes (MG) and accelerometers (MA).

Fig. 1. High-precision MG printed circuit board ADIS 16265.

* Corresponding author: mrv@security.tomsk.ru

Table 1. Summary table of systems for excavators.

Producer	System	System class
Topcon	Easy Control	1D
	X-22	2D
	X-62	2D
	X-32	2D
	X-33	3D
	X-63/X-63i	3D
Trimble	GCS 600	2D
	GCS 900	3D
Prolec	Level X	1D
	Digmaster X	2D/3D
	Digmaster Pro	2D/3D
	Digmaster PRO + RTK-GPS	3D
	PcX-3D	3D
Leica Geosystems	iCON iXE3	3D
Moba	Xsite Easy	1D
	Xsite Pro	2D
	Xsite Pro Advanced	3D

The system of continuous monitoring of the position in the space of functional elements of technological machines can be easily built with the aid (help) of an MG, an MA and a Madgwick filter uniting their readings. The application of the Madgwick filter does not require a large amount of processing power from the controller of the system and allows achieving sufficiently high-quality indications.

By combining data of these two MEMS sensors it is possible to calculate the position of the functional element of the processing machine in space at any time during any of its intervals.

The gyroscope operates in the mode of the angular velocity sensor [9], which for a certain initial orientation can be integrated with the passage of time to obtain the current orientation of the platform of the walking excavator.

The use of precise gyroscopes, for example, a fiber-optic or laser gyroscope is inexpedient in view of the dimensions and very high cost.

MG have significantly lower accuracy and cost, but the availability of digital filters and methods for correcting the indication significantly improves the quality of their work, which promotes active use in various fields of science and technology.

Direct integration of the angular velocity readings obtained with MG leads to a rapid accumulation of errors in the calculation of orientation due to MG errors. The main disadvantages of MG are the error of the conversion factor (the instability of the scale factor) and the displacement (drift) of zero. The drift of zero is a parasitic phenomenon, in which the gyroscope readings are different from zero when the gyroscope is at rest.

Thus, MG alone cannot provide an absolute orientation measurement. Figure 2 shows the graph of accumulation of the orientation error.

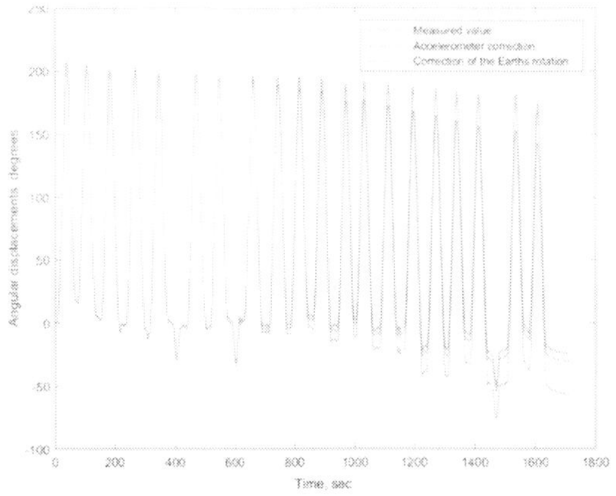

Fig. 2. Graphic accumulation of the orientation error as a result of integrating the MG readings.

The line of pink color is approximate real orientation values, the blue line is the measured orientation values.

It can be seen from the graph that a gradual accumulation of the error occurs due to the lack of additional correction by the digital filter.

Figure 3 shows the zero drift of the angular velocity of the MG ADIS16265 at rest.

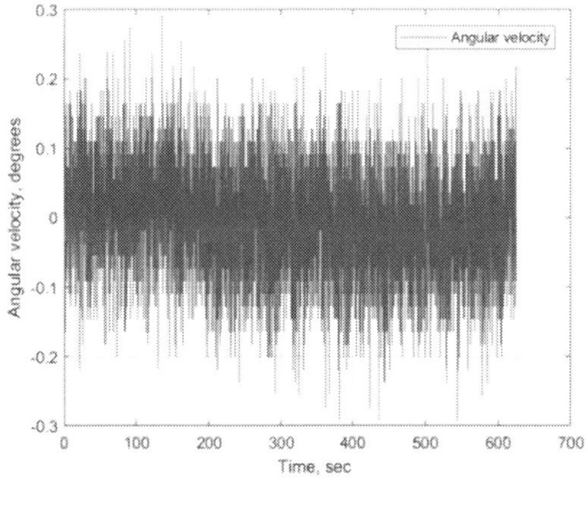

Fig. 3. Drift of the zero point of the angular velocity of the gyroscope ADIS16265.

The graph shows the apparent angular velocity during idle MG. The average value of the apparent angular velocity is 0.0830 degrees per second, which can lead to an absolute displacement of 4.98 degrees within a minute. A periodic harmonic signal with a frequency of 14 kHz from the sensor element is observed.

Apparent movements are formed as a result of integrating the apparent angular velocity and accumulation of integration errors.

Figure 4 shows the graph of moving the MG ADIS16265 at rest.

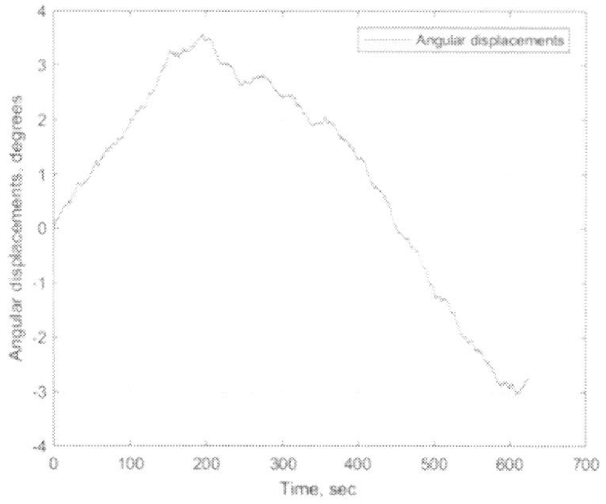

Fig. 4. Moving MG ADIS16265 at rest.

MA [10] – sensor measuring the projection of apparent acceleration. To determine the angle of the object's inclination, you can use MA, the indications of which determine the moment of rest of the object. MA is highly susceptible to various factors. Accumulation of the error as a result of integrating the readings of the MA is very large, determination of the orientation is impossible in a short time interval.

Combining the readings of MEMS sensors makes it possible to level out their essential shortcomings [11, 12]. For a gyro, this is zero drift and integration errors, and for the accelerometer - too much sensitivity to vibrations and other external force effects. It is possible to perform such an aggregation using the Madgwick filter [13].

The Madgwick filter is a modern complex filter with very high speed, a small requirement for computational resources, and accuracy not inferior to the Kallman filter.

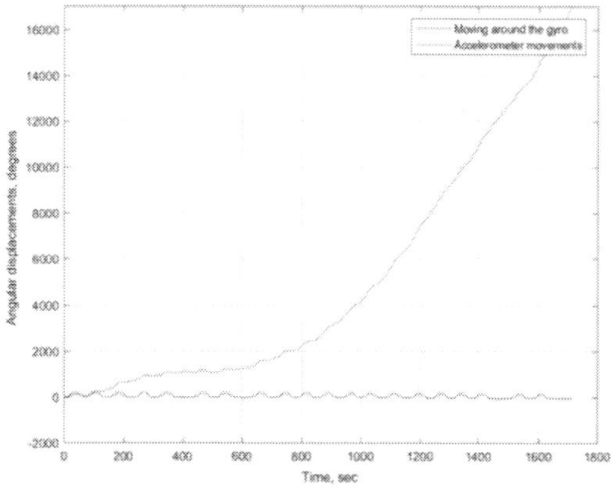

Fig. 5. Accumulating orientation error for separate use of MA and MG without correction by a digital filter.

The Madgwick filter is designed for mathematical processing of incoming data samples obtained from primary converters, eliminates the problem of errors and accuracy, as well as the filter parameter settings that are characteristic of Kallman filters. Adjustment of the readings of MA and MG is to calculate a single orientation estimate by measuring the accelerometer, gyroscope and, if necessary, a magnetometer.

The Madgwick filter uses a quaternion to represent the orientation to describe the position in space in three dimensions and does not contain problems related to the description of the position by the Euler angles (folding frames/folding frames). The accuracy of the filter is compared with existing industrial filters, similar to the Kallman filter. The filter can be realized not only with inertial sensors MA and MG but also with an additional sensor - a magnetometer (digital compass). The realization of a filter with a magnetometer implies compensation of magnetic distortions and compensation of gyroscope bias.

The main characteristics of the filter are two adjustable parameters β and ξ (one parameter for realization without a magnetometer) determined by observable system characteristics, analytical calculations, and gradient descent optimization, which improves accuracy at low sampling rates, magnetic distortion compensation and zero-zero gyro compensation time.

The gain of the filter β expresses all the errors associated with the operation of the MG and the digital filter, represented as a drift of the zero of the gyroscope, expressed as the magnitude of the quaternion derivative. The main sources of errors are sensor noise, filter smoothing operation, quantization process, calibration accuracy, sensor mounting and alignment, non-orthogonality of sensor axes and frequency characteristics. These errors are the displacement of the gyro.

The choice of the coefficient β depends on the drift of the zero of the gyroscope, the rate of accumulation of the calculation errors, and the conditions for using the walking excavator.

So, too much β value will result in the result of filter operation being greatly affected by the vibration of the machine body. Too small a value of β may not be sufficient to eliminate the drift of the gyroscope readings.

The initial coarse value of the Madgwick filter gain is determined at the stage of assembly and calibration of the AMPS. The MG readings are collected at rest for 8 hours, which corresponds to one working shift.

The obtained angular velocity data is processed by Allan's dispersion method with the construction of a graph on a logarithmic scale, along which time intervals and drift values of the angular velocity ω_β are determined. The angular value ω_β represents an estimate of the average error of zero measurements for each axis. After determining the value of the angular velocity, the value of the filter gain β is calculated from expression (1).

$$\beta = \left\| \frac{1}{2} \hat{q} \otimes \left[0 \; \tilde{\omega}_\beta \; \tilde{\omega}_\beta \; \tilde{\omega}_\beta \right] \right\| = \sqrt{\frac{3}{4}} \tilde{\omega}_\beta \qquad (1)$$

For each filter implementation, there is an optimal value of β, which should be high enough to compensate for the accumulation of errors and low enough so that unnecessary noise does not fall into the large steps of the gradient descent. For MG ADIS 16265, the optimum value in the steady state is 0.035.

The installation of global satellite navigation (GNSS) system will improve the accuracy of the orientation and record additional parameters of the technological process.

The main functional elements of AMPS functional units of mining technological machines are:

- sensors (primary converters), including MG and MA;
- navigation module;
- microcontroller;
- -jam-protected link; (noise-resistant communication line);
- main and backup memory module;
- information display device.

The system is designed according to the modular principle, where its nodes can be added depending on the type and function of the mining machine.

The use of the modular architecture in the creation of AMPS will allow assembling the system from unified modules depending on the operating conditions, the range of tasks to be performed and the logistics. Unified modules will allow you to quickly change the functionality of the system, to debug and implement new functionality without a complete system rebuild.

2 Software

2.1 Main software

The control microcontroller software, written in a middle-level C / C ++ language, sets and manages the logic of the entire system. The program provides: data collection, leveling of noise and errors by the Madgwick filter, leveling of the influence of vibration, minimization of the accumulated error due to the daily rotation of the Earth, the functioning of the navigation module, identification of the operating cycles of the walking excavator, recording of the event history in the black-box recorder, and the operation of the information display device.

Identification of work cycles is carried out by specialized subroutines that function as a "finite state machine".

2.2 Algorithms for identification of operating states

A distinctive feature of the proposed AMPS is that it does not require the inclusion of a walking excavator into the control and power contour. AMPS perform the registration of time, angular velocity, normal and tangential acceleration, angles of inclination relative to the horizon, spatial coordinates.

AMPS allows you to register the following operating conditions of the excavator:

- digging;
- transportation for unloading;
- deloading;
- transportation to scoop;
- idling;
- auxiliary operations;
- dredge (walking);
- readjustment (not available).

3 Graphical representation of the technological process of excavation

The operation of the excavator during the process can be represented in the form of a connected graph, shown in Figure 6. Each vertex of a connected graph represents a certain state of the technological process.

The interrelation between states can be represented by repeated transitions between the vertices of the state graph describing the technological state of the walking excavator.

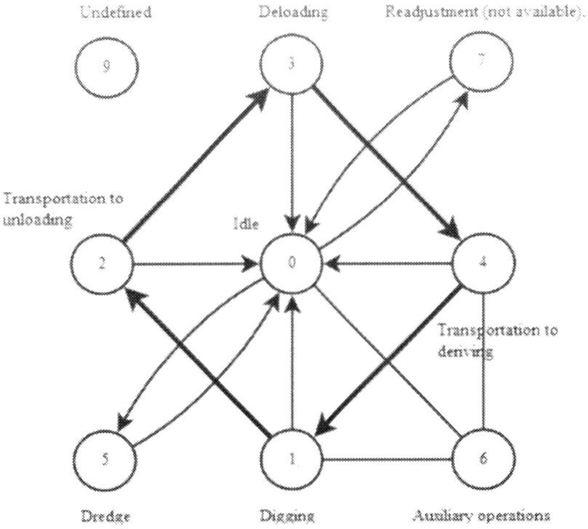

Fig. 6. The graph of the conditions of the technological process of excavation, registered by AMPS.

The graph of states describes the technological process of excavation and contains an oriented graph by points 1 2 3 4 1.

The idle state is connected by edges to all states, as a forced stop of the machine can occur at any time, interrupting the course of any operation.

Auxiliary operations that describe position 6 can begin and end with operations: idling, scooping, or transporting to scoop, so the vertex 6 is connected by edges to the corresponding vertices 0, 1, 4 of the graph.

The stepping operation begins and ends after the main mechanisms stop, so the vertex 5 is connected by an edge only to the point 0.

The state of "readjustment" or "unavailable" can occur only after the platform is completely stopped and all power drives.

The state "undefined" can occur from any state.

3.1 Software Development

Calculations of the cycle duration and corresponding energy costs can be performed only with a special identification procedure, which makes it possible to distinguish the beginning and the end of the corresponding technological state of the excavation process.

To assess the effectiveness of the operation of the dragline, it is necessary to perform the determination of the main cycles, broken down according to a precise deterministic state: "calculation of the new duty cycle", "continuation of the cycle", "idle cycle", "completion of the duty cycle," "not available."

The judgment about such a state should be deductible on the basis of statements about the values available for registration, the current variables: angular velocity, normal and tangential acceleration, the fact of absence or presence of an onboard power supply, spatial coordinates.

3.2 Development of identification algorithms

Algorithm for identification of ACMP work cycles is developed on the basis of analysis of the typical technological process of the excavator operation at the transshipment of rock mass. It consists of three main parts: the main cycle, the control subroutine, and the calculation subroutines.

Figure 7 shows a block diagram of the main loop, which describes the current state of the system and prescribes the program's further behavior, depending on the state of the excavator platform. The subroutine is identified to calculate the cycles broken down by a clear deterministic state: "calculation of a new duty cycle", "idling cycle", "and completion of the duty cycle".

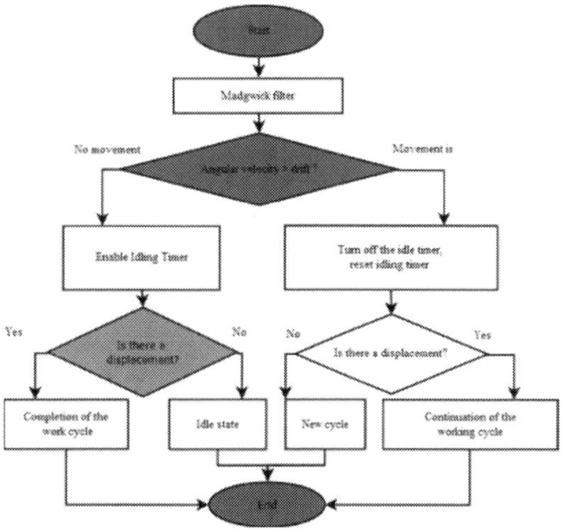

Fig. 7. Block diagram of the algorithm of the main loop of the AMPS.

The main cycle describes the current state of the system and determines the further state of the state machine, depending on the current variables. The main cycle initiates the start and stops of the idling timer; it makes a transition to the state of the work cycle identification.

Figure 8 shows a block diagram of the subroutine for calculating a new work cycle with a subroutine for monitoring the drift of the angular velocity of the MG. The subroutine is designed to track the correct operation of the Madgwick filter.

The subroutine initiates the calculation of a new work cycle with a preliminary check of the angular velocity drift [14, 15]. If the drift of the angular velocity has reached a certain threshold, the reset flag of the MG is set to reset the accumulated errors.

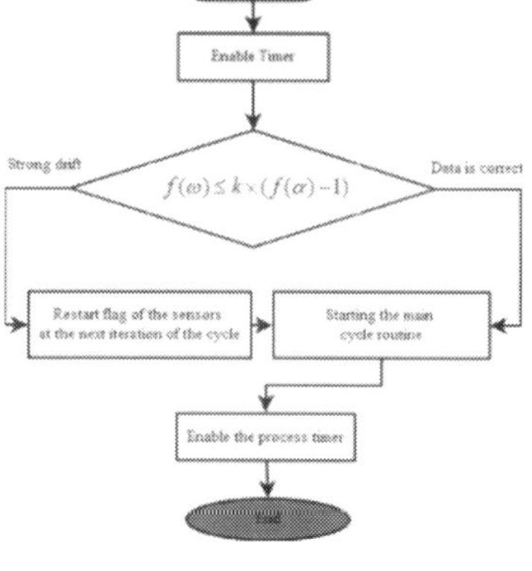

Fig. 8. Block diagram of the subroutine for calculating a new work cycle with the subroutine for monitoring the drift of the angular velocity.

The logic of the control subroutine is described by the expression 2.

$$f(\omega) \le k \times (f(\alpha) - 1) \quad (2)$$

Expression 2 describes the ratio of the angular velocity X to the total acceleration Y with the projection of the acceleration due to gravity falling. If at the moment of rest of the platform of the walking excavator the angular velocity is greater than the linear acceleration, then there is a large drift of the zero angular velocity. The subroutine can be run as part of another subroutine or run autonomously.

Direct identification of cycles "scooping", "walking", "transportation for unloading", "disruption" is carried out separately on the server to enable rapid changes in identification conditions [16-18].

4 Tests of AMPS

Industrial tests of AMPS were carried out under the conditions of the Kedrovsky section. The AMPS unit was installed in the cab of the ESh-11/70 excavator driver, who works at the transshipment of the rock mass.

After the installation of the AMPS unit, a self-diagnostic system and internal self-calibration of the orientation in space was launched. The calibration consisted in obtaining initial data from the sensors - the accelerometer SCA3100-D04-1 and the gyroscope - ADIS16265.

The calibration data were accelerations projections that were recalculated into the angles of the initial orientation of the AMPS block and the angular velocity readings for adjusting the gain of the Madgwic filter beta.

After receiving the calibration data, the communication with the server was established for the operation of the module.

After connecting to the server, the control program and the hardware part continued working in the normal mode. The data arriving at the server in automatic mode was processed and placed on the developer's site. The complex worked continuously for ten days. Throughout the time of testing showed high accuracy and stability of operation. Thus, the monitoring system accurately recorded the working cycles of the ESH excavator - 11/70 the time for changing the workers' crews of the drivers. The array of received data corresponds to the events recorded in the journal of working hours.

Figure 9 shows the hourly chart of the workers. In this diagram, the number of short cycles is shown in a dark green color, and the number of long cycles of the excavator's operation is light green.

The results of the tests of the automated monitoring and positioning system of the functional elements of mining machines in laboratory and real production conditions at one of the coal enterprises of Kuzbass confirm the possibility of its application for remote continuous monitoring of the angular orientation of the excavator platform and automated recording of the operating cycles of walking excavators for a fixed period of time through the use of microelectromechanical sensors MA and MG with combination Madgwick filter. The use of the Madgwick filter does not require large computational powers, fine tuning, and increased data polling frequencies, which are inherent in various implementations of the Kallman filter.

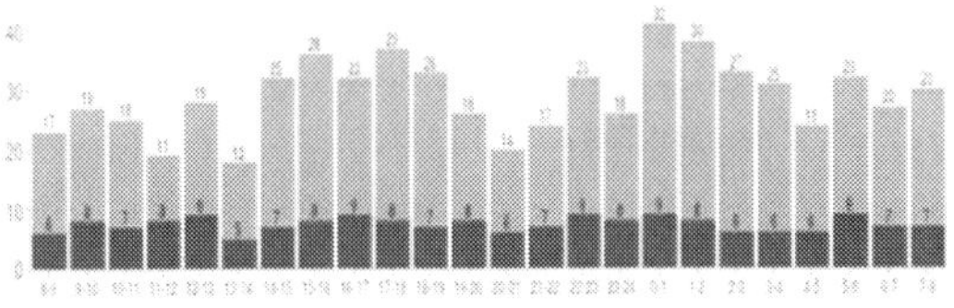

Fig. 9. The hourly graph of work cycles.

5 Conclusion

Thus, the work presents an automated system for monitoring and positioning functional units of mining technological machines for coal-mining enterprises. The system had based on the method of leveling the errors of MEMS sensors by mean of combining the Madgwick filter.

The analysis reports show that system makes it possible to significantly simplify software for identifying the operating cycles of walking excavator, allows achieving sufficiently good indications to the total operating time of technological machines in various operating conditions and angular orientation parameters during operation.

In the future, it is planned to finalize and introduce software with elements of self-learning and dynamic automatic correction of the AMPS operation.

The given paper is completed with the support of the Ministry of Education and Science of the Russian Federation within the limits of the project part of the state assignment of TUSUR in 2017 and 2019 (project 2.3583.2017).

This work was financially supported by the grant of the Russian Federation President for supporting the leading scientific schools in Russian Federation No. NSh-3070.2018.8.

References

1. M.Y. Drygin, N.P. Kuryshkin, Y.E. Meshcheryakov, O.V. Lyubimov, Bulletin of KuzSTU, **5**, 28–31 (2014)
2. Y.E. Meshcheryakov, A.M. Korikov, Actual problems of modern engineering: a collection of works of the International Scientific and Practical Conference, 38–42 (2014)
3. Y.E. Meshcheryakov, Modern technology and technology: Sat. works of XVIII International Scientific-Practical. Conf. students, graduate students and young scientists, **3** (2), 359–360 (2012)
4. Y.Mescheryakov, A. Korikov, Applied Mechanics and Materials, **770**, 612–616 (2015)
5. Y.Mescheryakov et. al., Proceedings in Electronic International Interdisciplinary Conference. The 2nd Electronic International Interdisciplinary Conference, 401–403 (2013)
6. Y.E. Meshcheryakov, A.M. Korikov, Natural and Intellectual Resources of Siberia: reports (conference materials) of the 22nd International Scientific and Practical Conference (2016)
7. Y.E. Meshcheryakov, A.M. Korikov, N. Kuryshkin, Systems of automation and education, science and production: Proceedings of the 10th All-Russian Scientific and Practical Conference (2015)
8. Y.E. Meshcheryakov, A.M. Korikov, Scientific session of TUSUR-2015: materials of the All-Russian scientific and technical conference of students, graduate students and young scientists, **3**, 169–172 (2015)
9. Site of the manufacturer of industrial radio electronic components Murata. Available at: www.murata.com/en-global/products/sensor.
10. Site of the manufacturer of high-precision radio-electronic components Analog Devices. Available at: www.analog.com/en/products/mems.html
11. O.S. Salychev, *Applied inertial navigation: problems and solutions* (M.: BMSTU Press, 2004)
12. M. A. Basarab, A.V. Fomichev, The bulletin of MGTU of N.E, **3**, 80–96 (2014)
13. S.O.H. Madgwick, A.J.L. Harrison, R. Vaidyanathan, Proceedings of the IEEE International Conference on Rehabilitation, 1–7 (2011)
14. Power spectral density. Available at: http://ocw.mit.edu/courses/electrical-engineering-andcomputer-science/6-011-introduction-to communication-control-and-signal-processing-spring-010/readings/MIT6_011S10_chap10.pdf
15. Angle random walk. Available at: http://www.moogcrossbow.com/Literature/Application_Notes_Papers/Angle_Rom_Walk_Estimation_for_Rate_Gyros.pdf
16. R.V. Meshcheryakov, V.P. Bondarenko, Cybernetics and Systems Analysis, **44** (2), 175–184
17. V Budkov, M Prishchepa, A Ronzhin, Pattern Recognition and Image Analysis, **21** (3), 458 (2011)
18. A Ronzhin, A Saveliev, O Basov, S Solyonyj, International Conference on Interactive Collaborative Robotics, 32–39 (2016)

Blockchain-based framework for ontology-oriented robots' coalition formation in cyberphysical systems

Nikolay Teslya* and *Alexander* Smirnov

SPIIRAS, 199178, 14th line, 39, Russia

Abstract. The creation and functioning of intelligent robots coalition is impossible without the organization of an information interaction environment between them. To solve this problem, it is proposed to use the concept of cyberphysical framework, which involves the unification of physical and cyber (virtual) environments. Intelligent robots can interact with each of the environments. The paper proposes methodology of the cyber physical smart space creation for the intelligent robots' coalition formation and functioning, based on the concept of the "blackboard" with the support of smart contracts over the blockchain technology. It provides a new approach to the distribution of sensory, computational, information-control and service tasks between intelligent robots, embedded devices, fixed service equipment, cloud computing and information resources.

1 Introduction

The development of information technology and robotics in particular, achieved to date, allows to develop robots that are able to make independent decisions in unknown situations. It becomes available by equipping robots with a large number of various sensors to check the physical environment parameters, actuators to make actions in it and the development of computational algorithms that allow making decisions based on information received from robots' sensors and external information sources.

At the same time, the task remains open to organize the joint work of various robots to solve a common problem under current conditions. To date, there are a large number of various solutions dedicated to the joint work performed by a group of robots. Among them, such models of joint work can be distinguished, like swarms, schooling and collective (coalition) work. The first two models are based on imitating swarms or schools peeped in wildlife, i.e. fish schools on insect swarms. Robots in such models have simple instructions that allow them to make decisions based on a limited amount of information provided by nearest neighbors.

In a collective robot interaction model, each robot is viewed as a separate entity or agent, which allows the use of approaches developed for multi-agent systems [1-3]. Agents can form a coalition in which joint actions will be planned for the solutions of the problem facing the robots. Each agent in the coalition operates independently of the others, but within the scope of the developed plan and taking into consideration the current context, which provides possibility to make maximum use of the capabilities of individual robots to achieve a common goal. The robots' coalition formation with the distribution of tasks among all coalition members is an NP-complete task. To date, many solutions have been proposed that provide an approximate solution to the coalition forming task [2, 4-7]. Another important task in the robots coalition formation is the organization of an information interaction environment between them, which allows information exchanging about the current situation and the distribution of tasks among the participants in the coalition.

The purpose of this paper is to develop a methodology for creating a platform where robots are viewed as separate independent entities – agents interacting with each other in cyberphysical space and able to unite in a coalition to solve a common complex problem. During a coalition creation, robots negotiate among themselves spreading knowledge they know about the task being solved, their current state and the state of the environment. The negotiation process is carried out in a common information space, and the result of negotiations is the coalition in which the common complex problem is divided to tasks distributed among the coalition participants. The tasks are bond to the robots, as well as the required resources, and the remuneration for solving the problem. Assignment and binding of tasks, resources and remuneration is proposed to be carried out using a smart contract that is based on the blockchain technology.

The methodology presented in this paper is described in the following sections. Section 2 includes the description of creating a smart space for robots interaction. Section 3 provides the basis of smart contract and blockchain technologies as well as the way of interaction organization through smart contracts. Section 4 describes the integration of smart space and smart contract technologies for organizing trusted interaction between robots.

* Corresponding author: teslya@iias.spb.su

2 Organization of a smart space for robots' interaction

The formation and functioning of an intelligent robots coalition is impossible without the organization of an information interaction environment between them. To date, many solutions have been proposed in the field of collective decentralized robot control systems. In works [2, 5, 6, 8, 9] models and algorithms for solving problems of optimal distribution of goals between robots of the group are presented, as well as algorithms for collective control of robots in conditions of active counteraction from the environment. Original methods for solving the problem of controlling large groups of robots are proposed. Also presented is the development of a decentralized production management model in which the dispatcher is distributed among program agents and solves the task of scheduling a plan for the performance of a production task [3, 4, 8, 10].

To organize an information interaction environment between robots, it is proposed to use the concept of cyberphysical systems that is actively developed now. It involves the integration of physical and cyber (virtual) environments. Intelligent robots can interact with each of the environments using their hardware and software. In the physical environment, the interaction is effected through sensors and actuators that robots are equipped. With their help, the robot can track the development of the current situation (its context) by measuring the physical characteristics of the environment and objects in it, moving around in environment and performing various operations (for example, measuring, capturing, moving) over physical objects in the environment. To exchange data received from the sensors, current actuators' status and robots' roles, and actions in the coalition it is proposed to use the smart space concept. This concept is based on the "blackboard" information sharing model [11, 12], which assumes the availability of a public information object where it is possible to write, read, and search for information without any restriction. This approach makes it possible to inform all coalition members about the context of the current situation in order to access the extent to which the overall goal of the coalition has been achieved and to analyze the change in the environment for a timely response. The computation power over the cyberphysical framework at all can be expanded by adding powerful computing devices on which additional processing of information can be done in order to accelerate the decision-making on the formation and functioning of the coalition.

The semantic interoperability between robots during the interaction through the smart space is supported by using an ontology that contains the main characteristics of the robot with their current values, robot resources such as time to the maintenance, current battery charge, type and amount of consumable resources that is depended on the problem area (for example, in medicine – the amount of available drugs, in agriculture – the supply of fertilizers, and in aerospace – the fuel supply). In addition, for each robot, the competences are defined, such as tasks that can be handled by the robot, and patterns of its behavior. The description of the robot competences of is necessary for the formation of the list of tasks with which it is able to cope.

It should be noted that in spite of all the advantages of the cyberphysical systems concept while ensuring interaction of robots, the problems associated with ensuring the control of the process of performing tasks by individual robots remain topical. This control includes the assignment of specific tasks for robots, the description and monitoring of the implementation of the norms of the coalition, the distribution of scarce resources, and the distribution of remuneration for the solved task.

3 Smart contracts in the robots interaction

The requirements presented in the end of previous section can be satisfied by using the blockchain technology and smart contract concept. The properties of the blockchain and smart contract that allows meeting the requirements are presented below.

3.1 Blockchain

Blockchain was originally considered as a distributed transaction ledger for keeping records of operations with the cryptocurrency [13], but it can also be used for other purposes. It is a chain of transaction blocks containing a header and a list of transactions. The header specifies its own hash, hash of the previous block, hash of transactions and additional service information. To calculate transaction hashes, hashing algorithm based on the Merkle Tree is using [13, 14]. Thus, due to linking with the previous block by hash and calculating of the total hash for several transactions this technology can keep the data in block immutable (Fig. 1). The earlier block was created the harder to change data in it.

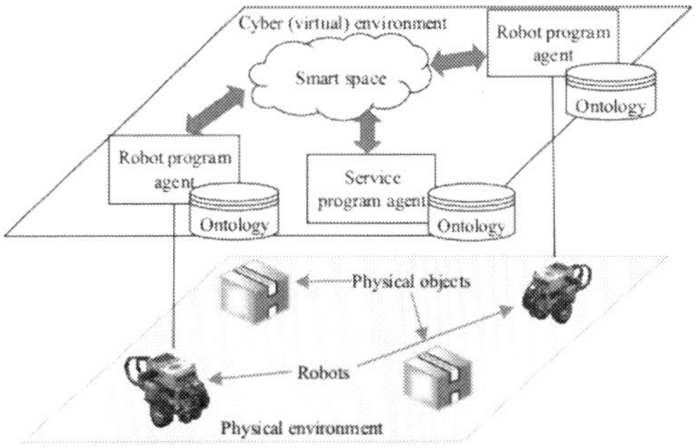

Fig. 1. Cyberphysical space for robots' interoperability.

In the simplest case, a transaction is a record that specifies the operation id and type, the operation itself, and the users participating in the specified operation [15]. For each user, an open-private key pair is usually formed on the registration stage, which is used to sign a transaction to unambiguously establish the ownership of the operation [16]. On the next step, to form a block in the blockchain, a hash function is calculated over all transaction information, and the hash value is then used to calculate the hash of the block. Thus, information about the transaction becomes also unchangeable and ensures the safety and durability of data storage in the blockchain.

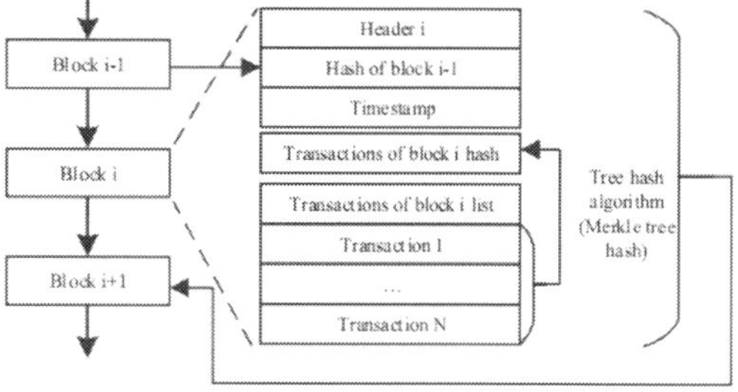

Fig. 2. Structure of the blockchain.

It should be noted that the transaction ledger is designed to be distributed in peer to peer way. This means that each blockchain user has access to the entire transaction log and can check any entry that it contains. Thus, it is possible to achieve a common consensus when carrying out operations, since each user can check the hash of the new block and determine the correctness of the transaction.

The built-in consensus mechanism allows to organize all published information, so that any protocols can be implemented on top of the blockchain, and any distributed business processes can be managed. For example, ownership of virtual resources with support for their transmission and atomic exchange can be realized. To do this, the following form of statements can be used:

1) "Participant X is an owner of resource A";
2) "Participant Y proposes to exchange resource A to their resource B from participant X (or any other, or from the list)";
3) "Participant X is agreed to the exchange of resource B to A from participant Y";
4) "Participant Y cancel the offer of exchange of resource A to B from participant X".

Statement "1" can be used for the transmission and creation of a resource. For atomic exchange, statement "2" can be used and followed by "3" or "4" in case of cancel the offer. It should be noted that users could publish statements without having the necessary resource. Such statements should be ignored, and it is necessary to track who owns what resources, considering the entire history of the transactions to determine the invalid statements. The above conceptual process is typically implemented based in a smart contract, whose ideas are presented below (Fig. 3).

To the date, the blockchain is mainly used as a basis for cryptocurrencies, for example, in platforms like Blockchain, Ethereum, LiteCoin, etc. However, during recent years there have been published works, the main purpose of which is the investigation of blockchain usage in other areas, such as supply chain [17], medicine [18], and IoT [19].

3.2 Smart contract

The idea of smart contract was proposed first in 1994 by Nick Szabo, even before the development of blockchain technology. N. Szabo had defined a smart contract in the following way: "A smart contract is a set of promises, specified in digital form, including protocols within which the parties perform on these promises." [20]

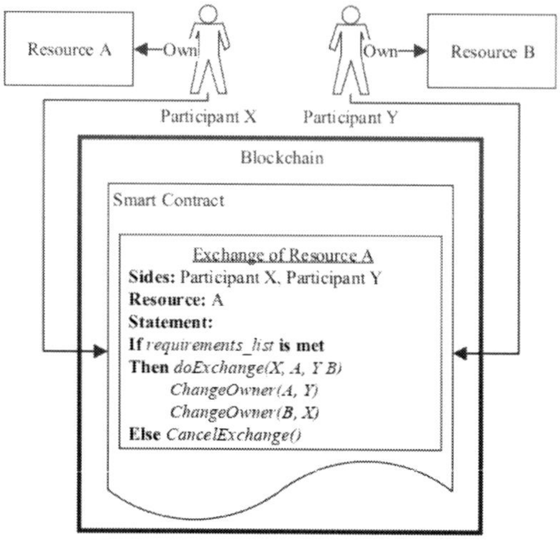

Fig. 3. Smart contract usage example.

Smart contract in scope of the blockchain technology can be viewed as a decentralized application available to all users of the blockchain. Due to the use of the Turing-complete language, the description of the contract code allows implementing of rather complex algorithms. At the same time, it is mandatory to have conditions under which the contract must be executed, and the list of actions assigned to the submitted conditions. All conditions of a smart contract must have a mathematical description and a clear execution logic. In this regard, the first smart contracts in the chain of blocks are created to formalize the simplest relationships, and consist of a small number of conditions.

Users sign a contract by using their private key [16] and send signed contract as a transaction that is written to the chain of blocks. After signing by all contract sides, the smart contract comes into force. To ensure the automated performance of contract obligations, an environment of existence is required that allows fully automating the execution of contract items. This means that smart contracts can only exist within an environment that has unrestricted access to executable code of smart contract objects. Having unimpeded access to the objects of the contract, the smart contract monitors the specified conditions of achievement or violation of the points and makes independent decisions based on the programmed conditions. Thus, the main principle of a smart contract is the complete automation and reliability of the performance of contractual relations between people.

Some blockchain platforms that provide environment for smart contracts are listed below:

1) Bitcoin [13]. It should be noted that this platform uses not a Turing-complete language and has strict restrictions on the format of the contract

2) Ethereum [21]. Contracts can be created using the internal program language – Solidity.
3) Hyperledger Fabric [22]. Platform leverages container technology to host smart contracts called "chaincode" that comprise the application logic of the system.

3.3 Robots interaction through smart contracts

The main way of robots interaction through blockchain technology, is the use of smart contracts. Since in the previous section it was noted that the chain of smart contracts always starts with a basic contract describing the simplest relationship, then for the interaction of robots such a contract will provide the basic principles of interaction, such as transferring a resource for use, paying a reward, receiving and performing a task. During the negotiation process robots can form additional contracts, using various combinations of basic contracts (Figure 4).

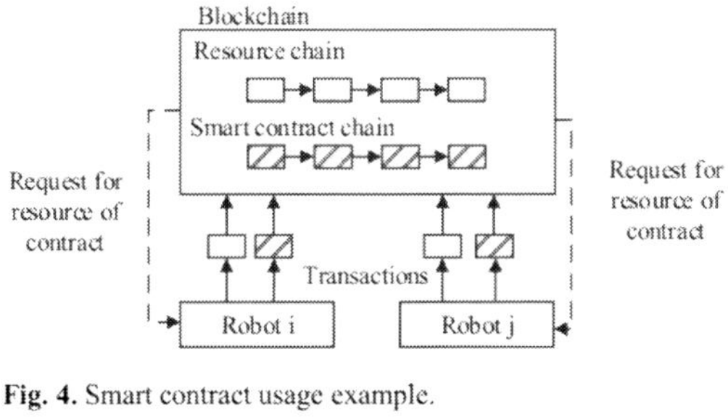

Fig. 4. Smart contract usage example.

When the robot is connected to the blockchain network, its main function is to the transfer of the transaction, which contains its requirements and competencies, drawn up in the form of a smart contract code. A new contract is created when forming the coalition based on combination of robots contracts and chain's base contracts. The condition of the new contract is the fulfillment of a certain task, and the body – the expected reward for the robot.

Resource allocation can be carried out in a separate chain of blocks in the blockchain. The resource is represented in the form of tokens that plays a role of a replacement for the resource in blockchain. The number of tokens is coincides with the available amount of the resource, while one token exactly corresponds to the resource unit and can not be divided into parts. The allocation of resources between robots can be carried out in a similar manner to the cryptocurrency in the Ethereum or Bitcoin system, but using the Proof-of-Stake consensus algorithm that does not require a "mining" procedure [25].

Contracts recorded in the blockchain remain immutable, which allows to track the sequence of robots' actions in the coalition and, for example, to correct the behavior of the robot to improve the effectiveness of its actions in the coalition. Also, by monitoring the time of the robot operation, it is possible to estimate the time to maintenance and pre-contract with the service organization for repairing the robot, which in the future will reduce the downtime of the robot due to breakdowns. Tracing the resources used by the robot will allow distribution of the resources according to the needs of the robot and to adjust the algorithm for forming a coalition to increase the efficiency of resource use.

4 Framework for robots interaction based on the smart space & smart contract concepts

The creation of a cyberphysical framework for the interaction of intelligent robots is based on the integration of the methods presented in the previous sections. This allows to create a system that combines advantages and closes the shortcomings of each of the concepts used. The conceptual scheme of the cyberphysical framework for the interaction of robots is presented in Figure 5. The framework is based on the organizing a smart cyberphysical space. It provides the ability to organize basic interaction of robots in the physical and cyber (virtual) spaces. The basic interaction includes solo and joint manipulations of robots with physical objects, information exchange about the current state of robots and objects for planning further joint actions during the coalition formation.

In the organization of information interaction of robots, heterogeneity and distribution of the cyberspace should be taken into account that is raised due to a large number of different robots and sensors, from different manufacturers and with different software, connected through a wireless data network. To provide interoperability, the information exchange between robots is represented using ontologies. Such approach allows to expand the context by providing additional information about the semantic meaning of individual context parameters or robot characteristics. Moreover, within the main idea of the approach, a common ontology can be developed for the whole system, or ontologies of individual components can be used, and information exchange between them is carried out using ontology matching methods [26, 27]. The concept of "blackboard" in this case allows providing all robots a common information space through which they can distribute information. All robots have equal access rights to read and write information in cyberphysical space.

The use of blockchain technology in this method is explained by the need of information exchange that requires long-term storage and the impossibility of unauthorized changes. These properties of the blockchain are required to solve several tasks at once, namely: monitoring the implementation of the basic norms of coalition interaction and agreements recorded in the smart contracts, as well as control over the expenditure of resources, including the basic resources of the system, specific to the problem area of the problem being solved. To do this, several independent chains are created in the blockchain, storing the distribution state of each of the available resources. In addition, a separate chain of blocks uses smart contracts to store the basic norms of behavior in the coalition and agreements formed based on the norms for the distribution of tasks among the coalition members.

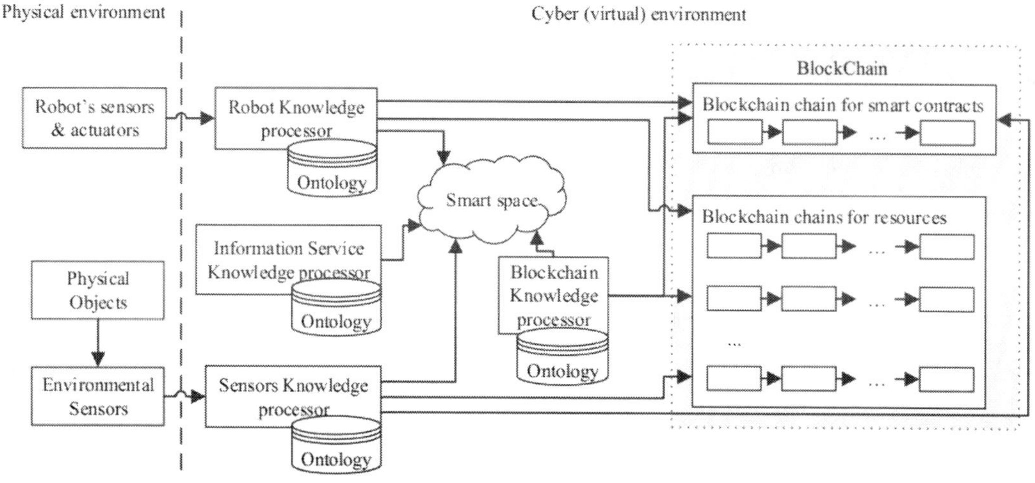

Fig. 5. Cyberphysical framework with blockchain support.

When an agreement is concluded between robots in a coalition, its conditions are transformed into a smart contract that is placed in a new block of a separate contract-oriented chain. The terms of the contract are the fact that the task or action is performed on a particular resource within the designated time range. In the case of this contract, the robot receives a bonus, which is to increase the competence parameter associated with the task being solved. In case of non-fulfillment of the contract, the robot gets a lowering of the relevant competence, and the other participants receive an automatic notification about faults to adjust the plans for performing their tasks.

Robots interact with the smart space and the blockchain by using of special software called knowledge processor (KP) that is located in the robot control block. KP realize the functions of processing information from physical components of robots, as well as their mapping into cyber space. In addition, with the help of KP, robots interact with each other during the formation of a coalition. KP also has access to the blockchain to obtain information about the amount of available resources required to solve the problem and obtain the required amount, according to the corresponding smart contract.

In addition to robots KP in cyber (virtual) space, there are KPs available for physical and cyber resources that perform control and timely informing all participants of cyberphysical system about changes in available resources and their characteristics. Resource characteristics changes are displayed in the smart space, while a change in the available amount of resources is displayed in the corresponding blockchain chain, which allows tracking its consumption and arrival. In order to ensure trust when changing the amount of resource, access to the blockchain is provided by all KPs directly with using the private-public key pair.

To solve the service tasks related to processing and storing large data sets from sensors and robots in the cyber space, the Information service and Sensors KPs are allocated. In addition, it is possible to interoperate the blockchain and smart space networks with blockchain KP. This KP is necessary for automatic control over the performance of smart contract conditions based on information located in smart space.

5 Conclusion

In the work the methodology of creation of the cyberphysical smart space, allowing formation and functioning of coalitions of intelligent robots is proposed. The methodology is based on the use of the concept of a "blackboard" to ensure the prompt exchange of information between coalition members and smart contracts in blockchain technology for the distribution of sensors, network, computing, information management and service tasks between intelligent robots, embedded devices, and equipment. To ensure the interoperability of heterogeneous participants in cyberphysical space, an ontology can be used that represents knowledge and competencies of robots in the system.

The use of blockchain technology and smart contracts in the proposed approach has allowed to provide immutable distributed storage of transactions about robot actions and resource allocation. This kind of storage is important on the negotiation stage when the coalition is formed to bind separated tasks with concrete coalition participants as well as on the execution stage to check the tasks execution process.

The present research was supported by the projects funded through grants # 17-29-07073, 17-07-00247 and 17-07-00327 of the Russian Foundation for Basic Research.

References

1. A. Kalyaev, Procedia Comput. Sci., **18**, 2193–2202 (2013)
2. J. Guerrero, G. Oliver, O. Valero, PLoS One, **12**, 1–26 (2017)
3. A. Kalyaev I. Kalyaev, M. Khisamutdinov, I. Korovin, G. Schaefer, 5th Int. Conf. on Informatics, Electronics and Vision (ICIEV), 1140–1142 (2016)
4. E. Ayari, S. Hadouaj, K. Ghedira, 18th Int. Conf. on Advanced Robotics (ICAR), 250–255 (2017)
5. C. Mouradian, J. Sahoo, R.H. Glitho, M.J. Morrow, P.A. Polakos, 13th Int. Wireless Communications and Mobile Computing Conf. (IWCMC), 1909–1914 (2017)
6. A. Dukeman, J.A. Adams, Auton. Agent. Multi. Agent. Syst., **31**, 1424–1466 (2017)
7. M. Agarwal, N. Agrawal, S. Sharma, L. Vig, N. Kumar, Expert Syst. Appl., **42**, 7797–7811 (2015)
8. M. Padmanabhan, G. R. Suresh, 3rd Int. Conf. on Signal Processing, Communication and Networking (ICSCN), 1–5 (2015)
9. A. Smirnov, A. Kashevnik, M. Petrov, V. Parfenov, Interactive Collaborative Robotics (ICR 2017), **10459**, 229–238 (2017)
10. J. Guerrero, G. Oliver, Rob. Auton. Syst., **60**, 1295–1307 (2012)
11. S. Boldyrev, I. Oliver, J. Honkola, 1st Int. Work. Manag. Data with Mob. Devices (MDMD 2009)(Milan, Italy), 1–10 (2009)
12. J. Honkola, H. Laine, R. Brown, O. Tyrkkö, Proc. IEEE Symp. on Computers and Communications, 1041–1046 (2010)
13. S. Nakamoto, www.bitcoin.org, 9 (2008)
14. R.C. Merkle, Conference on the Theory and Application of Cryptographic Techniques, 369–378 (1987)
15. D. Drescher, *Blockchain basics: a non-technical introduction in 25 steps.* (Apress, 2017)
16. O. Goldreich, *Foundations of cryptography I: Basic Tools,* (Cambridge: Cambridge University Press, 2001)
17. S.A. Abeyratne and R. P. Monfared, Int. J. Res. Eng. Technol., **5** (9), 1–10 (2016)
18. M.B. Hoy, Med. Ref. Serv. Q., **36** (3), 273–279 (2017)
19. X. Jia et al., Framework of blockchain of things as decentralized service platform, 1–23 (2017)
20. N. Szabo, *Smart Contracts: Building Blocks for Digital Markets Copyright,* alamut.com, (1996). URL: http://www.fon.hum.uva.nl/rob/Courses/InformationInSpeech/CDROM/Literature/LOTwinterschool2006/szabo.best.vwh.net/smart_contracts_2.html (accessed: 16-Jan-2018)
21. Ethereum, URL: https://www.ethereum.org/ (accessed: 16-Jan-2018)
22. Hyperledger Fabric, URL: https://hyperledger-fabric.readthedocs.io/ (Accessed: 16-Jan-2018)
23. W. Xie, Y. Shi, G. Xu, Y. Mao, Proceedings. Fourth IEEE International Conference on Multimodal Interfaces, 429–434 (2002)
24. D.G. Korzun, S.I. Balandin, V. Luukkala, P. Liuha, A.V. Gurtov, *Proc. Int. Conf. Artif. Intell. Intell. Syst.*, **4**, 64–67 (2011)
25. I. Bentov, C. Lee, A. Mizrahi, M. Rosenfeld, Cryptol. ePrint Arch., **452**, 1–19 (2014)
26. A. Smirnov et al., Smart Spaces and Next Generation Wired/Wireless Networking, **6294**, 72–83, (2010)
27. J. Euzenat, P. Shvaiko, *Ontology Matching,* (Springer Berlin Heidelberg, 2013).

Active stabilization in robotic vision systems

Vladimir Serebrennyi, Andrei Boshliakov and Georgii Ovsiankin*

Bauman Moscow State Technical University, Mechanical Engineering Department, 105005 ul. Baumanskaya 2-ya, 5, Moscow, Russia

Abstract. In the article considered prospective approaches to the design of active systems for stabilizing systems based on the parallel kinematics mechanism and possible applications of such systems. Attention is drawn to the fact that not only object fluctuations are an important object for stabilization, but it is also important to compensate for the body vibrations, along with its vibrations. Based on the analysis, it was concluded that it is perspective to use mechanisms with parallel kinematics for the design of active stabilization systems. Was obtained a mathematical model of the hexapod, according to which a computer model in the Simulink package was designed. Its analysis confirmed the possibility of using a mechanism with parallel kinematics in designing an active stabilization system and presented requirements to the actuators of the system.

1 Introduction

At present, optical systems are widely used to analyse and assess the current situation on the ground. At the same time, the requirements imposed on the quality of the image obtained from the equipment on such systems have increased. The requirements for the accuracy of optical systems can reach units of angular minutes. As a consequence, the task of stabilization comes first. The field of application of systems, where high demands are placed on the quality of the information received, is quietly wide.

2 Existing approaches

In remote earth sensing systems, high demands are placed on the quality of the resulting image. This is due to the growing popularity of automated intelligent analysis of aerospace information. The moving parts of the spacecraft create structural vibrations, which, acting on the mirror, degrade the quality of the resulting image [1].

In the aero cartography problem perspective approaches are (Fig. 1):

- Laser scanning
- Creation of digital models of terrain and objects
- Photographing

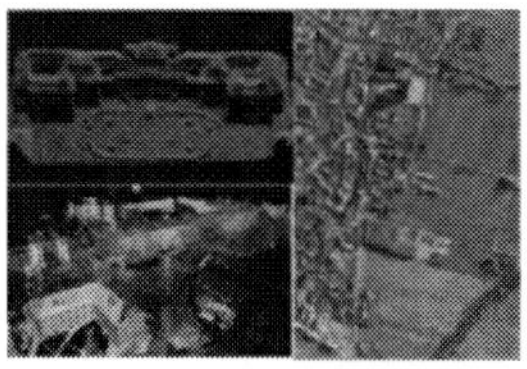

Fig. 1. Laser scanning. Creation of digital models of terrain and objects.

A serious impact on the quality of the resulting image is due to the housing vibrations generated by the engine operation. Vibration data due to the high altitude of the flight of the aircraft (100 ... 2500 m) lead to a shift in the pixels of the CCD matrix of photo equipment installed on board the aircraft. According to GOST 26382-84, the permissible levels of vibro-displacement lie in the interval 63...437 um at frequencies of 20 ... 100 Hz. Therefore, it is possible to form the following requirements for advanced high-precision stabilization systems

- accuracy of stabilization - units of angular seconds
- range of operating frequencies - up to a hundred hertz
- stabilization in three axes

Actual question is increasing the accuracy of existing systems, both by designing new ones, and increasing the accuracy of the existing ones. An important part of the stabilization problem is not only compensation of the body oscillations, but also of its vibrations. For this task, a perspective approach is to use the mechanisms of parallel kinematics, which make it possible to design an active system of three-dimensional stabilization. At the moment, studies are underway to build stabilization systems that differ in the principle of operation, the number of stabilized axes, the limiting values of deviations and other characteristics.

Design of a stabilization system for radio telescope with segment mirrors [1] (Fig. 2).

The diameter of each mirror is about 3 meters, on each of them is installed a system of passive vibration damping. Tuned mass dampers (TMDs) were selected to mitigate the effects of the Segmented Mirror Telescope segment vibration modes because of their ability to add significant damping to resonant frequencies as add-on passive devices without changing the original optical load path. These devices have previously been shown to be effective on telescope structures with image jitter due to resonant

* Corresponding author: george.ovciankin@gmail.com

vibration. Figure 3 shows the effect in the frequency domain of a typical TMD on a resonant vibration mode of an idealized structure, along with the associated time domain response with and without the TMD.

Fig. 2. Segmented Mirror Telescope.

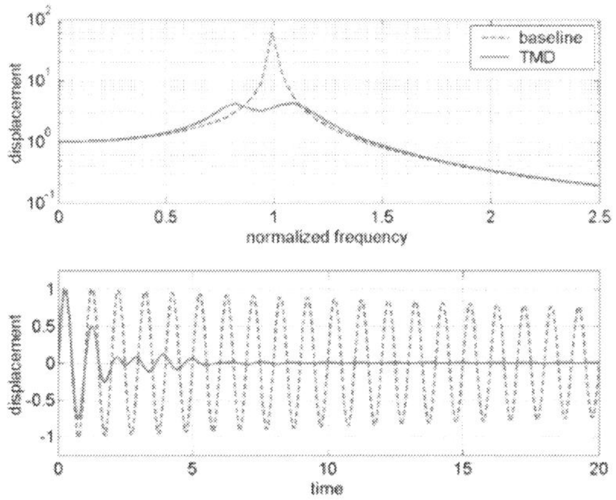

Fig. 3. Comparison of frequency and time domain responses with and without TMD.

Within work [2] a model of a 6 DOF (degree-of-freedom) vibration isolation system with semi-active control, using magnetorheological (MR) technology is investigated. The geometry of the parallel platform is determined in a way which maximizes the Quality Index. Parallel platform mechanisms are ideal candidates for 6 DOF positioning and vibration isolation. A 6 DOF parallel platform which utilizes semi-active vibration control has not received as much attention as its passive and active counterparts have. Each leg of the platform is modelled as a two DOF system with a magnetorheological (MR) damper for adjustable damping. The class of semi-active drives is considered in detail in [3].

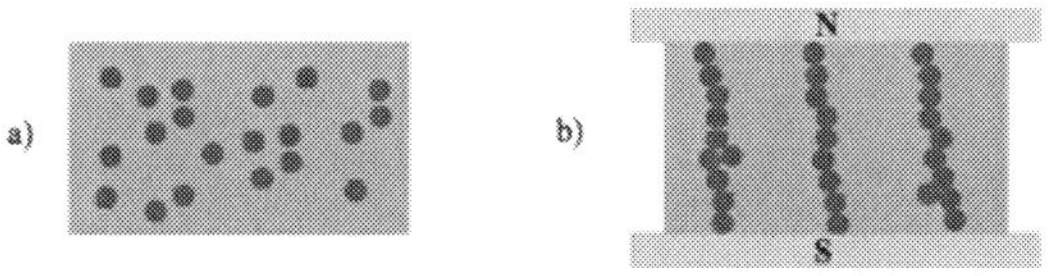

Fig. 4. Magnetorheological fluid: a) no magnetic field, b) with magnetic field.

The authors analysed the design of a parallel manipulator and selected a model for the legs (Fig. 5) and analysed the stabilizing properties of the received leg and semi-active control (Fig. 6).

a)

b)

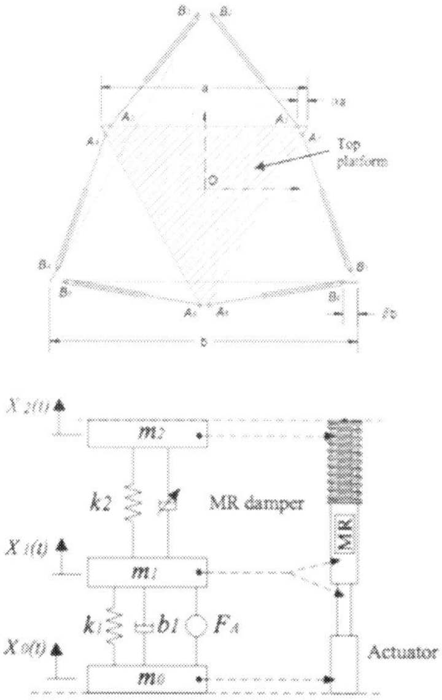

Fig. 5. Two DOF platform leg model with MR damper.

The MR damper was modelled in Simulink. The input to the model is the relative velocity between the two masses which the damper connects and the control current. In the simulations, the MR damper is assumed to respond very fast and the time delay associated with the damper's response time is ignored.

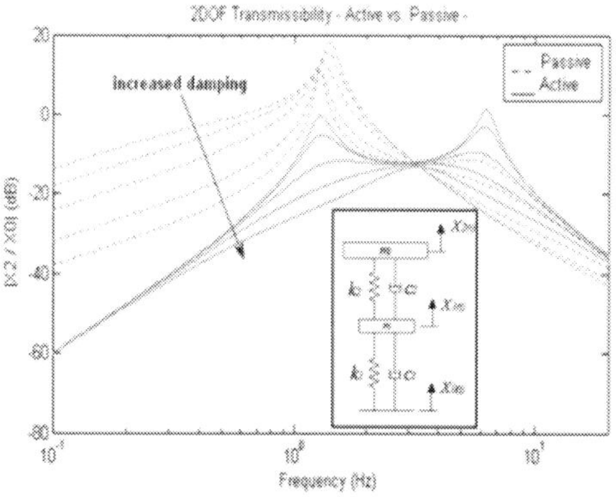

Fig. 6. Transmissibility of the two DOF mass-spring- damper system with passive and active skyhook damper control

Papers [4, 5] disclose a stiff active interface wherein a six degree-of-freedom Stewart platform, a standard hexapod with a cubic architecture, is used to actively increase the structural damping of flexible systems attached to it. It can also be used to rigidly connect arbitrary substructures while damping them.

Each leg of the active interface consists of a linear

piezo electric actuator, a collocated force sensor and flexible tips for the connections with the two end plates. By providing the legs with strain or elongation sensors, this active interface can also be used as an interface with infinite stiffness at low frequency (i.e. for machine tools), a 6 DOF positioning and steering device for space applications as well as a micro vibration isolator (Fig. 7).

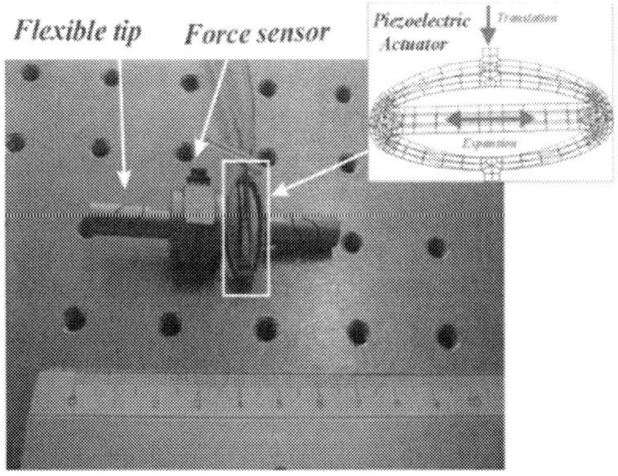

Fig. 7. The active leg assembly of Stewart platform.

To avoid the problems of friction and backlash in the joints, flexible tips were used instead of spherical joints.

These flexible tips have zero friction, zero backlash, high axial stiffness and relatively low bending stiffness. It will be shown that the existence of this bending stiffness makes a limitation for the control authority because it shifts the transmission zeroes to a higher frequency, which will decrease the damping effect expected from each closed-loop pole. (Fig. 8 a, b).

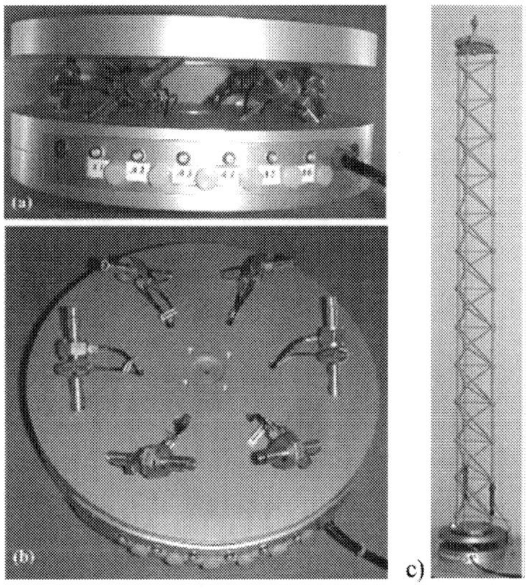

Fig. 8. The Stewart platform (a): complete hexapod, (b): the hexapod with the upper plate removed, (c): experimental setup.

In order to evaluate the damping performances of the interface, the hexapod is connected to a flexible payload, a 150 cm long steel truss structure (Fig.8 c). A decentralized integral force feedback control scheme was applied to the experimental system. The six independent controllers have been implemented on a DSP board. The six loops have been closed separately and, although the control loops were independent, the feedback gains used in all the loops are identical. Figure 9 presents some experimental results. The time response shows the signal from one of the force sensors located in the legs, the truss is subjected to an impulse at middle height, without and with control.

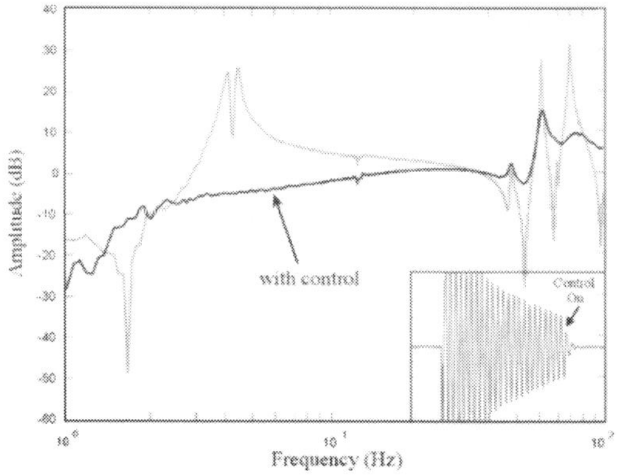

Fig. 9. Experimental time response and frequency response function of the truss mounted on the active interface.

In work [6] the authors propose approach to designing a semi-active stabilization system for space vehicles. In the design of space telescopes, massive primary mirrors are very often used, which can be damaged during the launch phase of the spacecraft. The authors propose a multilevel version of the design of a stabilization system, for stabilizing the structural resonances of the system. Several damping and insulation approaches are discussed, including damping of the viscoelastic restrained layer and damping of the magnetic tunable mass of the mirror segment, passive isolation of spacecraft disturbances and control of the indicators of the active optical telescope. As the actuator of the semi-active stabilization system, the Stuart platform was chosen, and a drive was developed to control the system (Fig. 10).

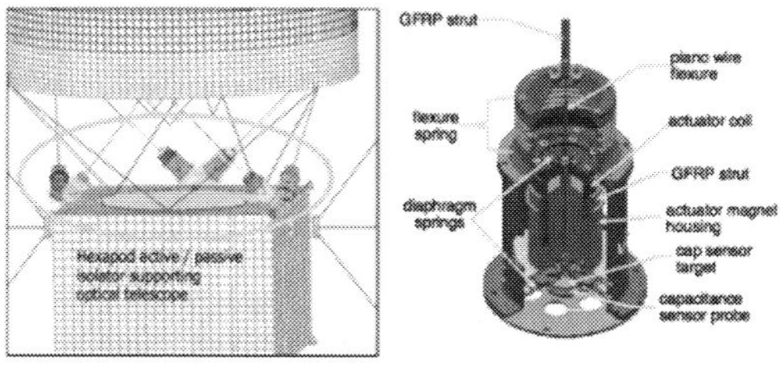

Fig. 10. Implementation of a semi-active stabilization system.

Designed long-stroke actuator for this isolator study has demonstrated stroke capability of +/- 7mm, which results in significant dynamic range performance and payload steering capability. The actuator design features a voice coil as the active element, diaphragm springs to align the voice coil, and a flexure spring that provides tunable passive isolation capability (see Fig. 11). This design can accommodate a capacitance sensor probe as the feedback element, which can be used for telescope positioning in addition to or instead of optical instruments or inertial sensors at the telescope.

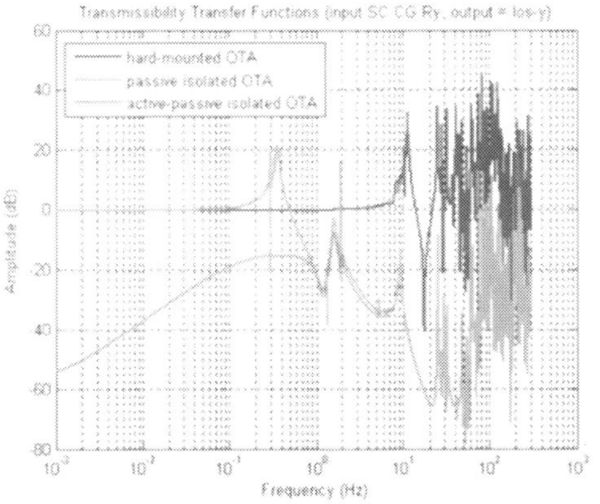

Fig. 11. Transmissibility plots for inputs at spacecraft CG and outputs at telescope LOS for hard- mounted, passive-isolated, and active-passive isolated optical telescope assembly (OTA).

3 Stabilization system model

A survey of promising studies on the designing of active stabilization systems has shown that the basic approach to designing is the use of mechanisms with parallel kinematics [7, 8]. Such mechanisms perceive the load as the spatial trusses, which increases their accuracy characteristics, and the carrying capacity of the entire system. At the given values of the load, the legs of the mechanism are much easier to obtain than the links of a similar mechanism with an open kinematic chain. Thus, a rigid structure with light links makes it possible to significantly increase the speed and accuracy of the system. Therefore, based on the capabilities of such mechanisms, a conclusion was made about the possibility of using such a manipulator as an executive mechanism of the system. In [9], mathematical calculations were carried out, as well as modelling in the MatLab environment, which shows the possibility of fulfilling the requirements for a system of three-dimensional stabilization of optical equipment.

3.1 Mathematical model of the system

Let the generalized coordinates of the legs $q_1, q_2, q_3, q_4, q_5, q_6$ be the length $Aa, Bb, \dots, Ff$ coordinates, respectively. Suppose also that in a fixed Cartesian coordinate system $OXYZ$ the coordinates of the points A, B, C, D, E, F are equal (X_1, Y_1, Z_1), and the points $- a, b, c, d, e, f$ $- i(X_1, Y_1, Z_1)$, where $I \in \{A, B, C, D, E, F\}, i \in \{a, b, c, d, e, f\}$ (Fig. 12).

In the above notation, the solution of the inverse kinematics problem for hexapod (with both four and six degrees of freedom) is given by the following system of equations (1)

$$q_i = (X_{A_i} - X_{B_i})^2 + (Y_{A_i} - Y_{B_i})^2 + (Z_{A_i} - Z_{B_i})^2 \qquad (1)$$

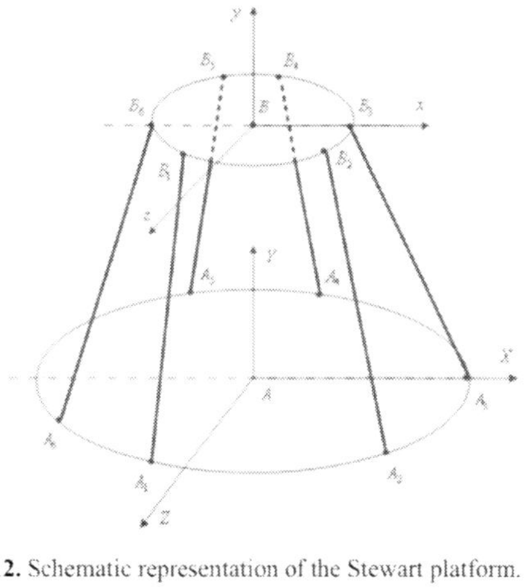

Fig. 12. Schematic representation of the Stewart platform.

The solution of the forward kinematics problem. Let the weightless legs A_iB_i $i = 1,2,\dots,6$, AB consist of two half-stems connected by translational kinematic pairs. The legs A_iB_i $i = 1,2,\dots,6$ are attached to the platform at points B_i with the help of spherical hinges, and to the base in points A_i - using universal joints; The leg AB is connected with the base motionless, and with the platform - with the help of a spherical hinge. Points B_i lie on a circle of radius r with centre at point B, in which the centre of mass of the platform is located. The base A_i is horizontal and the points are located on a circle of radius R with the centre at point A. The legs A_iB_i have lengths $l_i (l_i > 0)$ and are inclined to the plane of the base at angles γ_i $(0 < \gamma_i < \frac{\pi}{2})$. The distance between the point B and the base (the plane) is denoted h.

Position of joints $A_1, A_2, A_3, A_4, A_5, A_6$, $B_1, B_2, B_3, B_4, B_5, B_6$ in the coordinates system $AXYZ$ (Fig. 12) is determined by vectors (2)

$$A_i = (A_{i,1}, A_{i,2}, A_{i,3}),\ B_i = (B_{i,1}, B_{i,2}, B_{i,3}), i \in [1:6]. \qquad (2)$$

The position of the platform relative to the base is determined by the Euler angles $\varphi_1, \varphi_2, \varphi_3$, and vector $B = (X_B, Y_B, Z_B)$. Geometric attitudes between coordinates system $AXYZ$, $Bxyz$ are given in the transformation matrix$(4x4)$ - (3).

$$T = T(X_B, Y_B, Z_B, \varphi_1, \varphi_2, \varphi_3) = \begin{bmatrix} a_{1,1} & a_{1,2} & a_{1,3} & X_B \\ a_{2,1} & a_{2,2} & a_{2,3} & Y_B \\ a_{3,1} & a_{3,2} & a_{3,3} & Z_B \\ 0 & 0 & 0 & 1 \end{bmatrix} \qquad (3)$$

Where components $a_{i,j}$, $i,j = 1,2,3$ of matrix T are expressed in terms of Euler angles as follows (4)

$$a_{1,1} = \cos\varphi_1 \cos\varphi_3 - \sin\varphi_1 \cos\varphi_2 \cos\varphi_3 ;$$

$$a_{1,2} = \sin\varphi_1 \cos\varphi_3 + \cos\varphi_1 \cos\varphi_2 \sin\varphi_3 ;$$

$$a_{1,3} = \sin\varphi_2 \sin\varphi_3 ; \qquad (4)$$

$$a_{2,1} = -cos\varphi_1 cos\varphi_3 - sin\varphi_1 + sin\varphi_1 cos\varphi_2 cos\varphi_3;$$

$$a_{2,2} = -\sin\varphi_1 \sin\varphi_3 + \cos\varphi_1 \cos\varphi_2 \cos\varphi_3 ;$$

$$\begin{aligned} a_{2,3} &= \sin\varphi_2 \cos\varphi_3\,; \\ a_{3,1} &= \sin\varphi_1 \sin\varphi_2\,; \\ a_{3,2} &= -\cos\varphi_1 \sin\varphi_2\,; \\ a_{3,3} &= \cos\varphi_2. \end{aligned} \qquad (4)$$

Thus, position of joint $B_i, i \in [1:6]$ in coordinates system $AXYZ$ is determined by vector (5)

$$[\,B_i\,] = \begin{pmatrix} [\,B_{i,1}] \\ [\,B_{i,2}] \\ [\,B_{i,3}] \\ 1 \end{pmatrix} = T \begin{pmatrix} B_{i,1} \\ B_{i,2} \\ B_{i,3} \\ 1 \end{pmatrix} = T\,B_i \qquad (5)$$

Consequently, generalized coordinates l_i, as a function of quantities $X_B, Y_B, Z_B, \varphi_1, \varphi_2, \varphi_3$ determined by (6)

$$l_i = l_i(X_B, Y_B, Z_B, \varphi_1, \varphi_2, \varphi_3) = \\ = \sqrt{\sum (A_{i,j} - [B_{i,j}])^2}, \qquad (6)$$

$$j = 1,2,3\,, i \in [1:6].$$

The notation introduced, the angles between the leg A_iB_i, $i \in [1:6]$ and axes of the coordinate system $AXYZ$ is determined by (7)

$$\cos\gamma_{i,j} = \cos\gamma_{i,j}\,(X_B, Y_B, Z_B, \varphi_1, \varphi_2, \varphi_3\,) = \\ = \frac{[B_{i,j}] - A_{i,j}}{l_i}, \qquad (7)$$

$$j = 1,2,3,$$

where angles $\gamma_{i,1}$, $\gamma_{i,2}$, $\gamma_{i,3}$ - angles between leg A_iB_i and axes AX, AY, AZ .

3.2 Designing model in Matlab.

Simulation of the kinematics and dynamics of our system will be performed using Matlab. The Simulink model consists of 3 blocks (Figure 13). In the first block, "Platform", a model of the system is created, which consists of 2 platforms and 6 bars (Fig. 14).

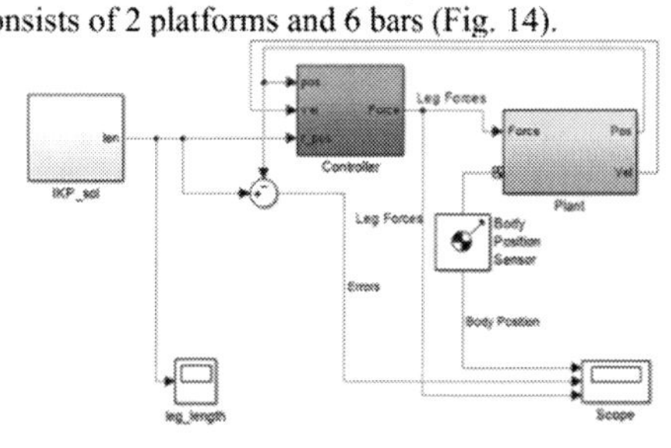

Fig. 13. General model of the system.

The platforms are implemented by the "body" block, in which are given the data on the system mass and coordinates of the points where the hinges connecting the platforms with the links (Fig. 14).

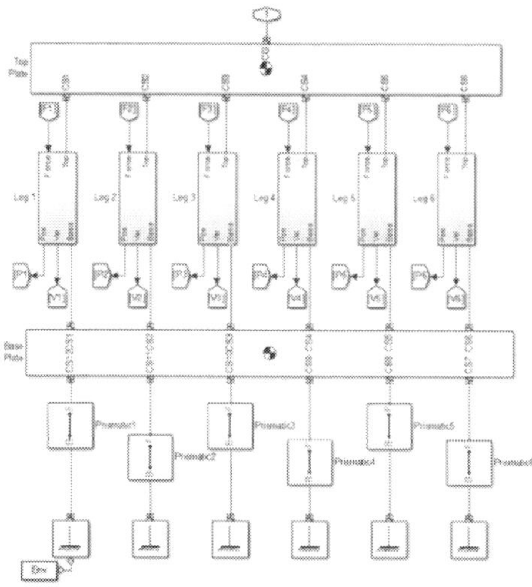

Fig. 14. Hexapod model in Matlab.

As a result of the simulation of the system, we obtain data of the position of the links, the speed, the difference between the desired and actual position of the platform (Fig 15).

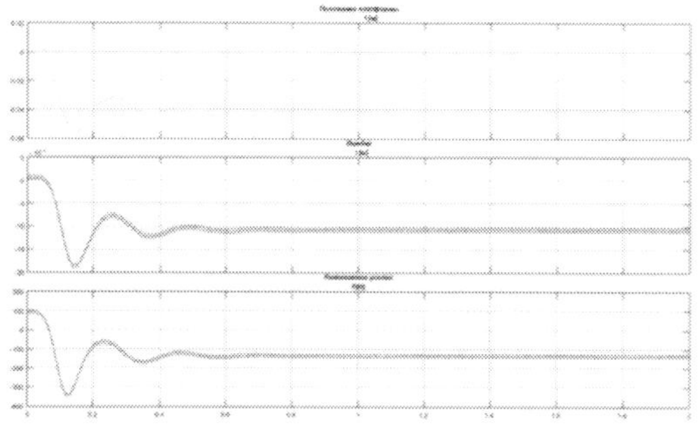

Fig. 15. Modelling results.

The first graph shows the position of each joint of the system. The second graph shows the error generated. The third graph shows the efforts that occur in each boom of the system. The results allowed to put forward demands on the executive modules of the stabilization system and conclude that the hexapod can be used as an executive device for the stabilization system.

4 Conclusion

The analysis of the development of high-precision active stabilization and guidance systems shows that it is promising approach to designing such systems is the use of mechanisms with parallel kinematics. It provides the possibility of an even load distribution to the executive modules of the stabilization system. The designed mathematical model of the system made it possible to make demands to the executive modules of the system that would provide the required quality of work.

References

1. J.R. Maly et al., Proc. SPIE 8450, Modern Technologies in Space- and Ground-based Telescopes and Instrumentation **II**, 845004 (2012)
2. M. Unsal, C. Crane, C. Niezrecki, Florida conference on recent advances in robotics (FCRAR), 25–26 (2006)
3. M. Elahinia, C. Ciocanel, S. Wang, Smart Materials Research, **2013**, Article ID 831017 (2013)
4. C. Wang, X. Xie, Y. Chen, Z. Zhang, 12th International Conference on Recent Advances in Structural Dynamics (2016)
5. Ir. A. Abu Hanieh, A. Preumont, proceedings of the 9th European Space Mechanisms and Tribology Symposium (2001)
6. V. Camelo, A. Bronowicki, S. Simonian, R. Hejal, S. Brennan, 50th AIAA/ASME/ASCE/AHS/ASC Structures, Structural Dynamics, and Materials Conference, **2009**, 2637 (2009)
7. Y. Liu, X. Han, J. Han, International Journal of Advanced Robotic Systems, **9** (2) (2012)
8. Y. Song, X. Sun, Appl. Sci. **7**, 393 (2017)
9. J.T. Kaganov, A.P. Karpenko, Science and Education of the Bauman MSTU, **11** (2009)
10. J.N. Artemenko, Science and Education of the Bauman MSTU, **01**, (2013)
11. J.N. Artemenko, S.N. Sajapin, Science and Education of the Bauman MSTU, **06**, (2013)
12. S.V. Volkomorov, J.T. Kaganov, A.P. Karpenko, Informacionnye tehnologii. Prilozhenie, **5**, 1–32 (2010)
13. S.V. Volkomorov, A.P. Karpenko, Science and Education of the Bauman MSTU, **12** (2010)
14. S.V. Volkomorov, A.P. Karpenko, A.M. Leletko, Science and Education of the Bauman MSTU, **08** (2010)
15. A.I. Galushkin, S.N. Sajapin, A.V. Sinev, 7 All-Russian Conference "Neirokompjutery i ih primenenie", 184–192 (2001)
16. A.L. Lapikov, V.N. Pashhenko, Science and Education of the Bauman MSTU, **12**, (2014)

Experiments on mobile robot stereo vision system calibration under hardware imperfection

Ramil Safin[1,*], *Roman* Lavrenov [1], *Subir Kumar* Saha[2] and *Evgeni* Magid[1]

[1] Laboratory of Intelligent Robotic Systems, Higher Institute for Information Technology and Information Systems (ITIS), Kazan Federal University, 35 Kremlyovskaya street, Kazan, 420008, Russian Federation
[2]Mechanical Engineering, Indian Institute of Technology Delhi, Hauz Khas, New Delhi-110 016, India

Abstract. Calibration is essential for any robot vision system for achieving high accuracy in deriving objects metric information. One of typical requirements for a stereo vison system in order to obtain better calibration results is to guarantee that both cameras keep the same vertical level. However, cameras may be displaced due to severe conditions of a robot operating or some other circumstances. This paper presents our experimental approach to the problem of a mobile robot stereo vision system calibration under a hardware imperfection. In our experiments, we used crawler-type mobile robot «Servosila Engineer». Stereo system cameras of the robot were displaced relative to each other, causing loss of surrounding environment information. We implemented and verified checkerboard and circle grid based calibration methods. The two methods comparison demonstrated that a circle grid based calibration should be preferred over a classical checkerboard calibration approach.

1 Introduction

Stereo vision is widely used in robotics for navigation, obstacle avoidance and path planning [1], object tracking, 3D scene reconstruction, simultaneous localization and mapping [2]. Calibration of a stereoscopic vision system is a necessary step in 3-D computer vision in order to achieve high accuracy in deriving metric information from 2D images such as distance between objects and their dimensions [3]. Calibration is used in order to estimate specific parameters of cameras, which help establishing proper correlations between 2D images and 3D world, and to determine lens distortion models.

Ideally, stereo vison system sensors should be perfectly aligned along the same horizontal line passing through the principal points (image plane center) of both images of the stereo cameras. Otherwise, misalignment could lead to the error growth in measurements [4]. However, in practice, we have to deal with initially presented or appeared later due to multiple reasons hardware imperfection issues, for example, due to severe operation conditions. As in general case it is not possible to solve this issue operatively, the calibration process becomes more complicated and error-prone.

This paper presents our experimental approach to the problem of a mobile robot stereo vision system calibration under a hardware imperfection. In our experiments, we used crawler-type mobile robot «Servosila Engineer» (Fig. 1). The left camera of the stereo pair was displaced after the robot had felt from a significant height during an experiment. Our goal was to verify whether it is possible to achieve acceptable stereo system calibration results in a presence of such hardware imperfection. We implemented and verified checkerboard and circle grid based calibration methods. The two methods comparison demonstrated that a circle grid based calibration should be preferred over a classical checkerboard calibration approach.

Fig. 1. «Servosila Engineer» crawler robot.

The rest of the paper is organized as follows. Section 2 presents mobile robot stereo vision system. Section 3 is dedicated to camera calibration approaches. Experiments on calibration with a checkerboard and a circle grid based methods are presented in Section 4. Finally, we conclude in Section 5.

2 Robot's vision-related features

Russian crawler-type mobile robot «Servosila Engineer» [5] is equipped with four cameras:

* Corresponding author: safin.ramil@it.kfu.ru

- a forward-view optical zoom camera;
- a wide angle of view rear-view camera;
- a pair of cameras for stereo vision.

Stereoscopic vision system consists of two monocular cameras which are located on the front of the robot's head (Fig. 2). Supported pixel formats and frame resolutions of the stereo system are described in Table 1.

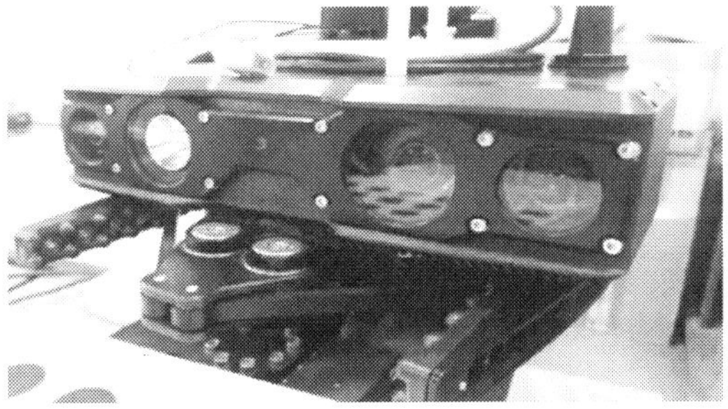

Fig. 2. «Servosila Engineer» robot stereo vision system.

Table 1. Stereoscopic vison system featuresof «Servosila Engineer».

Camera model	Image format	Supported camera resolution
DFM 22BUC03-ML	Bayer RGB	744x480 and 640x480

3 Camera calibration

3.1 Problem statement

Camera calibration is used in order to estimate extrinsic and intrinsic parameters and distortion coefficients of the camera systems (i.e., monocular or stereo) [6, 7].

Intrinsic parameters, which specify a camera itself, include internal optical characteristics of the camera:
- a principal point (location of an image center)
- a focal length (a distance between the camera lens and an image plane)
- a skew coefficient
- a radial and a tangential distortion coefficients of the camera lens

Extrinsic, or external, parameters relate the world coordinate system to the camera centered coordinate system. It is defined by a translation vector and a rotation matrix. Both intrinsic and extrinsic parameters enable us to transform objects of 3D space into a 2D image and vice versa. This process is referred as a perspective projection and it maps each 3D point to a 2D image plane [6].

For stereoscopic vision systems, intrinsic parameters are calculated separately for both cameras. However, external parameters are estimated for an entire stereo vision system, and thus spatial relations between the cameras are determined [8]. Obtained parameters are used to align images from both cameras along a same plane. This process is called stereo rectification and it is a crucial step in a case of misalignment of the cameras.

3.2 Calibration using 2D apparatus

To calculate camera parameters, a proper mapping between 3D world points and their corresponding 2D image points is required. This mapping could be obtained applying several images of a calibration pattern (e.g., a checkerboard). With regard to calibration object dimensions, approaches could be classified into four groups [3]:
- self-calibration, or zero-dimensional calibration
- one-dimensional object-based calibration
- two-dimensional plane-based calibration
- three-dimensional object-based calibration

The process of calibration using a 2D plane includes tracking of a planar pattern, which is located at different orientations relatively to the camera [9], and various patterns could be used (Fig. 3).

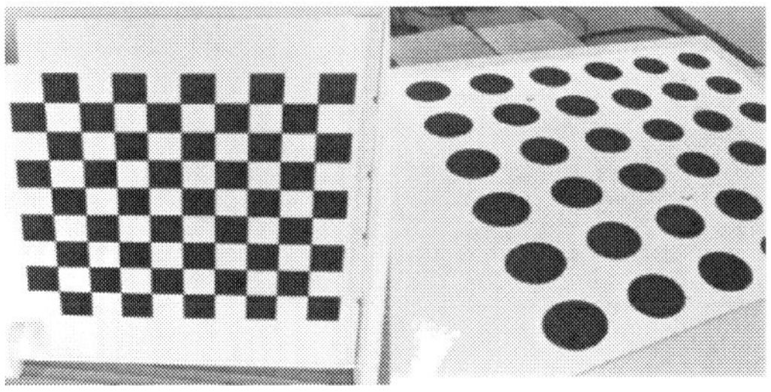

Fig. 3. Examples of 2D calibration patterns: a checkerboard (left) and a circle grid (right).

Comparing to other calibration approaches, calibration with 2D apparatus is easier to implement as 2D calibration pattern is easy to create. For example, self-calibration techniques are generally less accurate and require solving more complicated mathematical problem resulting in a more difficult calibration process. In contrast with 3D object-based calibration (Fig. 4), which uses up to three orthogonal to each other planes, 2D object-based calibration does not require an elaborate setup and has a relatively good calibration quality.

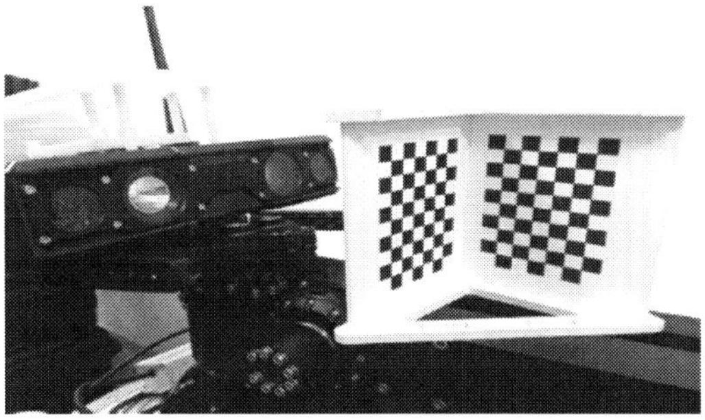

Fig. 4. 3D reference object-based calibration pattern.

3.3 Pinhole camera model

Most calibration methods are based on a pinhole camera model (Fig. 5) and deal with radial distortion model. The pinhole camera parameters are represented in a matrix that is referred as camera matrix.

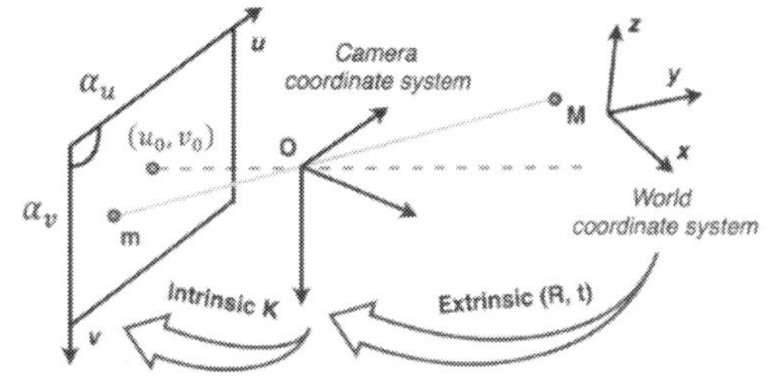

Fig. 5. Pinhole camera model.

In Figure 5 an optical center of the camera O denotes the origin of the camera's coordinate system and the axes of this coordinate system define an image plane. Position of a 3D point **M** and the coordinates of its corresponding projection to the image plane **m**, representing a 2D point, are denoted as

$$\mathbf{M} = [\mathbf{X}, \mathbf{Y}, \mathbf{Z}]^{\mathbf{T}}, \tag{1}$$

$$\mathbf{m} = [\mathbf{u}, \mathbf{v}]^{\mathbf{T}}. \tag{2}$$

The ray, which connects point **M** and the camera's optical center O, intersects the image plane at point **m**. The corresponding augmented vectors of 2D and 3D points are denoted respectively as

$$\widetilde{\mathbf{M}} = [X, Y, Z, 1]^T, \tag{3}$$

$$\widetilde{\mathbf{m}} = [u, v, 1]^T. \tag{4}$$

Intrinsic, or internal, parameters of the camera are defined as the following 3-by-3 matrix (intrinsic matrix):

$$\mathbf{K} = \begin{bmatrix} \alpha_u & \gamma & u_0 \\ 0 & \alpha_v & v_0 \\ 0 & 0 & 1 \end{bmatrix}, \tag{5}$$

where α_u and α_v denote focal length in u and v axes directions respectively, $(u_0,\ v_0)$ are the coordinates of the center of the image plane, and γ is a skew coefficient. If the image plane axes are orthogonal to each other, then the skew coefficient is equal to zero.

Extrinsic, or external, parameters of the camera represent a transformation from a 3D world coordinate system to a 3D coordinate system of the camera. It includes of rotation and translation:

$$[\mathbf{R} \quad \mathbf{t}] = \begin{bmatrix} r_{11} & r_{12} & r_{13} & t_1 \\ r_{21} & r_{22} & r_{23} & t_2 \\ r_{31} & r_{32} & r_{33} & t_3 \end{bmatrix}, \tag{6}$$

where **R** is a rotation matrix, **t** is a translation vector.

A 3-by-4 matrix **C**, consisting of both extrinsic and intrinsic parameters (Eq. 5, 6), is called camera projection matrix, or camera matrix, and it could be defined as follows:

$$\mathbf{C} = \mathbf{K}\,[\mathbf{R} \quad \mathbf{t}]. \tag{7}$$

Using Eq. 3, 4, 7 it is possible to define the relationship between 3D point **M** and its image projection **m**:

$$s\widetilde{\mathbf{m}} = \mathbf{K}\,[\mathbf{R} \quad \mathbf{t}]\,\widetilde{\mathbf{M}} \equiv \mathbf{C}\,\widetilde{\mathbf{M}}, \tag{8}$$

where s is a arbitrary scale factor.

This way, 3D world points are transformed into 3D camera model coordinates using the external parameters. Given 3D points camera coordinates are projected into the image plane using the internal camera parameters (Eq. 5). To estimate the camera parameters, 3D world points and their corresponding 2D image points are used. In order to get those correspondences (samples), multiple images of a calibration pattern could be used, such as a checkerboard or circle grid (Fig. 3). Thus, the camera parameters can be determined.

4 Calibration experiments

4.1 Hardware imperfection issues

In our experimental work the goal was to calibrate the stereo vision system of Russian crawler-type mobile robot «Servosila Engineer» (Fig. 2). Even though initially the stereo pair was mounted correctly by the manufacturer, the left camera was displaced from its original position as a result of the robot turning upside down and falling from a significant height during an experiment. Figures 6-7 demonstrate that the images from the left and from the right cameras are displaced relative to each other (the left camera's frame is higher), so their principal points are not on the same horizontal line; no cropping or any other image processing were applied to the original images. Thus, approximately 20% of visual information will be discarded during the rectification procedure. In addition, barrel lens distortion affects the images.

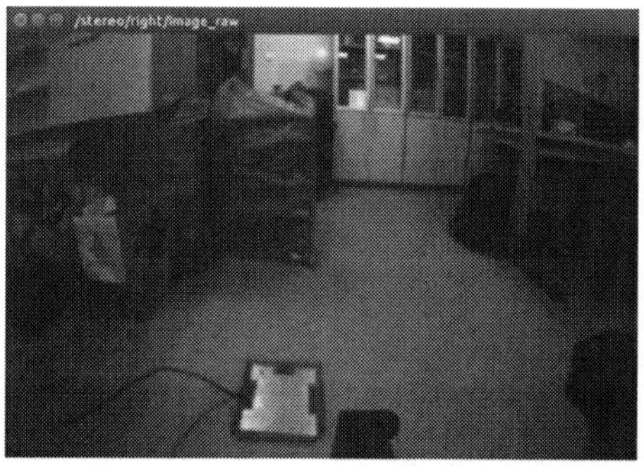

Fig. 6. An image from the right camera of «Servosila Engineer» robot stereo vision system.

Fig. 7. An image from the left camera of «Servosila Engineer» robot stereo vision system.

4.2 Calibration with a checkerboard

The first calibration method that we used in our experiments was based on checkerboard calibration pattern approach [13]. This method requires the dimensions of the pattern to be known and to be constant during the calibration process. The algorithm detects chessboard corners (as an intersection of two black squares) from left and right images being captured by the left and the right cameras of the stereo vision system respectively. In order to estimate camera system parameters, images are obtained at different orientations of the calibration pattern.

Calibration was conducted using robot operating system (ROS). To capture video stream from the stereo system of the robot we employed our own ROS package for simultaneous video streaming from several different cameras [14]. Stereo camera calibration ROS package was used for calibration purposes [15]. It required a checkerboard template with known dimensions (number of squares and their size in meters). It should be noted that the pattern must be hold horizontally and the number of its columns should exceed the number of rows. When the calibration process successfully terminates, the camera system parameters are displayed and applied.

Four different checkerboard patterns were tested in order to obtain an optimal solution (Table 2). Calibration pattern with dimensions of 8 by 6 and a square side of 26 mm showed acceptable results (Fig. 8, 9). The process of calibration took approximately 5 minutes on our machine.

Table 2. Checkerboard patterns performance analysis.

Square size (one side), mm	Columns x rows	Epipolar error (pixels)	Results
26	8x6	0.33	good
26	9x6	0.35	good
46	8x6	1.09	poor
46	11x7	0.48	fair

Fig. 8. Calibration with the checkerboard (8x6, cell size is 26 mm, epipolar error is 0.33).

Fig. 9. Undistorted images after camera calibration with 8x6 26 mm cell checkerboard.

In an attempt to achieve more accurate calibration results, we tried increasing the number of checkerboard columns by one column. However, the pattern of 9 by 6 with cells demonstrated almost the same results (Fig. 10).

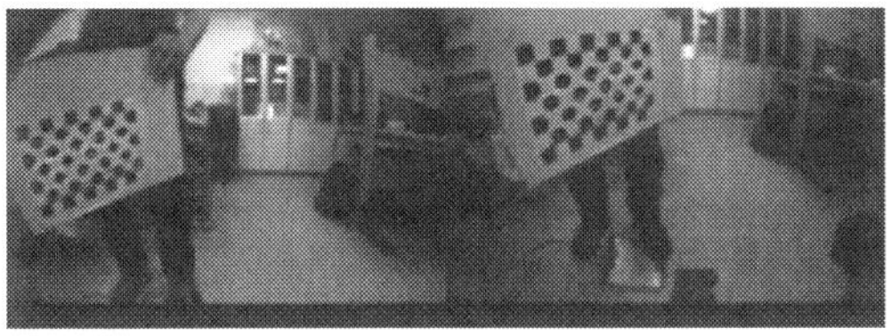

Fig. 10. Calibration with the checkerboard (9x6, cell size – 26 mm, epipolar error – 0.35).

In contrast to the calibration pattern with a cell size of 26 mm, a checkerboard with a cell size of 46 mm and dimensions of 8 columns by 6 rows demonstrated poor calibration quality results, producing an epipolar error that exceeded one pixel (Fig. 11). Some of the experimental trials did not converge at all and the obtained distortion coefficients were zero.

Fig. 11. Calibration with the checkerboard (8x6, cell size - 46 mm).

In the next experiment, we used a checkerboard with 46 mm cells and 11 columns by 7 rows and successfully estimated distortion (Fig. 12). The obtained epipolar error was greater than the one for the 26 mm cell size pattern (0.48 vs 0.33) and camera parameters estimation time also increased with the size of the pattern.

Fig. 12. Calibration using 11x7 checkerboard pattern (cell size - 46 mm, epipolar error – 0.48).

When the pattern size was increased (number of rows and columns), it become harder to accomplish the calibration process. Nevertheless, after successful calibration, images had less common pixels due to cameras displacement (Fig. 13) and some information about the surrounding environment was lost.

Fig. 13. Loss of information due to left camera displacement.

4.3 Calibration with a circle grid

Circle grid based calibration is another option of camera calibration [16], which could be more accurate than classical checkerboard based approaches. We experimented with robot stereo vision system calibration applying a 6 by 6 circle grid pattern with a circle diameter of 23 mm (Fig. 14). ROS stereo camera calibration package was used. Calibration resulted in estimated distortion coefficients and rectified images (Fig. 15). The obtained epipolar error was 0.18 pixels, which is almost two times better than the result of the best checkerboard based calibration.

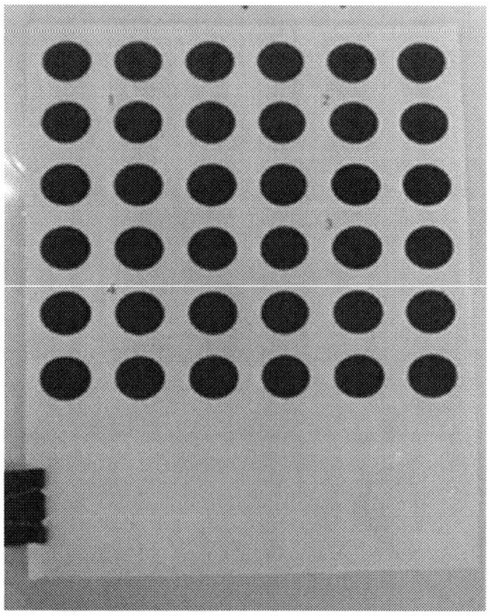

Fig. 14. Circle calibration grid (6x6, circle diameter – 23mm).

Fig. 15. Calibration with the circle grid (6x6, circle diameter – 23 mm).

4.4 Monocular calibration experiments

In addition to main set of stereo vision system calibration experiments, we attempted to calibrate only one of the robot cameras and to apply the estimated distortion directly on the other camera. For calibration we selected the right camera as it had not been displaced from its original location. Based on our previous conclusion on a better calibration approach, the circle grid was used. As a result, the error remained, approximately, the same, as if stereo calibration with circle grid pattern was used (Fig. 16).

Fig. 16. Calibration of the right camera of the stereo vision system (with a circle grid).

5 Conclusions

In this paper, we reviewed and experimentally compared two popular stereo vision system calibration approaches, the checkerboard and the circle grid based approaches. Moreover, the experiments were performed for a mobile robot stereo vision system calibration under a hardware imperfection. We used a crawler-type mobile robot «Servosila Engineer». Stereo system cameras of the robot were displaced relative to each other, causing loss of surrounding environment information. We implemented and verified checkerboard and circle grid based calibration methods. The two methods comparison demonstrated that a circle grid based calibration should be preferred over a classical checkerboard calibration approach. Although fairly good calibration results were achieved, the rectification process discarded almost 20% of the surrounding environment information, which decreases robot's stereo vision perception information.

As a part of our future work, we plan to implement calibration methods, which are based on fiducial markers. These approaches allow camera calibration in the presence of marker occlusions and rotations [17] and such features are especially important for autonomous self-calibration of a robot in real world operation conditions.

This work was partially supported by the Russian Foundation for Basic Research (RFBR) and Ministry of Science Technology & Space State of Israel (joint project ID 15-57-06010). Part of the work was performed according to the Russian Government Program of Competitive Growth of Kazan Federal University.

References

1. C. Brand, M. J. Schuster, H. Hirschmüller, M. Suppa, In IEEE/RSJ Int. Conf. on Intelligent Robots and Systems, 1846–1853 (2014)
2. A. Buyval, I. Afanasyev, E. Magid, Ninth Int. Conf. on Machine Vision, Int. Society for Optics and Photonics, **10341** (2017)

3. Z. Zhang, IEEE Transactions on pattern analysis and machine intelligence, **26** (7), 892–899 (2004)
4. W. Zhao, N. Nandhakumar. Pattern Recognition, **29** (12), 2115–2126 (1996)
5. N. Alishev, R. Lavrenov, Y. Gerasimov, Int. Conf. on Artificial Life and Robotics, 204–207 (2018)
6. J. Suriansky, M. Cmarada, Annals & Proceedings of DAAAM International, (2012)
7. M. Fiala, C. Shu. *Fully automatic camera calibration using self-identifying calibration targets* (Techn. Rep. 48306, 2005)
8. J. Heikkila, O. Silven, Proc. IEEE Computer Society Conf. on Computer Vision and Pattern Recognition, 1106–1112 (1997)
9. P.F. Sturm, S.J. Maybank, Proc. IEEE Computer Society Conf. on Computer Vision and Pattern Recognition, 432–437, (1999)
10. J. Wang, F. Shi, J. Zhang, Y. Liu, Pattern Recognition, **41** (2), 607–615 (2008)
11. C.B. Duane. Photogramm. Eng, **37** (8), 855–866 (1971)
12. T.A. Clarke, J.G. Fryer, The Photogramm. Record, **16** (91), 51–66 (1998)
13. Z. Zhang, IEEE Transactions on pattern analysis and machine intelligence, **22** (11), 1330–1334 (2000)
14. R. Safin, R. Lavrenov. Int. Conf. on Artificial Life and Robotics, 220–223 (2018)
15. Camera calibration at ROS Wiki. Ros.org, Available at: http://wiki.ros.org/camera_calibration
16. P. F. Luo, J. Wu. Optical Engineering, **47** (3), 033607 (2008)
17. A. Sagitov, K. Shabalina, L. Sabirova, H. Li, E. Magid. Int. Conf. on Informatics in Control, Automation and Robotics, 182–191 (2017)

Issues of physical interaction of unmanned aircraft manipulators with ground objects

Vinh Nguyen[1,*], *Oksana* Solenaya[1] and *Petr* Smirnov[2]

[1]SUAI, Department of Electromechanics and Robotics, 190000, Saint-Petersburg, Bolshaya Morskaya Str., 67, Russia
[2]SPIIRAS, Laboratory of Autonomous Robotic Systems, 199178, Saint-Petersburg, 14-th line of V.I., 39, Russia

Abstract. Adding an onboard manipulation system to an unmanned aerial vehicle (UAV) significantly complicates framework, functioning algorithms, and leads to an increase in overall dimensions. The physical interaction of the manipulator with objects influences to unstabilization of UAV, which in turn leads to difficulties in positioning the UAV and reduces the accuracy of gripper motion. In addition, the physical interaction of the manipulator with objects requires increased power resources of UAVs. The article analyzes modern research of UAVs with a manipulator, including flight control problems, avoidance of contact with the earth, surrounding space, as well as manipulations with the captured object. On the basis of the analysis, a list of new problems arising in the physical interaction of UAVs with objects through an embedded manipulator is formulated.

1 Introduction

At present, unmanned aerial vehicles (UAVs) are increasingly being used to solve problems of monitoring various types of objects and territories. Also, a promising area of research is the transportation of goods through UAVs, the main advantage of which in the urbanized area is high speed, while strict security requirements are imposed. UAVs should avoid obstacles, do not slow down or interfere with the movement of land vehicles. Unlike fixed-wing aircraft, rotorcraft (FWA), such as helicopters and multicopters, can travel through narrow aisles. The FWA can also hang and observe what is useful when interacting with stationary objects. Some FWA can land in the course of the task to prolong battery life and reduce power consumption, for example, when viewed from a stationary site. In work [1] approaches to the solution of the problem of increasing the operating time of unmanned aerial vehicles in autonomous missions based on charge or replacement of on-board batteries on the accompanying robotic platform are analyzed.

Recent advances in global and local navigation, localization, detection and prevention of obstacles have significantly increased the autonomy of UAVs. Stabilization of flight parameters and constant monitoring of surrounding objects are important aspects in autonomous operations performed by UAVs. Most of the research to date has focused on developing robust autopilot systems, automated workplaces for pilot-to-aircraft interaction using 3D modeling and virtual reality systems, and analyzing safety issues, including preventing the drones from collapsing.

The ability of aerial unmanned vehicles to manipulate or carry surrounding objects greatly expands the types of missions and applications, enhancing the operator's capabilities in performing dirty, dangerous or monotonous operations such as: manipulation of objects, building structures, repair of building infrastructures, assistance in natural disasters, loading cargo, inspection and maintenance of industrial plants, taking samples of materials from hard-to-reach areas, inspection of transport, agricultural land and other uses. Such typical tasks can be: collection of soil samples, maintenance on the bridge or on the facade of the building, cleaning of debris in the danger zone. In such tasks, the UAV must be able to manipulate in its environment, including removing obstacles that block movement to the target point. However, at present only ground-based robotic vehicles are equipped with manipulators [2, 3].

The control systems of the multi-rotor platforms equipped with a manipulator with a grip are also called MM-UAV (Mobile Manipulating Unmanned Aerial Vehicles). The development of unmanned aircraft manipulators (UAM) is accompanied by a much greater number of difficulties than the creation of traditional ground-based robots that perform the tasks of movement together with manipulation. The most difficult issue is the lack of a stable platform in the air. Ground vehicles can stop to perform operations and in a stable state to perform precise manipulations, while UAMs in most cases do not have this capability. In fact, even with the use of robust flight stabilization systems, the UAM will be in a certain area of space, without having precise fixed coordinates, especially when operating in an open space. The lack of stability of the air platform can be partially compensated by manipulator control. The range of manipulator movements and the field of vision sensors also depend on the side with which the platform

* Corresponding author: nguyenvanvinhhvkt@gmail.com

flies to the object. The workspace of the manipulator is limited by the body of the air platform itself, so to increase it in most cases the manipulator is mounted directly to the lower part of the platform [4].

Thus, the focus of previous research was the solution of the problem of flight control, avoidance of contact with the earth and the surrounding space. The purpose of this study is the physical interaction of UAVs with objects through an integrated manipulator. The following is an overview of the main problems encountered by unmanned aircraft manipulator developers and the analysis of published results. It should be noted that experiments on full-scale models are still rarely carried out, mainly computer or combined modeling, where the model of the manipulator is suspended in a special frame that simulates the free movement of the UAV. In the last section, a list of tasks that arise in the development of unmanned aircraft manipulators, not studied previously in the design of UAVs and ground-based manipulators, is formulated.

2 Problems of development of unmanned aerial manipulators and analysis of existing solutions

Interacting with objects using a manipulator mounted on unmanned aerial vehicles is an important task, since UAVs can access many places that are in some cases inaccessible to other ground transport robotic platforms [5]. However, adding an on-board manipulation system to a flying robot causes serious problems, as operations of physical interaction with ground objects affect the overall stability of the entire air platform. During the flight of an UAV with a moving mobile manipulation system, the mass distribution changes and additional dynamic reaction forces arise [6]. The design of the mechanism for capturing and manipulating objects affects the permissible mass of the payload, the inertial and dynamic characteristics of the entire UAV. It is important that the mass of this mechanism is low and its center of mass is closer to the base of the UAV, in this case the emerging reaction forces and torque that will arise during the UAV movement will not significantly affect its stability. The effect of propagating contact forces from the final grab to the UAV can be minimized through the connection of the air manipulator with the UAV base [7]. Such aspects are usually not taken into account in standard algorithms of platform stabilization, so a new algorithm is needed that solves such problems. It should also be noted that manipulators perform various types of movements, translational, rotating, with a payload and without, which causes various effects on UAVs that violate its stability [6].

It is also necessary to analyze a number of other serious problems before it becomes possible to manipulate objects through UAVs, among which [5] are distinguished: 1) the UAV platform constantly has micro-propulsion and, accordingly, it is extremely difficult to accurately position the links of the UAV manipulator relative to the object of interest for manipulation; 2) manipulator movements affect the stability of the UAV platform, which further complicates positioning.

In work [4], the authors have constructed a mini-gantry test rig and built a simulation environment in OpenRAVE for a dual arm UAV manipulator system. The authors envision Mobile Manipulator UAV system performing similar tasks as existing ground robots, but with a higher degree of autonomy, speed, and mobility. The design consists of two manipulators that are mounted to the bottom of the mini-gantry crane. The authors focus on dual arm manipulation and grasping tasks. Attached to the arms are 1-DOF Melissa hands consisting of four fingers and a thumb. As a result, the forces of reaction that the UAV experiences during manipulation when pointing to an object were investigated. A model for compensating this force and ensuring the quasi-stability of UAVs at contact with the object is created.

In work [5], a multi-link manipulator was installed on a platform with 3 degrees of freedom to simulate the linear motions of the Ascending Technologies Pelican quadrotor. Angles by rolls, pitch and yaws were not applied in emulated movements due to the mechanical complexity of the implementation. The manipulator is assembled from servomotors and passive elements. Reconfigurable capture with two degrees of freedom and one drive allows you to take a variety of objects with minimal movements. A preliminary demonstration of the manipulator's capabilities was performed on an example of simulating the task of removing and installing insulators from high-voltage power lines.

In work [7], has presented the design and experimental validation of a compliant and low weight (40 grams) anthropomorphic finger module specifically designed for aerial manipulation and grasping. The three joints of the finger are under actuated and driven by a nylon tendon that is rolled up on a small section reel attached to a micro motor shaft The anthropomorphic finger consists in three 8 mm U-shape aluminum profile sections of 45 mm (proximal phalange), 20 mm (middle phalange) and 15 mm (distal phalange) length. The finger module is capable of grasping objects of different shapes and forms in a stable way by itself, without requiring any other fingers. If the finger is not grasping any object, it can be used for detecting and reacting against collisions of the manipulator with the environment in a safe way as joint compliance will cause the passive contraction of the finger. This may be especially useful in the navigation of the aerial platform in narrow areas. For this purpose, the finger would remain extended and the metacarpophalangeal joint position controller disabled.

In work [8], The MM-UAV prototype is constructed using a low-cost quadrotor-type aircraft, a GAUI 330, and three manipulator arms. The GAUI quadrotor uses four equally positioned brushless motors equipped with 8-inch propellers to provide lift and maneuverability. The total vehicle diameter is 330 mm and the total weight is approximately 400 g with a maximum flying weight of 1100 g. The overall weight of the aircraft is kept bellow 950 g, thus making the aircraft capable of lifting small objects bellow 200 g. The final construction and important construction points (i.e. Center of Assembly—COA, Center of Mass of manipulator arm—COM, and Center of Gravity of aircraft—COG) are appropriately marked.

Each arm of the manipulator has 2 degrees of freedom. For additional manipulation capabilities at the end, the manipulator is equipped with a hook-to-hook, push / pull objects from the floor. The simulation results showed that the manipulator creates the least amount of UAV interference in the case when all the links move simultaneously. Nevertheless, experimental results show that the multicopter has oscillations.

In work [9], a multicopter equipped with a manipulator weighing 200 g, capturing a payload of up to 200 g, is described. The multicopter itself can carry a payload in excess of 500 g (greater than the total weight of the manipulator with payload). In the developed controller, three control methods were used to move the UAV: the linear feedback method (FL-PID), the direct fuzzy control method (DFLC) and the fuzzy model control method (FMRLC). The methods were tested in the capture of the payload and release from it. The FL-PID method did not ensure the stability of the UAV when increasing the payload and there were delays in performing real-time operations. The DFLC and FMRLC method allowed to maintain the stability of the UAV when adding / removing from the payload, while the FMRLC method showed a higher positioning accuracy.

In work [10] the problem of the accuracy of navigation and motion control of several connected UAVs with manipulators was investigated, a general structure of modeling of aviation vehicles with several fuselages was developed. To demonstrate the proposed models as a separate element used a hexagonal UAV, equipped with a simple manipulator.

In work [11] the helicopter SST-Eagle2-GS Long Tail, released by HIROBO, equipped with a manipulator, is described. During the tests, the effect of the added weight (manipulator) on the parameters of the helicopter control system was studied. Taking into account that the added weight of the manipulator changed the total weight of the helicopter by 25%, the MP2128Heli autopilot equipped with PID controllers did not provide stability and when testing shortly after takeoff the helicopter heared into lateral fluctuations, therefore the algorithm for calculating the PID controller values was modified and with new parameters was provided stable helicopter flight with additional weight. The testing was carried out with the helicopter hanging autonomously at a height of 3 m.

In work [14] a successful application of a quadrotor with a two-DOF robot arm attached for picking up and delivering an object. Kinematic and dynamic models for the combined system are analyzed and an adaptive sliding mode controller is designed to control them all together. This is demonstrated in an experiment with a scenario consisting of picking up an object, moving, and releasing it inside of a shelf. The experimental results demonstrates satisfactory performance to perform the designed tasks. The proposed aerial manipulation using the adaptive sliding mode controller is validated by an experiment.

The first mission for the combined system is picking up an object. The object used in the experiment is an wood block of dimension (H) 7.5x (D) 5x (W) 1.5 cm. Given that the gap of the end effector is only 3.2 cm, it is clear that grabbing the wood block requires accurate control of the quadrotor and robot arm concurrently. Even though no information such as mass and moment of inertia about the wood block is given to the adaptive sliding mode controller, overall manipulation is successful because of robustness of the sliding mode controller. During the pick-up phase, root-mean-square error of the quadrotor position is only 2.08 cm. The second mission is releasing the wood block inside a shelf. The size of the gap of shelf is about (H) 16 x (W) 22 cm. To lay the wood block down inside the shelf, most part of the second link with gripper needs to be put into the shelf. Again, given the size of the second link with respect to the size of the shelf, it is a difficult task that demands exact control of the combined system.

From the entire trajectory histories, it is shown that the quadrotor and robotic arm are controlled as desired during the full flight including manipulations of an object. These results also means that the provided approach can be utilized in expanded manipulation and transportation applications.

In work [15] presented a method for planning a time-optimal trajectory for a quadrotor equipped with an onboard gripper with the goal of grasping a moving target while traveling between generic initial and final states. We introduced and used two complementary set of trajectories able to continuously keep the quadrotor gripper in contact with the moving target. The method takes into account upper/lower bounds on the propeller speeds and the need of a finite time for the gripper in order to successfully perform the grasping task. Extensive physically-based simulations demonstrate the effectiveness of the proposed strategy.

With respect to the state of the art, the contributions of this paper can be summarized as follows: i) we explicitly consider the duration of the trajectory in order to minimize it; ii) we allow the quadrotor to attain generic, and in particular non-hovering, states for starting and ending the grasping phase; iii) we take into account the fact that the target is moving; iv) we include additional practical constraints such as limited actuation capabilities and the requirement of a finite time to actually lock the gripper and perform the grasp. Although the main focus is on the planning stage, we also discuss how to implement our proposed strategy by closed loop tracking control for the quadrotor motion.

3 Analysis of new tasks arising in the management of UAVs with a manipulator

On the basis of the above analysis of existing modern solutions of physical interaction of UAVs with surrounding ground objects, we further formulate the spectrum of new tasks that arise when the UAV is jointly controlled and the manipulator with gripper is installed on it during contact operations with the transported object. In Table 1, new tasks are listed in accordance with the main stages of development and operation of UAVs when manipulating a ground object. At the design stage, it is necessary to calculate the increased values of the UAV parameters and the expansion of the sensor system for manipulating and transporting the object.

Table 1. New tasks in the implementation of manipulation of UAVs with ground objects.

No.	Task name	demand
At the design stage		
1.	Selecting the installation location of the manipulator with the gripper on UAVs	1. Provide the necessary angles of motion of the manipulator and grip for working with objects; 2. The absence of intersection of the workspaces of the manipulator with moving elements of the UAV.
2.	Selection and location of the sensor system	1. Arrangement of additional sensors and video cameras to control the manipulator and grip when manipulating the object.
3.	The calculation of the increased values of parameters UAV for transporting objects	1. Providing increased flight performance for a flight with a manipulator and payload; 2. Providing additional energy resources for flight with a manipulator and payload.
On the take-off phase		
4.	Selecting the position of the manipulator	1. Providing minimal aerodynamic drag of the manipulator; 2. The provision of compact parking position of the manipulator with minimal increase in the dimensions of the UAV; 3. Ensure the stable position of UAVS when moving the manipulator to the parking position; 4. Providing fixation of the grip on the body of the UAV.
5.	The synchronization of the work of UAV and gripping, used for landing on ground stationary objects	1. Providing a synchronized take-off and compression of the grip with the ground object, based on the analysis of data from the sensors; 2. Providing the transfer of the manipulator to the parking position.
At the stage of object grasp		
6.	Approach to the manipulated object	1. Navigate to the object close enough for translation of the manipulator in working condition; 2. Maintaining a stable position of UAVS when moving the manipulator from the parking position to the working position.
7.	Pointing the manipulator on the object	1. Providing sensory data capture position relative to the object and the surrounding objects based on the variability of the position UAV; 2. Providing direct line of sight of the manipulated object with the joint movements of the UAV and the manipulator; 3. The provision of a sufficient distance from the UAV to the surrounding objects; 4. Maintain a stable position of the UAV when hovering the manipulator.
8.	Capture an object	1. Synchronization of object capture and stabilization of UAVs; 2. Providing takeoff UAV with an object in the capture; 3. Transfer of the manipulator to the optimal position for the flight, taking into account the parameters of the transported object.
At the stage of object transportation		
9.	Transportation	1. Ensuring the safe navigation of UAVs taking into account the dimensions of the transported object.
At the stage of unloading the object		
10.	The approach to the unloading site	1. Navigation UAV to the unloading site taking into account the dimensions of the transported object; 2. Transfer of the manipulator to the optimal position for unloading the transported object.
11.	Placement of the object on the landing site	1. Ensure the stable position of UAVS hovering manipulator; 2. Installation of the object in the given area of the unloading site; 3. Release of the grip and retraction of the manipulator / UAV from the object; 4. Transfer of the manipulator to the parking position.

At the take-off phase, it is required to select a compact fixed parking position of the manipulator, which provides minimum aerodynamic resistance, and also to maintain a stable position of the UAV when moving the manipulator to the parking position. In the event that the manipulator was used to land a UAV on a ground facility, it is required to ensure synchronization of the operation of the UAV and the capture during undocking and take-off.

At the stage of grasp an object, the UAV is navigated to the object at a sufficiently close distance to move the manipulator to the working state, pointing the manipulator on the object, capturing and taking off the UAVs with the object. The main difficulty at this stage is the provision of a stable position of the UAV during the movement of the manipulator and accurate capture guidance in the dynamically varying position of the UAV.

At the stage of object transportation, it is necessary to ensure safe navigation of UAVs taking into account the dimensions of the transported object.

At the stage of unloading the object, an approach is made to the unloading area, the manipulator is moved to the optimal position for unloading the transported object and placing the object on the landing site. After the installation of the object in the given area of the unloading area, the grip is extended and the manipulator / UAV is withdrawn from the object. Ensuring a stable position of the UAV allows for more accurate manipulation of the manipulator and installation of the object. The above tasks and requirements allow defining the UAV hardware and software parameters necessary for manipulating and transporting ground objects, as well as interaction with ground-based service robotic platforms and other collaborative robots [1, 12, 13].

4 Conclusion

Historically, the investigation of control systems of UAVs was focused on preventing falling and contact with surrounding objects. This was mainly due to the inadequate use of the payload available to the UAV. Until now, UAVs have been used mainly for surveillance and monitoring tasks, for example, in search and rescue missions. Nevertheless, the ability of air vehicles to manipulate and move objects they encounter can significantly expand the types of missions performed by unmanned systems. Flying robots equipped with manipulators can lead to significant changes in transport logistics in near-Earth environments.

To date, the use of UAVs for direct interaction with the environment is still limited due to its instability. A number of studies on this topic have been carried out, but most of them have been performed at the modeling level. Undoubtedly, working with objects using a manipulator mounted on unmanned aerial vehicles and studying the control features of multi-rotor platforms equipped with manipulator grip are promising directions for further research.

The present research was partially supported by state research № AAAA-A16-116033110095-0 and project № 16-29-04101-ofi_m of the Russian Foundation for Basic Research.

References

1. Q.T. Ngo, O.Ya. Solenaja, A.L. Ronzhin, Trudy MAI, **95**, (2017)
2. A.A. Ivanov, O.A. Shmakov, SPIIRAS Proceedings. **49**, 190–207 (2016)
3. V.G. Gradetsky, I.L. Ermolov, M.M. Knyazkov, E.A. Semyonov, A.N. Sukhanov, Mechanism and Machine Science, Springer, Heidelberg, 275–281 (2014).
4. C.M. Korpela, T.W. Danko, P.Y. Oh, J Intell Robot Syst., **65**, 93–101. (2012)
5. T.W. Danko, P.Y. Oh, J Intell Robot Syst., **73**, 709–723 (2014)
6. S. Gardecki, A. Kasiński, A. Bondyra, P. Gąsior, International Conference Automation, 357–366 (2017)
7. A. Suarez, G. Heredia, A. Ollero, Robot 2015: Second Iberian Robotics Conference, 543–555 (2015)
8. M. Orsag, C. Korpela, P. Oh, J Intell Robot Syst,. **69**, 227–240 (2013)
9. A. Khalifa, M. Fanni, International Journal of Control, Automation and Systems **15** (3), 1434–1446 (2017).
10. M. Kobilarov, J Intell Robot Syst., **73**, 679–692 (2014)
11. G. Chmaj, T. Buratowski, T. Uhl, K. Seweryn, M. Banaszkiewicz, Aerospace Robotics, 109–119 (2013)
12. A.S. Kodyakov, N.A. Pavlyuk, V.Yu. Budkov, R.A. Prakapovich, Journal of Physics: Conference Series, **803** (1), 012074 (2017)
13. N.E. Shlyakhov, I.V. Vatamaniuk, A.L. Ronzhin, Journal of Physics: Conference Series, **803** (1), 012146 (2017)
14. S Kim., S. Choi, H.J. Kim, 2013 IEEE/RSJ International Conference on Intelligent Robots and Systems (IROS), 4990–4995 (2013)
15. R. Spica, A Franchi., G. Oriolo, H.H. Bülthoff, P.R. Giordano, 2012 IEEE/RSJ International Conference on Intelligent Robots and Systems (IROS), 4985–4992 (2012)

The model of UAV formation based on the uniform allocation of points on the sphere

Mikhail Khachumov, *Vyacheslav* Khachumov

Federal Research Center "Computer Science and Control" of the Russian Academy of Sciences, 117312, Moscow, Russia

Abstract. We consider an approach to constructing the desired virtual structure, which should be formed by unmanned aerial vehicles (UAVs). The proposed model is based on the principle of quasi-uniform allocation of points, previously used by the author in clustering problems. The mathematical apparatus for solving the problem of forming the desired structure is given: necessary theorems are proved; the hypothesis on the uniform distribution of points on the boundary of the circle and the sphere is put forward and partially proved. The method for optimal assignment of UAVs to goal positions in the desired formation is considered.

1 Introduction

Due to the active use of unmanned aerial vehicles (UAVs) in various spheres of human activity, the problem of forming UAV coalitions and using them in solving practical tasks becomes urgent. The general problem of forming a desired formation contains the following steps:

1) Forming the desired topology of homogeneous UAVs in a certain area, i.e. the structure of goal positions. We note the existence of a huge number of possible structures, but we focus on the bounded regular geometric shapes. As a rule, these are regular polygons on the plane and Platonic solids realizing the principle of equidistance for adjacent vertices. Another approach assumes the uniform allocation of points in a given space. The problem is not yet solved and requires the development of algorithms for calculating coordinates for an arbitrary number of points.

2) Planning the motion of UAVs to the goal positions from a given arbitrary location. This problem is called formation or reconfiguration of a given shape. The problem of planning optimal (in terms of time, distance, energy costs, etc.) and at the same time safe UAVs routes with respect to size and velocities is posed. It is needed to define a sequence of controls that lead a group of homogeneous UAVs from the initial position to the desired one [1]. The task is complicated by the presence of wind disturbances that affect the velocities, trajectories, distances between UAVs and, ultimately, the safety of group motion.

The problems of formation and reconfiguration of a given structure for a group of mobile vehicles (robots, drones, aircrafts, etc.) are considered in many state-of-art research papers.

In paper [2] the authors provide an extensive overview of the methods of UAVs allocation, based on the concept of uniform distribution of points on different surfaces. Control and navigation algorithms are proposed for a swarm of autonomous homogeneous mobile robots involved in the formation of a given spatial convex surface. Calculation of robots trajectories for swarm reconfiguring is carried out with minimizing time and energy costs (computation time and total length of trajectories) and with controlling collisions. The developed algorithm has quadratic computational complexity.

In paper [3] the authors proposed a new method for topology design of a group of robots based on a multi-level structure, which is employed to present different shapes of swarm robots in the desired region. A potential function-based controller makes it possible to control the robots in forming the desired formation shape and avoid collisions. The local minima problem that causes the robots to get stuck at undesired positions is solved by using new shape regulation control force. Experiments are performed to demonstrate the effectiveness of the proposed approach.

Formation control strategies for UAVs with fully available group trajectory information are considered in [4]. A unified, distributed architecture for controlling vehicles' layers requires only local neighbour-to-neighbour information exchange. A group of robots should form a certain spatial shape. It is proposed to decompose the virtual shape into parts and divide the swarm into several subsets (groups) responsible for the formation of its own part of the shape.

In paper [5] the authors propose formation control of nonholonomic mobile robots avoiding obstacles in a distributed manner for cluttered environments. The asymptotic convergence of follower robots to the position and orientation of the reference robot is ensured using the Lyapunov function. Simulations results are

* Corresponding author: khmike@inbox.ru

provided demonstrating formation control and obstacle avoidance for multirobots using the proposed scheme.

The problem of forming large groups of individuals moving together toward a common target location is considered in [6]. The possibility of multi-agent technologies application to this problem is studied. Consensus-based algorithm to control a swarm in dynamically changing and noisy environments is proposed.

Basically, all considered papers solve the problem of reaching already established goal points. Coordinates of those points are assumed to be known. In this paper, we solve two particular, but theoretically interesting problems of structure formation when UAVs must form a regular flat or spatial shape that meets the requirements of uniform allocation of goal points.

Earlier we obtain theoretical results in cluster analysis that may be utilised for solving UAV formation problems. The methods and models of uniform distribution of points presented in [7] are applicable to solving the problem of safe formation of desired shapes for UAVs. This article provides an interpretation of these results.

2 The model of UAVs formation

We consider the geometrical model of a UAV formation based on the distribution of n points in a sphere. We introduce the following definitions.

Definition 1. The condition of the uniform distribution m_i (i=1,2,…,n) of points on a p-sphere in R^p is fulfilled if the points are the vertices of regular polyhedrons that are circumscribed by the sphere. The regular N dimensional polyhedron is a convex polyhedron (i.e., the distance between any nearest points is constant and equal to the edge of polyhedron).

Definition 2. .The condition of the uniform distribution m_i (i=1,2,…,n) of points on a sphere of radius r in R^p ($n>p$) is fulfilled if for each point m_i exist p other points m_{jp} such that $d(m_i, m_{jp})$ = const(n,p,r).

Definition 3 (Thomson's uniformity). Let us allocate n identical free charges on a sphere. We consider that the distribution is quasi-uniform if it corresponds to the minimum of the potential energy of the system [8].

It follows from Definition 2 that the number of solutions of the problem is bounded by the number of regular polyhedra. Definition 3 does not cover the cases when $n \leq p$, as well as leads to difficulties in constructing an algorithm for calculating the coordinates of uniformly spaced points.

2.1 The hypothesis of quasi-uniformity of the distribution of points

Let $m_i \in R^p$ (i = 1, 2, ⋯, n) be the points on the p-sphere of radius r; let S be the boundary of the p-sphere and $d(m_i, m_j)$ be the distance between the points m_i and m_j. Let us propose the following hypothesis about the quasi-uniformity of their distribution in a space bounded by a sphere.

Hypothesis. The expression $\sum\limits_{1 \leq i < j \leq n} d(m_i, m_j)$ takes the maximum when the points m_i are quasi-uniformly located on the border S of a p-sphere.

Unfortunately, the proof of Hypothesis for a p dimensional sphere causes some difficulties; thus, it is proposed to consider some particular cases, whose solution has an independent and important value in practice.

2.1.1 The problem of optimal allocation of points on a plane bounded by a circle

Let us show that the given Hypothesis is valid for the planar case. Consider model problem 1 (packing problem): how to allocate the points in a circle of radius r such that the sum of the distances is maximal.

Lemma 1 (about the quasi-uniform allocation of points within the circle). Let m_i (i=1,2,…,n) be the points that lie on the circle $K(r)$ of radius r; let L be the boundary of the circle $K(r)$ and $d(m_i, m_j) = d_{i,j}$ be the distance between the points m_i and m_j. The expression $\sum\limits_{1 \leq i < j \leq n} d(m_i, m_j)$ takes the maximum values when the points m_i are quasi-uniformly located on the boundary of the circle L. The proof of this lemma is given in [9].

2.1.2. The problem of optimal allocation of points in a sphere

The original statements of problems on the sphere are given in [10]. Consider model problem 2: how can one allocate n points on the p-sphere of radius r such that the sum of the distances is maximal.

The solution of the problem is the subject of actual research and underlies the predictable geometric allocaton of points [7, 9].

Theorem. Let $m_i \in R^p$ (i = 1, 2, ⋯, n) be the points on the p-sphere $K(r)$ of radius r, let S be the boundary of the sphere $K(r)$ and $d(m_i, m_j)$ be the distance between points m_i and m_j. Then exist points $m_i' \in S$ (i = 1,2,⋯,n) such that $\sum\limits_{1 \leq i < j \leq n} d(m_i, m_j) \leq \sum\limits_{1 \leq i < j \leq n} d(m_i', m_j')$.

Proof.

Let some allocation of the points m_i (i=1,2,…,n) exist in p sphere $K(r)$. Show that any point $m_i \in K(r)$ can be shifted to the sphere S so that the value $\sum\limits_{1 \leq i < j \leq n} d(m_i, m_j)$ does not decrease. We select the arbitrary point m_i and plot a chord through it. Introduce a new coordinate system, wherein one of the coordinate axes will coincide with the chord. Since m_i moves along the chord, its coordinates are (t; 0; …; 0), while the coordinates of the point are m_j = (a_1; a_2; ⋯; a_p). Then, the function of the parameter t, which is equal to the distance from the point $m_j \in K(r)$ to the point m_i that lies on the chord is $d(m_i, m_j) = f(t) = \sqrt{(a_1 - t)^2 + a_2^2 + ... + a_p^2}$ .The function $f(t)$ depend only on the variable t, as the point $m_j \in K(r)$

is fixed. The function $f(t)$ is convex since its second derivative is positive.

Let us consider the analogous function for other points. The sum of convex functions $\sum_{j=1}^{n} d(m_i, m_j)$ is convex. The maximum value of a convex function is in one of the ends of a segment, i.e., a shift of the point to the sphere does not reduce the values of $\sum_{1 \le i < j \le n} d(m_i, m_j)$.

Theorem is proved.

Let us consider the case of a three-dimensional sphere.

Lemma 2. Let m_i $(i = 1,2, \ldots,4)$ be the points on a three-dimensional sphere. The expression $\sum_{1 \le i < j \le 4} d(m_i, m_j)$ takes the maximum when the points m_i matches the vertices of the regular tetrahedron.

The proof of this lemma is given in [9].

Lemmas 1 and 2 are important and give partial results that confirm the Hypothesis. We note that in both concidered special cases the conditions of Definition 2 are also satisfied, e.i., the distribution of points is uniform.

2.2 The Thomson model of quasi-uniform placement of points.

Definition 3 relates to the theoretical justification of the optimality of the configuration of charges considered by the physicist Thomson [8]. The problem is to determine the global minimum potential energy configuration of equal charges constrained to the surface of a sphere (without external influences): $\min U = \sum (1/(r_i - r_j))$, $i<j$, where $r_1, r_2, \ldots, r_n$ are radius-vectors of charges (unit charges and vectors are considered).

An iterative algorithm for practical allocation of points on a sphere is developed. Visualization of the results for seven charges is shown in Figure 1.

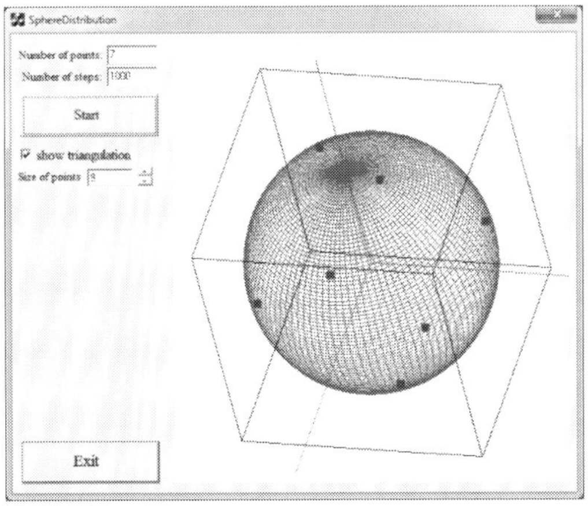

Fig. 1. Quasi-uniform distribution of seven points on the sphere.

If necessary, the algorithm additionally constructs the triangulation mesh. By solving the triangulation problem a reasonable distribution of points can be achieved.

Definition 4. Triangulation is called optimal if the sum of the lengths of all edges is minimal among all possible triangulations constructed on the same initial points. The mesh constructed by the methods of exhaustion, as a rule, have good quality. By optimizing the position of nodes one can achieve better quality.

The Thomson problem in three-dimensional space is solved exactly only for the cases of 2, 3, 4, 6 and 12 charges on a sphere. When the number of charges is 2,3 or 4, in extreme configuration all pairwise distances between them are equal. For other cases, the optimality of any disposition is not proved.

Note that, for these particular cases the results of placement of points coincide according to Definitions 2 and 3. That confirms the validity of the chosen criterion.

3 The problem of assignment of UAVs to goal positions

3.1 Formulation of the problem

Assume that we know the goal positions in the desired structure and positions of all UAVs. The problem is to optimally assign UAVs to available points in the formation. The problem can be presented in the form of a bipartite graph (Fig. 2), where vertices of one layer correspond to UAVs $\{p_1, p_2, \ldots, p_n\}$ and vertices of another layer correspond to the positions of the formation $\{q_1, q_2, \ldots, q_n\}$.

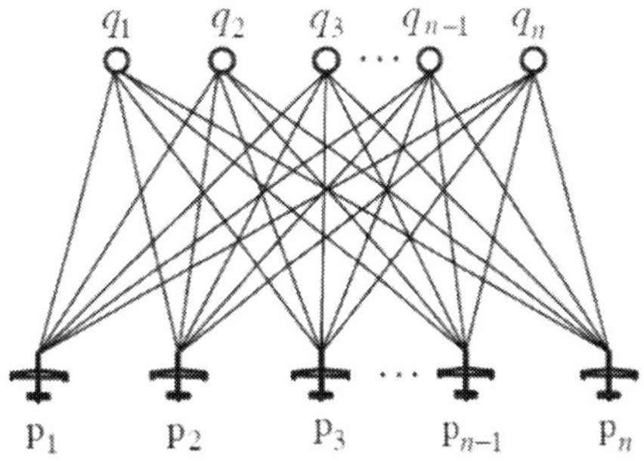

Fig. 2. Bipartite graph of relations "UAV-formation".

The edges of the graph correspond to the elements of the distance matrix $D=\|d_{ij}\|$, containing information about the distances between UAVs and goal positions. In addition, we assign $d_{ij} = \infty$, if the flight of UAV i to position j is impossible or undesirable.

Costs that correspond to the flight of UAV i to point j, we denote as c_{ij}. Here, we assume $c_{ij} = d_{ij}$. Alternatively one can use time costs that correspond to the flight of UAV i to point j, i.e. $c_{ij} = d_{ij} / (v_i + v_{ij})$ $(i, j = 1,2,\ldots,n)$, where v_i is an average speed of UAV i; v_{ij} is the speed of the air flow between position of UAV i and goal position j. We introduce x_{ij} $(i, j = 1,2,\ldots,n)$, such that $x_{ij} = 1$ if UAV i is assigned to point j, and $x_{ij} = 0$ otherwise.

A typical assignment problem is formulated as a linear programming problem [11]:

$$c = \sum_{i=1}^{n}\sum_{j=1}^{n} c_{ij}x_{ij} \rightarrow \min,$$

where: $\sum_{i=1}^{n} x_{ij} = \sum_{j=1}^{n} x_{ij} = 1 \ \forall i, j = 1,2,...,n.$

To solve the problem the Hungarian method can be used [11]. As a result of solving the problem a table is obtained with marked zeros that correspond to the assigned items. A similar approach is concidered in [12, 13].

3.2 Flight safety assurance

When all UAVs are assigned to goal positions it is necessary to plan their flight trajectories guaranteeing safety. Collision avoidance can be ensured by constructing for each vehicle of a tube (3D curve) [14] determined by circular object. All trajectories within such a tube are safe. The main criteria for optimal selection of tubes are: time of construction and length of the tube. To plan the optimal route one can use special software T-FLEX CAD 3D, which implements the basic operations of 3D modeling. Alternatively to detect collisions one can use the matrix of the current distances between UAVs. In that case decisions are made by certain rules adopted in aviation. It is planned to carry out further research on the construction of safe trajectories for collisions avoidance.

4 Experimental researches

We simulate the problem of forming a structure in a group of UAVs, taking into account real mathematical models of objects and wind disturbances [15]. Figure 3 shows formation flight of four UAVs and demonstrates the effect of wind loading on the motion trajectories.

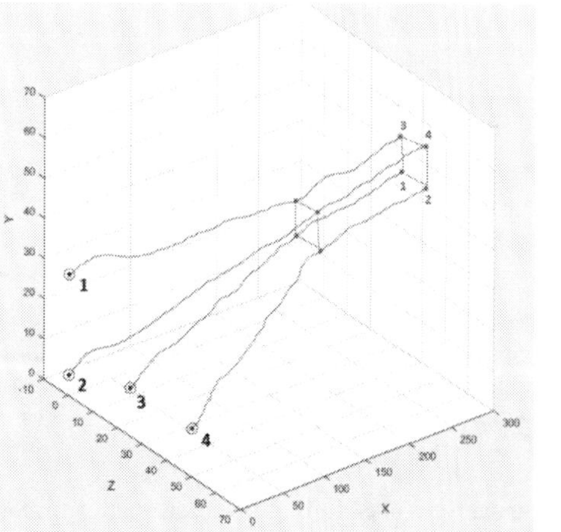

Fig. 3. Formation flight of four UAVs.

It can be seen that vehicles (grey markers) are assigned to uniformly distributed positions (red markers) in the desired formation. The developed simulating system contains a special module of intelligent control, realizing strategies and control rules for prompt response to changes in the external environment.

5 Conclusion

The proposed definitions and hypothesis of quasi-uniformity of the distribution of points in a circle and a sphere, as well as the lemmas and theorem given in the paper, are of practical importance for solving the particular problems of forming a structure in group of UAVs. It can be assumed that quasi-uniformity in the general case also ensures the safety of the formation in the presence of wind loads. Thus, the proposed models of points' allocation can be applied for planning of UAV formations. The problem of assignment of UAVs to goal positions is formulated and considered.

This research was supported by the Russian Science Foundation (Project No. 17-71-10163).

References

1. D.Y. Ivanov, XII All-Russian conference on control problems, 1971–1978 (2014)
2. I.V. Vatamaniuk, G.Y. Panina, A.L. Ronzhin, Large Scale Syst. Control, **58**, 285–305 (2015)
3. X. Yan, D. Sun, *Handbook of Research on Design, Control, and Modeling of Swarm Robotics*, 233–261 (IGI Global, 2016)
4. W. Ren, N. Sorensen. J. Rob, Autom. Syst., **4**, 324–333 (2008)
5. M.U. Khan, S. Li, Q. Wang, Z. Shao, ACM Trans. Auton. Adapt. Syst., **2**, 1–22 (2016)
6. V.A. Erofeeva, O.N. Granichin, V.I. Kiyaev, Russ. SC Days, 285–305 (2016)
7. V.V. Atamanov, M.A. Kozachok, V.V. Trushkov, M.V. Khachumov, J. Neurocomp., **1**, 73–76 (2009)
8. E.L. Altschuler, T.J. Williams, E.R. Ratner, R. Tipton, R. Stong, F. Dowla, F. Wooten, Phys. Rev. Lett., **78**, (1997)
9. M.V. Khachumov, Sci. Tech. Inf. Process. **6**, 310–316 (2012)
10. N.J.A. Sloane, R.H. Hardin, T.S. Duff and J. H. Conway, Discrete Comput. Geom., **14**, 237–259 (1995)
11. H. Kuhn, Nav. Res. Logist. **2**, 83–97 (1955)
12. A.A. Dyachenko, Izvestiya SFedU. Engineering sciences, **3**, 22–30 (2012)
13. I.A. Kalyaev, A.R. Gayduk, S.G. Kapustyan, *Models and algorithms of collective control in groups of robots* (Phismathlit, Moscow, 2009)
14. Y. Zhou, A. Raghavan, J.S. Baras, IEEE Conference on Decision and Control (2016)
15. N.S. Abramov, D.A. Makarov, M.V. Khachumov, Autom. Remote Control, **76**, 1070–1080 (2015)

MATEC Web of Conferences 161, 03023 (2018) https://doi.org/10.1051/matecconf/201816103023
13th International Scientific-Technical Conference on Electromechanics and Robotics "Zavalishin's Readings" - 2018

Classification of robotic battery service systems for unmanned aerial vehicles

Tien Ngo[1*], *Mehmet* Guzey [2] and *Vladimir* Dashevsky[3]

[1]SUAI, Department of Electromechanics and Robotics, 190000, Sankt-Petersburg, Bolshaya Morskaya Str., 67, Russia
[2]Erzurum Technical University, Department of Electrical and Electronics Engineering, 25700, Erzurum, Turkey
[3]SPIIRAS, Laboratory of Autonomous Robotic Systems, 199178, Sankt-Petersburg, 14-th line of V.I., 39, Russia

Abstract. Existing examples of prototypes of ground-based robotic platforms used as a landing site for unmanned aerial vehicles are considered. In some cases, they are equipped with a maintenance mechanism for the power supply module. The main requirements for robotic multi-copter battery maintenance systems depending on operating conditions, required processing speed, operator experience and other parameters are analyzed. The key issues remain questions of the autonomous landing of the unmanned aerial vehicles on the platform and approach to servicing battery. The existing prototypes of service robotic platforms are differed in the complexity of internal mechanisms, speed of service, algorithms of joint work of the platform and unmanned aerial vehicles during the landing and maintenance of the battery. The classification of robotic systems for servicing the power supply of multi-copter batteries criteria is presented using the following: the type of basing, the method of navigation during landing, the shape of the landing pad, the method of restoring the power supply module. The proposed algorithmic model of the operation of battery power maintenance system of the multi-copter on ground-based robotic platform during solving the target agrarian problem is described. Wireless methods of battery recovery are most promising, so further development and prototyping of a wireless charging station for multi-copter batteries will be developed.

1 Introduction

Automation of the process of recovering the energy resources of the multi-copters on the ground-based robotic platform in the target solution area will allow for increased the duration of the multi-copters operation and the scope of the tasks in the autonomous mission. Increased autonomy for the operation of multi-copters in the target area by the automatic charging system is the subject of many scientific studies. There are two types of station recharging system and systems with battery replacement. Systems with battery replacement can significantly shorten the preparation time of the multi-copters for a new flight and increase the total number of multi-copters that are simultaneously in the autonomous mission. The recharging system has a lower cost compared to the battery replacement system due to the minimization of the mechanical components of the structure.

The development of charging robotic ground stations is relevant for solving the problems of monitoring remote territories, in which most of the energy resources is used on reaching the specified area. In this case, multi-copters have the opportunity to replenish their energy resources in the territory of operation and work in a continuous mode until the end of the platform resources. The developing charging system should solve the problem of automating the battery maintenance process and could be used in fixed locations. In the framework of this study, models and prototypes of the battery charging system is developed, characterized by wireless transmission of energy, mechanization of the manipulation process with the battery on the multi-copter and the service platform.

Ground-based robotic systems and unmanned aerial vehicles, including multi-copters are actively used in various industries both for solving applied specialized tasks and for the entertainment industry. Considering the wide range of training of operators and functional capabilities of the multi-copters, it is necessary to distinguish three most popular applications of multi-copters: 1) researchers; 2) farming; 3) security [1-3].

Table 1 lists the main requirements for service systems of multi-copter power supply systems for the three above categories [4]. The potential uses for security multi-copters are reconnaissance, target spotting, riot control and border patrol, among others. Uses for farming multi-copters are checking cattle, checking fences, monitoring crops and spraying crops, among others. Uses for researchers are recreational play, pilot training and data collecting, among others.

Different customers also have different design constraints (budget, weight, complexity), but it is possible to identify some major common constraints; they are summarized in Table 2.

* Corresponding author: quoctienbn@gmail.com

Table 1. Customer needs summary [4].

Security	Farming	Researchers
1. Charge UAV batteries without human assistance		
2. UAV charging process should not damage the UAV		
3. Charging platform's geometry which has high probability of successful landing		
4. Shape/color of the charging platform is easy to discern from the surroundings, either by humans or navigation system		
5. Erroneous landing on the charging platform inflicts no more damage than erroneous landing in normal landing pads		
6. Parts of charging platform which are potentially harmful to users are out of reach of humans		
7. Simple system set-up		
8. Indicates when the process is done, to humans and/or navigation system		
9. The charging platform endures regular usage (steady landing, assembly/ disassembly)		
10. Can be carried out and assembled/disassembled by one average human		
11. The charging platform can operate without physical connection to the power grid or main base		The charging platform does not use much energy from the power grid
12. Endures non-ideal weather conditions (light rain, light snow, wind)		Endures being wet without breaking, but need not operate wet
13. Fits different sizes of UAVs	Designed for one UAV	
14. Allows more than one UAV to charge at the same time on the platform	Charge one UAV at a given time	

When developing automatic battery power maintenance system of the multi-copter should also take into account the following functional and technical requirements:

- Provide identifiable landing space: 1) Provide sufficient area to land (bigger than 1.5 times the position error of navigation system and proportional to the number and size of skids/footprint/tail); 2) Provide means to communicate current position of platform to navigation solution.
- Charge batteries: 1) Provide safe electrical interface between battery on multi-copter, multi-copter electronics, charger on platform, and multi-copter detection system when appropriate on appropriate linkage; 2) Identify that multi-copter has landed in correct position; 3) Charge battery; 4) Identify charge needs.
- Provide power to the system: 1) Acquire power; 2) Adapt power to be used on the platform.
- Provide portability: 1) Provide easy setup; 2) Provide way to transport.

Table 2. Design constraints

1	Modifications to the helicopter, if any, should add as little weight as possible to prevent reduction in flight duration due to increase of body mass
2	Battery disabling systems must guarantee that the UAV will not be disabled unintentionally
3	UAV dimensions and physical properties
4	The battery is very sensitive to recharging voltage/current
5	UAV electronics should not be connected to battery during recharging
6	Pilot skills/auto-pilot skills
7	Human strength/skills

Above functional and technical requirements will be taken into account when developing a ground-based robotic platform equipped with landing sites and multi-functional mechanisms for capturing unmanned aerial vehicles (UAVs) and servicing their embedded power systems, taking into account previously developed technological solutions [5-7].

2 Electromechanical maintenance systems for unmanned aerial vehicles

Autonomous landing of unmanned aerial vehicles in modern studies is considered not only on a fixed platform but also on a mobile platform that carries out traffic in various environments. Landing UAVs on a service charging station is realized with the help of various systems of navigation and analysis of the surrounding area. The paper [8] describes the vision system developed for target detection and tracking. Adopted an optimized solution to save computation time ensuring a real time image processing supported by the robot operating system environment. The vision system works in indoor / outdoor environments even in absence of GPS or similar localization systems.

A new decentralized method for controlling the joint functioning of a UAV and a mobile platform was considered in [9]. The presented experimental results for the small Aeryon Scout and the Clearpath Robotics A200 Husky mobile platform confirm the possibility of landing both in enclosed spaces with high-quality navigation data and outdoors in windy conditions.

In the paper [10] proposes a vision-based target following and landing system for a quadrotor vehicle on a moving platform. The system employs a vision based landing site detection and locating algorithm using an omnidirectional lens. The latest smartphone was attached to the UAV and served as an on-board image acquisition and process unit. Measurements from the omnidirectional camera are combined with a proper dynamic model in order to estimate position and velocity of the moving platform. An adaptive control scheme was implemented on the flight computer to deal with unknown disturbances in outdoor environments. The system was validated on a quadrotor UAV and the vehicle successfully landed on the moving platform during outdoor flight tests.

In [11] present the design and implementation of a vision based autonomous landing algorithm using a downward looking camera. To demonstrate the efficacy of algorithms emulate the dynamics of the ship-deck, for various sea states and different ships using a six degrees of freedom motion platform. Present the design and implementation of robust computer vision system to measure the pose of the ship-deck with respect to the vehicle. A Kalman filter is used in conjunction with vision system to ensure the robustness of the estimates.

The paper [12] suggests analysis of unmanned ground vehicle component power requirements followed by recommendations for the use of low power sensors that do not affect vehicle functionality. Design specifications for an order of magnitude higher endurance unmanned ground vehicle compared with the existing one, based on a hybrid battery configuration. Design itself of an energy-efficient (solar cell) gimballed landing platform serving as a transport and charging base for small unmanned helicopters. An indirect benefit of this design is that it also allows the unmanned vertical takeoff and landing vehicle to take off from one unmanned ground vehicle and land on another, maximizing operational range.

To increase the autonomy of an unmanned aerial vehicle, it is necessary to recharge its power source and replenish other consumables based on automated recharge systems. For systems of UAVs to achieve near autonomy, the automation of ground tasks is required. To address this problem we first conducted an economic evaluation of two solutions concepts: refill and exchange of consumable reservoirs. We demonstrated that refilling stations are economically superior for low target coverage. For high coverage systems, exchange stations are better [13].

In work [13] three types of charging stations for the UAV are proposed: Rollin 'Mat, Concentric circles, Honeycomb. They differ in cost, features and functions. Power stations such as Rollin 'Mat and Concentric circles are simple in design, easy to integrate and have a relatively low cost. However, the location and size of the terminals at the station depend on the type of UAV, which naturally affects the size of the landing pad at the station. In particular, if the UAV is rather small, the landing accuracy provided by the navigation systems may not be sufficient for docking with the charging sites. The Honeycomb power station has many advantages - the system is easily expandable, because a large number of UAVs can be charged at the same time by adding additional elements and chargers. Another feature is that the wireless IR emitter/sensor communication system can be readily replaced with another wireless system. Honeycomb can be used in almost any situation, when you need to recharge. But this solution is more expensive, therefore the Honeycomb platform is recommended in case when exact landing is required on a small site in difficult weather conditions.

Also [13] proposed a system for changing the UAV battery. Using an exchange process could increase significantly the ratio of maximum possible flight time (from take-off to landing) per ground time, therefore, decreasing the total number of UAVs. On the other hand, implementation cost will increase, since exchanging an empty reservoir for a full one is a more sophisticated approach than simply charging a waiting UAV.

The paper [14] presented designs options for an autonomous UAV battery replacement system ground station. Station is to automatically swap the depleted batteries of a UAV for a recharged battery without human intervention. Here, instead of tackling navigation control algorithm problems, we work with design parameters to solve issues such as: guiding the UAV to the battery replacement site, orienting the UAV in a desired direction, locking the UAV position on the station. UAV-battery connections: extracting and placing a battery in the UAV, Battery transportation and Battery array recharging.

The design options presented in [14] assume that the ground stations are outdoors, where weather conditions cannot be predicted. The focus here is not on landing algorithms, but on how to address UAV positioning after landing with small error. While UAV navigation systems are assumed to exist, allow landing position error. The goal is to allow the UAV to reach the location where battery swapping will be conducted, even if its landing position is non ideal due to navigation errors, weather conditions, UAV damage, etc. Normally, batteries are connected to UAVs via wire plugs and placed inside the UAV, thereby, ensuring terminal connection and battery stability. However, in a platform that has to rapidly swap a UAV's depleted battery for a recharged one, such a connection/holding system is not suitable. The method by which the battery is secured and physically connected to the UAV is of great importance because it will effect the repeatability, complexity, and the time that the ground station will take to swap batteries. Also, the added weight will effect the UAV's payload and flight time. In order to create the interface between UAV and platform, mechanical and magnetic couplers are considered and devices that can easily hold and release batteries while ensuring terminal connections in a UAV are proposed.

When the service station functions, [14] uses several modules to replace the batteries with high accuracy: orientation-fixing module, UAV Locking/unlocking module, battery extraction module, battery swap module. The station can compensate for the orientation and positioning errors of the UAV during landing. The complete station is able to compensate for orientation and positioning errors. It addresses navigation imprecision, weather conditions and/or UAV damage, which are common issues in outdoor missions and combat environments. The ground station is also capable of handling heterogeneous UAV fleets with not only different shapes and sizes, but also different number of battery cells per battery pack.

For the recharge platform, the duration the time UAV spends at the service station should be at least as long as a battery charging time; the UAV remains with the platform until its battery is fully charged. For the replacement platform the time UAV spends at the service station will be small relative to the charging time a battery requires to achieve full charge when fully depleted [14].

Figure 1 presents a classification of robotic battery service systems for multi-copter batteries. The battery service systems can be divided into the following features: 1) by type of basing: ground and water surface; 2) by the way of navigation during landing: local navigation systems, vision systems and global navigation systems; 3) by form of a landing site: flat, toroidal, system with mechanical guides to the center of the platform; 4) by method of battery recovery: battery charge system, battery replacement system. In the battery charge system there are two types: contact and non-contact. In the battery replacement system, depending on the way the battery is attached, there are magnetic and mechanical systems; by the type of placement of batteries there are linear and disc stores.

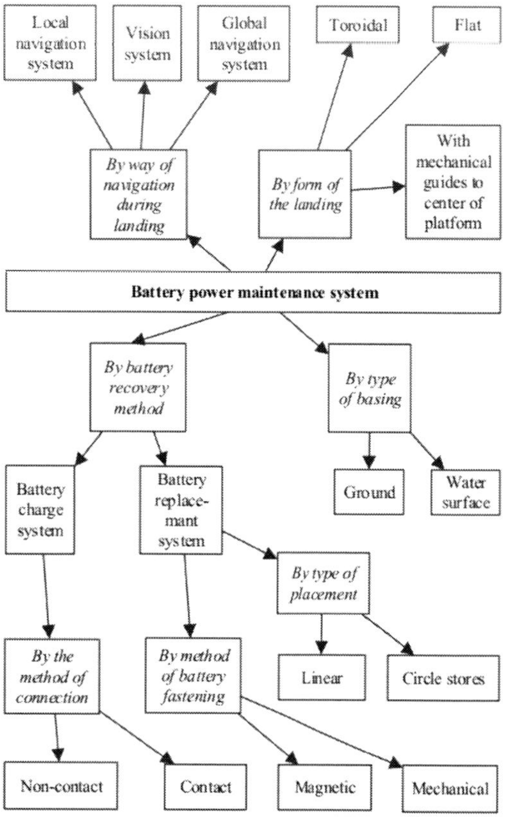

Fig. 1. Classification of robotic battery service systems for multi-copter batteries.

The design of battery maintenance systems for multi-copters with battery charge is simpler and less expensive. At the same time, the speed of service is not great and the number of simultaneously charged multi-copters is limited. Systems with battery replacement have more complicated design, but the speed of their maintenance is much faster, and therefore the number of overcharged multi-copters also increases.

With the development of wireless charging technology and battery replacement service stations, the UAV developed an algorithmic model for the interaction between multi-copter and service ground robotic station. Figure 2 shows the algorithmic model of the battery service systems when solving the target agrarian problem. The first step is monitor the current battery charge of UAV. If the battery is discharged, the UAV flies to the platform.

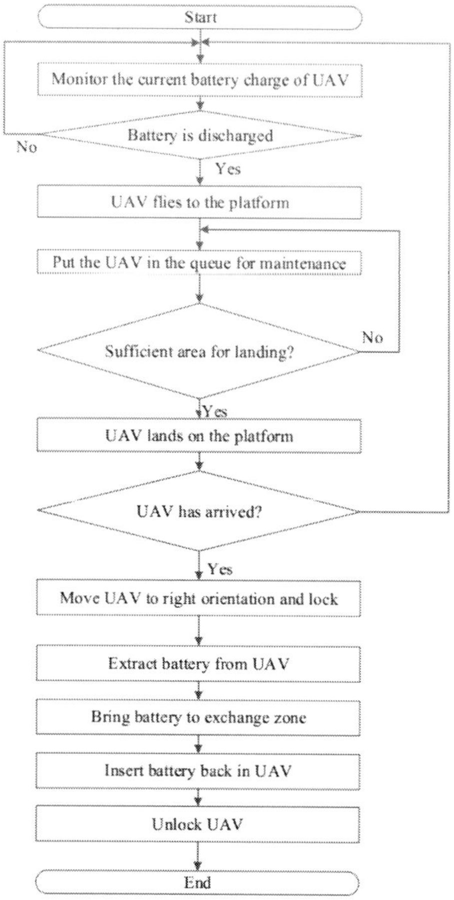

Fig. 2. Algorithm for servicing the UAV on terrestrial service platform.

The next step is to put the UAV in the queue for maintenance in accordance with the current battery charge. After receiving the charging requirement signal from the UAV, the system notifies the UAV of the presence of sufficient area for landing. Upon receiving the signal, the UAV lands on the platform with an acceptable error. After landing, the system positions drag UAV to the desired position and locks the UAV. Follows the step of extracting the battery from the UAV. The discharged battery is removed from UAV and moved to the charging area, the full battery is inserted back in UAV. Next UAV is unlocked and take off.

3 Conclusion

Now unmanned aerial vehicles are actively used for monitoring land, mapping the yields of land, planning zones for fertilizer application. Already the first prototypes appear, physically interacting with surrounding objects, which requires the consumption of even greater energy resources. The joint work of robotic ground platforms and multi-copters allows significantly increase the duration of work in the autonomous mission [15-18].

The article analyzes the requirements of the UAV battery maintenance systems for three main target areas: entertainment, farming, security. Analyzing the requirements and limitations of UAV battery charging stations. They were distinguished in the speed of processing and the complexity of the servicing mechanism.

Robotic systems for servicing power supply modules UAV can be divided into two main categories: battery charge system, battery replacement system. The battery replacement system have more complex designs, but provide a high speed of service, which increases the number of UAVs that have replenished their energy resources.

The algorithm of UAV service on a terrestrial service platform is proposed, which includes the main stages of interaction between UAV and service ground robotic station when solving the target agrarian problem.

The most promising method of battery recovery is wireless methods. The main problem of existing commercially available solutions for wireless transmission systems is low transmission power. The battery capacity of autonomous robots far exceeds the capacity of power sources for smartphones and other low-power devices whose extensive market dictates requirements to developers and manufacturers. Charging robotic systems requires capacities that are tens of times greater than chargers of existing standards. As future work, the development and prototyping of wireless charging station for UAV will be carried out.

This work was supported by RFBR № 18-58-76001_ERA.Net.

References

1. O.A. Lipa, L.V. Belyaeva, S.I. Kopylov, E.V. Khromov, Questions of electromechanics. **144**, 16–20 (2015)
2. V.A. Prince, B.V. Vishnyakov, Yu.V. Vizilter, V.S. Gorbatsevich, O.V. Vygolov, SPIIRAS Proceedings, **45**, 26–44 (2016)
3. V.A. Bogushevskaya, et al., MAI Proceedings, **51** (2012)
4. P.F. Kemper, K.A.O. Suzuki, J.R. Morrison, Journal of Intelligent & Robotic Systems, **61**, 369–397 (2011)
5. V. Nguyen, Q. Vu, O. Solenaya, A. Ronzhin, 12th International Scientific-Technical Conference on Electromechanics and Robotics "Zavalishin's Readings" – 2017, MATEC Web of Conferences, **113** (02009) (2017)
6. Q. Vu, V. Nguyen, O. Solenaya, A. Ronzhin, International Conference on Interactive Collaborative Robotics, **10459**, 260–267 (2017)
7. A.S. Kodyakov, N.A. Pavlyuk, V.Yu. Budkov, R.A. Prakapovich, Stability Study of Anthropomorphic Robot Antares under External Load Action. IOP Conf. Series: Journal of Physics: Conf. Series, **803**, 012074 (2017).
8. F. Cocchioni et al., Journal of Intelligent & Robotic Systems, **84**, 511–528 (2016).
9. J.M. Daly, Y. Ma, S.L. Waslander, Autonomous Robots, **38**, 179–191 (2015)
10. J.W. Kim, Y.D. Jung, D.S. Lee, D.H. Shim, Journal of Intelligent & Robotic Systems, **84**, 529–541 (2016)
11. Sanchez-Lopez J. L., Pestana J., Saripalli S., Campoy P. An Approach Toward Visual Autonomous Ship Board Landing of a VTOL UAV. Journal of Intelligent & Robotic Systems, **74**, 113–127 (2014)
12. S. Ioannou, K. Dalamagkidis, K.P. Valavanis, E.K. Stefanakos, Journal of Intelligent & Robotic Systems, **53**, 399–416 (2008)
13. P.F. Kemper, K.A.O. Suzuki, J.R. Morrison, Journal of Intelligent & Robotic Systems, **61**, 369–397 (2011)
14. K.A.O. Suzuki, P.K. Filho, J.R. Morrison, Journal of Intelligent & Robotic Systems, **65**, 563–586, (2012)
15. A. Zalevsky, O. Osipov, R. Meshcheryakov, International Conference on Interactive Collaborative Robotics, 268–274 (2017)
16. N. Pavluk, A. Ivin, V. Budkov, A. Kodyakov, A. Ronzhin, Interactive Collaborative Robotics, 113–123 (2016)
17. A. Ronzhin, I. Vatamaniuk, N. Pavliuk, 2016 International Conference and Exposition on Electrical and Power Engineering (EPE), 675–680 (2016)
18. A.A. Ivanov, O.A. Shmakov, SPIIRAS Proceedings. **49** (6), 190–207 (2016)

A decentralised solution for coordinating decisions in large-scale autonomic systems

Olga Melekhova[1,*], *Jacques* Malenfant[1], and *Roman* Mescheriakov[2] and *Aleksandr* Chueshev[3]

[1]Sorbonne Universités, UPMC Univ Paris 06, CNRS, UMR 7606 LIP6, 4 place Jussieu, 75005, Paris, France
[2]Tomsk State University of Control Systems and Radioelectronics, 634050, pr. Lenina, 40, Tomsk, Russia
[3]Polzunov Altai State Technical University, 656038, pr. Lenina, 46, Barnaul, Russia

Abstract. In this paper, we address the large-scale coordination of decisions impacting the consumption of a common shared resource of limited capacity by managed elements. We propose a decentralised token-based scheme, where each token represents a share of the resource. Our token-based protocol is meant to provide statistical guarantees on the average total resource usage and the average lateness of node actions due to the coordination. Experiments with the coordination of 10.000 autonomic managers have shown very good results for large spectrum of parameter values and system's regimes.

1 Introduction

The enormous growth in complexity of today's computer systems and applications has pushed IBM to propose a comprehensive effort, called *autonomic computing*, to automate their management [1]. In IBM's vision, an autonomic system is composed of *autonomic elements i.e.*, a *managed element* observed and adapted by an *autonomic manager* that operates a closed-loop control in the sense of control theory. Autonomic elements implement the well-known MAPE-k loop from [1]. As today's systems are more and more large-scale, new schemes are needed that allow a large number of autonomic managers to coordinate effectively with each others [2].

In this paper, we address the problem of coordinating decisions where the dependency among them is system-wide and implicit, namely the level of consumption of a central shared resource. The objective is to make sure that autonomic managers take decisions at the pace of changes in the need of their managed elements while the total required resource never exceeds its total capacity. More precisely, the coordination is meant to provide for statistical guarantees both on the global resource usage and on the fair distribution of the resource during the whole execution of the system. Coordination can be made to *emerge* [3–5] from local decisions, but making the current levels of resource usage of all managed elements available to every autonomic manager would inevitably leads to breach the time constraints of local control loops as the system scales.

We rather propose a token-based scheme, inspired from decentralised mutual exclusion algorithms [6]. The total amount of available resource is divided up in tokens that represent a right to use the corresponding amount of resource. Autonomic managers are organized into an overlay network [7] over which tokens circulate using random walks [6]. When an autonomic manager takes a decision that implies a raise in resource consumption, it first limits it to the amount represented by tokens it holds and then try to acquire new tokens to raise it further. When the autonomic manager takes a decision that lowers the consumption, tokens are reinjected in the overlay network for other autonomic managers.

Such a coordination scheme must exhibit both a fair distribution of the resource among autonomic managers, and the best possible global usage of the shared resource. These properties are highly influenced both by the quality of the interconnection provided by the overlay network and by the fluidity in the flow of tokens. In this paper, we propose an algorithm for the overlay network and we study a token-based coordination scheme exposing different parameters and show how their values can be chosen to balance the above objectives and scale. We report on experiments on systems of 10.000 autonomic managers that validate the approach and provide insights to tune the parameters to the application to get the best possible coordination.

2 Coordinating decisions in autonomic computing

Large-scale coordination of autonomic managers can be seen as a dynamic distributed decision-making problem. Centralised solutions, where all of the information is collected and decisions taken by a central entity do not scale. To scale, decisions must be made locally, and then coordinated by disseminating the information to every decision-making sites. But collecting locally all of the information at decision-time does not scale when the pace of decision-

*e-mail: melekhova@gmail.com

making is faster than the time required for the collection. Approaches to decouple the collection of information from the decision-making have been proposed, yet they do not scale further when the time for a required information to travel from the site where it is generated to the site where it is needed exceeds the deadlines of local decisions it impacts. In autonomic computing, where decisions must be made within the strict time limits imposed by the local control, the decoupling approach must also ensure the local availablility and freshness of data within these strict time constraints.

We presente our general token-based approach to large-scale coordination decomposes into three steps: (1) define a suitable semantics of tokens with regards to the local decision-making processes ; (2) link autonomic managers through a specific overlay network and adopt a suitable token communication policy ; (3) adopt a local decision-making process able to execute within the decision deadlines.

2.1 A use case in geotracking

Our use case considers the geotracking of large fleets of trucks coming from the French ANR project SALTY [8], however the scope can be different: robot/sensor networks for global resource charing, for example, communications, etc.. Typically, geotracking is used to follow trucks as to provide users with alerts as soon as they exit a predefined corridor or, perhaps more commonly, when approaching logistic bases to trigger resource allocation (doors, personal) for their arrival. Our industrial partner offers value-added services by making its customers receive their alerts through a unique centralised position processor called GeoHub[1]. Operators of fleets have trucks with GPS that send positions to the GeoHub through GSM networks.

2.2 Related work

Today, perhaps the most promising approaches are bio-inspired algorithms where global behavior *emerges* from purely local decisions [3, 9]; some applications have been conceptually explored [5] and only recently experimented for the first time in autonomic computing [10]. Crucially depending on information flows [11], emergence approaches for large-scale autonomic computing are also limited by the convergence delays in the broadcasting of information and in reaching an overall stability [12, 13]. Indeed, emergence approaches still require a lot of experiments, to which our work claims to contribute.

Louleiro *et al.* [14] address a large-scale resource allocation problem closely related to ours. Autonomic managers gather each workloads, then its autonomic manager uses its utility function to find the right decision. The authors use an *epidemic protocol* [12] to broadcast the individual workloads to all entities. Simulated on overlay networks of 500 entities, with degrees between 5 and 7, their approach requires more than 10 cycles of message exchanges between all of the nodes to propagate the information over the whole network. The number of cycles can be diminished by augmenting the degrees, but the number of messages per cycle would also augment. Picard *et al.* [15] propose a similar solution for distributed constraint satisfaction, without experimenting a dynamic version of the problem though. Other works on protocols for resource discovery in grids have shown that broadcasting approaches have an exponential complexity in the number of messages [16, 17].

Few related works address decision-making in the context of decentralised control where the system has strict timing constraints. Decentralised, networked control is still an open problem in control theory [18], even at the small-scale. Nevertheless, efforts to coordinate decisions in distributed autonomic systems have been made. IBM's teams have tackled the problem of energy management in data centers, but for small-scale systems only [19]. Work from Oliveira *et al.* [20] on synchronising autonomic control loops also targets the small-scale and doesn't take into account the timing issues of control. Many of the other works published recently are in the same vein [21–23].

Coordination in general has been studied in many fields, among which multi-agents systems (MAS), to such an extent that it is impossible to cover all of them. Compared to our proposal, most of these works address relatively small-scale systems or do not consider the timing issues of control [24]. For example, Gerber and Jung [25] consider a solution where local decisions are taken without delays, but use a global reasoning coordinating them that would hardly scale. Rustogi and Singh [26], as us, use local decision-making and show that giving locally more global information does not always lead to better decisions, concurring to more decoupling between local decision-making and information dissemination.

In the field of decision-making, reinforcement learning and *collaborative reinforcement learning* (CRL) has also been studied as a coordination mechanism, but most of the works are for a few decision-makers [27–30]

In conclusion, our contribution is the design and the implementation of new approaches and techniques in the field of decentralized decision-making subject to time constraints, more or less strict.

3 Overlay network

The coordination decisions in large-scale systems demands all nodes be interconnected to respect the fluidity in the information exchanging and the management of the dynamic network. We propose a self-organization approach based on decentralised algorithms for the construction and the maintenance of overlay network with the very weak distribution of node degrees. In more detail, the overlay network management algorithms and the results of experimentations are introduced in [31] and are excluded from our actual discussion.

[1] The term "GeoHub" belongs to Deveryware.

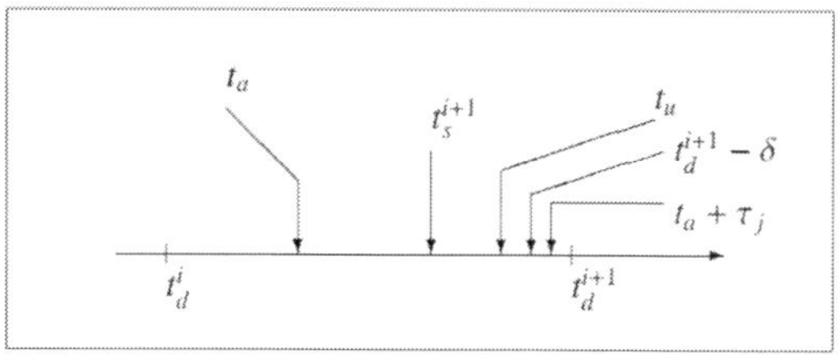

Fig. 1. Main instants when deciding and handling arriving tokens.

4 Token-based coordination

Based on the overlay network management algorithms, nodes exchange tokens controlling the (quantitative) access to the central shared resource, the GeoHub. Tokens circulate on the overlay network by random walks *i.e.*, token messages do not identify their destination but are rather sent to a randomly selected neighbour which either captures it for its own usage or immediately releases it and sends it again to a randomly selected neighbour among its own ones. Each token corresponds to the same rate of position sendings. To enforce its rate, the token has a period p_j and an earliest time of next utilisation t_u; each time $t \geq t_u$ a token is used to enable a position sending, its t_u is reset to $t + p_j$.

4.1 Coordination algorithms

The coordination decomposes in two main coordinated decision-making: handling position receptions and handling token arrivals. The coordinated decision-making occurs first when a position is received from the GPS of the truck through GeoHub, at which time the local decision-making process proposes a delay to the next position sending according to the needs of the geotracking applications and tries to plan it if the necessary token is available. When a token arrive, the autonomic manager must decide if it can capture or reemit it, and in the case of capture, plan a new position sending according to the most recently required time for it..

A central goal of the coordination is to minimise the waiting time for a token when an autonomic manager needs one and one is available somewhere. This goal is achieved by favouring, everywhere in the algorithms, the fluidity in the circulation of tokens against their retention. Considering the events introduced in the previous paragraph, the issue boils down to the conditions under which tokens can be kept after usage (at position reception time) or captured (at token arrival time). Our proposal is to use a unique parameter to control the fluidity, namely the time period τ_j during which a token can be kept before using it. The larger τ_j is, the more likely a token can be kept or captured. This criteria is therefore used both at token arrival time, to decide if it can be captured, and after a decision to decide if it can be kept.

Figure 1 illustrates the different important instants for some illustrative scenario. At time t_d^i, a position is received and the local decision-making proposes a delay for the next position sending that should occur at t_s^{i+1}. The coordinated decision is then to plan the next sending at $t_d^{i+1} \geq t_s^{i+1}$ that corresponds to the least t_u of a token the autonomic manager holds. Figure 1 shows a scenario where the autonomic manager does not have a token matching t_s^{i+1}, so $t_d^{i+1} > t_s^{i+1}$. When a token arrives at t_a, the autonomic manager captures it only if the conditions over τ_j hold and if it can use it to advance or better match t_d^{i+1}. The presented scenario illustrates the case where $t_a \leq t_s^{i+1} \leq t_a + \tau_j$ and $t_u \leq t_a + \tau_j$, which implies that the token can be used to advance t_d^{i+1}, hence it can be captured. Pragmatically, modifying t_d^{i+1} involves sending an order to the GPS through the GeoHub, hence the change must worth this effort (and respect a minimal delay). This requirement is formulated as the constraints $t_s^{i+1} + \epsilon \leq t_d^{i+1} - \delta$ and $t_u \leq t_d^{i+1} - \delta$ for some predefined ϵ and δ representing the shortest delay for modification and the smallest allowed one.

Algorithm 1 defines more precisely the coordinated decision at position reception time. First, the token that was previously captured to "cover" the current sending is marked as used by resetting its t_u to $t_d^i + p_j$. Next, the token is looked up to see if it must be released. Again, it will be kept only if it can be used for the next required sending, *i.e.* if the token can be used and the next sending planned within the next τ_j period of time, or more formally if $t_{u'} \leq t_d^i \wedge t_s^{i+1} \leq t_d^i + \tau_j$. If the token is kept, the next position sending can be planned (line 4). Otherwise, the autonomic manager will have to wait for the arrival of another, capturable token (line 7). Line 4 shows how the pragmatical parameter ϵ is used to force the next position sending to occur in at least ϵ seconds. Algorithm 2 implements the coordinated decision at token arrival time. To capture a token, the autonomic manager must need one and/or be able to use it during the next τ_j period. From the above discussion about Figure 1, the following conditions must be observed to (re)plan the next sending:

- $t_s^{i+1} \leq t_d^{i+1} - \delta$ *i.e.*, t_d^{i+1} can be advanced;
- $t_a + \epsilon \leq t_d^{i+1} - \delta$ *i.e.*, advancing it is feasible;
- $t_u \leq t_a + \tau_j$ *i.e.*, the token can be used;
- $t_u \leq t_d^{i+1} - \delta$ *i.e.*, and be used for that;
- $t_s^{i+1} \leq t_a + \tau_j$ *i.e.*, and the next sending would occur before τ_j.

Common data (algorithms 1 and 2)

f_j, tokens' position sending frequency
$p_j = 1/f_j$, the period of tokens
ϵ, minimal delay between a decision and the actual sending
δ, minimal modification in the time of the next sending justifying its change
t_u^j, earliest time for the next utilisation of token j
t_d^i, time of the ith local decision
t_s^{i+1}, requested time for the next position sending
t_d^{i+1}, time for the next position sending as imposed by
the coordination

Algorithm 1: Position reception time t^i_d

```
let t be the token captured for this sending in
    // mark the token for the current sending
    set t_u' ← t^i_d + p_j
    // release the token, if required
    if t_u' ≤ t^i_d ∧ t^{i+1}_s ≤ t^i_d + τ_j then
        keep the token t
        // choose the time of next sending t^{i+1}_d
        set t^{i+1}_d ← max(t^{i+1}_s, t_u', t^i_d + ε)
    else
        release the token t
        // no token left, so no next sending
        set t^{i+1}_d ← ∞
end
```

Algorithm 2: Token t arrival time t_a

```
if t^{i+1}_s ≤ t^{i+1}_d − δ ∧ t_a + ε ≤ t^{i+1}_d − δ ∧
    t_u ≤ t_a + τ_j ∧ t_u ≤ t^{i+1}_d − δ ∧ t^{i+1}_s ≤ t_a + τ_j
 then
    if a token has been captured since t^i_d then
        release the previous token
    keep the new token
    set t^{i+1}_d ← max(t^{i+1}_s, t_u, t_a + ε)
else
    if first token since t^i_d then
        release the token
    else
        let t' be the previously seized token in
            if PREFER(t^{i+1}_d, t, t') then
                keep t and release t'
            else
                keep t' and release t
        end
```

Again to augment the fluidity of tokens, only one can be captured between two position sendings. The first to arrive will be if close enough to t^{i+1}_s. If another token has been captured before, then the autonomic manager will keep the one that it prefers. Two criteria for the function PREFER have been tested: always keep the previously captured token t' or keep t if its t_u is closer to t^{i+1}_d, in which case $t^{i+1}_d = t^{i+1}_s$ and then $t'_u < t_u \leq t^{i+1}_s$

5 Experimental results

Our experimental setup uses a discrete-event simulation of the GeoHub to receive each request for position sendings from autonomic managers and to plan a push of data at the requested moment. Autonomic managers are implemented as concurrent and distributed components in Java connected through RMI. 10.000 autonomic managers were distributed among 50 JVM (Oracle Java 7 update 51 on Mac 0s X 10.9), themselves running by group of 10 on 5 computers (Mac mini late-2012, with 2.6GHz Intel Core i7 and 16Gb 1600MHz DDR3) connected through Ethernet. Another computer (Mac pro mid-2012, with 2×2.4GHz Core Intel Xeon and 24Gb 1333MHz DDR3 ECC) runs the GeoHub simulator.

The overlay network topology was very stable, with each autonomic managers having an average of 15.5 neighbours, with a standard deviation of 1.5.

5.1 The core experiments

The core experiments were targeting two major questions: is the token-based coordination effective and fair enough, and can we characterise to some extent the spectrum of parameters of the algorithms that works well and in what situations? As a measure of effectiveness, we have looked at the average global resource usage as well as its standard deviation. To evaluate the fairness among the different autonomic managers, we have measured for each autonomic manager the average lateness in each of its position sendings (*i.e.* the delay between the requested time and the actual time the sending was done), and then we have looked at the distribution, the mean and the standard deviation of these average latenesses.

To cover different and meaningful situations, we have evaluated three main system's regimes: overloaded, even and underloaded. In each of them, the capacity of the Geohub has been fixed at 50 positions processed per second (*pps*).

As the number of tokens can influence the coordination, we have experimented with 400, 800, 1200 and 1600 tokens, giving token periods $p_j = 8, 16, 24$ and 32 seconds. Our previous work [31] has shown through simulations that the coordination works well when $\tau_j \approx p_j$. So experiments used as values of τ_j roughly $0.75p_j$, p_j, $1.125p_j$, $1.25p_j$ and $1.5p_j$. Each run lasted 1.5 hours. To compute the distribution of the rate of positions actually processed, we have measured it by periods of 30 seconds and then computed the average and standard deviation of these values. To compute the mean of lateness means, we have computed the mean for each autonomic manager, and then we have computed the mean of these means and the standard deviation.

As expected from our previous simulations, these experiments show that good results are obtained for values of τ_j between p_j and $1.5p_j$, but slightly dropping as τ_j increases over $1.25p_j$. The influence of the number of tokens appears less important, but a too low or a too large number of tokens doesn't seem to help, especially for small τ_j. Under the overloaded scenario, the global usage of the resource can be very close to the optimal 50 pps. The distribution of the mean latenesses also show a very good fairness, though it should be enhanced and finding new criteria for capturing tokens that would decrease the standard deviation and increase the fairness is certainly a goal of future research. For larger τ_j the distributions are very similar and relatively grouped, while for the smaller ones they tend to be less grouped, with the one for $\tau_j = 16$ having a somewhat long tail, hence favouring $\tau_j > p_j$ The variants on the PREFER function (Algorithm 2) did not show significant differences, so they are not included here.

The even and underloaded scenarios make it easier to assess the loss of resource imposed by the algorithm. In the even scenario, we can see a larger standard deviation in global resource usage and a larger relative standard deviation in the mean of mean latenesses. This is due to the fact that when the pressure on the central resource is no longer controlled, when under 50, the full randomness of the requested positions shows directly. This is even more the case in the underloaded scenario, where basically no control is imposed by the tokens, but only a lower performance due to the potential unavailability of tokens at the right moment. Hence the means in these two cases are more important than the standard deviations. And we can observe that in the underloaded scenario, the mean resource usage and the mean of mean lateness are close to 33.3 and 0, showing a low impact of the token-based scheme *per se*.

6 Conclusion

In this paper, we have presented a new token-based scheme for large-scale coordination in decision-making for autonomic systems that provides for statistical guarantees on the quality and fairness of the coordination. Our main contributions lie in the quantitative nature of tokens experimented in our coordination scheme, in a novel simple local rule to capture and release tokens, and in large-scale experiments on systems of 10.000 autonomic managers. Experiments on our geotracking use case have shown that, on an overloaded scenario, our scheme can achieve as much as 98–99% in the global usage of the shared resource with a standard deviation of less than 1%. Fairness is also very good. In the same overloaded scenario, the ratio standard deviation over average is kept under 25%. Overall, very good coordination results are obtained for a wide spectrum of parameter values and system's regimes.

References

[1] IBM, White paper, IBM Corporation (2003), 1st Edition, 4th in June 2006

[2] J.O. Kephart, *Engineering decentralized autonomic computing systems*, in *SOAR'10* (ACM, New York, NY, USA, 2010), 1–2, ISBN 978-1-4503-0087-2

[3] L. Steels, *Towards a Theory of Emergent Functionality*, in *From Animals to Animats (First Int. Conf. on Simulation of Adaptive behaviour)* (MIT Press, 1990), 451–461

[4] T.D. Wolf, T. Holvoet, Tech. Rep. CW384, Department of Computer Science, KU Leuven (2004)

[5] T.D. Wolf, T. Holvoet, *Emergence versus self-organisation: different concepts but promising when combined*, in *Engineering Self Organising Systems: Methodologies and Applications* (Springer-Verlag, 2005), Vol. 3464 of *Lecture Notes in Computer Science*, 1–15

[6] A. Israeli, M. Jalfon, *Token Management Schemes and Random Walks Yield Self-Stabilizing Mutual Exclusion*, in *ACM Symposium on Principles (PODC'90)* (1990), 119–131

[7] E.K. Lua, J. Crowcroft, M. Pias, R. Sharma, S. Lim, IEEE Communications Surveys and Tutorials **7**, 72 (2005)

[8] O. Melekhova, M.A. Abchir, P. Châtel, J. Malenfant, I. Truck, A. Pappa, *Self-Adaptation in Geotracking Applications: Challenges, Opportunities and Models*, in *ADAPTIVE 2010* (IEEE, 2010), 68–77, ISBN 978-1-61208-001-7

[9] A. Ronzhin, A.I. Saveliev, O. Basov, S. Solyonyj, *Conceptual Model of Cyberphysical Environment Based on Collaborative Work of Distributed Means and Mobile Robots*, in *Interactive Collaborative Robotics - First International Conference, ICR 2016, Budapest, Hungary, August 24-26, 2016, Proceedings* (2016), 32–39

[10] T.D. Wolf, G. Samaey, T. Holvoet, D. Roose, *Decentralised Autonomic Computing: Analysing Self-Organising Emergent Behaviour using Advanced Numerical Methods*, in *IEEE ICAC'05* (2005)

[11] T.D. Wolf, T. Holvoet, *Designing Self-Organising Emergent Systems based on Information Flows and Feedback-loops*, in *IEEE SASO 2007* (2007)

[12] M. Jelasity, A. Montresor, O. Babaoglu, ACM Transactions on Computer Systems **23**, 219 (2005)

[13] I. Hegedűs, R. Ormándi, M. Jelasity, *Gossip-based Learning under Drifting Concepts in Fully Distributed Networks*, in *IEEE SASO'12* (2012), 79–88

[14] E. Loureiro, P. Nixon, S. Dobson, ACM Transactions on Autonomous and Adaptive Systems **7**, article 14:1 (2012)

[15] G. Picard, M.P. Gleizes, P. Glize, *Distributed Frequency Assignment Using Cooperative Self-Organization*, in *IEEE SASO'07* (2007)

[16] E. Jeanvoine, C. Morin, D. Leprince, *Un protocole de découverte de ressources optimisé pour l'allocation de ressources dans les grilles*, in *CFSE 2006* (2006), 49–59

[17] A. Iamnitchi, I.T. Foster, D. Nurmi, *A Peer-to-Peer Approach to Resource Location in Grid Environments*, in *IEEE HPDC-11)* (2002), 419

[18] V. Kumar, N. Leonar, A.S. Morse, eds., *Cooperative Control*, Vol. 309 of *Lecture Notes in Control and Information Sciences* (Springer-Verlag, 2005)

[19] J. Kephart, H. Chan, R. Das, D. Levine, G. Tesauro, F. Rawson, C. Lefurgy, *Coordinating Multiple Autonomic Managers to Achieve Specified Power-Performance Tradeoffs*, in *ICAC'07* (2007)

[20] F.A. de Oliveira Jr., R. Sharrock, T. Ledoux, *Synchronization of Multiple Autonomic Control Loops: Application to Cloud Computing*, in *Coordination'12* (Springer-Verlag, 2012), Vol. 7274 of *LNCS*, 29–43

[21] P. Vromant, D. Weyns, S. Malek, J. Andersson, *On Interacting Control Loops in Self-Adaptive Systems*, in *ACM SEAMS'11* (2011), 202–207

[22] D. Weyns, S. Malek, J. Andersson, *On Decentralized Self-Adaptation: Lessons from the Trenches and Challenges for the Future*, in *ACM SEAMS'10* (2010), 84–93

[23] L. Baresi, S. Guinea, G. Tamburrelli, *Towards Decentralized Self-adaptive Component-based Systems*, in *ACM SEAMS'08* (2008), 57–64

[24] G. Mahdi, A. Gouaïch, F. Michel, *Towards an Integrated Approach of Real-Time Coordination for Multi-agent Systems*, in *KES-AMSTA'10* (Springer-Verlag, 2010), Vol. 6070 of *LNAI*, 253–262

[25] C. Gerber, C. Jung, *Resource Management for Boundedly Optimal Agent Societies*, in *ECAI'98* (John Wiley & Sons, 1998), 1–6

[26] S. Rustogi, M. Singh, *The Bases of Effective Coordination in Decentralized Multi-agent Systems*, in *ATAL'98* (Springer-Verlag, 1999), Number 1555 in LNAI, 149–161

[27] L. Busoniu, R. Babuska, B. De Schutter, Systems, Man, and Cybernetics, Part C: Applications and Reviews, IEEE Transactions on **38**, 156 (2008)

[28] J.R. Kok, N. Vlassis, Journal of Machine Learning Research **7**, 1789 (2006)

[29] I. Dusparic, V. Cahill, *Research Issues in Multiple Policy Optimization Using Collaborative Reinforcement Learning*, in *ACM SEAMS'07* (2007), 18:1–7, ISBN 0-7695-2973-9

[30] J. Dowling, S. Haridi, in *Reinforcement Learning: Theory and Applications* (I-Tech Education and Publishers, 2008), chap. 8, 143–166

[31] O. Melekhova, Technique et science informatiques **33**, 31 (2014)

MATEC Web of Conferences **161**, 03025 (2018) https://doi.org/10.1051/matecconf/201816103025
13th International Scientific-Technical Conference on Electromechanics and Robotics "Zavalishin's Readings" - 2018

Robust anti-windup control for marine cyber-physical systems

Mikhail Kakanov, *Mikhail* Kaliuzhny, *Oleg* Borisov, *Vladislav* Gromov, *Sergey* Vlasov, *Sergey* Somov and *Anton* Pyrkin

ITMO University, Department of Computer Science and Control Systems, 197101 49 Kronverksky Pr., St. Petersburg, Russia

Abstract. In this paper the robust output control with anti-windup compensation and its implementation to the robotic boat are addressed. The detailed control design and stability analysis of the closed-loop systems are provided in the work. Extensive experimental verification of the dynamic positioning system based on various modifications of the basic controller is carried out by means of robotic boat. The corresponding experimental results are presented and analysed.

1 Introduction

Development of cyber-physical systems (CPS) is a brand new ongoing scientific trend in applied control theory. It assumes design of sophisticated systems comprised of multiple smart devices interconnected between each other and able to make decisions on their own and adapt to external changes without human intervention. There is a wide range of applications of such systems including for instance manufacturing and transportation.

This research is focused on development of robust output control algorithm with anti-windup compensation for MIMO plants with uncertain parameters and bounded inputs. In particular a solution suggested in this paper can be used in dynamic positioning (DP) problem to control a surface vessel being a component of marine CPS.

If an integral term being a part of the internal model is included in the control system, input saturation might cause various adverse effects when implementing regulator to real plants. Anti-windup modification first mentioned in [1] can be used to resolve this issue. There is a number of scientific works developing anti-windup approaches. Overview of modern anti-windup design is consider in [2]. Application of the toolkit of linear matrix inequalities (LMI) to synthesis is presented in [3-8]. The anti-windup is combined with internal control mode (ICM) in [9-11]. In the application to discrete-time systems, the anti-windup scheme is considered in [12]. The consecutive compensator approach described in [13, 14] and based on the pacification approach [15] is used as a basis for control design for marine CPS. The relevant problem of output regulation for MIMO systems with unknown disturbance compensation and application of the suggested control approach to surface vessels to control its velocities is addressed in [16]. Detailed theoretical background on ship control is provided in [17]. Some results on saturated control design being applied to the research setup of the robotic boat is presented [18, 19] and to the quadcopters in [20-22].

The consecutive compensator approach is augmented with an auxiliary integral loop and anti-windup scheme. Such solution is needed, on the one hand, to eliminate the steady state error and, on the other hand, to compensate integral windup.

This paper is organized as follows. The problem of this research is formulated in Section 2. Control design is given in Section 3. Stability analysis of the closed-loop system is provided in Section 4. Experimental verification and the corresponding test results are described and discussed in Section 5. The paper is summarized in Conclusions.

2 Problem formulation

Consider the plant

$$\dot{\boldsymbol{x}} = A\boldsymbol{x} + \boldsymbol{b}u + R\boldsymbol{\omega}, \tag{1}$$

$$y = \boldsymbol{c}^T\boldsymbol{x}, \tag{2}$$

where $\boldsymbol{x} \in \mathbb{R}^n$ is the state, $\boldsymbol{\omega} \in \mathbb{R}^{n_\omega}$ is the disturbance, which derivative and initial conditions are

$$\dot{\boldsymbol{\omega}} = 0, \qquad \boldsymbol{\omega}(0) \neq 0, \tag{3}$$

$y \in \mathbb{R}^1$ is the measurable output, $u \in \mathbb{R}^1$ is the input satisfying

$$u = sat(v) = \begin{cases} u_{max} \; if v \geq u_{max}, \\ u \; if u_{min} < v < u_{max}, \\ u_{min} \; if v \leq u_{min}, \end{cases} \tag{4}$$

where u_{max} and u_{min} are the input saturation limits, v is the control signal generated by the nominal linear controller, A, R, $\boldsymbol{b}$, $\boldsymbol{c}$ are matrices and vectors of corresponding dimensions.

Assumption 1: The plant (1), (2) is minimum phase.

Assumption 2: The relative degree of the plant (1), (2) $\rho \geq 1$ is known.

Assumption 3: The input saturation limits u_{max} and u_{min} have different signs, but the same magnitude.

Assumption 4: The disturbance is bounded $\omega \in \mathcal{L}_\infty$ and the nominal control signal u_0 needed for its compensation at steady state satisfies

* Corresponding author: kakanov.mikhail@niuitmo.ru

$$u_{min} \leq |u_0| \leq u_{max}.$$

The objective is to asymptotically stabilize the output under conditions of external disturbance effects and presence of the integral windup.

3 Robust control design

The plant model (1), (2) can be represented as

$$y(s) = \frac{b(s)}{a(s)}\bar{u}(s) + \frac{r(s)}{a(s)}\omega(s), \quad (5)$$

where $b(s)$ is the Hurwitz polynomial due to Assumption 1.

Rewrite the plant model (5) as

$$\frac{a(s)}{b(s)}y(s) = \bar{u}(s) + \frac{r(s)}{b(s)}\omega(s),$$

$$\left(c(s) + \frac{d(s)}{b(s)}\right)y(s) = \bar{u}(s) + \left(r_2(s) + \frac{r_1(s)}{b(s)}\right)\omega(s), \quad (6)$$

$$c(s)y(s) = \left(\bar{u}(s) - \frac{d(s)}{b(s)}y(s) + \frac{r_1(s)}{b(s)}\omega(s)\right) + r_2(s)\omega(s).$$

and express (6) as two interconnectyed subsystems as

$$z(s) = \frac{d(s)}{b(s)}y(s) - \frac{r_1(s)}{b(s)},$$

$$y(s) = \frac{1}{c(s)}\left(\bar{u}(s) - z(s)\right) + \frac{r_2(s)}{c(s)}\omega(s),$$

which is equivalent to the state-space model

$$\begin{bmatrix}\dot{z}_1\\ \dot{z}_2\end{bmatrix} = \begin{bmatrix}A_{11} & A_{12}\\ A_{21} & A_{22}\end{bmatrix}\begin{bmatrix}z_1\\ z_2\end{bmatrix} + \begin{bmatrix}0\\ b_2\end{bmatrix}\bar{u} + \begin{bmatrix}R_1\\ R_2\end{bmatrix}\omega, \quad (7)$$

$$y = \begin{bmatrix}0 & c^T\end{bmatrix}\begin{bmatrix}z_1\\ z_2\end{bmatrix}, \quad (8)$$

where A_{11} is the Hurwitz matrix due to Assumption 1, if the matrix A_{22} is chosen in companion form, then $\boldsymbol{b}_2^T = [0 \ \cdots \ 0 \ \ b_0]$, $\boldsymbol{c}_2^T = [1 \ \ 0 \ \cdots \ 0]$.

Choose the control law

$$v = -\kappa\left(c_q^T\xi + y\right) - \gamma\eta, \quad (9)$$

$$\dot{\xi} = A_q\xi + b_q y, \quad (10)$$

$$\dot{\eta} = \kappa\left(c_q^T\xi + y\right) + v\psi(v), \quad (11)$$

$$\psi(v) = v - sat(v), \quad (12)$$

where $\psi(v)$ is the memoryless nonlinearity, $\kappa > 0$, $\gamma > 0$, $v > 0$, A_q, $\boldsymbol{b_q}$, c_q are given as

$$A_q = \begin{bmatrix} q'_\rho\sigma & 1 & 0 & \cdots & 0\\ -q'_{\rho-1}\sigma^2 & 0 & 1 & \cdots & 0\\ \vdots & \vdots & \vdots & \ddots & \vdots\\ -q'_2\sigma^{\rho-1} & 0 & 0 & \cdots & 1\\ -q'_1\sigma^\rho & 0 & 0 & \cdots & 0\end{bmatrix},$$

$$b_q^T = \begin{bmatrix}q'_\rho\sigma & q'_{\rho-1}\sigma^2 & \cdots & q'_2\sigma^{\rho-1} & q'_1\sigma^\wedge\rho\end{bmatrix},$$

$$c_q^T = \begin{bmatrix}q_1 & q_2 & \cdots & q_{\rho-1} & q_\rho\end{bmatrix},$$

where $\sigma > 0$ and $q'_i \ (i = \overline{1,\rho})$ are chosen for the system (10) to be Hurwitz, $q_i \ (i = \overline{1,\rho})$ are coefficients of an arbitrary polynomial of the degree $\rho - 1$.

Interconnecting the plant (7), (8) and the control law (9)–(12) obtain the model of the closed-loop system (13).

Perform the change of coordinates $\chi = z_2 - \xi$ and compute the derivative

$$\dot{\chi} = \dot{z}_2 - \dot{\xi} = A_{21}z_1 + \left(A_{22} - I_0 - \kappa b_2\left(c_q^T + c_2^T\right)\right)z_2 + \left(A_q + \kappa b_2 c_q^T\right)\chi - b_2\gamma\eta - b_2\psi + R_2\omega,$$

where

$$I_0 = A_q + b_q c_2^T = \begin{bmatrix}0 & 1 & 0 & \dots & 0\\ 0 & 0 & 1 & \dots & 0\\ \vdots & \vdots & \vdots & \ddots & \vdots\\ 0 & 0 & 0 & \dots & 0\end{bmatrix}.$$

This change of coordinates puts the system (13) into the form (14).

Perform the second change of variables $\zeta = \left(\boldsymbol{b}_2^T\boldsymbol{b}_2\right)^{-1}\boldsymbol{b}_2^T z_2 + \eta$ and compute the derivative

$$\dot{\zeta} = \left(\boldsymbol{b}_2^T\boldsymbol{b}_2\right)^{-1}\boldsymbol{b}_2^T\dot{z}_2 + \dot{\eta} = \left(\boldsymbol{b}_2^T\boldsymbol{b}_2\right)^{-1}\boldsymbol{b}_2^T A_{21}z_1 + \left(\boldsymbol{b}_2^T\boldsymbol{b}_2\right)^{-1}\boldsymbol{b}_2^T(A_{22} + \gamma I)z_2 - \gamma\zeta + (v-1)\psi + \left(\boldsymbol{b}_2^T\boldsymbol{b}_2\right)^{-1}\boldsymbol{b}_2^T R_2\boldsymbol{\omega},$$

where I is the identity matrix of the corresponding dimension.

For the sake of simplicity denote

$$\bar{b}_2^T = \left(\boldsymbol{b}_2^T\boldsymbol{b}_2\right)^{-1}\boldsymbol{b}_2^T. \quad (15)$$

This change of coordinates taking into accound (15) puts the system (14) into the form (16) :

$$\begin{bmatrix}\dot{z}_1\\ \dot{z}_2\\ \dot{\xi}\\ \dot{\eta}\end{bmatrix}=\begin{bmatrix}A_{11} & A_{12} & 0 & 0\\ A_{21} & A_{22}-\kappa\boldsymbol{b}_2\boldsymbol{c}_2^T & -\kappa\boldsymbol{b}_2\boldsymbol{c}_q^T & -\boldsymbol{b}_2\gamma\\ 0 & \boldsymbol{b}_q\boldsymbol{c}_2^T & A_q & 0\\ 0 & \kappa\boldsymbol{c}_2^T & \kappa\boldsymbol{c}_q^T & 0\end{bmatrix}\begin{bmatrix}z_1\\ z_2\\ \xi\\ \eta\end{bmatrix}+\begin{bmatrix}0\\ -\boldsymbol{b}_2\\ 0\\ \nu\end{bmatrix}\psi+\begin{bmatrix}R_1\\ R_2\\ 0\\ 0\end{bmatrix}\omega. \tag{13}$$

$$\begin{bmatrix}\dot{z}_1\\ \dot{z}_2\\ \dot{\chi}\\ \dot{\eta}\end{bmatrix}=\begin{bmatrix}A_{11} & A_{12} & 0 & 0\\ A_{21} & A_{22}-\kappa\boldsymbol{b}_2(\boldsymbol{c}_q^T+\boldsymbol{c}_2^T) & -\kappa\boldsymbol{b}_2\boldsymbol{c}_q^T & -\boldsymbol{b}_2\gamma\\ A_{21} & A_{22}+I_0-\kappa\boldsymbol{b}_2(\boldsymbol{c}_q^T+\boldsymbol{c}_2^T) & A_q+\kappa\boldsymbol{b}_2\boldsymbol{c}_q^T & -\boldsymbol{b}_2\gamma\\ 0 & \kappa(\boldsymbol{c}_q^T+\boldsymbol{c}_2^T) & \kappa\boldsymbol{c}_q^T & 0\end{bmatrix}\begin{bmatrix}z_1\\ z_2\\ \chi\\ \eta\end{bmatrix}+\begin{bmatrix}0\\ -\boldsymbol{b}_2\\ -\boldsymbol{b}_2\\ \nu\end{bmatrix}\psi+\begin{bmatrix}R_1\\ R_2\\ R_2\\ 0\end{bmatrix}\omega. \tag{14}$$

$$\begin{bmatrix}\dot{z}_1\\ \dot{\zeta}\\ \dot{\chi}\\ \dot{z}_2\end{bmatrix}=\begin{bmatrix}A_{11} & 0 & A_{12} & 0\\ \bar{b}_2^TA_{21} & -\gamma & \bar{b}_2^T(A_{22}+\gamma I) & 0\\ A_{21} & -\boldsymbol{b}_2\gamma & A_{22}+I_0-\kappa\boldsymbol{b}_2(\boldsymbol{c}_q^T+\boldsymbol{c}_2^T)+\gamma\boldsymbol{b}_2\bar{b}_2^T & A_q+\kappa\boldsymbol{b}_2\boldsymbol{c}_q^T\\ A_{21} & -\boldsymbol{b}_2\gamma & A_{22}-\kappa\boldsymbol{b}_2\left(\boldsymbol{c}_q^T+\boldsymbol{c}_2^T\right)+\gamma\boldsymbol{b}_2\bar{b}_2^T & \kappa\boldsymbol{b}_2\boldsymbol{c}_q^T\end{bmatrix}\begin{bmatrix}z_1\\ \zeta\\ \chi\\ z_2\end{bmatrix}+\begin{bmatrix}0\\ (\nu-1)\\ -\boldsymbol{b}_2\\ -\boldsymbol{b}_2\end{bmatrix}\psi+\begin{bmatrix}R_1\\ \bar{b}_2^TR_2\\ R_2\\ R_2\end{bmatrix}\omega. \tag{16}$$

As noted above, the matrix A_{11} is Hurwitz due to Assumption 1. The matrix

$$A_1=\begin{bmatrix}A_{11} & 0\\ \bar{b}_2^TA_{21} & -\gamma\end{bmatrix}$$

is Hurwitz due to the choice of the parameter $\gamma>0$.

There exists a number κ_0 such that for $\forall\kappa\geq\kappa_0$ the matrix

$$A_2=\begin{bmatrix}A_{11} & 0 & A_{12}\\ \bar{b}_2^TA_{21} & -\gamma & \bar{b}_2^T(A_{22}+\gamma I)\\ A_{21} & -b_2\gamma & A_{22}-\kappa b_2\left(c_q^T+c_2^T\right)+\gamma b_2\bar{b}_2^T\end{bmatrix}$$

is Hurwitz, since only the element (3, 3) depends on the parameter κ, which can be set to be sufficiently large.

The overall state matrix of the closed-loop system (16) can be forced to be Hurwitz due to the structure of the matrix A_q and sufficiently large value of the parameter σ, which is included only in the block (4, 4).

Statement 1: Consider the plant model given at state-space (7), (8) and the control law (9)–(12). Hurwitzness of the state matrix of the resultant closed-loop system (16) follows from Hurwitzness of all its diagonal blocks, which is forced by the parameters $\gamma,\ \kappa,\ \sigma$.

4 Stability analisys of closed-loop system

Consider the case without taking into account the external disturbances, i.e. $\omega=0$. Denote

$$F=\begin{bmatrix}A-\kappa\boldsymbol{b}\boldsymbol{c}^T & -\kappa\boldsymbol{b}\boldsymbol{c}_q^T & -\gamma\boldsymbol{b}\\ \boldsymbol{b}_q\boldsymbol{c}^T & A_q & 0\\ \kappa\boldsymbol{c}^T & \kappa\boldsymbol{c}_q^T & 0\end{bmatrix},\boldsymbol{g}=\begin{bmatrix}-\boldsymbol{b}\\ 0\\ \nu\end{bmatrix},\boldsymbol{h}=\begin{bmatrix}-\kappa\boldsymbol{c}\\ -\kappa\boldsymbol{c}_q\\ -\gamma\end{bmatrix},\boldsymbol{\varsigma}=\begin{bmatrix}\boldsymbol{x}\\ \xi\\ \eta\end{bmatrix}. \tag{17}$$

Represent the closed-loop system (16) using notations (17) as follows

$$\dot{\boldsymbol{\varsigma}}=F\boldsymbol{x}+\boldsymbol{g}\psi(\upsilon), \tag{18}$$

$$\upsilon=\boldsymbol{h}^T\boldsymbol{\varsigma}. \tag{19}$$

According to the Popov criterion [22], the closed-loop system consisting of the linear part (18)–(19) with Hurwitz state matrix F and the memoryless nonlinear function $\psi(\upsilon)$ is absolutely stable for all initial conditions if there exists a constant $\bar{\omega}\geq0$ such that $(1+\lambda_i\bar{\omega})\neq0$ for each eigenvalue λ_i of the matrix F and the transfer function $W(s)=1+(1+s\bar{\omega})W_l(s)$ is strictly positive real, where $W_l(s)$ is the transfer function of the system (18)–(19).

The transfer function $W_l(s)$ can be derived as follows

$$W_l(s)=\boldsymbol{h}^T(sI-F)^{-1}\boldsymbol{g}$$
$$=\begin{bmatrix}-\kappa\boldsymbol{c}^T & -\kappa\boldsymbol{c}_q^T & -\gamma\end{bmatrix}\mathcal{F}\begin{bmatrix}-b\\ 0\\ \nu\end{bmatrix}, \tag{20}$$

where

$$\mathcal{F}=(sI-F)^{-1}=\begin{bmatrix}sI-A+\kappa\boldsymbol{b}\boldsymbol{c}^T & \kappa\boldsymbol{b}\boldsymbol{c}_q^T & \gamma\boldsymbol{b}\\ -\boldsymbol{b}_q\boldsymbol{c}^T & sI-A_q & 0\\ -\kappa\boldsymbol{c}^T & -\kappa\boldsymbol{c}_q^T & s\end{bmatrix}^{-1}.$$

Represent the matrix $\mathcal{F}$ in the block form

$$\mathcal{F}=\bar{\mathcal{F}}^{-1}=\begin{bmatrix}\bar{\mathcal{F}}_{11} & \bar{\mathcal{F}}_{12}\\ \bar{\mathcal{F}}_{21} & \bar{\mathcal{F}}_{22}\end{bmatrix}=\begin{bmatrix}\mathcal{F}_{11} & \mathcal{F}_{12}\\ \mathcal{F}_{21} & \mathcal{F}_{22}\end{bmatrix}, \tag{21}$$

where

$$\bar{\mathcal{F}}_{11}=sI-A+\kappa\boldsymbol{b}\boldsymbol{c}^T,\quad \bar{\mathcal{F}}_{12}=\begin{bmatrix}\kappa\boldsymbol{b}\boldsymbol{c}_q^T & \gamma\boldsymbol{b}\end{bmatrix},$$
$$\bar{\mathcal{F}}_{21}=\begin{bmatrix}-\boldsymbol{b}_q\boldsymbol{c}^T\\ -\kappa\boldsymbol{c}^T\end{bmatrix},\qquad \bar{\mathcal{F}}_{22}=\begin{bmatrix}sI-A_q & 0\\ -\kappa\boldsymbol{c}_q^T & s\end{bmatrix},$$

and use the Frobenius formula for block matrices inversion.

Calculate the inverse matrix of the block $\bar{\mathcal{F}}_{22}$

$$\bar{\mathcal{F}}_{22}^{-1}=\begin{bmatrix}sI-A_q & 0\\ -\kappa\boldsymbol{c}_q^T & s\end{bmatrix}^{-1}=\begin{bmatrix}\mathcal{A}_q & 0\\ \frac{\kappa}{s}\boldsymbol{c}_q^T\mathcal{A}_q & \frac{1}{s}\end{bmatrix},$$

where $\mathcal{A}_q = (sI - A_q)^{-1}$.

Calculate the block $\mathcal{F}_{11}$

$$\mathcal{F}_{11} = (\bar{\mathcal{F}}_{11} - \bar{\mathcal{F}}_{12}\bar{\mathcal{F}}_{22}^{-1}\bar{\mathcal{F}}_{21})^{-1} = \left(sI - A + \boldsymbol{b}\frac{\kappa b_q(s)(s+\gamma)}{sa_q(s)}\boldsymbol{c}^T\right)^{-1},$$

where

$$\frac{b_q(s)}{a_q(s)} = \boldsymbol{c}_q^T\mathcal{A}_q(s)\boldsymbol{b}_q + 1. \quad (22)$$

Remark 1: Note that, Hurwitzness of the numerator and denominator of the transfer function (22) can be achieved by choosing matrix A_q and vectors $\boldsymbol{b}_q$ and $\boldsymbol{c}_q$.

Calculate the block $\mathcal{F}_{12}$

$$\mathcal{F}_{12} = -\mathcal{F}_{11}\bar{\mathcal{F}}_{12}\bar{\mathcal{F}}_{22}^{-1} = \left[-\frac{\kappa(s+\gamma)}{s}\mathcal{A}_{11}\boldsymbol{b}\boldsymbol{c}_q^T\mathcal{A}_q(s) \quad -\frac{\gamma}{s}\mathcal{A}_{11}\boldsymbol{b}\right].$$

Calculate the block $\mathcal{F}_{21}$

$$\mathcal{F}_{21} = -\bar{\mathcal{F}}_{22}^{-1}\bar{\mathcal{F}}_{21}\mathcal{F}_{11} = \begin{bmatrix} \mathcal{A}_q(s)\boldsymbol{b}_q\boldsymbol{c}^T \\ \frac{\kappa}{s}(\boldsymbol{c}_q^T\mathcal{A}_q(s)\boldsymbol{b}_q + 1)\boldsymbol{c}^T \end{bmatrix}\mathcal{F}_{11} = \begin{bmatrix} \mathcal{A}_q(s)\boldsymbol{b}_q\boldsymbol{c}^T\mathcal{F}_{11} \\ \frac{\kappa b_q(s)}{sa_q(s)}\boldsymbol{c}^T\mathcal{F}_{11} \end{bmatrix}.$$

Calculate the block $\mathcal{F}_{22}$

$$\mathcal{F}_{22} = \bar{\mathcal{F}}_{22}^{-1} - \mathcal{F}_{21}\bar{\mathcal{F}}_{12}\bar{\mathcal{F}}_{22}^{-1} =$$
$$\begin{bmatrix} \left(1 - \frac{\kappa(s+\gamma)}{s}\mathcal{A}_q(s)\boldsymbol{b}_q\frac{\beta(s)}{\alpha(s)}\boldsymbol{c}_q^T\right)\mathcal{A}_q(s) & -\frac{\gamma}{s}\mathcal{A}_q(s)\boldsymbol{b}_q\frac{\beta(s)}{\alpha(s)} \\ \left(1 - \frac{\kappa(s+\gamma)b_q(s)}{sa_q(s)}\frac{\beta(s)}{a_q(s)}\right)\frac{\kappa}{s}\boldsymbol{c}_q^T\mathcal{A}_q(s) & \left(1 - \frac{\gamma\kappa b_q(s)}{sa_q(s)}\frac{\beta(s)}{\alpha(s)}\right)\frac{1}{s} \end{bmatrix},$$

where

$$\frac{\beta(s)}{\alpha(s)} = \boldsymbol{c}^T\mathcal{F}_{11}\boldsymbol{b} = \frac{b(s)sa_q(s)}{sa_q(s)a(s) + \kappa b_q(s)(s+\gamma)b(s)}.$$

Substitute derived expressions of the blocks $\mathcal{F}_{11}, \mathcal{F}_{12}, \mathcal{F}_{21}$ and $\mathcal{F}_{22}$ into the matrix $\mathcal{F}$ (for the sake of brevity the middle column is omitted)

$$\mathcal{F} = \begin{bmatrix} \mathcal{F}_{11} & * & -\frac{\gamma}{s}\mathcal{F}_{11}\boldsymbol{b} \\ \mathcal{A}_q(s)\boldsymbol{b}_q\boldsymbol{c}^T\mathcal{F}_{11} & * & -\frac{\gamma}{s}\mathcal{A}_q(s)\boldsymbol{b}_q\frac{\beta(s)}{\alpha(s)} \\ \frac{\kappa b_q(s)}{sa_q(s)}\boldsymbol{c}^T\mathcal{F}_{11} & * & \left(1 - \frac{\gamma\kappa b_q(s)}{sa_q(s)}\frac{\beta(s)}{\alpha(s)}\right)\frac{1}{s} \end{bmatrix}.$$

Substitute the matrix $\mathcal{F}$ in (21) and obtain

$$W_l(s) = \beta_\kappa(s) - \frac{\gamma v a_q(s)a(s)}{sa_q(s)a(s)} + \beta_\kappa(s),$$

where

$$\beta_\kappa(s) = \kappa b_q(s)(s+\gamma)b(s). \quad (23)$$

Set $\bar{\omega} = 0$, then the sufficient condition of the absolute stability of the system is strictly positive realness of the transfer function $W(s) = 1 + W_l(s)$. Calculate $W(s)$

$$W(s) = W_l(s) + 1 = 2\beta_\kappa(s) + \frac{(s - v\gamma)a_q(s)a(s)}{sa_q(s)a(s) + \beta_\kappa(s)} = \frac{2\kappa b_q(s)(s+\gamma)b(s) + (s - \gamma v)a_q(s)a(s)}{sa_q(s)a(s) + \kappa b_q(s)(s+\gamma)b(s)}.$$

It easy to show, that there exists a number κ_0 such that for $\kappa \geq \kappa_0$ both the numerator and denominator of the transfer function $W(s)$ are Hurwitz. The relative degree of $W(s)$ is zero. As a consequence, strictly positive realness follows, i.e. $\mathrm{Re}W(j\omega) > 0$, $\forall\omega \in [-\infty, \infty]$or equivalently $\mathrm{Re}W_l(j\omega) > 1$, $\forall\omega \in [-\infty, \infty]$ and the absolute stability of the system (18), (19) follows in accordance to the Popov criterion [22].

Consider the case taking into account the external disturbances, i.e. $\omega \neq 0$, and analyze the steady error. Obviously, at steady state the relations $v = \mathrm{sat}(v)$ and $\psi(v) = v - \mathrm{sat}(v) = 0$ hold.

In order to determine steady state behavior use the Sylvester equation applied to the model (13) and focus on the fourth row

$$\Sigma S = \kappa(\boldsymbol{c}_2^T\Pi_2 + \boldsymbol{c}_q^T\Pi_\xi),$$

which shows that at steady state

$$\boldsymbol{c}_2^T\Pi_2 + \boldsymbol{c}_q^T\Pi_\xi = 0. \quad (24)$$

Find the relation between $\boldsymbol{c}_2^T\Pi_2$ and $\boldsymbol{c}_q^T\Pi_\xi$. Consider the auxiliary variable z_0

$$\boldsymbol{z_0} = \boldsymbol{y} + \boldsymbol{c}_q^T\boldsymbol{\xi}, \quad (25)$$

taking into account (24), the steady state value of which is zero.

From (10) follows

$$\xi(s) = (sI - A_q)^{-1}\left(\boldsymbol{b}_q y(s) + \xi(0)\right),$$

where $\xi(0)$ is a vector of initial conditions. Rewrite (25)

$$z_0(s) = \left(\boldsymbol{c}_q^T(sI - A_q)^{-1}\boldsymbol{b}_q + 1\right)y(s) + \epsilon(s),$$

where $\epsilon(s) = \boldsymbol{c}_q^T(sI - A_q)^{-1}(\xi_1(0) + \xi_2(0))$ corresponds to the exponentially decaying function $\epsilon(t)$. If $\boldsymbol{c}_q^T$ is chosen so that the numerator of the transfer function $\left(\boldsymbol{c}_q^T(sI - A_q)^{-1}\boldsymbol{b}_q + 1\right)$ is Hurwitz and the relative degree is zero, then from

$$y(s) = \left(\boldsymbol{c}_q^T(sI - A_q)^{-1}\boldsymbol{b}_q + 1\right)^{-1}(z_0(s) - \epsilon(s))$$

find that the steady-state error of y as well as $\boldsymbol{c}_q^T\xi$ converges to zero.

5 Experimental verification

The practical verification of this research has been carried out using the robotic boat shown in Figure 1. It is intended to test DP systems of marince CPS under laboratory conditions.

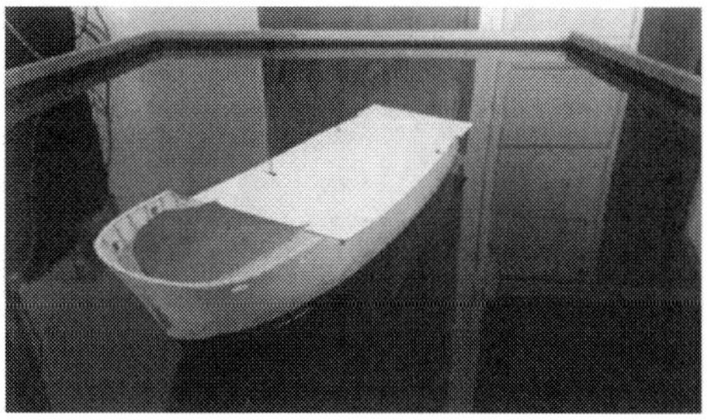

Fig. 1. Robotic boat.

The boat is equipped with the main engine, bow and stern tunnel thrusters and heading servo drive (Fig. 2). Dimensions of the boat are (0.432×0.096×0.052) *m*. Its workspace is represented by the specific basin made of plywood sheets with dimensions (1.50×1.10×0.1) *m*. The computer vision system emulating satelite navigation systems is based on the RGB video signal provided by the digital camera placed above the basin. The control signals transmitted via radio channel saturate at the rates $[-127;\ 127]$ due to the pulse-width modulation technique.

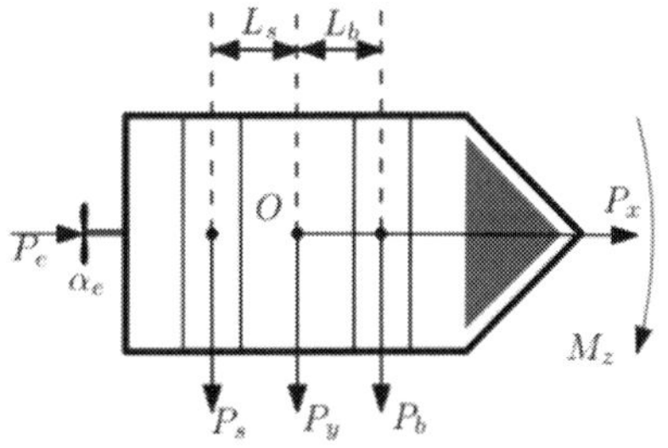

Fig. 2. Actuator configuration of the robotic boat.

The particular form of the control law is (9)–(12) applied to the boat model with the relative degree being equal to 2 is

$$A_q = \begin{bmatrix} -q_2'\sigma & 1 \\ -q_1'\sigma^2 & 0 \end{bmatrix},\quad \boldsymbol{b}_q = \begin{bmatrix} q_2'\sigma \\ q_1'\sigma^2 \end{bmatrix},\quad \boldsymbol{c}_q = \begin{bmatrix} q_1 \\ q_2 \end{bmatrix},$$

Three experiments have been carried out. The first one focuses on the simple robust output control with the internal model (11) deactivated (set $\gamma = 0$). The second one is the robuts output control with the integral term and excluding the anti-windup scheme (12) (set $\gamma \neq 0, \nu = 0$). The third one is robuts output control with the integral term and anti-windup scheme, which is the suggested algorithm (set $\gamma \neq 0, \nu = 1$).

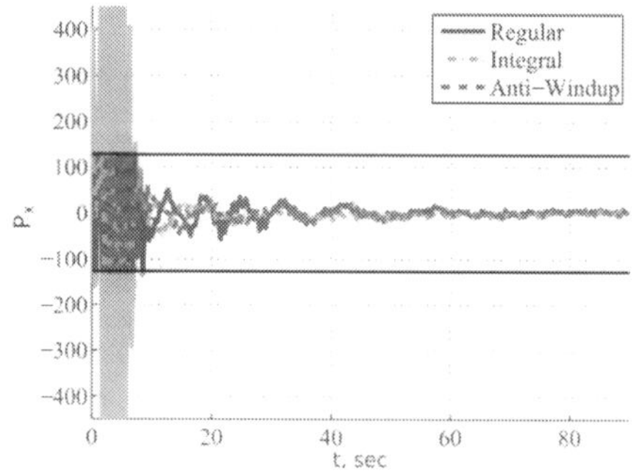

Fig. 3. Plots of the saturated control x-axis.

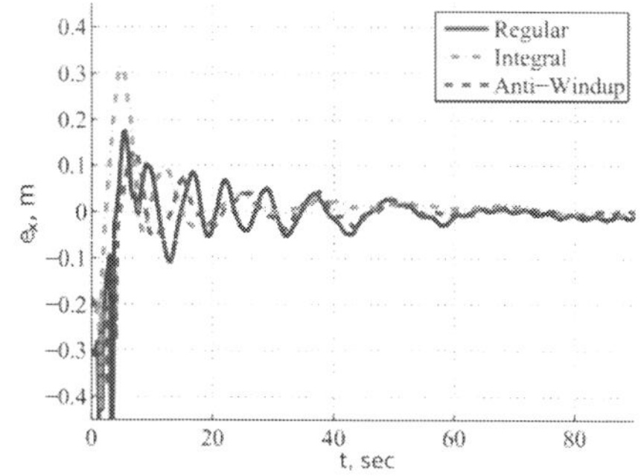

Fig. 4. Plots of the error signals along x-axis.

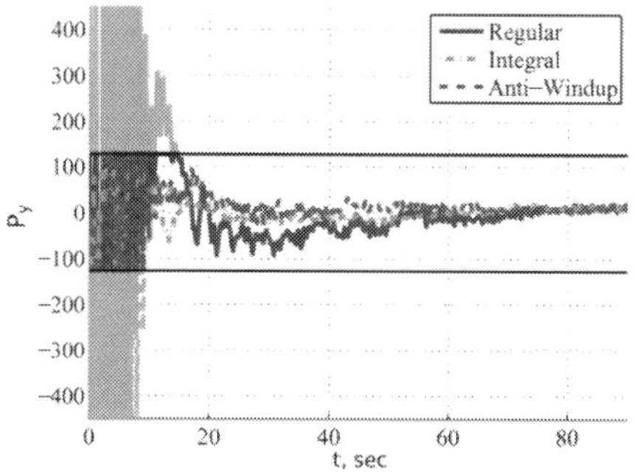

Fig. 5. Plots of the saturated control along y-axis.

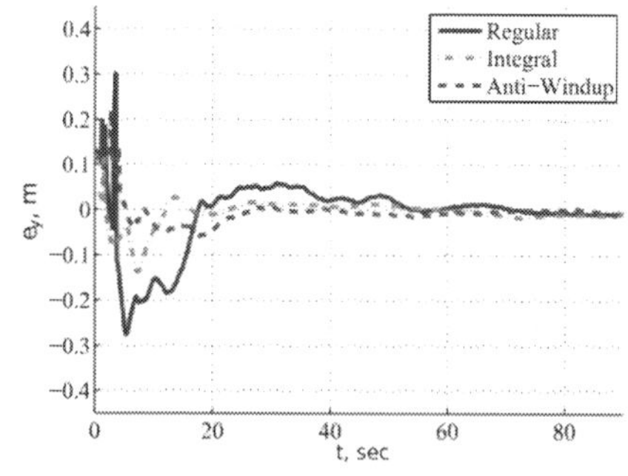

Fig. 6. Plots of the error signals along y-axis.

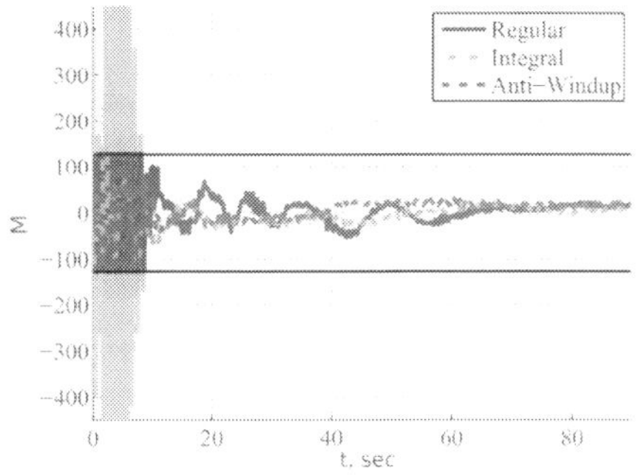

Fig. 7. Plots of the saturated control along ψ-axis.

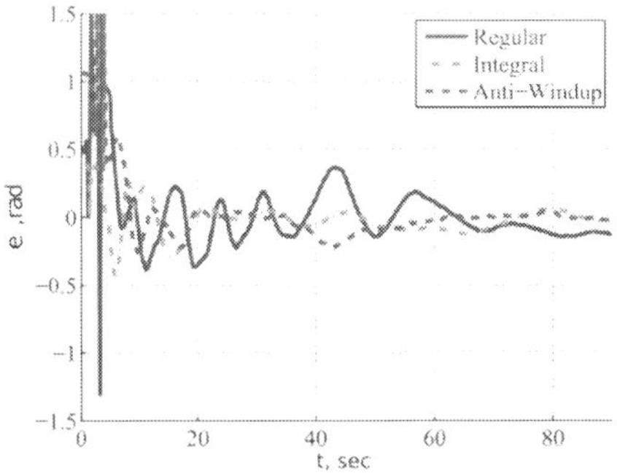

Fig. 8. Plots of the error signals along ψ-axis.

The experimental results are given in Figures 3-8. Cyan curves illustrate the control signals generated by the nominal controller without saturation. Blue curves are assigned to the simplest controller with the integral term deactivated. Green curves correspond to the controller augmeted with the integral term. Red curves refer to the controller augmented with the itegral term and anti-windup scheme.

The experimental data shows that the first controller leads to the slight steady-state error. It is compensated by means of the integral term which accumulates the error value and increase the control signal. In this case integral windup occurs and leads to the overshoot increase. This effect can be fixed using the anti-windup approach.

7 Conclusions

This research focuses on the design the robust output control algorithm with anti-windup compensation for DP systems. The theoretical background is based on the consecutive compensator approach, which uses the high-gain principle to suppress plant uncertainties. The paper provides control design strategy, strict stability analysis and experimental verification by means of the robotic boat setup.

This article is supported by the Russian Federation President Grant 14.Y31.16.9281-НШ, the Government of the Russian Federation (grant 074-U01), the Ministry of Education and Science of the Russian Federation (goszadanie no 2.8878.2017/8.9, project 14.Z50.31.0031) and Russian Foundation for Basic Research (project 17-58-53129).

References

1. H.A. Fertik, C.W. Ross, ISA transactions, **6**, 317 (1967)
2. S. Galeani, S. Tarbouriech, M. Turner, L. Zaccarian, ECC 2009, 306–323 (2014)
3. G. Grimm, J. Hatfield, I. Postlethwaite, A.R. Teel, A.C. Turner, L. Zaccarian, IEEE Trans. Autom. Control, **48** (9), 1059–1525 (2003)
4. E.F. Mulder, M.V. Kothare, M. Morari, Automatica, **37**, 1407–1416 (2001)
5. M. Hussain, N. Us Saqib, M. Rehan, ICET 2016, 7813237 (2016)
6. T. Hu, J. A.R. Teel, L. Zaccarian, Automatica, **44**, 512–519 (2008)
7. J.M. Gomes Da Silva, S. Tarbouriech, R. Reginatto, 2002 IEEE Int. Symp. Comp. Aided Control Systems Design Proc., **2**, 1106–1111 (2002)
8. M.V. Kothare, M. Morari, 1997 ECC, 2156–2161 (1997)
9. A. Zheng, M. V. Kothare, M. Morari, Int. J. Control, **60**, 1015–1024 (1994)
10. Z. Liu, J. Wang, Q. Jin, L. Qi, B. Jiang, DDCLS 2017, 512–517 (2017)
11. W. Wu, 2011 50th IEEE CDC-ECC, 4783–4788 (2011)
12. J.M. Gomes Da Silva, R. Reginatto, S. Tarbouriech, Controle and Automacao, **15** (1), 3–9 (2004)
13. A.A. Bobtsov, N.A. Nikolaev, Autom. Remote Control, **66**, 108–118 (2005)
14. A.A. Bobtsov, Automatica, **41**, 2177–2180 (2005)
15. A. Fradkov, Eur. J. Control, **9**, 577–586 (2003)
16. A. Pyrkin, A. Isidori, **50**, 5498 – 5503 (2017)
17. T. Fossen, *Marine control systems: Guidance, Navigation, and Control of Ships, Rigs and Underwater Vehicles* (2002)
18. O. Borisov, V. Gromov, A. Pyrkin, A. Bobtsov, I. Petranevsky, A. Klyunin, MMAR 2016, 13–18 (2016)
19. O. Borisov, V. Gromov, A. Pyrkin, A. Bobtsov, N. Nikolaev, IFAC-PapersOnLine, **49**, 256–261 (2016)
20. O. Borisov, A. Bobtsov, A. Pyrkin, V. Gromov, AIP Conference Proceedings, **1798** (2017)
21. O. Borisov, V. Gromov, A. Pyrkin, A. Bobtsov, N. Nikolaev, IFAC-PapersOnLine, **49**, 287–292 (2016).
22. S. Tomashevich, O. Borisov, V. Gromov, A. Pyrkin, A. Bobtsov, MED 2017, 1029–1034 (2017).
23. H. K. Khalil, *Nonlinear systems (*2002).

Smartphone-based tourist trip planning system: a context-based approach to offline attraction recommendation

Sergei Mikhailov[1,2,*] and *Alexey* Kashevnik[2]

[1]ITMO University, Saint Petersburg, Russian Federation
[2]SPIIRAS, Saint Petersburg, Russian Federation

Abstract. The paper presents a tourist trip planning system that is aimed at offline attractions recommendation for a tourist based on current situation in the region and user profiles. Authors present the related work analysis in this area and based on this analysis develop the reference model of the proposed system and describe implementation that is based on previously developed Tourist Assistant — TAIS.

1 Introduction

According to the UNWTO World Tourism Organization, the total number of international tourist trips increased from 25 million in 1950 to 674 million in 2000, and to 1.235 million in 2016. Income from international tourism increased from $ 2 billion in 1950 to $ 495 billion in 2000 and to $ 1.220 billion in 2016. It is expected that the number of international travel will increase by 3.3% per year until 2030 and will reach 1.8 billion. The tourism industry began to actively use Internet technologies since the mid-1990s. At the current time, the planning of tourist trip and support for the decisions of the tourist is an object of considerable scientific interest [1].

Over the past ten years there has been an explosive growth in the use of mobile technologies among tourists. E-tourism systems provide a good opportunity for mobile services that help tourists by offering recommendations based on their preferences and context [2]. Presented in the paper Tourist Trip Planning System is based on developed earlier Tourist Assistant — TAIS [3] that is a system that proactively recommends the tourist attractions around, which is currently better to attend personally for him\her. The recommendation is based on tourist's preferences and current situation in the location region. The collaborative filtering technique is used to determine tourists preferences and generate recommendations based on these preferences. Information about attractions is extracted from different Internet sources on the fly (e.g., Wikipedia, Wikivoyage). The main idea of the approach is that TAIS consists of several services, which are united by the smart space for providing the tourist with information according to personal preferences and current situation in the location.

The Tourist Trip Planning System is a service-oriented system which provide attraction visiting route and information about attraction to a tourist. Authors offers to use a context-oriented approach for constructing tourist attraction visiting route. The proposed system consist of a several services which are interacting with each other. Also, the system provides a region databases for offline attraction details processing from accessible Internet source on the user smartphone.

The paper is structured as follows. Section 2 describes the related work in the area of attraction recommendation systems. Section 3 offers concept of the tourist trip planning system based on context approach and describers system services. Section 4 provides a detailed description of system services and section 5 gives a conclusion.

2 Related work

The paper [4] describes personalized location-based mobile tourism application. This application is based on the iBike system in Taichung City, Taiwan and uses a hybrid filtering technique to collect tourist information, to gather tourism information and adapts the ant colony optimization algorithm for more precise adjustment of geolocation recommendations to tourists. The authors use the technology acceptance model and the information systems success model for users intentions investigating to further using in the developed system. The technology acceptance model has been used to interpret users acceptance of information technology. It comprises three dimensions: perceived ease of use, perceived usefulness, and usage behavioural intention. The information systems success model contains a six factors: system quality, information quality, system use, user satisfaction, the effect on the individual user, and the effect on the organization.

Authors of the paper [5] develop geolocation services for the multi-agent information collection of Dr.What-Info system [6]. This system uses Google Maps and image recognition technologies as a source of tourist information and a route planner for users. Users requests to system are sent in the format of the canonical user request representa-

*e-mail: saboteurincave@gmail.com

tion language, which represents information as knowledge frames. The user request is processed by several agents: the ontological data mining agent is used to provide current knowledge in the field of data mining techniques and cluster classification; the ontological case-based reasoning agent uses a reasoning mechanism based on use cases to provide a user-centric and information-related solution. If these agents can not complete the task, then the ontology information agent shell searches for solutions from external mechanisms by crawling information with OntoCrawler tool, retrieving data using OntoExtractor tool, classifying data using OntoClassifier tool and the recommended OntoRecommander agent.

The authors of the paper [7] are develop a personalized travel planning system which uses the time framework concept with regulated recommendations to users. The personalized travel planning system assumes that users can setup their personal requirements through the Internet platform. Through the travel planning module a preliminary travel schedule is created, and the configuration functions help the user to obtain the necessary travel plan. The travel planning module uses a schedule determination method to create a time framework-based travel schedule. This method includes distance calculations, popularity calculations, and time calculations. The information of all matching locations and time block categories are the input parameters for the schedule determination method. The travel requirements module cross-referenced the user's requirements and the travel destination to select the destination that best meets the user requirements. Places that meet the user's requirements are connected and formatted in time framework depending on the time for planning a preliminary trip schedule.

In the paper [8] the authors propose a prototype of a tourist system that offers a method based on multi-criteria collaborative filtering. This method improves the accuracy of forecasting tourist guidance systems by using clustering, reducing the size of samples and forecasting methods. The authors used an adaptive network based on the adaptive neuro-fuzzy inference system and support vector regression methods for predictions. The principal component analysis was used as a dimensionality reduction technique. The authors used the self-organizing map and the expectation maximization algorithm, because they are well-known methods of clustering in scientific community. To improve the accuracy of multi-criteria collaborative filtering recommendations the hypergraph partitioning algorithm are applied to the clustering results. The authors evaluated the accuracy of the recommendation method in the TripAdvisor data set and their experiments indicate that cluster ensembles have greater prediction accuracy than single-source clustering techniques.

In the paper [9] the authors represent the PlanTour system, which creates personalized tourist routes using the information collected from the travellers social network MINUBE[1]. The system uses an automated planning approach to create a multi-day plan with the most interesting sights of the visited city\region. In particular, the system collects information about users and points of interest from the MINUBE, groups these points using clustering methods to divide the problem into subtasks. After these steps the system uses an independent from the tourist area automated scheduler, which finds tourist plans. Unlike other tourist recommendation systems, the PlanTour planner can organize the relevant objects of interest, taking into account the expected travel of the user and user estimates from a real social network.

The paper [10] uses geo-tagged photos in social networks to build routes to attractions. Also, the authors use a separate segment of the road as the basic routing unit instead of the GPS path segment and have created a system of recommendations that provides users with the most popular places in the city\area, as well as the best travel routes between attractions. Using Flickr photos with geotags, it is possible to determine the best places to travel in the city and the best travel routes between popular tourist destinations. The authors apply the spatial clustering method to determine the main tourist reference points and the subsequent ranking. Using a machine learning method, the system calculates the tourist popularity of the road in terms of the relevant parameters, for example, the number of users and the number of points of interest. The best route recommended to the user should have a minimum distance and maximum tourist popularity.

The authors of the paper [11] solve the widespread issue of information congestion in advisory systems and the problem of incomplete data in areas with low level of tourism and non-standard routes of attractions. To solve these problems, the authors propose a new personalized travel recommendations scheme that uses explicit user interaction and multimodal travel information. At first, the system collects heterogeneous sights information through a multi-user exchange, which can be considered as a collective intelligence used to provide reliable references by other travellers. Secondly, valuable knowledge is extracted from the collective intelligence to filter out noise from the data and to structure travel information. The model of Personalized Attraction Similarity is designed to offer attractions through the merger of heterogeneous information with weighted adaptation and simultaneous study of explicit interaction with the user. A contextual information, such as a user location, is used to generate the recommendations that can affect the user choice at a certain point in time.

The paper [12] offers the «Filter-first, tour-second» approach for creating personalized recommendations on tourist routes, based on information from social networks and other online data sources. The collaborative filtering technique is used to define a subset of additional points of interest that maximize potential user satisfaction from the route. Further, the orienteering problem is solved using the iterative tabu search algorithm, which creates tourist routes that must contain all the mandatory places and maximize the overall satisfaction from the route when using additional daily visited places, taking into account access times and hours of operation, restrictions on tour time, user's budget and other restrictions.

[1] http://www.minube.com/

The paper [13] presents the eCOMPASS multimodal tourist tour planner. This tourist tour planner provides a navigational aid and incorporates multimodality within its routing logic aiming at deriving near-optimal sequencing of points of interest (POI) along recommended tours to utilize the time available for sightseeing and minimize waiting time at transit stops. eCOMPASS allows users to define arbitrary start\end locations (e.g. the current location of a mobile user) rather than choosing among a fixed set of locations. The core engine of eCOMPASS is based on a novel cluster-based heuristic approach, the SlackRoutes. The main incentive behind approach is to motivate visits to topology (i.e. touristic) areas featuring high density of «good» (i.e. highly profitable) candidate vertices, even if those are located relatively far. SlackRoutes takes into account time dependency (i.e. multimodality) in calculating travel times from one vertex (i.e. POI) to another.

In the paper [14] authors describes a tourist preferences based recommendation system. The system can build a route taking into account the use of various types of vehicles and context information (weather restrictions and attraction schedule). The authors uses an ant colony optimization algorithm with using an attraction weighted categories as a combinatorial optimization method for a global route constructing. The attractions was stored into a light-weighted ontology.

All the solutions considered provide either a set of attractions that a tourist can visit, or a route of attractions, which does not take into account the current situation in the city. The authors of papers [4, 7, 9, 12] does not work with context information during attraction route visiting at all, authors of paper [11] formalizes context only as user location (city and coordinates), [14] context does not contain traffic information, [10] context contains only a topological cafeterias, [5, 6] does not give many details about route planning and more focusing on ontological method for attraction processing and the authors of [8] describes only a recommendation tourism system without route planning. The proposed method in system will take into account the situation on the road, including traffic jams, current weather and season, attractiveness of sights and their schedules. In another words, the constructing route method should use the current contextual information. It is also proposed to use the user profile to build a route with remembering the history of visits and taking into account the assessments and preferences of users on the attractions.

3 Tourist trip planning system approach

The tourist trip planning system is a service-oriented system which provide attraction visiting route and attraction information to a tourist. This system offers to use a context-oriented approach for constructing tourist attraction visiting route in his\her region. Context situation in the region and user profile are formalized accordingly to an ontology. The system can store a region information in sqlite-database and provides it to user for offline attraction processing on a smartphone. For building the attraction visiting route, methods and algorithms of graph theory are used.

A reference model of tourist trip planning system is presented on Figure 1. The tourist trip planning system provides content to an user and consists of a set of interacting services. The user uses a smartphone for viewing attractions on current region map and information about them and can plan the route in the region. The smartphone stores a user profile, which includes information about the user and his\her preferences. The presented reference model consists of the following services.

Information aggregation service is responsible for gathering information about attractions at the region using the accessible information sources in offline mode. For the information sources at the moment the OpenStreetMaps and Wikipedia are used. OpenStreetMaps is used as geodata source for fetching information about attractions in the region. This information includes attraction name, position (latitude and longitude) and tag (attraction category such as museums, parks, etc.). After fetching an attraction list, service tries to gather detailed information from open-source databases. Usually, the service is looking for an attraction description and images.

The main purpose of the information aggregation service is gathering attraction information in selected regions, extracting data into database and providing databases to the content delivery service. Process of gathering information can be divided into two parts: bringing geo-data about attractions geolocation services and searching attraction details in in accessible information services. As input, the service requires a region list. For each region, the service performs the following operations: district processing, attraction data searching and data writing into the database.

The service uses the developed searching module for information processing and aggregation. The reference model of search module is presented on Figure 2. The input of the attraction information search module is formed by the attraction details, such as title and coordinates given from external geolocation servers. As the search module output, the attraction article with description, keywords and images is given. For improving the quality of searching the following set of measures has been identified. At first, unnecessary words are removed from the attraction title, such as «monument», «tomb», etc. Secondly, module performs geo-search by coordinates, if it existed, and basic search by using attraction title. Main criterion for the attraction article choice correctness is the correspondence of the chosen article with the selected one, computed by using fuzzy search algorithms (with Levenstein distance metric).

Article candidates from geo- and basic- search are compared with each other and the most suitable variant is chosen. After comparison, the selected article is checked for correctness according to the following criteria: whether the article object is in the selected region, whether the content of the article of the original sight is suitable (for example, is the article describes person in case if as input module has monument of that person) and so on. If the check was successful, the module writes information

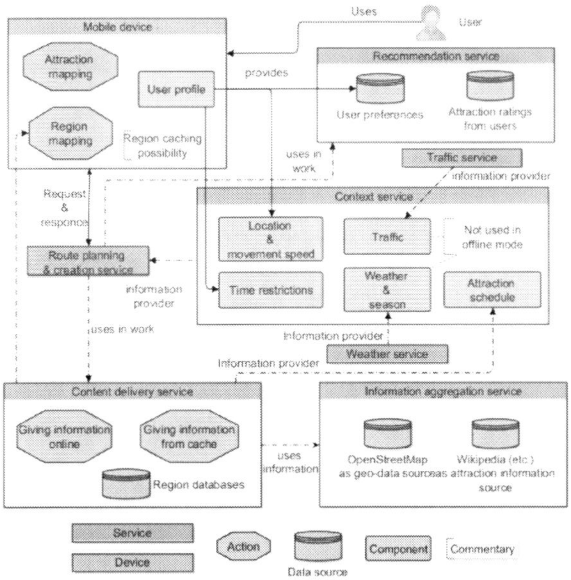

Fig. 1. Tourist trip planning system reference model

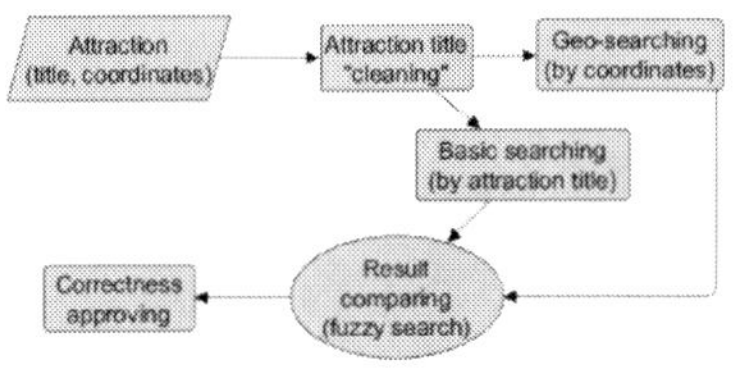

Fig. 2. Reference model of attraction searching module

about the article (title, short description, keywords, links to images) to the list and at the end of the work transfers data to the information aggregation service. At the end, information aggregation service stores information into a local region database.

Content delivery service acts as a storage of information and takes attraction region data from *information aggregation service*. User can download offline region cache from service for the selected region. Also, the content delivery service provides attraction information for the user in case if users does not download region cache for offline mode and Internet connection is available. The *trip planning service* uses a region map in the form of a graph from this service and information about the sights for constructing a route for the user. The *context service* takes information about attraction schedule for its work.

Recommendation service is responsible for ranking attractions, images, and descriptions for providing the tourist the best attractions to see and the best images and description of chosen attraction for acquaintance. To improve the quality of information presented to the tourist the list of attractions found should be ordered with respect to a predicted degree of interestingness for the tourist as well as reachability (taking into account the current situation in the area). The service takes tourist ratings associated with each attraction by all tourists as an input and performs user-based collaborative filtering.

Trip planning service is interacting with *recommendation service*, *context service* and *content delivery service* for creating an attraction route for the user. The trip planning service is interacting with the following services: constructed region graph are gained from *content delivery service*, user and region context information are provided from *context service* and information about user preferences and attraction ratings are gathered from *recommendation service*.

The purpose of the context-oriented tourist trip planning service development is to increase the tourist attractiveness and total saving of the tourist time when visiting the attraction. The developed method provides attraction visiting plan for the tourist. The context includes the tourist location, travel speed and preferences, temporary tourist restrictions, season and current weather, city traffic and a schedule of attractions. For building the attraction visiting route, methods and algorithms of graph theory are used, since the set of attractions and the roads between them can be represented as a graph. Region graph can be described as a indirection graph (V, E) where every node $v \in V$ represents an attraction and every edge $e \in E$ represents a path between an attractions. Every path has a function f which is representing a cost (e.g. travel time between attractions). The context is described as a set of modifiers that can be applied to the path cost. It assumes the use of algorithms for finding the minimum path using the average path time between the selected sights, taking into account the current context situation (for example, traffic, preferred attraction types, weather situation, etc.).

The tourist route can be described as a sequence of visited attractions for the developed method. It is possible to apply parallel calculation of the best route for each segment for the route constructing acceleration. The context-oriented constructing route method involves re-evaluating of the region graph using different metrics for applying the new graph to various methods for finding the shortest route, such as ant colony optimization algorithm, tabu search algorithm, brunch and cut algorithm, etc. The method assumes the utilization of knowledge in machine-readable form, as well as the formalization and modelling of the current situation in the region.

Context service is designed for storing and working with a context information. The context has been split into two parts: user context and region context. User context contains information about location and movement of user; user time restrictions (e.g. tourist arrived to a region at the morning and have free-time until evening). Region context contains information about the current traffic situation; attractions schedule of region and information about current weather and season. Traffic and weather information can be obtained via appropriate services — *weather service* and *traffic service*. To create a tourist profile and a contextual information model of the region, it is suggested to use an ontological approach. It is needed to create an tourist profile that will allow describing the tourist himself, his preferences and the situation in the region. The approach used to describe the profile is well established in recent years and allows to precisely formalize the area under investigation.

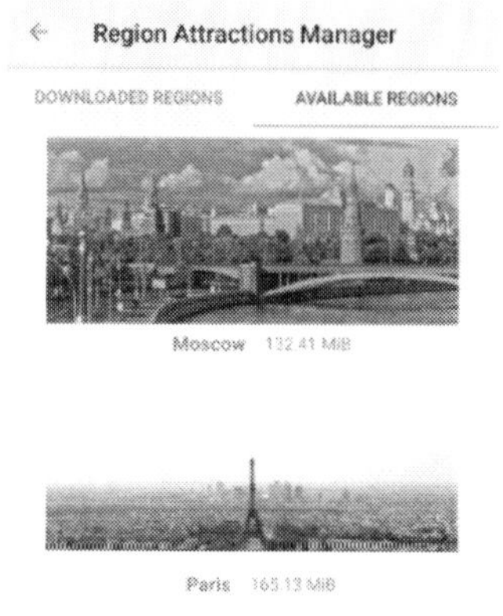

Fig. 3. Regions download and selection in the application

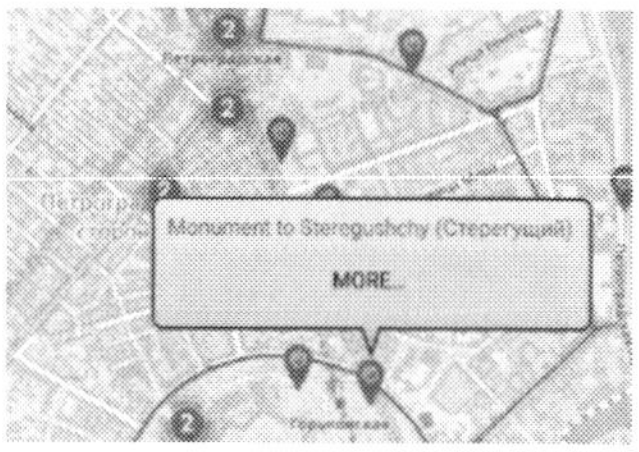

Fig. 4. Map viewer in the application

4 Implementation

The prototype implementation is based on Tourist Assistant — TAIS [15]. The main feature of the implemented application is providing the information to users based on offline source such as sqlite database in contrast the TAIS that searches the information using online resources. The application provides information to user based on the content delivery service region list file developed from the information aggregating service. The user can download desired regions in a background and select one for utilization (Fig. 3).

The region sqlite database contains information about attractions (title, coordinates, keywords, images, etc.) and districts (name and polygon coordinates). The developed application searches the closet to user attractions and provide information about them; track and display the current region district and provided the full map of selected region with attractions. Map viewing (Fig. 4) is organized by using *osmdroid* library and using OpenStreetMaps data for displaying the current region. For improving the perception of a large number of attraction markers on the map, the clustering technique was used.

The information aggregation service is implemented using the Python language with additional packages for working with network, database, images, etc. As geo-data source the OpenStreetMaps is chosen and for attraction information extraction the Wikipedia is used. Input configuration file for the service is coded as JSON-file and contains list of regions, which are needed to be processed. Each region in the configuration file contains name (both original and english versions), region identification number in OpenStreetMaps servers, language code and types of attraction for search. The service accesses to OpenStreetMaps server data by using the Overpass API for gathering information about districts and region attractions. The inner Wikipedia API is used for gathering information about attractions.

After collecting the attraction information from the thread pool the information aggregation service writes information about attraction to the database. For each attraction, the service checks possibility and insert into the database information about available attraction (name, description, keywords, language, and images).

The sqlite-databse models are presented on Figure 5. Table `Attraction` stores: english name, which is taken from information service (*name* field); name, which is taken from geo-service (*original_name* field); coordinates of place (*latitude* and *longitude* fields), general attraction description, which will be presented to user (*description* field); tag-information related to attraction (*tag* field); source type which is used for content generation (*source* field); translation flag notifier (*translated* field) and *url* field for storing information about url of content page. `Image` table contains: *attraction_id* for storing attraction identification number, *image* for keeping the image con-

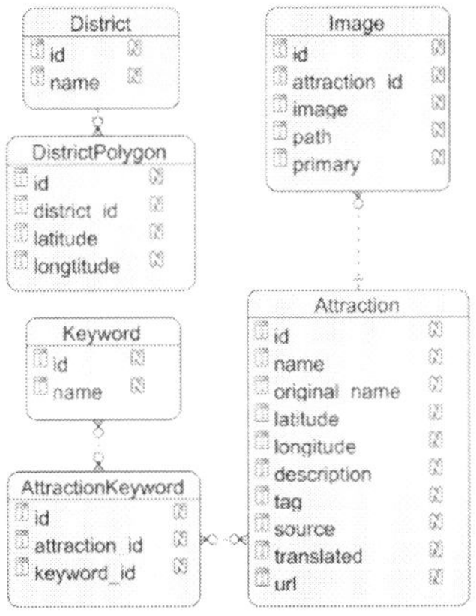

Fig. 5. Database models for information aggregation service

tent, *path* field contains the image url, and *primary* field is a flag for Android application for using particular image as title.

`Keyword` table works in pairs with `AttractionKeyword` table to provide keyword support for attractions. *Name* field of `Keyword` table contains the keyword; *attraction_id* and *keyword_id* from `AttractionKeyword` table keep information about identification number of attraction and keyword accordingly. `District` and `DistrictPolygon` tables stores information about region districts. *name* field of `District` table contains district title, which is gathered from geo-service and `DistrictPolygon` table uses the *district_id*, *latitude* and *longitude* fields for storing information about polygon points and district affiliation. All points in the polygon are inserted clockwise.

All gathered information from the search module of the information aggregation service are divided into parts. For each part the following operations are done: attraction entry creation and correctness checking, attraction keywords and image processing.

The content delivery service is written using Python and acts like a web-service. For building web-service *Flask* micro-framework is used, *Gunicorn* is used as a WSGI HTTP server and *nginx* works as proxy-server for a whole application. Web-service for the current moment contains only two available paths: root route for region list file providing (/) and region sqlite database downloading route from service (`/get/region/<id>`), where *id* acts as region identification number). The region list file is a JSON-file with data about available regions. Every region contains information about its name, identification number, image url for the Android application, current size of sqlite database with region and date of gathering region information, which acts like a version number for the Android application.

5 Conclusion

The paper presents the concept of the smartphone-based tourist trip planning system and describes implementation some parts of it. The architecture of this system is a set of services that interact with each other and provide content to the user. The information aggregation service is responsible for gathering attraction information in regions, such as descriptions, coordinates, keywords, etc. Also, this service save all information into sqlite database, which is used by the Android application for providing offline full-fledged work on finding and displaying tourist information. The trip planning service is interacting with recommendation service, context service and content delivery service for creating an attraction route for user.

The research is funded by the Russian Science Foundation (project # 16-11-10253).

References

[1] C.D. Huang, J. Goo, K. Nam, C.W. Yoo, Information & Management **54**, 757 (2017), smart Tourism: Traveler, Business, and Organizational Perspectives

[2] J. Borràs, A. Moreno, A. Valls, Expert Systems with Applications **41**, 7370 (2014)

[3] A.V. Smirnov, A.M. Kashevnik, A. Ponomarev, Personal Ubiquitous Comput. **21**, 297 (2017)

[4] C.C. Chen, J.L. Tsai, Future Generation Computer Systems (2017)

[5] S.Y. Yang, C.L. Hsu, Computers & Electrical Engineering **54**, 87 (2016)

[6] S.Y. Yang, D.L. Lee, K.Y. Chen, C.L. Hsu, 49–52 (2016)

[7] H.S. Chiang, T.C. Huang, Information Fusion **21**, 3 (2015)

[8] M. Nilashi, K. Bagherifard, M. Rahmani, V. Rafe, Computers & Industrial Engineering **109**, 357 (2017)

[9] I. Cenamor, T. de la Rosa, S. Núñez, D. Borrajo, Expert Systems with Applications **69**, 1 (2017)

[10] Y. Sun, H. Fan, M. Bakillah, A. Zipf, Computers, Environment and Urban Systems **53**, 110 (2015), special Issue on Volunteered Geographic Information

[11] J. Shen, C. Deng, X. Gao, Neurocomputing **173**, 789 (2016)

[12] S. Kotiloglu, T. Lappas, K. Pelechrinis, P. Repoussis, Tourism Management **62**, 76 (2017)

[13] D. Gavalas, V. Kasapakis, C. Konstantopoulos, G. Pantziou, N. Vathis, C. Zaroliagis, Expert Systems with Applications **42**, 7303 (2015)

[14] Novye informacionnye tehnologii v avtomatizirovannyh sistemah 75–80 (2016)

[15] A. Smirnov, A. Kashevnik, A. Ponomarev, N. Teslya, M. Shchekotov, S.I. Balandin, 40–51 (2014)

Ontology-driven approach for describing industrial socio-cyber-physical systems' components

Nikolay Teslya[1,*] and *Igor* Ryabchikov[1,2]

[1]SPIIRAS, 199178,14th Line, St.Petersburg, Russia
[2]ITMO University, 197101, 49 Kronverksky Pr., St. Petersburg, Russia

Abstract. Nowadays, the concept of the industrial Internet of things is considered by researchers as the basis of Industry 4.0. Its use is aimed at creating a single information space that allows to unite all the components of production, starting from the processed raw materials to the interaction with suppliers and users of completed goods. Such a union will allow to change the established business processes of production to increase the customization of end products for the consumer and to reduce the costs for its producers. Each of the components is described using a digital twin, showing their main characteristics, important for production. The heterogeneity of these characteristics for each of the production levels makes it very difficult to exchange information between them. To solve the problem of interaction between individual components this paper proposes to use the ontological approach to model the components of industrial socio-cyberphysical systems. The paper considers four scenarios of interaction in the industrial Internet of things, based on which the upper-level ontology is formed, which describes the main components of industrial socio-cyberphysical systems and the connections between them.

1 Introduction

The industrial Internet of things can be seen as a refinement of the concept of the Internet of things for the tasks of the industrial sector. It integrates industrial means of production into socio cyber-physical systems, including a description of business processes and people operating or using these processes and systems. This integration provides for the creation of a single information space in which an exchange takes place between the components of production and customers, which allows changing existing business processes in the direction of increasing the speed and quality of goods manufacturing, as well as increasing customization according to the customer's wishes [1].

The basis of the industrial Internet of things are smart "objects" — machines, people, processed materials and manufactured goods. In the production process, machines and people are able to analyze the current situation and make decisions based on available information about the status of production and the production facility [2, 3]. Interaction of objects occurs due to the development of software agents - digital twins, displaying properties of objects and information exchange with each other in the common information space, thus forming a smart factory. Unlike a conventional automated factory, a smart factory is characterized by the complete connection of all elements of production, which makes it possible to create a flexible system that can independently optimize productivity, adapt to new conditions and be trained in real or near-real time, and autonomously execute the production process [4].

In the production process, one or several smart factories can be involved. In the latter case, a network of smart factories is kept in mind that can adapt to complex production conditions, including the production of individual parts of the product, logistics and assembling in the assembly department [4].

Inside one smart factory as well as when they are combined, there is interaction of a large number of heterogeneous devices, as well as people, which requires interoperation and trust between all of them. Considering every element of the smart factory from the machine to the whole factory can be represented by its own software agent, their integration can be viewed as a multi-agent system in which each agent has certain characteristics, competencies and requirements. In multi-agent industrial systems, interoperability is most often provided by using ontologies [5, 6].

Ensuring trust and tracking the production process is relevant mainly for the case of combining several smart factories into a single production system. In this case, in the interaction between them, it is necessary to ensure the transparency of the ongoing transactions related to the production, supplying, reception and processing of products. Currently, a promising way to meet these requirements is the distributed decentralized registry technology or blockchain. With this technology, it is possible to organize decentralized distributed immutable storage, in which each record, called a block, will be linked to neighboring ones by computing complex hash functions, the arguments of which are transactions in the blockchain network and neighboring blocks [7].

* Corresponding author: teslya@iias.spb.su

This paper describes the ontologies used to provide interoperability of elements in smart factories, as well as the place of blockchain transactions in this ontology. The paper is structured as follows. Section 2 describes the application of ontologies in production processes. The main scenarios and ways of using ontologies in them are outlined. Section 3 provides an upper level ontology, combining the presented ontologies into one common. In conclusion, conclusions about the work done and plans for future work are determined.

2 Application of ontologies in the modern manufactory

One of the main trends of the modern manufactory are globalization and flexibility [8, 10]. Problems that arise in this case are ensuring interoperability between heterogeneous components of factories, both local (within the same factory) and distributed (distributed production); the consolidation of information from a variety of heterogeneous sources in order to obtain a new value; and the adaptation of production to rapidly changing customer requirements and production capabilities. The application of ontologies can contribute to the solution of these problems [9].

Analysis of publications and projects devoted to the fourth industrial revolution, the automation of production and the application of modern technologies in production makes it possible to single out the following main approaches/scenarios which can be implemented with the help of ontologies.

2.1 The application of the multi-agent system concept in smart factories

The concept of multi-agent system involves the creation of software agents for production components (production machines, schedulers, process analysers) interacting with each other in a common environment to coordinate actions and share information [11, 12]. For agents' interaction a common language is needed, which can be an ontology. The interaction environment can be a common smart space in which agents can publish their knowledge and query knowledge of others. The application of this concept allows providing modularity and simplifying the implementation of changes in production processes, thus providing the necessary flexibility.

Examples of the multi-agent system concept application in production are presented in many papers, including [13-15].

2.2 Collecting and analyzing of information about the products usage

One aspect of the Industry 4.0 concept is the closer integration of consumers with manufacturers and the receipt of a feedback about the products usage and state. Through the analysis of this information, manufacturers can improve the quality of their future products, as well as support users:

Paper [16] describes a scenario for collecting information on the use of the production equipment by its supplier in order to predict possible malfunctions and suggest repair/replacement services. This approach can reduce/prevent possible downtime. Service providers can also be third-party organizations.

Paper [14] describes solution for industrial IoT to enable pervasive monitoring of industrial machinery and products to provide predictive maintenance applications.

Thus, the problem of storage of information about the products state and functioning appears. Products can be divided into classes that have common and individual indicators. In addition, products can consist of parts supplied by different organizations (after assembling the final product, or as a result of the replacement during repair), and the indicators of interest can be considered for both the individual component and the whole product on the basis of the whole information. All this makes it expedient to use ontologies.

2.3 Collecting and provisioning of complete products information about supply chain to stakeholders

The availability of complete products information about supply chain (about materials used during production, its components, production methods, storage and transportation methods and etc.) can increase the confidence of the supply chain participants in a current state and location of their assets (for example, for long-distance deliveries), allowing early detection of errors and frauds. And for consumers to guarantee the quality of supplied products, for example, to confirm the license (that the product has not been forged, stolen and delivered illegally) or, in the event of a poor quality of materials/defects in the components/malfunction of production machines, to track which products were affected and where their current position.

This approach has become particularly relevant with the development of the blockchain technology [17], which allows to build a common decentralized information space of independent participants without a single point of trust and failure. Ontology in this case is necessary for presentation of comprehensive product related information.

The advantages of the increased transparency and awareness in supply chains are described in many works, including [18-20]. In [18], an overview of projects successfully using the blockchain technology for supply chains support is given, including tracking of fish in Indonesia to counter illegal activities; tracking the supply of products to counter the supply of low-quality products; tracking supplies by Walmart, etc.

2.4 The automation of enterprises interaction for participating in a single production and dynamic supply chain processes

Services provided by enterprises (supply of materials, production, transportation of goods, etc.) can be presented as web services with program interfaces for orders formation. This can increase the automation of interaction and, even, allow the automatic search and use of services. This becomes especially relevant in the context of availability of interchangeable service providers and a competition between them. In turn, for service providers, closer cooperation between industries can increase the efficiency of production resources and capacities utilization. Especially if production systems are flexibly configurable and can be used to produce a large class of products (like additive manufacturing).

The description of such services should include a service type, conditions of provisioning, an access interface (for example, REST or WSDL), as well as allow automatic searching and reasoning, which justifies the use of ontologies.

Services themselves can be implemented as REST, WSDL or other types of web services, but for their automatic search, a common registry containing the up to date information updated by the providers themselves is required. Such registry can be a common decentralized smart space, created, for example, based on the blockchain technology.

The automation of service search and execution of transactions between participants in supply chains is mentioned in a number of works, including [21-23].

In paper [21], the scenario of the automatic electricity supplier selection and purchase of the required amount of energy by a factory is described. The main criterion of choice is the current price. As a platform for transactions accounting a blockchain is used.

Paper [22] proposes the so-called Production as a Service framework for provision of production capabilities as services designed to help manufacturers use underutilized resources. Some of the shortcomings of the proposed solution are the low flexibility in description of orders and production capabilities, as well as the orientation towards manufacturing services, while supply chain services can include transportation, storage, etc.

In paper [23], the development of service-oriented manufacturing is also predicted, and a possible basic architecture of the solution is described.

3 Basic ontologies for solving of described tasks of modern industry

Scenarios in the solution of which ontologies can be applied are described above. This section presents specific ontologies that can be used as a base for developing final solutions.

The ontology required for the scenario 2.1 (the application of the multi-agent system concept in smart factories) should describe raw materials, manufactured products and components, production machines and their functionality.

The ontology required for the scenario 2.2 (collecting and analysing of information about the products usage) - should describe products and components, their state and functioning indicators.

The ontology required for the scenario 2.3 (collecting and provisioning of complete products information about supply chain to stakeholders) should also describe products and components, as well as an information on operations performed at different stages of the life cycle (for production, transportation, storage and etc.).

The ontology required for the scenario 2.4 (the automation of interaction of enterprises participating in a single production process, and the formation of dynamic supply chains) should describe services, particularly services itself (raw materials supply, food production, storage, transportation, etc.), terms of their provision (cost, geographic location, etc.) and interfaces for automatic requesting.

Thus, the required ontologies can be broken into:

- *Products and materials*. The ontology of types of products and materials;
- *State and functioning indicators*. The ontology of products state and functioning indicators;
- Structural relations. The ontology of structural relations between products and materials (structural components, used materials, etc.);
- *Supply chain operations*. The ontology of operations that were performed with products during supply chain passage (from what materials it was produced, in what way; what machines were used in production; how it was transferred and transported; in what environmental conditions it had been stored (temperature, humidity, illumination), etc.);
- *Manufacturing machines and capabilities*. The ontology of machines, parameters of functioning and other elements necessary for the development of multi-agent systems of smart factories;
- *Services*. The ontology of supply chain services (product manufacturing, supply of materials, transportation, storage, etc.) with a description of program interfaces for the orders formation.

Figure 1 depicts the ontologies relations. The products and materials ontology is central and is used by all other ontologies: the *state and functioning indicators* ontology and the *structural relations* ontology use it as a base; the *services* ontology use it for describing inputs and outputs; the *manufacturing machines and capabilities* ontology use it in the description of manufacturing processes; the *supply chain operations* ontology use it for the description of objects of operations. Also, the *supply chain operations* ontology uses the *manufacturing machines and capabilities* ontology and the *services* ontology for the description of subjects of operations and concretizes the structural relations ontology (it describes operations by which relations were formed). Also, some well-known standards and ontologies which can be used as parts of the described ontologies are shown in Figure 1, in particular GoodRelations [24], OWL-S [25], eCl@ss [26], eClassOWL [27], IEC 62264 [28]. In addition to the standard ontologies, there are many papers devoted to automation of production which describe their own ontologies that can be used in the development of the Manufacturing machines and capabilities, Supply chain operations and Structural relations ontologies, for instance, [8, 29, 30].

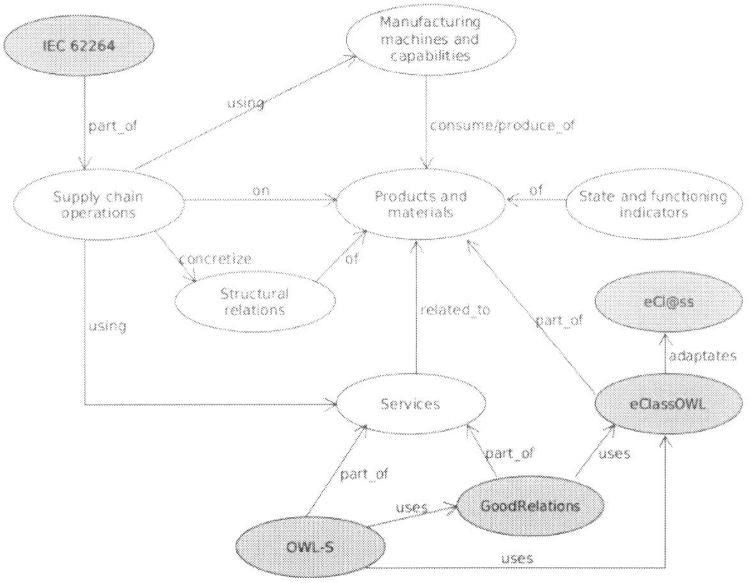

Fig. 1. Relations of the required ontologies.

As the basis for the implementation of scenario presented in section (2.4) GoodRelations [24], OWL-S [25], eCl@ss [26] (eClassOWL [27]) and other public ontologies of assets (for instance, see the list from http://wiki.goodrelations-vocabulary.org/Vocabularies) can be used. The relationship of these ontologies is shown on Figure 1.

GoodRelations is an ontology for describing commercial offers, for instance, on the sale/lease of assets or the provision of services. The ontology allows you to describe information about the provider, terms of the offer (cost, quantity, available methods of delivery, payment, etc.) and assets or services. To describe specific assets or services, the ontology relies on third-party standards, for instance, eCl@ss. However, this ontology is not enough to accept offers automatically because it misses a description of the service's program interfaces, the expected sequence of calls (protocol) and possible responses.

OWL-S is an ontology for describing Semantic Web Services. Its purpose is to provide the opportunity for automatic search, call and composition of web services. The ontology consists of three main parts: a service profile for search, a process model for operations description and grounding as a connection with physical interfaces. The profile contains a description of the service provider, a description of the service as a set of functions (with preconditions, input, output and effects) and a description of the service properties (service category, rating, etc.). It is assumed that the process model contains a more detailed description of the operations than presented in the profile, but they can also coincide. However, OWL-S contains only basic structural elements and recommendations for their addition. To specify the services provided, OWL-S can be integrated with GoodRelations.

The standard eCl@ss contains a description of many classes and properties of assets and services. eClassOWL is an ontology representing the standard eCl@ss in the OWL language.

OWL-S is an ontology for describing Semantic Web Services. Its purpose is to provide the opportunity for automatic search, call and composition of web services. The ontology consists of three main parts: a service profile for search, a process model for operations description and grounding as a connection with physical interfaces. The profile contains a description of the service provider, a description of the service as a set of functions (with preconditions, input, output and effects) and a description of the service properties (service category, rating, etc.). It is assumed that the process model contains a more detailed description of the operations than presented in the Profile, but they can also coincide. However, OWL-S contains only basic structural elements and recommendations for their addition. To specify the services provided, OWL-S can be integrated with GoodRelations. The integration point can be OWL-S Process referring to GoodRelations Offering. In this case, the process itself will describe the protocol of the offer usage, additional preconditions, results and possible responses.

The standard eCl@ss contains a description of many classes and properties of assets and services. eClassOWL is an ontology representing the standard eCl@ss in the OWL language.

Let's consider the application of these ontologies in the following example. Assume that in a certain region there are number of offers for steel sheets supply. Suppliers can be enterprises specialized in the steel supply or enterprises that use steel sheets in production and have unnecessary residues that take up space in the warehouse and are desirable to be utilized in short time (especially if materials are perishable). Offers can differ in cost, available quantity and method of delivery of the material. Consuming enterprises may have software agents whose task is to search for offers of required materials, to make deals for the acquisition of materials and to add actions on the handling of supplies to the schedule. The availability of unpredicted offers, more profitable than the offers of major suppliers, can increase the profit of enterprises.

The result of the offers selection by the enterprise agent can be presented as an ontology containing a description of the selected services and their operations for selling assets with specified parameters (quantity of assets, delivery address, etc.). This ontology can then be passed as input to a software library that will perform the required invocations (for example, via the REST interface of service provider) and return the result. Due to the standardization proposed by OWL-S, this library can be used for calls of any web services by their description.

To store the ontologies of the presented scenarios 2.1–2.4, a Smart Space shared by the stakeholders is required. However, the requirements for the Smart Space are different depending on the number of stakeholders. In scenarios 2.1 and 2.2 (when analysing the indicators of products of a single supplier) there is only one organization-stakeholder. In scenarios 2.3 and 2.4 (and, in some cases, in 2.2), there may be several organizations-stakeholders. In the first case, a centralized Smart Space platform can be used (for example, Smart-M3 [31]), but in the second case, a centralized solution is undesirable due to a single point of trust and failure - all organizations-stakeholders need to rely on the Smart Space provider, and to put on con their income. The solution can be a decentralized Smart Space, for example, developed based on the blockchain platform [32]. In this case, there is no single point of trust and failure, and the failure of a participant's node can only lead to a failure of the processes in which he took an active part.

4 Conclusion

The article describes the use of ontologies for solving the problems of interoperability between components of the industrial Internet of things. This task arises from the presence of a large number of heterogeneous components in production. In addition, the possible consolidation of smart factories in one production also requires interoperability.

In the work, four scenarios of using ontologies in the industrial Internet of things were revealed. All scenarios are related to different aspects of the interaction between production components and the support of final products: the interaction of devices within a single factory; collecting telemetry of products usage to identify possible malfunctions; collection of the complete productions information to support supply chain participants; and automation of production and distribution in several smart factories.

The individual ontologies used for each scenario can be combined into a common upper level ontology, providing interoperability between individual components of production and creating a link between scenarios. Based on the received ontology in the future, flexible production can be created, which allows to ensure high customization of the final product. At the same time, it is proposed to provide information support for intermediate production operations and the supply chain using blockchain technology, thanks to which it is possible to create an unchangeable distributed transaction log. This log can be used to ensure the transparency of intermediate production operations.

Further work will focus on expanding the resulting ontology and providing higher integration with the blockchain technology, particularly by creating an ontology of blockchain transactions to describe the contracts that can be concluded between the individual participants of the production chain. Contracts will describe such aspects as the transfer of components of the finished goods, resources and payment for services, etc.

The research is funded by the Russian Science Foundation (Project № 17-71-10223).

References

1. B. Sun, S.-L. Jämsä-Jounela,Y. Todorov, L. E. Olivier, I. K. Craig, IFAC-PapersOnLine, **50,** 65–70 (2017)
2. H. Bedenbender, A. Bentkus, U. Epple, T. Hadlich, *Industrie 4.0 Plug-and-Produce for Adaptable Factories: Example Use Case Definition, Models, and Implementation*, (Federal Ministry for Economic Affairs and Energy (BMWi), 2017)
3. J.R. Silva, S.Y. Nof, IFAC-PapersOnLine, **48,** 1628–1633 (2015)
4. R. Burke, A. Mussomeli, S. Laaper, M. Hartigan, B. Sniderman, *The smart Factory*. (2017). URL: http://www.smartfactory-owl.de/index.php/en/industry-4-0
5. S. Borgo, P. Leitão, Ontologies, 751–775 (2007)
6. M. Garetti, L. Fumagalli, IFAC Proc., **45,** 449–456 (2012)
7. S. Huckle, R. Bhattacharya, M. White, N. Beloff, Procedia Comput. Sci. **98,** 461–466 (2016)
8. H. Cheng et al., 2016 Third International Conference on Trustworthy Systems and their Applications (TSA), 42–47 (2016)
9. S. Lemaignan, A. Siadat, J.Y. Dantan, A. Semenenko, Proceedings - DIS 2006: IEEE Workshop on Distributed Intelligent Systems - Collective Intelligence and Its Applications, **2006,** 195–200 (2006)
10. S. El Kadiri et al., Comput. Ind., **79,** 14–33 (2016)
11. Y. Zou, T. Finin, L. Ding, H. Chen, Proc. 5th Int. Conf. Electron. Commer, 95–101 (2003)
12. D.D. Corkill, Proc. Int. Lisp Conf., **3,** 23–118 (2003)
13. R.Y. Zhong, X. Xu, E. Klotz, S.T. Newman, Engineering, **3,** 616–630 (2017)
14. F. Civerchia et al., J. Ind. Inf. Integr., **7,** 4–12 (2017)
15. K. Upasani, M. Bakshi, V. Pandhare, B.K. Lad, Comput. Ind. Eng., **108,** 1–14 (2017)
16. A. Braune, *et al. Exemplification of the Industrie 4.0 Application Scenario Value-Based Service following IIRA Structure*, (Federal Ministry forEconomic Affairs and Energy (BMWi), 2017)
17. L.W. Cong, Z. He, J. Zheng, SSRN Electron. J., **w24399** (2017)

18. N. Kshetri, Int. J. Inf. Manage., **39**, 80–89 (2018)
19. A. Bahga, V.K. Madisetti, J. Softw. Eng. Appl., **9**, 533–546 (2016)
20. S.A. Abeyratne, R.P. Monfared, Int. J. Res. Eng. Technol., **5**, 1–10 (2016).
21. J.J. Sikorski, J. Haughton, M. Kraft, P. Street, P.F. Drive, Appl. Energy, **195**, 234–246 (2016)
22. E.C. Balta et al., 2017 13th IEEE Conf. Autom. Sci. Eng., 382–389 (2017)
23. M. Moghaddam, J.R. Silva, S.Y. Nof, IFAC-PapersOnLine, **48**, 828–833 (2015)
24. M. Hepp, International Conference on Knowledge Engineering and Knowledge Management, **5268**, 329–346 (2008)
25. D. Martin et al., W3C member submission, **22** (4) (2004)
26. O. Gräser et al., *White paper AutomationML and eCl@ss integration* (2015)
27. M. Hepp, P. de Leenheer, A. De Moor, Y. Sure, *Ontology Management* (Springer Science & Business Media, 2007).
28. IEC. *IEC 62264-1 Enterprise-control system integration – Part 1: Models and terminology*. (2003). URL: https://webstore.iec.ch/preview/info_iec62264-1%7Bed1.0%7Db.pdf>
29. Z. Usman, R.I. Young, K. Case, J. Harding. Enterp. Interoperability IV Mak. Internet Futur. Futur. *Enterp.* 147–155 (2010)
30. J.L. Martinez Lastra, I.M. Delamer. Advances in Web Semantics I, **4891**, 276–289 (2008)
31. J. Honkola, H. Laine, R. Brown, O. Tyrkkö, Proc. IEEE Symp. Comput. Commun. 1041–1046 (2010)
32. N. Teslya, I. Ryabchikov, Proceeding 21st Conf. Fruct Assoc., 321–329 (2017)

MATEC Web of Conferences **161**, 03028 (2018) https://doi.org/10.1051/matecconf/201816103028
13th International Scientific-Technical Conference on Electromechanics and Robotics "Zavalishin's Readings" - 2018

Reinforcement learning and convolutional neural network system for firefighting rescue robot

Tien Kun Yu, *Yang Ming* Chieh and *Hooman* Samani*

Electrical Engineering, National Taipei University, 23741 Sanxia New Taipei City, Taiwan

Abstract. In this paper, we combine the machine learning and neural network to build some modules for the fire rescue robot application. In our research, we build the robot legs module with Q-learning. We also finish the face detection with color sensors and infrared sensors. It is usual that image fusion is done when we want to use two kinds of sensors. Kalman filter is chosen to meet our requirement. After we finish some indispensable steps, we use sliding windows to choose our region of interest to make the system's calculation lower. The least step is convolutional neural network. We design a seven layers neural network to find the face feature and distinguish it or not.

1 Introduction

As the advancement of robot technologies, people use robot to execute the high risk tasks for human beings. Hence, it is usual for robots entering the high-risk area. Take the earthquake for example, it is unavoidable for people to enter the collapsed building or burning house to search survivor. However, the structure for the building is too weak, when aftershock strikes, it may cause huge lost. To prevent this situation, people try to build robot to keep the rescue team from danger.

For this paper, we focus on fire situation. We want to design a hexapod robot with color sensors and infrared image sensors for firemen to check how many survivors was trapped and the environment. In our opinion, four legs robot would be unstable if it losses any leg of it. However, eight legs robot's locomotion is too complex to move fast. Without a doubt, hexapod robot is the best choice.

This paper has two parts of robot design, locomotion control and human detection. For the controlling part, we used Q-Learning, a model-free method base on the reinforcement learning technique, achieve it's control motivation by interacting environment and maximize the reward. The reinforcement learning to provide maximize reward based on the action. Make the hexapod learned the gait pattern by Q-Learning, with the reward given by the speed and direction. In natural, insect with hexapod has two different gait tripod and bipod, tripod, in every moment there are three legs on the ground which is much slower than bipod every moment there are only two legs on the ground. The hexapod in this paper is used to search survivor in the fired area, so the searching must be done as quickly as it can, with the heavy load it has, the speed by tripod will be unreliable, but compare the tripod and bipod, tripod has better stability than bipod which is also necessary for the robot. With the Q-Learning, hexapod can find the best gait for speed and stability, not only limited by bipod and tripod. Besides learning the gait for robot, robot might stranded by the unknown terrain and obstacle cause by fire, to conquer unpredictable situation, when hexapod detect the position was stranded, Q-Learning will updating the movement immediately, keeping the searching go smoothly. On the other side, we try to search the trapped in the burning house by the near infrared sensors and the color sensors. We use kalman filter to combine the information to get better result than the data only be gotten by one kind of sensor. Although we train our convolutional neural network (CNN) by IIIT-Disguise Face Database [1, 2], which provides lots of faces pictures with infrared and color version, with 130*150 pixels. But our images' size is different from the form. Thus, we would choose suitable region of interested (ROI) to meet the input limitation of CNN. To find out the ROI, we use the sliding windows concept. It would cut image into a lot of pieces. But every sub image has some overlapping part. It is necessary to avoid the face part from being cut. It also can make the detection part faster. The traditional sliding windows only move one pixel or one column in a loop. However, we consider it is too similar to change the CNN's result. We set the window's moving rules that the window would move 13 columns or 15 rows. It is obvious that the calculation can be lower.

In this paper, backgrounds would be mentioned at section two. There are some different methods other researcher done. Section three's content is about our method. We design some experiment for the system and make sure it is work. At section four, we would show the result and explain the experiments. At the final chapter, there is the conclusion.

* Corresponding author: hooman@mail.ntpu.edu.tw

2 Background

There are so many people tried to make the dream, which we can save everyone even we don't need to go into a crisis section, comes true. Like "Control Architecture Design for a Fire Searching Robot using Task Oriented Design Methodology" [3], which is published at 2006, it focuses on three parts on its robot. The first one is Stable mobility in a hazardous and uncertain environment. The second one is providing accurate information regarding the fire and the people in danger. The last one is Safety of the operator, and convenience in operation. To meet the requirement, the robot is designed based on task oriented design (TOD). The robot will be controlled by wireless controller. The operator could watch real time environment by the images which taken by the camera. It means the robot is not an automatic or semi-automatic system. We still need a body to control it. Using a robot to search whole environment maybe is not a good solution for a rescue mission. So "Multi-Robot Exploration and Fire Searching" [4] tries to solve the efficiency issue by mult-robot. It uses improved A* algorithm to get the target way-points. It applies employs a decentralized frontier based exploration method according to these way-points. How to avoid obstacle is practiced by potential field method. The communication between all robots is by decentralized control way. These robots share information and make a complete and accurate map. By simulation, it can get a good result in an unknown environment `To make the rescue speed faster, unmanned aerial vehicles (UAV) also are good choices. "Semi-Autonomous Indoor Firefighting UAV" [5] improved it can finish the entire searching mission with some help of human beings. Moreover, it could carry a fire extinguisher to cool the trapped in order to prevent them from the hurt caused by high-temperature. With carbon fiber and G10, its structure is fireproofed, low-density and strong. To keep the inner equipment from broken, it also has a thermoelectric cooling system. And the communication system between remote controller with UAV is done by radio control. According to the experiment the UAV can work steadily when the environment temperature is less than 1300 degrees. Moreover, the link between the controller with UAV can reach 500 meters. But it still need to somebody to check the monitor.

3 Method

In this section, we show the details of all the process including the movement part, the detection part and sliding windows. At first, we would like to show that how the robot moves by Q-learning. After then, we will explain the convolutional neural network and sliding windows concept detailed.

3.1 Locomotion

The joint circled by red circle is the joint we control in this paper. For this paper, we only control one joint to learn the walking pattern, we will finish the rest later.

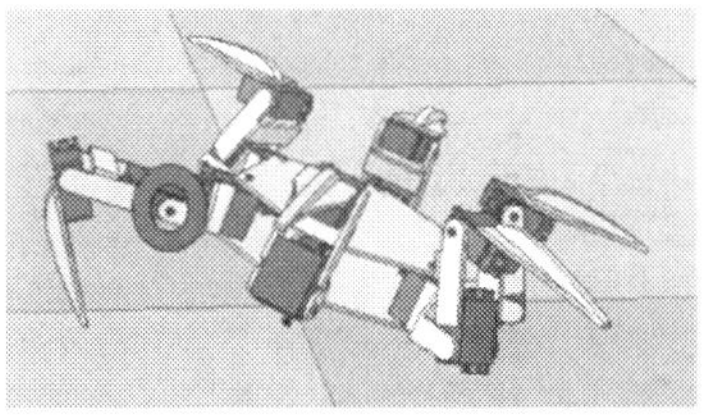

Fig. 1. Joint for Hexapod.

Control the joint roatation for the hexapod movement by the reinforcement of Q-learning. The Q-learning algorithm.

$$a_t) \leftarrow (1 - \alpha) \cdot (r_t + \gamma \cdot \max Q(s_{t+1}, a_t)) \qquad (1)$$

We define the state: θ ,by the rotate for the joint, set the horizontal from previous object to θ = 0.

Fig. 2 Define theta.

Every state have two action increase 20° or decrease 20°. a_t is the action for every state.Each state has two action, each action has it's own Q value ,which updated during the learning by algorithm and store inside the Q-table. Q-table[state,action]=[Q-value], θ+20° is countercloclwise rotation -20° is clockwise rotation.

Table. 1 Q-table for State-Action Action (Next State).

state	40°	20°	0°	-20°	-40°
20°	State20 - action2		State20 - action1		
0°		State0 - action2		State0 - action1	
-20			State-20 - action2		State-20 - action1

The Q-value is used to determine the action that which is best for the movement, the value will rapidly update during learning, when learning beacome stable, the value will converge. Like the above table, state0-action1 and state0-action2 will have different value. When the joint position was set to 0°, program will choose the higher Q-value value from one of them, and set the position to the choosen one.

The α from the algorithm is the learning rate, determine the importance for the new information, α should be low at the beginning, and increase the value along with the learning

become stable. γ the discount factor, determine the importance of future reward, if set gamma=0, the Q-learning will be short sight, may not learn the perfect movement, because it only choose the easy way to get the reward, which may cause bad result, if set γ=1, the result might be divergent. Every parameter change the learning result more or less. Adjusting the value will be neccisarily.
The most important part for Q-learning is the reward, the reward cause very different result for learning. The reward for the learning is determine by the loaction for hexapod change by every action. When the θ change, matlab will pass the θ to V-rep and Hexapod will rotate the joint order by the θ, and read the current location for Hexapod and pass back to matlab, compare the current location and previous location to set the reward. The forward directioin is set to positive x-aixs. The reward is given by below. First location is the starting location when simulation begin, if current position less than current position means the gait is still backward so give the largest negative reward.

Table. 2 Define reward.

Situation	Reward
Current location > previous location	Reward = 20
The action makes hexapod move forward.	
Current location < previous location	Reward = -5
Also compare to first location, if current location < first location reward = -100.	
Current location = previous location	Reward =0
If the action is converge to regular, but the location didn't increase the reward will change negative. Reward = -10.	

To find the fastest gait, there is also a reward is given by the velocity. Each action will have current location and previous location, substract them, it can seem to be the velocity for the action, velocity=distance/time, time for every action can approximate to the same. If current location larger than previous location, the reward for the velocity will add into the reward for the action.

To read the Q-table and each value easier, we use matlab to compile the moving module, use V-rep to run the simulation, the controlling value given by matlab will pass to V-rep and return the data back to matlab, and pass to program to calculate the data and update the Q-value.
When the joint increase or decrease 20°, matlab will take the location for hexapod from V-rep, and determine the location and update the Q-value, as the program started, first rotate the joint from 0° to 120° and decrease to -120° and back to 0°, and record the movement and update Q-value, it will have a initial Q-table, these help the learning much faster than only by random action, after the schedule rotation, the program start to have random action, the action will compare the random action with every action's Q-value and choose the largest, beside the current action, it also compare the next action for next position, which can get the highest reward, and make it under consideration for current action.

3.2 Vision

3.2.1 Sensor fusion

Because we use two kinds of sensors, we need to combine the color images with infrared images. We use extended kalman filter as our solution. Kalman filter is a kind of efficient recursive filter. It always is used on denoising or predicting. There are some functions for kalman filter to finish its mission. Kalman filter has two parts. The first one is predicting, the second one is updating. For the predicting part, we use follow two formulas [6]

$$\hat{x}_k^- = A\hat{x}_{k-1}^- + Bu_{k-1} \quad (2)$$

$$P_k^- = AP_{k-1}A^T + Q \quad (3)$$

There are some definitions as bellowed. $\hat{x}_k^-$ means the uncorrected state matrix at time k. u_{k-1} means the control matrix at time k-1. The formula (2) can get the state from the previous moment. A is the state transfer function. B is the control gain. Q is the noise covariance P_k^- is the uncorrected estimated covariance matrix. By formula (1) and (2), we can predict the following state by previous state, but we hope we can get a more precise estimation. There are three functions to correct the variable X^- and P^-

$$K_k = P_k^- H^T(HP_k^- H^T + R)^{-1}, \quad (4)$$

$$\hat{x}_k = \hat{x}_k^- + K_k(Z_k - H\hat{x}_k^-), \quad (5)$$

$$P_k = (1 - K_k H)P_k^-. \quad (6)$$

According to formula (4), we calculate the variable kalman gain (K_k). It is an important value for system to correct the final output. If the kalman gain is low, it means the output is closer to predicting measurement than the real situation. Otherwise, the output is closer to the actual measurement. After updating the measurement, updating the estimated covariance would be done. Although kalman filter is a time-domain filter, we transfer the pixels from a 2D image to an array to make our images fit the form.

3.2.2 Convolutional Neural Network

After sensor fusion, we get some image information for convolutional neural network. We train the convolutional neural network by the face pictures which provided by IIIT-Disguise Face Database [1, 2]. But its images' size is 130*150. So, we use sliding windows to found out the face shown in any place in the pictures. Why the CNN structure is chosen is because CNN has the feature that it can simplify the calculation.

The first layer is input layer. It allows us to get the test data and training data. The second layer is a convolutional layer. It has twenty kernels. Every kernel is a 5*5 filter to find out the features in the input layer. The formula (7) would be calculated by the weights are showed at bellowed:

$$F_2(x) = \sum_{i=1}^{5} \sum_{j=1}^{5} x_{ij} \times \omega_{ij}. \quad (7)$$

ω_{11}	ω_{12}	ω_{13}	ω_{14}	ω_{15}
ω_{21}	ω_{22}	ω_{23}	ω_{24}	ω_{25}
ω_{31}	ω_{32}	ω_{33}	ω_{34}	ω_{35}
ω_{41}	ω_{42}	ω_{43}	ω_{44}	ω_{45}
ω_{51}	ω_{52}	ω_{53}	ω_{54}	ω_{55}

Fig. 3. Filter weight position.

The third layer is a Rectified Linear Units (ReLU) layer. Our data will not be changed if the data is positive. However, the negative data is replaced by zero just like function (8) because the negative data means the feature is too different to use. We delete the information to lower system loading and speed up the calculation:

$$F_3(x) = \begin{cases} x, if\ x > 0 \\ 0, if\ x < 0 \end{cases} \quad (8)$$

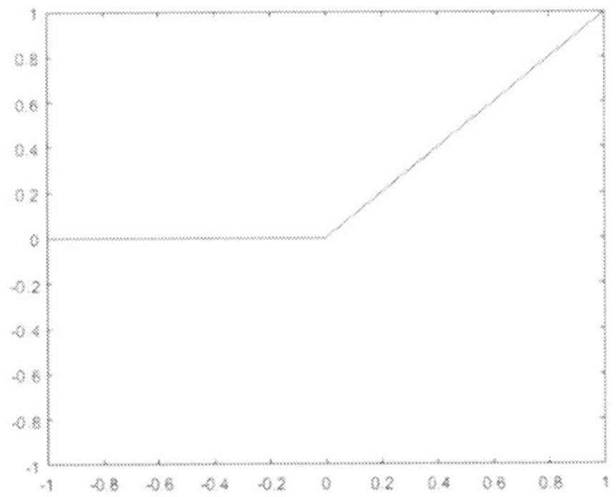

Fig. 4. ReLU function.

After then, the next layer is pooling layer. This pooling layer is max pooling. It is another method to relieve the system loading. We use 2*2 maximum filters to choose the maximum value as the layer's output. The image only gives quarter information to fifth layer.

$$F_4(x_{i,j}) = \max(x_{i,j}, x_{i+1,j}, x_{i,j+1}, x_{i+1,j+1}). \quad (9)$$

At fifth layer, it is a fully connected layer. It turns the image into an array and connects by the perception structure. The last two layers are softmax layer and classification layer. Its missions are calculating weights and finish detection separately. Classification layer is a standard to detect the image is face image or not. If the value is higher than the standard, it is a face image. If not, the image doesn't have face. The weight is from softmax layer. Softmax layer uses softmax function to calculate the final value. Softmax function is showed at bellowed. The range of softmax function is from zero to one. The function (10) is softmax function.

$$F_6 = \frac{e^{Z_j}}{\sum_{k=1}^{K} e^{Z_k}}, j = 1 \sim K. \quad (10)$$

Because our image is an array now, we show the all data as Z. K means how many pixels in an image. We are able to get the weight for every pixel by the function. The weight means how similar between the pixel and face. So the final layer is trying to detect any face by the result the previous six layers calculated.

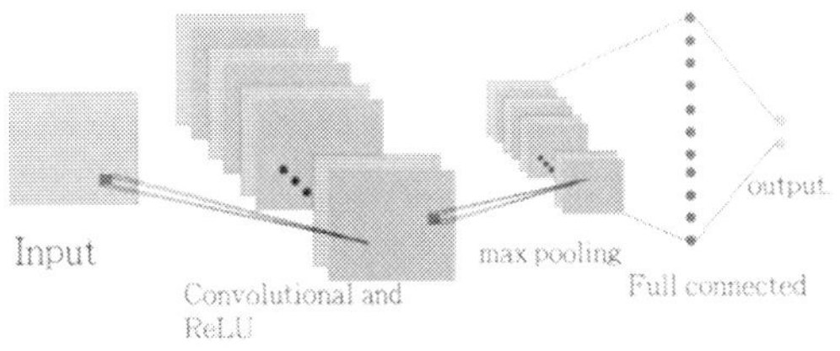

Fig. 5. Convolutional neural network structure.

3.2.3 Sliding windows

Because the input image's size is different from the first layer, we use sliding window to choose region of interested (ROI). Sliding window is a common concept for image processing and signal processing. It sets a 130*150 rectangle at the upper left corner of images. It is the first ROI for CNN input. Our system will detect human being by the selected ROI with the convolutional neural network we trained. Different from the traditional sliding windows, the rectangle doesn't move a column by a column because neural network is a time-cost technology. We hope our system can make a good balance between time-cost and accuracy. We choose move 13 columns (10% of rectangle's length). If the ROI is at the rightest part, the ROI would move 15 rows downward and start from the leftest side. Just like the Figure 6 showed.

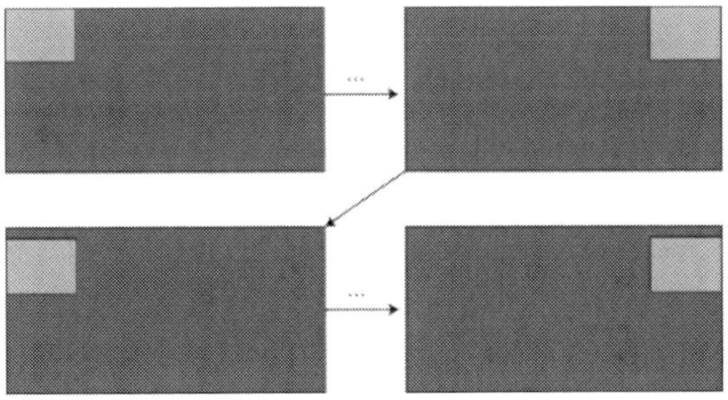

Fig. 6. Sliding windows filter moving steps.

4 Results

In order to prove our system is work, some experiments are designed. Experiment results about locomotion part will be showed as followed. After that, face detection function's experiment result will be displayed.

4.1 Locomotion

For the value form the table below, column represent the state (θ) and row is the action (next θ), set column=7 when rotation=0, column=6 when rotation=-20 and so on. For the result, When θ=-20 and column=6, row 5 and row 7 has two different value, each represent the Q-value of action decrease to rotation=-40° and increase to rotation=0°, row 5 is 20 that is much higher than row 7, so the action will be row 5, decrease to rotation=-40°, and θ change to column 5, and compare the value between the two action, makes the θ back to rotation=-20°. So the gait of the rotation will alternate between -20° and -40°.

Table. 3 Q-table for the result.

	1	2	3	4	5	6	7	8	9	10	11	12	13
1	0	-100	0	0	0	0	0	0	0	0	0	0	0
2	-100	0	-100	0	0	0	0	0	0	0	0	0	0
3	0	-100	0	-100	0	0	0	0	0	0	0	0	0
4	0	0	-100.0000	0	-100.0000	0	0	0	0	0	0	0	0
5	0	0	0	-49.9878	0	50	0	0	0	0	0	0	0
6	0	0	0	0	20	0	-49.3750	0	0	0	0	0	0
7	0	0	0	0	0	-37.8125	0	-98.5156	0	0	0	0	0
8	0	0	0	0	0	0	-49.5313	0	-95	0	0	0	0
9	0	0	0	0	0	0	0	-76.1500	0	-78.0938	0	0	0
10	0	0	0	0	0	0	0	0	-76.2375	0	-62.5000	0	0
11	0	0	0	0	0	0	0	0	0	-57.1875	0	-75	0
12	0	0	0	0	0	0	0	0	0	0	-86.0250	0	-87.5000
13	0	0	0	0	0	0	0	0	0	0	0	-119.5375	0

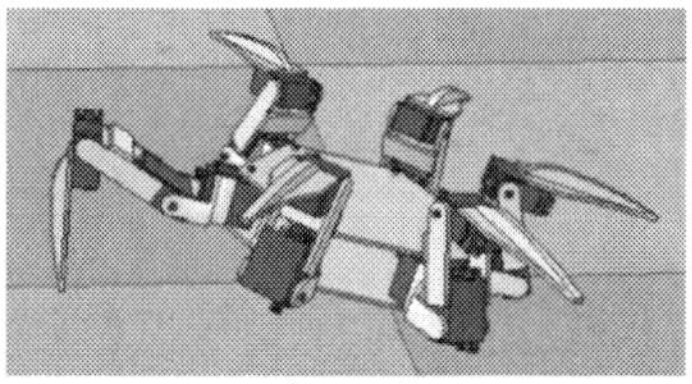

θ=-20°

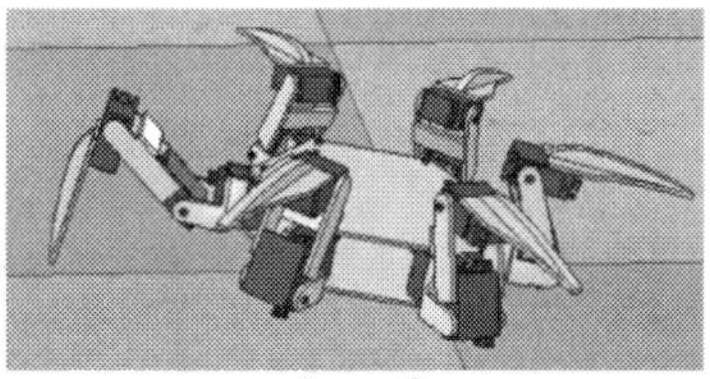

θ=-40°

Fig. 7. Rotation for the one joint hexapod gait.

The discount factor γ is very important at this learning, when θ larger than 0, joint rotate counterclockwise, hexapod will go forward so reward will be positive, and rotate clockwise the hexapod will get negative reward, on the contrary counterclockwise makes backward movement, clockwise makes forward movement. It makes the action stock at θ=100° and 120° or -100° and -120°, because the action for -120° or 120° only have one option is increase or decrease to θ=100° and -100°, and θ=100° and -100° will select the highest Q-value that will makes them go back to θ=-120° and 120°. So the discount factor need to set very high to makes the algorithm know there is a higher reward.

Table. 4 Wrong result from wrong define γ.

Theta=-120 and -100

	1	2	3	4	5	6	7	8	9	10	11	12	13
1	0	-0.6000	0	0	0	0	0	0	0	0	0	0	0
2	52.4692	0	-0.6000	0	0	0	0	0	0	0	0	0	0
3	0	46.8109	0	-0.6000	0	0	0	0	0	0	0	0	0
4	0	0	35.0227	0	-0.6000	0	0	0	0	0	0	0	0
5	0	0	0	10.4640	0	16.8109	0	0	0	0	0	0	0
6	0	0	0	0	16.8000	0	16.8000	0	0	0	0	0	0
7	0	0	0	0	0	1.1520	-0.6000	14.7658	0	0	0	0	0
8	0	0	0	0	0	0	-0.8400	0	18.4416	0	0	0	0
9	0	0	0	0	0	0	0	-0.8400	0	12.3034	0	0	0
10	0	0	0	0	0	0	0	0	-0.8400	0	9.4330	0	0
11	0	0	0	0	0	0	0	0	0	-0.8400	0	5.7776	0
12	0	0	0	0	0	0	0	0	0	0	-0.8400	0	5.9705e-...
13	0	0	0	0	0	0	0	0	0	0	0	-1	0

Theta=100 and 120

When I first started the simulation I expect the gait will be the joint rotate 120° to 120° regularly, but no matter how I rewrite the program the result can't be what I expect, than I discovered the reason is the velocity for the backward and forward, for example, joint regularly rotate between θ=0° and -120°, when joint rotate from -120° to 0°, the distance will increase the most, but when rotate 0 to -120, the location will also decrease the most. Compare the movement than the gait rotate between -40° and -20°, the velocity for the gait between -40° and -20° is higher than gait between 0° to -120°.

4.2 Face detection

To determine our kalman filter is work or not, we use thermal images and RGB images which are provided by IIIT- Disguise Face Database. There are some images showed bellowed.

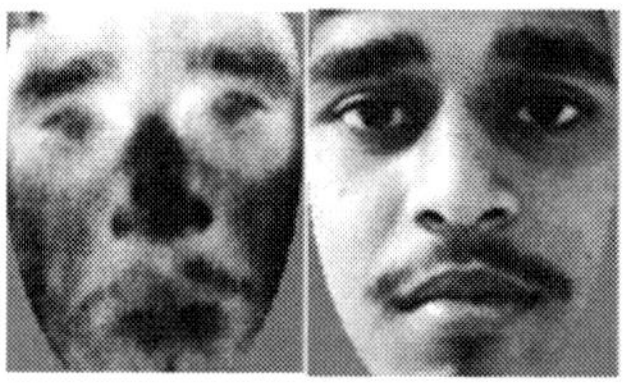

Fig. 8. Infrared and RGB Image from IIIT-Disguise Face Database.

We use kalman filter which mentioned at chapter three to combine two kinds of image to get the result.

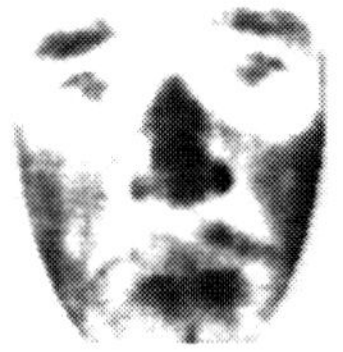

Fig. 9. Composite Image by kalman filter.

According to the image, we can keep the important information, like nose and contour. The most important thing is that the image includes the temperature information.

We detect human not only with shape but temperature. The image also helps us to understand our method is work.

Though image fusion is work, we still need to make sure the CNN is useful. It is very important that CNN has a good performance with high noise. For this reason, we do the experiment with different salt and pepper noise. We use matlab to make some noise.

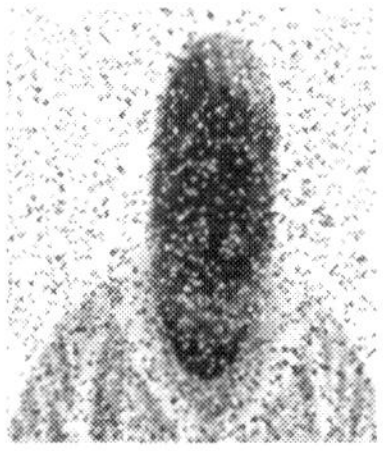

Fig. 10. Face image with 30% salt and pepper noise.

We use some images which are like the above image and some noise images to check the accuracy of convolutional neural network. Because the difference between non-face images with face images is too huge, our convolutional neural network's performance is very well. Actually, it reaches 100%. However, it is not for real situation. We will continue to change and correct our experiment [7].

5 Conclusion

We use Q-learning to learn the gait for robot locomotion. By the simulation we, it works at one-leg situation. On the other hand, we use kalman filter and CNN to finish face detection as well. It provides a trusted result for the system. In other words, we made robot move efficiently and detect human accurately with infrared and RGB sensor. Which is necessary for fire rescue robot to work at fire ground in short time. Although we only get one-leg situation and face detection with normal temperature environment, we will continue to improve it. Our final system would be tested by hexapod situation and real fire ground situation. Even we can't fulfil the dream now, we still provide a new rescue concept that hexapod can go into fire ground and search people. We also finish the first step experiment to prove it is useful.

The authors gratefully acknowledge the financial support of the Ministry of Science and Technology of Taiwan through grant MOST 106-2221-E-305 -013.

References

1. T.I. Dhamecha, R. Singh, M. Vatsa, A. Kumar, PLoS ONE, **9** (7), e99212 (2014)
2. T.I. Dhamecha, A. Nigam, R. Singh, M. Proceedings of International Conference on Biometrics, 1–8 (2013)
3. P.H. Chang et al, SICE-ICASE International Joint Conference, Bexco, Busan, Korea 3126–3131 (2006)
4. A. Marjovi, J.G. Nunes, L. Marques, A. de Almeida, IEEE/RSJ International Conference on Intelligent Robots and Systems, 1929–1934 (2009)
5. A. Imdoukh, A. Shaker, A. Al-Toukhy, D. Kablaoui, M. El-Abd, 2017 18th International Conference on Advanced Robotics (ICAR), 310–315, (2017)
6. G. Welch, G. Bishop. *An introduction to the kalman filter* (Department of Computer Science, University of North Carolina //ed: Chapel Hill, NC, unpublished manuscript, 2006)
7. Samani H., SPIIRAS Proceedings, **56**, 56-75, (2018)

AUTHOR INDEX

AUTHOR INDEX